A COURSE IN
STATISTICAL THERMODYNAMICS

A COURSE IN

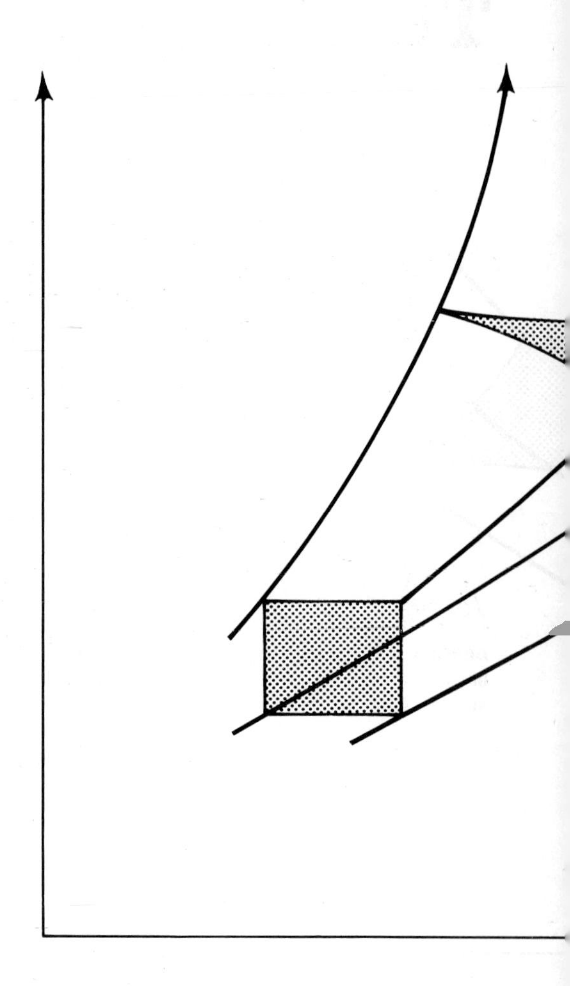

ACADEMIC PRESS

STATISTICAL THERMODYNAMICS

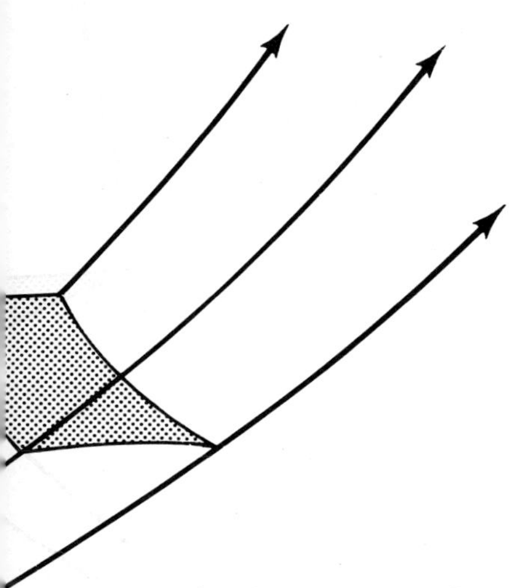

Joseph Kestin

DIVISION OF ENGINEERING
BROWN UNIVERSITY
PROVIDENCE, RHODE ISLAND

J. R. Dorfman

UNIVERSITY OF MARYLAND
COLLEGE PARK, MARYLAND

NEW YORK and LONDON

ACADEMIC PRESS, INC.
111 Fifth Avenue, New York, New York 10003

United Kingdom Edition published by
ACADEMIC PRESS, INC. (LONDON) LTD.
Berkeley Square House, London W1X 6BA

LIBRARY OF CONGRESS CATALOG CARD NUMBER: 70-137595

PRINTED IN THE UNITED STATES OF AMERICA

CONTENTS

Chapter 3. Quantum Mechanics

Chapter 4. Topics in Mathematics

Chapter 5. Foundations of Statistical Thermodynamics

Chapter 14. **Fluctuations**

PREFACE

It is becoming increasingly accepted that students of physics, chemistry, engineering, or biology should be introduced into the methodology of statistical thermodynamics early in their careers. Hence it follows that a difficult subject must be presented to them without the use of advanced mathematical methods. This book attempts to present such a course, putting all the emphasis on the physical aspects of the discipline. Being an introduction, it nevertheless sets out to give an accurate exposition, for there is nothing worse than misleading an inexperienced student. Otherwise, blocks are placed across the path of his future progress, blocks which he will have to remove at the cost of an avoidable mental effort. Our readers must judge to what extent we have succeeded in achieving this aim.

In the pursuit of this aim, great emphasis has been placed on a correct statement of the theory of Gibbsian ensembles couched in quantum-mechanical terms throughout. The kinetic theory, which, studied by itself, leads to incorrect expressions for the entropy and the specific heats of gases, is first conceived as the classical limit of quantum statistics. The book returns to this subject later and uses the kinetic theory directly to yield an understanding of transport properties and the Boltzmann equation. It is also thought that the effort of grasping statistical mechanics—of making the indispensable, and admittedly difficult, mental connections—is greatly reduced if the student has a prior, working knowledge of classical thermodynamics. In the authors' judgement, the recent attempts to introduce both aspects—the classical and the statistical—in a unified *first* treatment have not been crowned with patent success.

Although at least one semester's work in classical equilibrium thermodynamics is presupposed, Chapter 1 of this book contains a terse review of classical concepts and laws. These are so arranged that their connection with statistical thermodynamics can be easily seen later by the student, that is, when the essential identity of the two structures will have become apparent to him. This chapter emphasizes the concept of equilibrium, which is now commonly thought to constitute the central "primitive" concept of classical as well as of statistical thermodynamics, and contrasts it with that of a nonequilibrium state. Instructors and students who feel that they wish to grapple with statistical thermodynamics without diverting their attention, should leave this chapter until later. We have refrained from relegating it to an appendix, because we are of the opinion that a continuous awareness of the structure of classical thermodynamics is very helpful in grasping the underlying motivation for the statistical method. For these reasons, we have treated this chapter as an Introduction.

Starting with the concept of an equilibrium state, we are able to acquaint the student with the concept of a phase space in Chapter 2 as well as to explain the principle of its quantization by a heuristic argument. The chapter continues with a detailed physical discussion of the most important mechanical models of thermodynamic systems, and this leads naturally to the study of the rules for their quantization in Chapter 3 which is devoted to a simple presentation of quantum mechanics and spectroscopy. Once again, some readers will consider this to be an intrusion best handled by committing this review material to an appendix. On the other hand, many students of statistical thermodynamics lack fluency in quantum mechanics, and this is the point beyond which a clear understanding of the concepts of quantization, indistinguishability, and degeneracy should not be delayed. Readers who feel secure with these concepts may omit this chapter entirely.

Chapter 4 makes a digression into mathematics, concentrating, however, on a preparatory familiarization with certain indispensable results rather than on their rigorous derivation. In addition, care has been taken to develop in the student an intuitive "feeling" for *very* large numbers. Needless to say, students who are more mature in this respect may safely omit this chapter if they wish to accelerate their progress.

Chapter 5 can be said to contain the true focus of the book (and the subject). Starting with an exposition of the statistical method, it teaches that the structure of the physical theory is closely modeled on mathematical statistics (briefly reviewed in Chapter 4), and soon comes to grips with the concept of an ensemble. Particular, but not exclusive, emphasis is placed on stationary ensembles, the latter being firmly based on Liouville's theorem. This paves the way to a rigorous presentation of the method of the most probable distribution and the definition of the canonical distribution function. A mathematical study of the latter allows us to demonstrate the essential identity of its structure with that of the classical fundamental equation for the Helmholtz free energy. In this manner, the student is led to a restatement of the First Law of thermodynamics and of the First Part of the Second Law. After a brief discussion of the Second Part of the Second Law, the Third Law is formulated.

A final, summary section tersely reviews the essence of the subject and clarifies it with the aid of a histogram. A clear mastery of the material of this chapter on the part of the student should give him confidence in tackling any detailed topics in statistical thermodynamics that he may find necessary or useful in the future. On first reading, Sections 5.4, 5.9, and 5.18 may safely be omitted, as indicated in the text.

The simplest train of applications, that concerned with nondegenerate perfect gases, is contained in Chapter 6; it gives the student his first opportunity to test his newly acquired powers. Apart from the computation of the thermodynamic properties of perfect monatomic and polyatomic gases, inclusive of the contribution of electronic states, the student will encounter an extensive discussion of Gibbs's paradox and of the Sackur–Tetrode formula as well as an introduction to the kinetic theory of gases by the application of the Euler–Maclaurin approximation to quantal equations.

Chapter 6 concludes Part One of the book, and readers familiar with the two-volume work written by one of us† will recognize that Chapters 3–6 have been taken over almost verbatim from it. Both of us wish to express our deep appreciation to the Blaisdell Publishing Company for allowing us to do this as well as to revise the text without restriction. The authors are of the opinion that Part One contains the essential minimum recommended for a first study.

Chapters 7–14 contain a number of more specialized applications; they can be studied in any order that appeals to the student or to his instructor. Of necessity, the selection of topics for Part Two was somewhat arbitrary, reflecting, to a certain extent, the authors' personal interests and preferences. The topics range from treatments of real and degenerate gases (Chapters 7 and 8) through the theory of simple solids (Chapter 9), radiation (Chapter 10), and magnetic systems (Chapter 11), all treated, essentially, in terms of equilibrium properties, to an introductory exposition of nonequilibrium states and fluctuations. In all cases we have attempted to emphasize the essentials without venturing too far afield, but keeping close contact with the macroscopic form of presentation. In fact, here the two views—the macroscopic and the microscopic—are no longer kept separate in tight compartments, but are used discriminately following the demands of clarity in the physical discussion of phenomena. The authors hope that by now both views have been absorbed by the student enabling him to make transitions from one to the other with ease and fluency.

The treatment of transport phenomena (Chapter 12) is elementary, but special care has been taken to provide a rigorous derivation of Boltzmann's equation, a full discussion of the H-theorem and of the vexing paradox which

† J. Kestin, *A Course in Thermodynamics*, Vols. I and II, Blaisdell, Waltham, Massachusetts, 1966 and 1968.

arises when microscopic reversibility must be reconciled with irreversible behavior in the large (Chapter 13). The final chapter (Chapter 14) contains the bare essentials of fluctuation theory.

A set of problems has been designed to give an opportunity to the student to demonstrate his own creativity. The elementary problems in Chapters 1–6 are denoted by numbers set in boldface type; they are recommended to the instructor and the student for an introductory course. Selected problems are solved whenever it was felt that they amplified and supplemented the main text. A set of four tables has been added to facilitate numerical work, and all constants of physics, in the problems as well as in the text, have been quoted in accordance with the best-known, contemporary accuracy†; an edited extract is given in Table I. The fundamental constants of physics are distinguished in that they have been set in boldface type. However, care has been taken to avoid confusion with vectors, which are also, as is customary, set in boldface type. The authors feel that this compromise is preferable to the use of italics, in order to obviate confusion between Planck's constant (**h**) and enthalpy (h), the speed of light (**c**) and the specific heat (c), and so on.

Throughout the course, it has been assumed that the student has an understanding of classical mechanics, even though the more difficult results have been restated when required.

The book can be used for a leisurely course of two semesters in the junior or senior years, or as a first-year graduate course. With better preparation and later in the student's career, it can be covered in one semester as an accelerated course if some topics (at the instructor's discretion) are omitted from Part Two. Last, but not least, Chapters 2, 5, and 6 with the possible omission of Sections 5.4, 5.9, and 5.18, and with such additions from Chapters 3 and 4 as are deemed necessary, can be made the basis of a slow, *elementary* course restricted to fundamentals.

† B. N. Taylor, W. H. Parker, and D. L. Langenberg. *Reviews of Modern Physics* **41** (1969) 375. See also *The Fundamental Constants and Quantum Electrodynamics*. Academic Press, New York, 1969.

ACKNOWLEDGMENTS

The authors wish to discharge a pleasant duty and to acknowledge the support they received from several friends. Professor A. G. DeRocco made valuable suggestions for the composition of Chapter 7. Professor Y. Göğüş carefully proofread the draft of Chapter 10, evaluated all the constants given there, and prepared the corresponding graphs. Professors M. H. Ernst and P. C. Hemmer supplied us with a number of interesting problems. Professor E. G. D. Cohen provided hospitality to one of us (J.R.D.) at the Rockefeller University during most of the time when this book was being written. Professor E. A. Mason was always ready to act as a responsive sounding board. Mrs. J. Bardsley competently and uncomplainingly typed, and retyped the manuscript. Last, but not least, the editorial staff of Academic Press readily acceded to all our wishes and gave the book its pleasant outward appearance.

A COURSE IN
STATISTICAL THERMODYNAMICS

Introduction

CHAPTER 1

SUMMARY OF CLASSICAL THERMODYNAMICS

1.1. The Two Views of Matter

The science of thermodynamics is that branch of physics which deals with the effects of temperature changes and transformations of energy on a wide variety of natural phenomena. As the reader undoubtedly knows from his previous work, thermodynamics can be studied on two levels depending on the view taken regarding the structure of matter. In *classical thermodynamics* it is assumed that matter is continuous. By this we mean that homogeneous matter can be divided into arbitrarily small parts without limit and in a continuous way. This view is suggested to us by direct observation and by measurements performed with most instruments. In *statistical thermodynamics* we accept the view of modern physics which asserts that matter is particulate. By this we mean that all matter consists of molecules or atoms which, in turn, constitute complex structures of protons, neutrons, and electrons. We know, further, that such elementary particles simultaneously exhibit the properties of waves and that energy can be transferred between them only in discrete quanta. The impression of continuity and solidity is merely a scale effect. A typical atomic diameter is 1–10 Å, and, also typically, even a small quantity of such "thin" matter as a gas at low pressure constitutes a population of the order of 10^{23} molecules. Thus, when we imagine smaller and smaller volumes we conclude that the initial impression of solidity must, eventually, give way to an impression of random and fluctuating motion.

In spite of this seemingly radical divergence in the most fundamental concept, the two branches of thermodynamics—the macroscopically based classical branch and the microscopically based statistical branch—must lead to the same general results when they are applied to the description of a particular process. The difference between them resides in the fact that classical thermodynamics—based on a less detailed view of matter and a more directly phenomenological approach—can give us only general laws and equally general relations among the properties of substances. By contrast, the more elaborate picture of matter adopted by statistical thermodynamics succeeds in providing us with a more detailed description of physical phenomena. On the one hand, statistical methods allow us to retrieve all previous results which turn out to be those consequences that are independent of structure. On the other hand, statistical thermodynamics allows us to perform calculations of properties which are consequences of the details of the structure and thus to explain why different materials behave differently in similar circumstances. Beyond this, we gain an ability to deal with highly dilute systems which contain vastly smaller numbers of particles, that is, conglomerations for which the assumption of continuity ceases to be a good approximation.

At first sight it might appear that there is no point in studying classical thermodynamics at all because the more general science of statistical thermodynamics includes it as a special case. This, however, is not so for several reasons. First, being less detailed, the classical science is mathematically much simpler and hence easier to grasp. Secondly, since it is not concerned with structure, it expresses those relations which are independent of the latter. Thirdly, in many cases the details of structure are not known for important substances and systems, and we cannot, therefore, study them by other than direct, phenomenological methods. Finally, often we may know the molecular structure of a system and yet be unable to solve the problems of its mechanical motion owing to unsurmounted mathematical difficulties.

It follows that classical thermodynamics provides us with an easily grasped framework for statistical thermodynamics. For this reason, it is useful to preface a study of the latter with a terse review of the logical structure of the former. However, students familiar with classical thermodynamics† may turn

† H. A. Buchdahl, *The Concepts of Classical Thermodynamics*, Cambridge Univ. Press, London and New York, 1966; H. B. Callen, *Thermodynamics*, Wiley, New York, 1960; E. A. Guggenheim, *Thermodynamics*, North-Holland Publ., Amsterdam, 1957; J. Kestin, *A Course in Thermodynamics*, Vols. I and II, Blaisdell, Waltham, Massachusetts, 1966 and 1968; R. Kubo, *Thermodynamics*, Wiley, New York, 1968; A. B. Pippard, *The Elements of Classical Thermodynamics*, Cambridge Univ. Press, London and New York, 1957; A. Sommerfeld, *Thermodynamics and Statistical Mechanics* (J. Kestin, transl.), Academic Press, New York, 1956; M. W. Zemansky, *Heat and Thermodynamics*, 5th ed., McGraw-Hill, New York, 1968.

at this point directly to Chapter 2 which will introduce them to the fundamental, physical ideas of the subject of this book.

1.2. Definitions and Concepts

1.2.1. System

The concept of a *system* plays the same part in a thermodynamic analysis as the subject in a sentence. A specified collection of material particles enclosed in a specified, real or hypothetical, *boundary* constitutes a system. When the boundary is not crossed by matter, the system is called *closed*, otherwise it is called *open*. All principles of thermodynamics are enunciated for closed systems; they have to be modified when written for open systems, which must, therefore, be reducible to closed systems. For this reason, the present book will discuss only closed systems.

1.2.2. Subsystem

Any suitably defined portion of a system is called a *subsystem*. In many problems, such as those involving diffusion or chemical reactions, the different species regarded as subsystems, constitute open subsystems. However, the set of open subsystems under consideration represents a closed system. When inert or reacting mixtures are studied, it is useful to think of each species as filling the boundary of the whole system. Such subsystems are called *coexisting subsystems*. In subdividing systems into subsystems, it is possible to go further in subdivision than complete chemical species. Even in a homogeneous system, such as a specified mass of a polyatomic gas, it is sometimes useful to regard a specified set of molecular degrees of freedom† as an open or closed subsystem coexisting with the sets of other degrees of freedom. For example, the translational degrees of freedom may be regarded as one subsystem in contrast with rotational and vibrational degrees of freedom, each of which can be treated as a coexisting subsystem. Anticipating a concept from Chapter 6, we may say that each factor in a partition function may be thought of as relating to a separate subsystem.

1.2.3. Surroundings

Matter which affects the phenomena occurring in the system constitutes its immediate *surroundings*. In order to describe such effects, it is usual to say that the system *interacts* with its surroundings. A subsystem interacts with the other subsystems. The nature of such interactions is determined by the

† The concept of a molecular degree of freedom is explained in Section 2.3.6.

properties of the boundaries of systems and subsystems. A hypothetical rigid boundary which prevents all interactions is called *adiabatic*. A system or subsystem contained within, or constrained by, a rigid adiabatic boundary is called *isolated*. In any problem it is possible to choose a closed isolated system for the analysis by suitably defining its boundary.

1.2.4. Equilibrium

The most essential primitive concept in classical thermodynamics is that of *equilibrium*. A system is said to be in equilibrium when it ceases to interact with its immediate surroundings and when all conceivable subsystems within it cease to interact with each other. In other words, a system is said to be in equilibrium if it remains unchanged when its external boundary, or any internal boundary, suddenly becomes adiabatic. It is postulated that any thermodynamic system *can* be in equilibrium, and that any isolated system reaches equilibrium after a sufficient lapse of time.

On the molecular level of description, a system which is said to be in macroscopic equilibrium does not, of course, consist of noninteracting particles. We shall see later that it is precisely through such microscopic interactions that the system maintains a state of equilibrium in given circumstances. This dynamic picture of equilibrium allows us to see that the system must undergo fluctuations. The latter are ignored in the classical version of our subject because they can be understood only in statistical terms.

1.2.5. State

The *set* of measurable properties which are relevant for the analysis of a system when the latter is in equilibrium is called its *state*. It is found that of all relevant thermodynamic properties of a system only a limited number, n, characteristic of each system, can be given arbitrary values (within a domain) independently. Thus, the equilibrium state of a system is uniquely determined by n independent properties, $x_1, \ldots, x_n$, and any other property, y_k, is dependent on them. In other words, y_k satisfies a single-valued *equation of state*

$$y_k = y_k(x_1, \ldots, x_n). \tag{1.1}$$

The physical nature of the independent variables, x_i, or of the dependent variables, y_k, that are important for a particular system is suggested to us by direct observation and may vary very widely from system to system. For example, the state of a homogeneous fluid phase is described by its pressure, P, and volume, V, regardless of shape. In a solid, changes in shape described by its strain components, ε_{ij}, must be included. An electric capacitor's state will be described in terms of its potential difference, $\mathscr{E}$, and charge, z, whereas that of a paramagnet will employ the magnetic field intensity, **H**, and the magnetization, $\mathcal{M}$.

When a system is subjected to the action of a field (electrostatic, electromagnetic, gravitational, etc.), it is necessary to indicate the instantaneous spatial distribution $x_i(\mathbf{r})$ of each independent property, rather than a single value, in order to define a unique state of equilibrium. Such systems will be called *continuous* regardless of whether they are in equilibrium or not.† In the absence of fields, systems become *uniform.* A uniform system consists of a finite number of uniform and homogeneous subsystems. The equilibrium state of a uniform system can be represented as a point in a space of n dimensions with coordinates $x_1, \ldots, x_n$.

Equilibrium in a system is maintained by a set of internal and external *constraints,* that is, walls of different properties (rigid, deformable, semipermeable, adiabatic, etc.) as well as by a specified set of *internal constraints.*

1.2.6. Process

When the external or internal constraints are changed, the state of the system changes. The system is said to undergo a *process.* After the lapse of a sufficiently long period of time the system assumes a new state of equilibrium specified by the modified set of constraints. Since any process can be studied in terms of an isolated system, it is possible to say that thermodynamics is the study of processes which occur between two states of equilibrium in an isolated system after an initial set of constraints has been changed. Thus we can formulate the *fundamental problem of thermodynamics*: given an isolated system in an equilibrium state 1, determine that state 2 which will result after a specified number of internal constraints has been changed. The constraints which determine the equilibrium state of a system need not always be of a concrete nature, like a wall or partition. If a certain virtually possible process occurs at such a slow rate that it produces a negligible change in the interval of time of interest, it can be assumed that the system has developed a virtual internal constraint. For example, a stoichiometric mixture of oxygen and hydrogen at low temperature reacts to form water vapor at such a slow rate that we can regard the system as having been constrained by an ideal anticatalyst. Similarly, when a gas is compressed very fast in a metallic vessel, it is possible to assume that for short periods of time the external constraint is adiabatic and that the system is isolated at the end of the compression. The long-term process, however, occurs as if the adiabatic wall had

† Evidently, if the continuous system is not in a state of equilibrium, the number and nature of the set of fields $x_i(\mathbf{r})$ which describe it is not, in general, identical with the equilibrium set.

been exchanged for one which recouples the system to the surroundings (diathermal wall).

If the existence of internal constraints is postulated, there is no need to differentiate between stable, metastable, and unstable equilibrium. Metastable equilibrium differs from stable equilibrium merely in the manner by which constraints are changed: by direct manipulation in the former case, by a disturbance in the latter. As far as unstable equilibrium is concerned, we simply assume that it cannot be observed.

Any process which occurs in a system causes the system to depart from equilibrium, that is, to become continuous. Thus, a continuous system may be in equilibrium, but a system which has been caused to depart from equilibrium must be continuous. Any real process is called *irreversible.*

1.2.7. Reversible Process

In thermodynamics it is useful to consider idealized processes in the limit when their rates vanish. A process that occurs at a vanishingly small rate is called *quasi-static.* Certain quasi-static processes can be further idealized and regarded as a continuous sequence of equilibrium states. Such idealized quasi-static processes are called *reversible.*

Reversible processes in uniform systems can be represented as continuous lines in the space of n dimensions with coordinates $x_1, \ldots, x_n$. A real, irreversible process cannot be represented in such a simple way.

1.2.8. Intensive and Extensive Properties

A thermodynamic property is called *intensive* if its value in a uniform and homogeneous system in equilibrium is independent of mass. A property is called extensive if its value in such a system is proportional to mass.

1.2.9. Work

Work in thermodynamics is a natural extension of that in mechanics. Its definition can always be reduced to the evaluation of the product of a force and a distance, even though in thermodynamic equilibrium the quantities which appear in the appropriate expression have been transformed to represent thermodynamic quantities.

Work in a reversible process is a line integral evaluated along the curve which represents the process in the space $x_1, \ldots, x_n$. Moreover, the integrand is of the form

$$\mathrm{d}W^\circ = \sum_{\beta} Y_r(x_1, \ldots, x_n)\, \mathrm{d}z_r(x_1, \ldots, x_n), \tag{1.2}$$

that is, a sum of β products of intensive properties, Y_r, and perfect differentials of extensive properties, z_r. The properties z_r are often called *deformation variables*. The superscript $\circ$ in W° emphasizes the reversible nature of the process.

The work of a quasi-static or finite-rate irreversible process does *not* have the same mathematical properties as (1.2) even if its form is the same. In particular, in Equation (1.2) all quantities Y_r and z_r are thermodynamic properties. This is not necessarily the case for an irreversible process.

Work performed by the system is counted as positive, whereas that performed on the system is counted negative.

1.3. Equilibrium and Nonequilibrium Thermodynamics

The study of thermodynamics is naturally divided into two branches: equilibrium and nonequilibrium thermodynamics. Equilibrium thermodynamics deals with equilibrium states and reversible processes. Nonequilibrium thermodynamics deals with nonequilibrium states and irreversible processes in continuous systems. Equilibrium thermodynamics is, however, capable of making general statements about that final equilibrium state 2 which sets in after an irreversible process has been released in a system which was initially at an equilibrium state 1.

It is possible to take the view that equilibrium thermodynamics constitutes a closed, complete branch of physics. The same cannot be said about nonequilibrium thermodynamics.

1.4. The Laws of Thermodynamics

In stating the laws of thermodynamics we always have in mind a system which undergoes a reversible or irreversible process between an initial state 1 and a final state 2, *both of which are equilibrium states.*

1.4.1. Zeroth Law

The Zeroth law of thermodynamics asserts that two systems which are in equilibrium with a third across diathermal (that is, nonadiabatic) walls are in equilibrium with each other. It follows that all systems in equilibrium across diathermal walls have a common property called their *empirical temperature*, θ. An empirical temperature can be ascribed to every system, regardless of the nature of that system's independent properties. Thus, every system possesses a set of thermal equations of state

$$\theta = \theta(x_1, \ldots, x_n), \tag{1.3}$$

depending on the choice of the variables $x_1, \ldots, x_n$.

A preferred scale Θ for θ is defined with reference to the properties of real gases by postulating that

$$\Theta = \lim_{\rho \to 0} \frac{PV}{(PV)_3} \theta_3, \tag{1.4}$$

where P is the pressure, V is the volume, and ρ the density of the gas. The subscript 3 refers to measurements made when the gas is in equilibrium with the triple point of water, and by convention

$$\Theta_3 = 273.16 \text{ kelvin}. \tag{1.4a}$$

Here kelvin (K) denotes the internationally accepted unit of temperature, and the resulting scale is called the *perfect-gas temperature scale.*

1.4.2. First Law

The First law of thermodynamics states that: (a) given two arbitrary equilibrium states 1 and 2 of a system, it is always possible to release either process $1 \to 2$ or $2 \to 1$ connecting these two states *adiabatically*, that is, by the performance of work while the system is enclosed in a (usually deformable) adiabatic boundary; and (b) the amount of work performed in this way depends only on the two equilibrium states 1 and 2 and is independent of the details of the process.

It follows that there exists a potential $E(x_1, \ldots, x_n)$ in the n-dimensional space of states of any system whatsoever. This potential is called the *energy* of the system. In uniform systems at rest the energy is often referred to as the *internal energy* of the system, U. The energies of systems are defined except for a constant, and are usually normalized with respect to a reference state $x_1^*, \ldots, x_n^*$. The energies of subsystems, properly normalized, are additive and so energy is an extensive property.

When the system is not enclosed by an adiabatic boundary, the interaction with the surroundings involves the performance of work, W_{12}, and the *transfer of heat*, Q_{12} (positive into system and negative out of system), both quantities depending on the details of the process. However, their difference

$$Q_{12} - W_{12} = E_2 - E_1 \tag{1.5}$$

is equal to the difference in the potential E at the two states and is, therefore, independent of the details of the process.

In a reversible process

$$\begin{aligned} dE &= dQ^\circ - dW^\circ \\ &= dQ^\circ - \sum_\beta Y_r\, dz_r. \end{aligned} \tag{1.6}$$

1.4.3. First Part of Second Law

The first part of the Second law refers to reversible processes; it can be given a number of verbal formulations which all, eventually, lead to the statement that the linear differential form for reversible heat

$$\mathrm{d}Q^{\circ} = \mathrm{d}E + \sum_{\beta} Y_r \,\mathrm{d}z_r \tag{1.7}$$

is *integrable*. This means that there exists an integrating denominator $T(x_1, \ldots, x_n)$ which turns the form in Equation (1.7) into a potential $S(x_1, \ldots, x_n)$. Thus

$$\mathrm{d}S(x_1, \ldots, x_n) = \frac{\mathrm{d}Q^{\circ}}{T(x_1, \ldots, x_n)}$$
$$= \frac{\mathrm{d}E(x_1, \ldots, x_n) + \sum_{\beta} Y_r(x_1, \ldots, x_n)\,\mathrm{d}z_r(x_1, \ldots, x_n)}{T(x_1, \ldots, x_n)} \tag{1.8}$$

is a perfect differential. The potential S is called the entropy of the system. Like energy, entropy must be normalized and is additive; it is computed by integration along an arbitrary reversible path in the n-dimensional space of states. In thermodynamics it is assumed that such a reversible path *can* be indicated for any two equilibrium states 1 and 2 of any system whatsoever.

The integrating denominator, $T(x_1, \ldots, x_n)$, is a property of the system; it can be chosen so that it is made to be a unique function of the empirical temperature, θ, of every system:

$$T = T(\theta)\,. \tag{1.9}$$

For this reason, the integrating denominator is called the *thermodynamic temperature* of the system. If it is assumed that

$$T_3 = 273.16\ \mathrm{K}\,, \tag{1.9a}$$

it can be shown that this thermodynamic temperature, T, is identical with the perfect-gas temperature,

$$T \equiv \Theta\,. \tag{1.9b}$$

The sign of thermodynamic temperature is fixed by the convention adopted in Equation (1.9a) to be positive for so-called normal systems, that is, for systems whose energy has no upper bound. This is the case for the overwhelming majority of systems encountered in thermodynamics. Exceptions have been discovered (nuclear spin systems) when statistical methods have been applied to explain the results of experiments on the demagnetization of nuclei; they will not be discussed in this book.

The potential S defined by Equation (1.8) becomes singular for $T = 0$.

Consequently, the behavior of entropy as $T \to 0$ can only be inferred from a principle that is supplementary to the Second law.

1.4.4. Fundamental Equation of State

Equation (1.8) written as

$$dE = T\,dS - \sum_{\beta} Y_r\,dz_r \tag{1.10}$$

indicates that E can be regarded as a function of the $\beta + 1$ variables S, $z_1, \ldots, z_\beta$:

$$E = E(S, z_1, \ldots, z_\beta). \tag{1.11}$$

Hence it is seen that

$$n = \beta + 1. \tag{1.12}$$

Furthermore,

$$T = \left(\frac{\partial E}{\partial S}\right)_{z_i} \quad \text{and} \quad Y_i = -\left(\frac{\partial E}{\partial z_i}\right)_{S, z_j}, \tag{1.13a, b}$$

and

$$\left(\frac{\partial T}{\partial z_i}\right)_{S, z_j} = -\left(\frac{\partial Y_i}{\partial S}\right)_{z_j}, \quad \text{etc.} \tag{1.14}$$

Moreover, it can be shown that all equilibrium properties of any system can be derived from the equation of state (1.11) in which the energy is expressed as a function of the entropy and the deformation coordinates. For this reason, this equation is called a *fundamental equation*.

Additional fundamental equations can be obtained by performing Legendre transformations on Equation (1.10). To simplify matters, we suppose that (as for a gas)

$$Y_i = P, \qquad z_i = V, \qquad \beta = 1.$$

Hence

$$dU = T\,dS - P\,dV, \tag{1.15}$$

$$T = \left(\frac{\partial U}{\partial S}\right)_V, \qquad -P = \left(\frac{\partial U}{\partial V}\right)_S, \tag{1.16a, b}$$

and

$$\left(\frac{\partial T}{\partial V}\right)_S = -\left(\frac{\partial P}{\partial S}\right)_V. \tag{1.17}$$

The Legendre transforms of Equation (1.15) are most easily obtained by adding $\mathrm{d}(PV)$, $-\mathrm{d}(TS)$, and $\mathrm{d}(PV - TS)$ to both sides of the equation in turn. In this manner we obtain three additional forms

$$\mathrm{d}H = \mathrm{d}(U + PV) = T\,\mathrm{d}S + V\,\mathrm{d}P, \tag{1.18}$$

$$\mathrm{d}F = \mathrm{d}(U - TS) = -S\,\mathrm{d}T - P\,\mathrm{d}V, \tag{1.19}$$

$$\mathrm{d}G = \mathrm{d}(H - TS) = -S\,\mathrm{d}T + V\,\mathrm{d}P. \tag{1.20}$$

The potentials

$$H(S, P) = U + PV, \tag{1.21}$$

$$F(T, V) = U - TS, \tag{1.22}$$

$$G(T, P) = H - TS = U + PV - TS, \tag{1.23}$$

are called enthalpy, Helmholtz function, and Gibbs function, in that order. For each of those potentials, it is possible to write two *reciprocity relations* analogous to Equations (1.16a, b) and one *reciprocal* (or Maxwell) relation analogous to Equation (1.17). All such relations can be derived from the obvious property of the Jacobian

$$\frac{\partial(P, V)}{\partial(T, S)} \equiv 1. \tag{1.24}$$

The forms in Equations (1.21)–(1.23) represent alternative fundamental equations.

1.4.5. Second Part of Second Law

The second part of the Second law of thermodynamics can also be given several alternative formulations. They all lead to the statement that in an isolated system, the entropy, S_2, of a final state exceeds the entropy, S_1, of an initial state in an irreversible process that has occurred in it. Thus,

$$S_2 > S_1, \tag{1.25}$$

and the difference

$$\sigma = S_2 - S_1 \tag{1.26}$$

is called the *entropy produced* in the system during the irreversible process $1 \to 2$. Thus

$$\sigma \geq 0, \tag{1.27}$$

the sign of equality having been added to include the limiting case when process $1 \to 2$ is reversible.

If a system performs an *isothermal*, quasi-static, but irreversible process, the inequality (1.25) leads to

$$dS \geq \frac{dQ}{T} \tag{1.28}$$

or to

$$dS = \frac{dQ}{T} + d\sigma \qquad \text{with} \quad d\sigma \geq 0. \tag{1.29}$$

It can be shown that for such a process

$$T\,d\sigma = dW^{\circ} - dW. \tag{1.30}$$

1.4.6. Open Subsystem

When the first part of the Second law of thermodynamics is rewritten for an open subsystem, it takes the equivalent forms

$$dU = T\,dS - P\,dV + \sum_{\alpha} \mu_i\,dn_i, \tag{1.31}$$

$$dH = T\,dS + V\,dP + \sum_{\alpha} \mu_i\,dn_i, \tag{1.32}$$

$$dF = -S\,dT - P\,dV + \sum_{\alpha} \mu_i\,dn_i, \tag{1.33}$$

$$dG = -S\,dT + V\,dP + \sum_{\alpha} \mu_i\,dn_i. \tag{1.34}$$

Here n_i denotes the number of mols of a representative species of α chemical species present in the subsystem, and μ_i denotes the *chemical potential* of species i in the mixture. The fundamental equations for such a subsystem have the equivalent forms

$$U = U(S, V, n_1, \ldots, n_\alpha), \tag{1.35}$$

$$H = H(S, P, n_1, \ldots, n_\alpha), \tag{1.36}$$

$$F = F(T, V, n_1, \ldots, n_\alpha), \tag{1.37}$$

$$G = G(T, P, n_1, \ldots, n_\alpha). \tag{1.38}$$

Since the last form contains two intensive parameters, T, P, and α extensive parameters, n_i, and since G is extensive, it must be homogeneous of order 1 in the variables n_i. Consequently, we have

$$G = \sum_{\alpha} n_i \mu_i. \tag{1.39}$$

Taking the perfect differential of G and comparing the resulting expression with Equation (1.34), we derive the *Gibbs–Duhem–Margules relation*

$$-S\,\mathrm{d}T + V\,\mathrm{d}P - \sum_{\alpha} n_i\,\mathrm{d}\mu_i = 0. \tag{1.40}$$

Finally, from Equations (1.31)–(1.34), we see that

$$\mu_i = \left(\frac{\partial U}{\partial n_i}\right)_{S,V} = \left(\frac{\partial H}{\partial n_i}\right)_{S,P} = \left(\frac{\partial F}{\partial n_i}\right)_{T,V} = \left(\frac{\partial G}{\partial n_i}\right)_{T,P}. \tag{1.41}$$

There is no difficulty in writing reciprocity and reciprocal relations based on the four differential forms (1.31)–(1.34).

1.4.7. Third Law

The Third law of thermodynamics provides us with the independent principle, mentioned at the end of Section 1.4.3, which determines the behavior of entropy as $T \to 0$. The Third law asserts that the entropy difference in an isothermal process tends to zero as the thermodynamic temperature tends to zero. More precisely, we consider a system whose independent variables are T and x and assert that the entropy difference

$$\Delta S = S(T, x_2) - S(T, x_1) \tag{1.42}$$

must vanish as $T \to 0$, so that

$$\lim_{T \to 0} \Delta S = 0. \tag{1.43}$$

Here x_2 and x_1 denote the values of the second parameter at two states 1 and 2 which lie on the isotherm T. If more than two variables are independent, we write

$$\Delta S = S(T, x_2', x_2'', \ldots) - S(T, x_1', x_1'', \ldots).$$

Since the entropy of a system has the properties of a potential, it contains an arbitrary constant, S_0. The value of this constant depends on the normalization convention and is unaffected by the passage to the limit implied by Equation (1.43). Thus it is possible to assign a value zero to it, and to put

$$S(0, x_2) = S(0, x_1) = 0. \tag{1.44}$$

Entropy that is normalized in this manner is called *absolute entropy*. This particular convention does not alter the fact that entropy has the mathematical properties of a potential.

The application of the Third law, as expressed in Equation (1.43), to practical problems is made difficult sometimes by the fact that many systems develop internal constraints upon cooling and often exist in so-called metastable states. It is, therefore, necessary to remember that the passage to the limit $T \to 0$ in Equation (1.43) must occur without altering or disturbing such internal constraints. If they existed at $T \neq 0$, they must continue to exist at all temperatures as $T \to 0$; if they become removed by the action of some disturbance at a higher temperature, care must be taken to avoid their development when experiments are extrapolated to $T = 0$.

The Third law leads to the conclusion that a temperature of $T = 0$ precisely is unattainable.

1.4.8. Equilibrium Principle

The Zeroth and First laws and the first part of the Second law assure us of the existence of three very important thermodynamic properties for the equilibrium states of all systems. These are: the energy, E, or the internal energy, U; the thermodynamic temperature, T; and the entropy, S. The Third law determines the behavior of entropy as $T \to 0$, and the second part of the Second law gives an indication of the direction of natural processes.

None of the preceding principles is capable of providing a unique solution to the fundamental problem of thermodynamics formulated in Section 1.2.5. This is achieved by the Equilibrium principle.

In order to grasp the physical meaning of the Equilibrium principle, it is necessary to realize two circumstances. First, we realize that in an isolated system we must have

$$E_1 = E_2. \tag{1.45}$$

Secondly, we must realize that the manipulation of internal constraints changes the number, and sometimes the character, of the independent variables in the functional relation for energy, that is, in the fundamental equation. This means that the independent variables in state 2 are not the same as those in state 1.

The Equilibrium principle asserts that *the entropy* S of the system conceived as a function of the independent variables of state 1 *must be a maximum.*

The Equilibrium principle can be expressed in several equivalent forms depending on the nature of the external constraints chosen for it. The following four forms are most useful in practice; in writing them, we introduce a representative deformation coordinate, z, and a representative conjugate intensive parameter, Y, but it must be remembered that the system's fundamental equation may contain γ parameters. The subscripts 1 and 2 refer to the initial and final equilibrium states of the whole system. The four forms are:

if $S_1 = S_2$ and $z_1 = z_2$, then E must be a minimum;
if $S_1 = S_2$ and $Y_1 = Y_2$, then H must be a minimum;
if $T_1 = T_2$ and $z_1 = z_2$, then F must be a miminum;
if $T_1 = T_2$ and $Y_1 = Y_2$, then G must be a minimum.

The principles are all so formulated as to render the *appropriate potential* a *minimum*. The system assumes this minimum value of the potential in the final equilibrium state, and the function which is to be rendered a minimum must contain all variables which characterize the initial equilibrium state and which are subject to change during the process.

1.5. Continuum Thermodynamics

Although the principles of equilibrium thermodynamics are sufficient to solve the fundamental problem (Section 1.2.6), the actual process which takes place between the initial and final states of equilibrium cannot be analyzed on this basis, except for the highly idealized case when a reversible process—a sequence of equilibrium states—can be substituted for it. During an irreversible process the system becomes continuous even in the absence of externally imposed fields, and velocity as well as diffusive fields are developed throughout the system. Clearly, a supplementary set of concepts and principles must be developed to handle such a sequence of nonequilibrium states. Although, as already stated, no general science of nonequilibrium thermodynamics has yet come into being, it is possible to develop an admittedly approximate formalism known as the *thermodynamics of irreversible processes*.†

1.5.1. Principle of Local State

The description of a nonequilibrium state is provided by the adoption of the principle of local state, which recognizes that each of the independent properties, x_i, of the equilibrium end-states must now be replaced by an instantaneous field $x_i(\xi_j)$, where ξ_1, ξ_2, ξ_3 denote the three Cartesian space coordinates. To these it is necessary to add the velocity field $v_i(\xi_j)$ and a set of $\alpha - 1$ composition fields, $v_i^\alpha(\xi_j)$, where α is the number of distinct chemical species present in the system. The principle of local state (also known as the *hypothesis of local equilibrium*) asserts that the continuous system can be thought of as a collection of an infinite number of infinitesimal, elementary (usually open) subsystems each of which undergoes a quasi-static irreversible process and that the fundamental equilibrium equations of state remain valid locally and instantaneously. In other words, the local spatial gradients of the properties x_i, v_i, v_i^α as well as their instantaneous temporal rates $\dot{x}_i$, $\dot{v}_i$, $\dot{v}_i^\alpha$

† See, for example, S. R. de Groot and P. Mazur, *Non-Equilibrium Thermodynamics*, North-Holland Publ., Amsterdam, 1962; J. Kestin, *A Course in Thermodynamics*, Vol. II, Chapter 24, Blaisdell, Waltham, Massachusetts, 1969; J. Meixner and H. G. Reik, Thermodynamik der irreversiblen Prozesse, in *Handbuch der Physik* (*Encyclopedia of Physics*), Vol. III/2 (S. Flügge, ed.), pp. 413–523, Springer, Berlin, 1959.

can be omitted from the description of the state of a continuous system. The process is fully described by indicating the functions

$$x_i(\xi_j, t), \qquad v_i(\xi_j, t), \qquad \text{and} \qquad v_i^{\alpha}(\xi_j, t)$$

of space coordinates and time. The underlying idea is identical with that used when Newton's equations are generalized to a continuum, and when Cauchy's equations of motion are derived.

1.5.2. Interaction Vectors

In contrast with an equilibrium state or a reversible process, during an irreversible process the elementary subsystems interact with each other. This interaction is described instantaneously by the vector field $\dot{q}_i(\xi_j)$ of heat flux, by the vector field $w_i(\xi_j)$ of work flux and by $\alpha - 1$ vectorial fields $j_i^{\alpha}(\xi_j)$ of diffusive fluxes. Adopting the Euler–Cauchy principle of continuum mechanics which states that the tensor field $\sigma_{ij}(\xi_k)$ is sufficient to describe the work interaction, that is, that distributed couple stresses are absent or unimportant, it is found that the stress tensor (in the absence of externally imposed fields) is symmetric and that

$$\dot{w}_j = -\sigma_{ji} v_i, \tag{1.46}$$

where the summation convention for repeated indices is implied.

1.5.3. Laws of Thermodynamics and Motion

The irreversible thermodynamic process which occurs in a continuous system is now governed by a set of field equations which are represented by the appropriately generalized laws of motion and laws of thermodynamics. In particular, the First law is

$$\rho \frac{\mathrm{d}e}{\mathrm{d}t} + \frac{\partial \dot{q}_i}{\partial \xi_i} + \frac{\partial \dot{w}_i}{\partial \xi_i} = 0, \tag{1.47}$$

and the second part of the Second law becomes

$$\rho \frac{\mathrm{d}s}{\mathrm{d}t} + \frac{\partial \dot{\eta}_i}{\partial \xi_i} = \dot{\sigma} \tag{1.48}$$

($\dot{\sigma} > 0$, irreversible; $\dot{\sigma} = 0$, equilibrium or reversible). For simplicity, both equations have been written for closed subsystems (absence of diffusion). Here $e = u + \frac{1}{2} v_i v_i$, and

$$\dot{\eta}_i = \frac{\dot{q}_i}{T} \tag{1.49}$$

denotes the entropy flux vector. The quantity $\dot{\sigma}$ represents the local volume-rate of *entropy production* which must be positive everywhere during an irreversible process and which must vanish in equilibrium.

1.5.4. Phenomenological Assumptions

The theory allows us to derive an explicit expression for the rate of local entropy production, $\dot{\sigma}$. This appears in the form of a sum of products

$$\dot{\sigma} = \sum J_\mu X_\mu \tag{1.50}$$

of generalized fluxes, J_μ, and generalized forces, X_μ. The former are divergences of extensive thermodynamic properties, and the latter are gradients of intensive thermodynamic properties. Since both vanish simultaneously at equilibrium, we use the phrase that the fluxes are driven by the forces and recognize that the relations between them must be homogeneous. It is clear that equilibrium thermodynamics provides no indication about this relation. In the case of most processes of practical interest, a linear phenomenological assumption of the form

$$J_\mu = L_{\mu\nu} X_\nu \tag{1.51}$$

proves to be adequate. The *transport coefficients*, $L_{\mu\nu}$, cannot be obtained from any fundamental equation, but must be measured separately in non-equilibrium states.

Regarding the transport coefficients, statistical considerations lead us to the adoption of the Onsager–Casimir principle, treated as a new postulate in continuum thermodynamics. This asserts that there exist two kinds of generalized forces. Forces of class α are even functions of the microscopic velocities of particles; they remain unchanged when all microscopic velocities change sign. Forces of class β are odd functions of the particle velocities; they change sign when all microscopic velocities change sign. The phenomenological coefficients satisfy the reciprocal relation (in the absence of a magnetic field or rotation):

$$L_{\mu\nu} = \varepsilon(X_\mu)\varepsilon(X_\nu)L_{\nu\mu},$$

where

$$\varepsilon(X_\mu),\quad \varepsilon(X_\nu) = \begin{matrix} +1 & \text{if } X_\mu \text{ or } X_\nu \text{ is of class } \alpha \\ -1 & \text{if } X_\mu \text{ or } X_\nu \text{ is of class } \beta. \end{matrix} \tag{1.52}$$

1.5.5. Limits of Validity of the Theory

The theory of continuum thermodynamics is based on the principle of local state and on linear phenomenological assumptions. Therefore, its validity is restricted to small departures from thermodynamic equilibrium. To be in a position to specify the degree of deviation from thermodynamic equilibrium which may be regarded as small, it is necessary to formulate a more general theory which would contain the present one as a special case. At present, a more general theory seems to be beyond our grasp other than in the context of statistical thermodynamics. Even though some progress in this direction has been made, no general theory of nonequilibrium states has yet evolved with the exception of one for gases, which is based on Boltzmann's equation (Chapter 13). In the case of gases, the formalism of continuum thermodynamics is, indeed, confirmed in first and second approximation, particularly with respect to the phenomenological hypotheses known as Fourier's law of conduction, Stokes's law of fluid friction, and Fick's law of diffusion (Chapter 12). These turn out to be valid under two restrictions: First, the temperature change over one mean free path (Section 12.4) must be small compared with the absolute temperature. Secondly, changes in velocity over the same distance or over a characteristic relaxation time must be small compared with the velocity of sound. Considering that the mean free path is of the order of 10^{-4} cm under normal conditions, we conclude that these restrictions still leave a wide margin which is only seldom transgressed. The preceding conclusions are supported by an impressive mass of experimental results.

PROBLEMS FOR CHAPTER 1†

1.1. Derive closed-form expressions for the fundamental equation $u = u(s, x)$ for the following systems:

(a) a mol of perfect gas with constant specific heats;
(b) a mol of a paramagnetic substance which obeys Curie's law $\mathscr{M} = C\mathbf{H}/T$ and whose specific heat $c_{\mathscr{M}}$ is a constant;
(c) an elastic rod which obeys Hooke's law and whose elastic constant E and specific heat c_ε are constants;
(d) a galvanic cell whose emf is the following function of temperature: $\mathscr{E} = \mathscr{E}^\circ + (aT - T_0)$ and whose heat capacity C is a constant.

Here x stands for the appropriate deformation variable: (a) $x = v$, specific volume; (b) $x = \mathscr{M}$, magnetization; (c) $x = \varepsilon$, strain; (d) $x = z$, charge.

SOLUTION: We denote the intensive parameter conjugate to x by Y and write $du = T\,ds + Y\,dx$ [(a) $Y = -P$; (b) $Y = \mathbf{H}$; (c) $Y = \sigma$; for this system we may assume that the volume, V_0, remains unchanged; (d) $Y = \mathscr{E}$]. Hence

$$\left(\frac{\partial u}{\partial x}\right)_T = T\left(\frac{\partial s}{\partial x}\right)_T + Y,$$

† Problems numbered in boldface type are considered to be relatively easy; they are recommended for assigning in an elementary first course.

but, by Maxwell's relation $(\partial s/\partial x)_T = -(\partial Y/\partial T)_x$, so that

$$\left(\frac{\partial u}{\partial x}\right)_T = -T\left(\frac{\partial Y}{\partial T}\right)_x + Y. \tag{1}$$

The right-hand side of this equation can now be evaluated explicitly with the aid of the thermal equation of state. In this manner,

(a) $$-P = -\frac{\mathbf{R}T}{v} \quad \text{and} \quad \left(\frac{\partial u}{\partial v}\right)_T = 0;$$

(b) $$\left(\frac{\partial u}{\partial \mathscr{M}}\right)_T = 0; \qquad \text{(c)} \quad \sigma = E[\varepsilon - \alpha(T - T_0)]$$

and $[\partial u/\partial \varepsilon]_T = E\varepsilon + \alpha E T_0$.

Integrating Equation (1), we obtain

$$u = -\int_{x_0}^{x} \left[T\left(\frac{\partial Y}{\partial T}\right)_x - Y\right] \mathrm{d}x + \psi(T) \tag{2}$$

and verify that $\psi'(T) = c_x$. Thus,

$$u = -\int_{x_0}^{x} \left[T\left(\frac{\partial Y}{\partial T}\right)_x - Y\right] \mathrm{d}x + c_x(T - T_0)$$

where T_0 is a reference temperature. In particular: (a) $u = c_v(T - T_0)$; (b) $u = c_{\mathscr{M}}(T - T_0)$; (c) $u = \frac{1}{2}E\varepsilon^2 + \alpha E\varepsilon T_0 + c_\varepsilon(T - T_0)$.

Next we compute the entropy from

$$\mathrm{d}s = \frac{\mathrm{d}u - Y\,\mathrm{d}x}{T}, \tag{3}$$

eliminating Y with the aid of the thermal equation of state.

(a) $$s = c_v \ln T/T_0 + \mathbf{R} \ln v/v_0 + s_0;$$

(b) $$s = c_{\mathscr{M}} \ln T/T_0 - (\mathscr{M}^2 - \mathscr{M}_0^2)/2C + s_0;$$

(c) $$s = c_\varepsilon \ln T/T_0 + \alpha E\varepsilon + s_0.$$

The elimination of temperature from Equation (2) and the integral form of Equation (3) yields the desired forms of the fundamental equation:

(a) $$u = c_v T_0\left[\left(\frac{v_0}{v}\right)^{\mathbf{R}/c_v} \exp\left(\frac{s}{c_v}\right) - 1\right]$$

with $s_0 = 0$ at v_0, T_0;

(b) $$u = c_{\mathscr{M}} T_0\left\{\exp\left[\frac{1}{c_{\mathscr{M}}}\left(s + \frac{\mathscr{M}^2}{2C}\right)\right] - 1\right\}$$

with $s_0 = 0$ at T_0, $\mathscr{M}_0 = 0$;

(c) $$u = \tfrac{1}{2}E\varepsilon^2 + \alpha E\varepsilon T_0 + c_\varepsilon T_0\left[\frac{\exp(s/c_\varepsilon)}{\exp(\alpha E\varepsilon/c_\varepsilon)} - 1\right],$$

with $s_0 = 0$ at $\varepsilon = 0$, $T = T_0$.

The same general procedure can be followed for system (d).

1.2. Derive the closed-form fundamental equation $h = h(s, P)$ for one mol of an incompressible fluid whose specific heat c is a constant.

1.3. Transform the fundamental equations in Problem 1.1 to the form $f = f(T, x)$, where f is the specific Helmholtz function and x is the appropriate deformation variable. Why can this not be done for the system in Problem 1.2? In the latter case, write the equation $g = g(T, P)$ for the Gibbs function.

1.4. Identify the canonical independent variables for the fundamental equation of a pure substance in which the dependent variables are: (a) the Massieu function $j = s - u/T$; (b) the Planck function $y = s - h/T$. List these variables for the following systems: (i) a pure substance; (ii) an incompressible fluid; (iii) a paramagnetic substance; (iv) a dielectric; (v) an elastic body.

1.5. Derive the fundamental equations $u = u(s, v, x)$, $h = h(s, P, x)$, $f = f(T, v, x)$, and $g = g(T, P, x)$ for a mixture of two perfect gases with constant specific heats. Here x is the mol fraction of one gas, that for the other being, evidently, $1 - x$.

1.6. Show that the thermal equation of state of a pure substance in a single phase, $F(P, v, T) = 0$, together with the function $c_p{}^\circ = c_p{}^\circ(T)$ which represents its specific heat at constant pressure and at an arbitrary standard pressure, P^*, as a function of temperature, are sufficient to determine the system's fundamental equation.

Indicate the steps required to derive the fundamental equation $g = g(P, T)$. [Note: See J. Kestin, *A Course in Thermodynamics*, Vol. I, Chapter 12, Blaisdell, Waltham, Massachusetts, 1966.]

1.7. It will be stated in Chapter 10 that black-body radiation (a perfect gas composed of photons in statistical equilibrium) can be treated as a macroscopic thermodynamic system. It can be shown that for such a gas, the energy per unit volume, u, is a function of temperature alone, $u = u(T)$, and that the pressure is $P = \frac{1}{3}u$. Hence, in a reversible process $dW^\circ = \frac{1}{3}u(T)\,\mathrm{d}V$.

(a) Show that the entropy of this system is given by

$$\mathrm{d}S = \frac{V}{T}\frac{\mathrm{d}u}{\mathrm{d}T}\,\mathrm{d}T + \frac{4}{3}\frac{u}{T}\,\mathrm{d}V.$$

(b) Show that

$$\frac{1}{T}\frac{\mathrm{d}u}{\mathrm{d}T} = \frac{4}{3}\frac{\mathrm{d}}{\mathrm{d}T}\left(\frac{u}{T}\right).$$

(c) Prove that the property (b) leads to $u = aT^4$, where a is an undetermined constant.

(d) By integration, show that $S = \frac{4}{3}aT^3V$, where, according to the Third law, we have put $S = 0$ at $T = 0$.
(e) Show that the equations of an isentropic process are

$$T^3V = \text{const} \qquad \text{or} \qquad PV^{4/3} = \text{const}.$$

(f) Show that the Helmholtz function is given by

$$F = -\tfrac{1}{3}aT^4V = -\tfrac{1}{3}U.$$

(g) Calculate the Gibbs function $G = U + PV - TS$ and comment on the result. Why is $G = 0$ identically? (If you cannot answer this question at this stage, revert to it after you have studied Chapter 10.)
(h) Finally, list which of the preceding equations are fundamental ones for the system.

SOLUTION: (a) Express the internal energy U as Vu and use the equation $dU = T\,dS - P\,dV$ to obtain dS.
(b) Use the Maxwell relation

$$\frac{\partial}{\partial V}\left(\frac{V}{T}\frac{du}{dT}\right) = \frac{\partial}{\partial T}\left(\frac{4}{3}\frac{u}{T}\right).$$

(c) Integrate the differential equation obtained in (b).
(d) Write $dS = (\partial S/\partial V)_T\,dV + (\partial S/\partial T)_V\,dT$ and use the result of part (a) to show that $S = Vs$, with $s = 4aT^3/3$.
(e) Along an isentropic process $S = \text{const}$, so that VT^3 must remain constant.

1.8. Assuming that a vessel containing a reactive mixture of α perfect gases in a single, gaseous phase is equivalent to α coexisting open systems, and that the species participate in a single independent chemical reaction, show that the rate of entropy production on approaching equilibrium is $\dot{\theta} = \mathscr{A}\dot{\xi}/T$ where $\dot{\xi}$ is the time-rate of change of the extent of reaction and

$$\mathscr{A} = -\sum_{i=1}^{\alpha} \mu_i \nu_i,$$

where ν_i are the stoichiometric coefficients.

Hence, show that the condition of equilibrium ("law of mass action") is $\sum \mu_i \nu_i = 0$.

Note that the extent of reaction is defined as $d\xi = dn_i/\nu_i$, with $\nu_i < 0$ for the reactants and $\nu_i > 0$ for the products.

Specializing to a perfect gas, show that this is equivalent to

$$\prod_{i=1}^{\alpha}\left(\frac{n_i^\circ + \nu_i\xi}{\sum_{i=1}^{\alpha} n_i^\circ + \xi\sum_{i=1}^{\alpha}\nu_i}\right)^{\nu_i} = \exp(-\Delta G/\mathbf{R}T)$$

$$= (P/P^*)^{-\Delta\nu}\exp(-\Delta G^\circ/\mathbf{R}T).$$

Here n_i° are the mol numbers of the species present in the reaction vessel at the beginning ($\xi = 0$),

$$\Delta\nu = \sum_{i=1}^{\alpha} \nu_i, \qquad \Delta G = \sum_{i=1}^{\alpha} \nu_i g_i(P, T),$$

$$\Delta G^\circ = \sum_{i=1}^{\alpha} \nu_i g_i^\circ(T), \qquad g_i^\circ(T) \equiv g_i(P^*, T),$$

and P^* denotes the standard pressure.

1.9. Perform the analysis of Problem 1.8 on the assumption that the vessel sustains r independent chemical reactions.

1.10. Two compartments, a and b, of an adiabatic cylinder contain different amounts of the same perfect gas. The gas constant is R and the specific heat c_v is a constant. In state 1 the left compartment contains m_a units of mass at temperature T_a in a volume V_a, whereas the right compartment contains m_b units of mass at temperature T_b in a volume V_b.

At a certain instant, the adiabatic slide S is removed, thus starting an irreversible process in the system which ends in an equilibrium state 2 of uniform temperature T_2.

(a) Determine the number of independent variables which describe state 1.
(b) Determine the number of independent variables which describe state 2. Determine the temperature T_2 and pressure P_2 explicitly.
(c) Write expressions for the energy and entropy of the system for a state which is intermediate between 1 and 2.

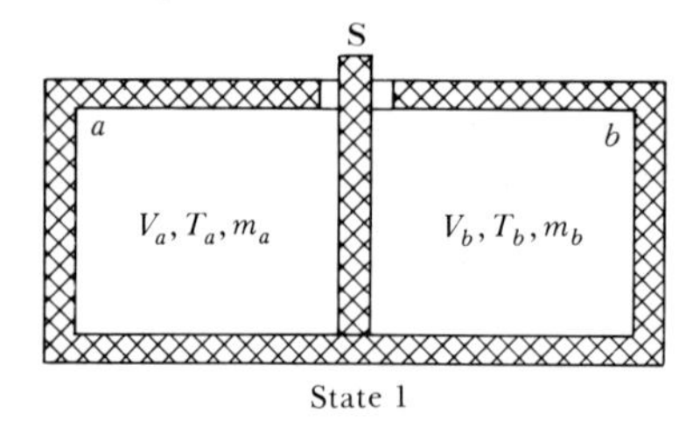

State 1

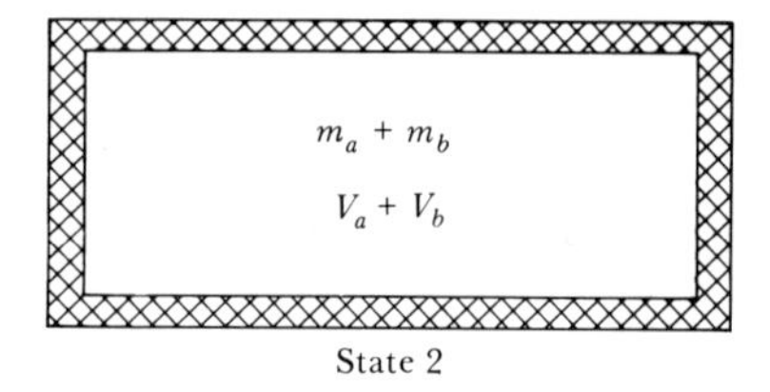

State 2

Note that part (c) of this problem must be solved in terms of several instantaneous fields, the velocity field $\mathbf{v}(\mathbf{r})$, the temperature field $T(\mathbf{r})$, etc. Indicate how these fields would be determined in a concrete case.

1.11. Solve Problem 1.10 on the assumption that the two gases are different, but inert chemically, that their gas constants are R_a and R_b and that their specific heats are c_{va} and c_{vb}, respectively.

1.12. Solve Problem 1.10 on the assumption that the two gases are different and chemically active. Denoting the chemical symbols of the two gases by A_m and B_n assume that they undergo the single reaction

$$a\mathrm{A}_m + b\mathrm{B}_n = \mathrm{A}_{am}\mathrm{B}_{bn}.$$

Specifically, solve for the case when $A_m = H_2$, $B_n = O_2$ and the reaction is

$$H_2 + \tfrac{1}{2}O_2 = H_2O\,.$$

1.13. A thin metallic rod connects two reservoirs, each maintained at a constant temperature, as shown in the sketch. In steady state, the temperature distribution along the rod may be assumed to be linear.

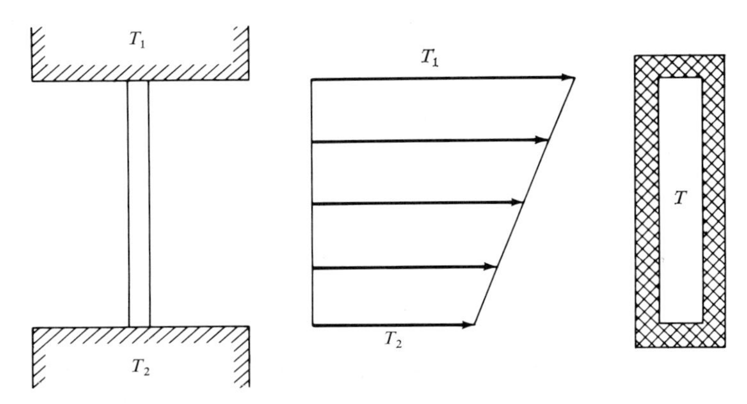

Enumerate the fields which are necessary to determine the state of the rod bearing in mind that it is not one of equilibrium, but assuming that the metal is incompressible.

At a certain instant, the rod is detached and insulated adiabatically, so that it reaches an equilibrium state after a sufficient lapse of time.

Determine the final temperature, T_f, and the change in entropy which is brought about by this process, assuming that the specific heat, c, of the metal is a constant. At which point did you invoke the Principle of local state?

1.14. Calculate the maximum amount of work that can be derived from the system performing the process described in Problem 1.13 if the surroundings are at temperature T_0.

LIST OF SYMBOLS FOR CHAPTER I

Latin letters

E	Energy
$\mathscr{E}$	Potential difference in a capacitor
e	Energy
F	Helmholtz function
G	Gibbs function
H	Enthalpy
$\mathbf{H}$	Magnetic field intensity

J	Generalized flux
j	Diffusive flux
L	Transport coefficient
$\mathcal{M}$	Magnetization
n	Number of independent properties of a thermodynamic system; number of mols
P	Pressure
Q	Heat
Q°	Reversible heat
q	Heat
$\mathbf{r}$	Vector with components x, y, z
S	Entropy
T	Temperature
U	Internal energy
V	Volume
$\mathbf{v}$	Velocity vector with components v_x, v_y, v_z
W	Work
W°	Reversible work
w	Work
X	Generalized force; deformation coordinate
x	Independent variable of a thermodynamic system; coordinate
Y	Intensive property
y	Dependent variable of a thermodynamic system; coordinate
z	Deformation variable; extensive property; charge on a capacitor; coordinate

Greek letters

α	Class of forces; number of independent components
β	Number of terms in the expression for the reversible work; class of forces
ε	Symbol having the values ± 1
ε_{ij}	Strain components
$\dot{\eta}$	Entropy flux
Θ	Temperature on perfect-gas scale
θ	Empirical temperature
μ	Chemical potential
ν	Composition variable
ξ_1, ξ_2, ξ_3	Cartesian spatial coordinates
σ	Entropy production
σ_{ij}	Stress tensor element

Subscript

3	Triple point of water

Special Symbol

$\cdot$	Time rate of change

PART 1

Fundamental Theory

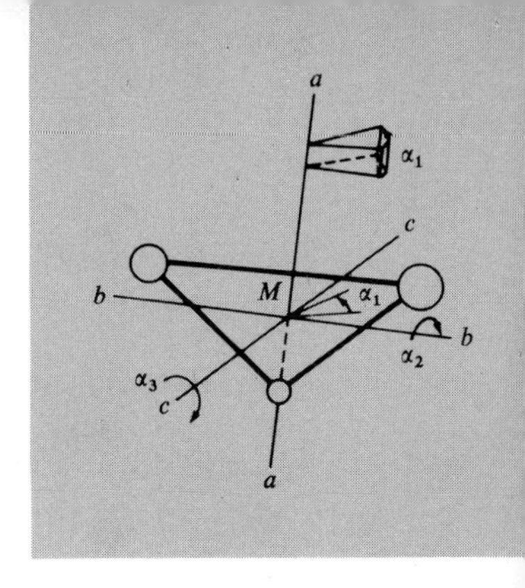

CHAPTER 2

INTRODUCTION TO STATISTICAL THERMODYNAMICS AND MECHANICAL MODELS

2.1. Prefatory Remarks

The theory of statistical thermodynamics is built with the aid of three blocks: the view that matter is composed of discrete elementary particles; the conviction that the motion of all elementary particles as well as that of systems comprising large collections of them is governed by the laws of *quantum mechanics*, that is, by *Schrödinger's wave equation*; and the idea that macroscopic thermodynamic properties constitute *statistical averages* of mechanical quantities taken over properly defined classes of microstates of the adopted model, that is, over a properly postulated *sample space*.

The First law of thermodynamics allows us to recognize that the motion of the mechanical model of a thermodynamic system must be conservative. This assures the existence of an internal energy which has the mathematical properties of a potential. The remaining laws of classical thermodynamics as well as the extremum principles emerge as natural generalizations of the statistical properties of large mechanical systems which are independent of the details of their structure. Beyond this, statistical thermodynamics leads to explicit expressions for the fundamental equation of state of every system whose mechanical model is known in sufficient detail.

The first modern successful steps toward the development of statistical thermodynamics were taken by Ludwig Boltzmann (1844–1906), Josiah

Willard Gibbs (1839–1903), and James Clark Maxwell (1831–1879). The decisive step toward the development of quantum mechanics was taken by Max Planck in 1900 when he discovered his empirical interpolation formula—the "lucky guess" as he described it in his Nobel Prize inaugural address in 1920—for the spectral distribution of energy in black-body radiation. The development of the concepts of quantum mechanics owes a great deal to the work of Albert Einstein who discovered the correct description of the photoelectric effect and who demonstrated the connection between the microscopic point of view and Brownian motion. The complete formulation of quantum mechanics emerged as a result of the work of Erwin Schrödinger (1926), Max Born, Werner Heisenberg (1925), Wolfgang Pauli, and P. A. M. Dirac. The idea that all forms of radiation possess particulate aspects and that all streams of elementary particles exhibit a wavelike ability to produce diffraction patterns was boldly formulated in 1924 by Prince Louis-Victor de Broglie; Niels Bohr pioneered these ideas in the study of the structure of atoms.

2.2. Microscopic Description of Thermodynamic Systems. Statistical Thermodynamics, Classical and Quantum Mechanics

In contrast with the macroscopic description, we can attempt to describe and to analyze the systems encountered in nature by accepting the view that matter consists of discrete particles, that is, of atoms and molecules. According to this point of view, for example, a certain quantity of gas enclosed in a vessel consists of an enormously large number of molecules which move chaotically in all directions colliding with one another and bouncing against the walls of the container. In developing this hypothesis, we shall find it necessary to make further assumptions regarding the nature of these collisions, to state, for example, whether or not they are elastic, and to specify whether the molecules interact during collisions only or whether they also exert long-range forces on one another. Finally, we shall be forced to speculate about the nature of the bonds which link atoms in a molecule, whether they are rigid, or otherwise. In other words, we shall be forced to *postulate* a suitable *molecular model* for every substance.

It is quite clear that no direct experiments to verify the correctness of such models are possible. Any instrument which we might use for the purpose would be enormously large compared with the size of a molecule and could not, therefore, serve to measure the forces acting on the latter. The only way in which we can judge whether any assumed model is adequate is to deduce from it the behavior of a large collection of molecules, that is, of a macroscopic system, and to compare the predicted behavior with large-scale experiments. Models which result in agreement with the experiments are retained, models which conflict with them are rejected or modified. Here, of course, agreement always leaves a margin of experimental error and refinements in the model usually accompany refinements in measuring techniques. By this means, the preceding *microscopic* description is made to agree with the macroscopic description in spite of the apparent, but superficial, discrepancy between them.

Adopting the microscopic point of view, we would say that two states of a system are identical if all molecules have identical positions and move with identical velocities in both cases. The natural way of describing such a *microstate* in the case of molecules which move like material points would be to introduce a system of coordinates x, y, z, and to specify for every molecule its position vector $\mathbf{r}(x, y, z)$ and its velocity vector $\mathbf{c}(u, v, w)$ by indicating its components u, v, w along x, y, z respectively. In actual calculations of this nature it has been found preferable to specify the components p_x, p_y, p_z of the linear momentum $\mathbf{p} = m\mathbf{c}$ of the molecules (of mass m) instead of the components of velocity. If the system contains N molecules, we would require $6N$ coordinates to define its state. The number of molecules, N, even in a minute macroscopic system is very large. From comparisons between calculations and experiment, for example, it is known that 1 cm^3 of a gas at atmospheric pressure and at the temperature of melting ice contains $N = \mathbf{L} = 2.687 \times 10^{19}$ molecules!† It is quite clear, therefore, that the preceding obvious scheme is impractical, even if methods for measuring the positions and momenta of molecules did exist. The number of coordinates would simply be impossibly large.

An alternative method consists in dividing the *physical space* in which the molecules perform their motion into cubical or rectangular cells whose volume is given by $\Delta x\, \Delta y\, \Delta z$, Figure 2.1a. We concentrate on a typical cell which encloses the portion of the physical space contained between the coordinates x, y, z and $x + \Delta x$, $y + \Delta y$, $z + \Delta z$. Instead of describing the momenta of all molecules contained in the cell, we again concentrate on those whose momenta are contained between the values p_x, p_y, p_z and $p_x + \Delta p_x$, $p_y + \Delta p_y$, $p_z + \Delta p_z$. If we imagine the three momentum components represented on the three axes of a Cartesian system of coordinates, Figure 2.1b, we obtain the *momentum space*. Hence, a molecule is represented by two associated points, one in each of the two spaces. In order to represent a molecule by one point, we can imagine, even if we cannot visualize, a six-dimensional space, the (classical) *phase space for a molecule* or μ-space,‡ whose Cartesian coordinates are x, y, z, p_x, p_y, p_z, and isolate from it a

† More precisely, $\mathbf{L} = (2.68719 \pm 0.00010) \times 10^{19}\ \text{cm}^{-3}$; the universal constant of physics **L** is called the Loschmidt number.

‡ This designation, first introduced by P. and T. Ehrenfest, suggests that the phase space relates to one molecule—hence the Greek letter μ.

typical cell of volume

$$\Delta\mathscr{V} = \Delta x \cdot \Delta y \cdot \Delta z \cdot \Delta p_x \cdot \Delta p_y \cdot \Delta p_z. \tag{2.1}$$

In this way the study of the distribution of momentum components over the molecules has been made equivalent to the study of the distribution of points in the phase space. For a given network of cells, it now suffices to indicate the number of molecules, n, contained in each cell, the *occupation numbers*, in order to complete the description.

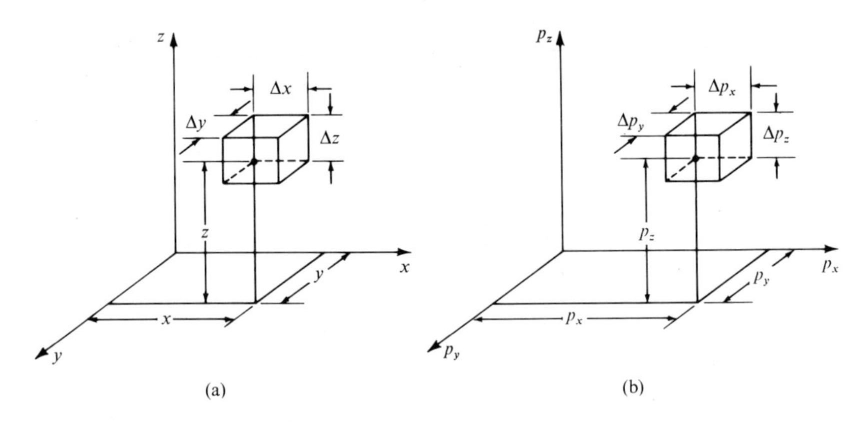

FIGURE 2.1. Cells in the physical and momentum spaces: a, physical space; b, momentum space.

Even this scheme is not entirely acceptable, since molecules cannot be counted and their momenta cannot be measured. Instead, we specify the *probable* occupation numbers, n, and this allows us to use the methods of probability and statistics in the theory. In fact, the preceding scheme for describing the state has been devised with that object in view. Hence, the branch of physics which uses these methods is called *statistical* thermodynamics.

On further examination it turns out that not all systems can be analyzed in terms of the μ-space and it is necessary to introduce a different phase space, the *phase space proper* or Γ-space,† whose coordinates represent the position and momenta of *all* molecules. An additional complication is introduced when polyatomic molecules are considered. Such molecules can accumulate additional energy by virtue of their rotation about an axis passing through them and by performing internal vibrations. This introduces further

† Here the Greek letter Γ stands for "gas."

coordinates which supplement the momenta considered so far. Hence, the Γ-space possesses $6N$ dimensions for a monatomic molecule and more for polyatomic molecules; and a particular microstate is represented by a single point in that space. We shall, however, refrain from pursuing this matter any further in this section. We shall be able to illustrate the main points sufficiently well, if inexactly, by confining our attention to the μ-space for a monatomic molecule alone.

Since to each cell in the μ-space there corresponds an average position in the physical space, x, y, z, and an average position in the momentum space, p_x, p_y, p_z, the average energy of a molecule is specified. Its potential energy depends on position and its kinetic energy is $\frac{1}{2}(p_x^2 + p_y^2 + p_z^2)/m$.† It follows, further, that one macroscopic state, or *macrostate*, of a system may correspond to a multitude of microstates, and that the behavior of a macroscopic system derived by this method will be determined by the details of the molecular model as well as by the coarseness of the lattice used for the cellular subdivisions in the μ-space. This is conditioned by the fact that large-scale instruments which measure thermodynamic macrostates are insensitive to rapid fluctuations caused by the motion of molecules. Such instruments are sensitive only to *average* forces, and it is intuitively evident that a given average or macrostate may be the same for a large number of microstates.

Comparing the two methods of analysis, we see that the science of classical thermodynamics leads to laws which are independent of any assumptions regarding the structure of matter, but from which the properties of substances, that is, their fundamental equations of state, cannot be derived. On the other hand, the science of statistical thermodynamics begins with suitable hypotheses from which equations of state as well as the laws of classical thermodynamics can be obtained. As already stated, these hypotheses must be subjected to subsequent verification by macroscopic experiments.

It might be supposed that the accuracy of the description of a thermodynamic system increases as the size of a cell in the phase space is made to decrease and that it is possible to carry out the analysis by passing to the limit of infinitely small cells with $\Delta\mathscr{V} \to 0$, and by integrating over the phase space. This is the method used in *classical statistical thermodynamics* (or *kinetic theory* in the case of gases). It leads to many results which agree with experiments but also to some which are in glaring contradiction with them. These very important contradictions can be removed only by postulating that the cells in the phase space have a finite volume. In the μ-space of a monatomic molecule this volume must be equal to $\Delta\mathscr{V} = \mathbf{h}^3$, where $\mathbf{h}$ is a universal

† In the presence of interactions, the potential energy of a molecule depends on its own position as well as on that of all the others. This feature forces us to introduce the Γ-space mentioned earlier.

constant of physics and is the same for all systems. It is easy to verify that its dimension (length × momentum), Equation (2.1), is

$$[\mathbf{h}] = [x][p] = \mathsf{ML}^2\mathsf{T}^{-1},$$

and is equal to that of the product of energy and time (or "action"). *Planck's constant*, or the "*quantum* of action," has the value

$$\mathbf{h} = (6.626196 \pm 0.000050) \times 10^{-34} \text{ J sec}. \tag{2.2a}$$

Often, Planck's constant appears as the ratio

$$\hbar = \frac{\mathbf{h}}{2\pi} = (1.0545919 \pm 0.0000080) \times 10^{-34} \text{ J sec}. \tag{2.2b}$$

The preceding postulate constitutes the starting point of *quantum statistics*. This is contrasted with *classical statistics* when the passage to the limit $\Delta\mathscr{V} \to 0$ is performed.

It should be realized that the process of passing to the limit $\Delta\mathscr{V} \to 0$ implies that the energy of a molecule, like its position, can change continuously, that is, that the difference between the energies associated with two neighboring cells in the phase space can be made arbitrarily small. The assertion that the volume of a cell in the phase space is finite means that this energy difference cannot be smaller than a certain quantum. As a molecule moves from the cell of the phase space to its neighbor, it is said to undergo a *quantum jump*.

Finally, it can be seen that the total number of microstates possible for a given number of molecules, N, is finite in quantum statistics. In classical statistics the number of possible microstates increases *ad infinitum* with increasing fineness of the lattice and becomes a continuum in the limit when the lattice is made infinitely fine. In quantum statistics this number is finite and the totality of microstates forms a *discrete* set.

As a matter of interest we may note here that quantum statistics puts a restriction on the *volume* of an element, $\Delta\mathscr{V}$, and not on all its sides. Assuming for the sake of simplicity that $\Delta x = \Delta y = \Delta z = \varepsilon$ and $\Delta p_x = \Delta p_y = \Delta p_z = \delta$, we see that the product $\varepsilon\delta = \mathbf{h}$. If the lattice is made finer in the physical space, it must be made correspondingly coarser in the momentum space. Consequently, we may assert broadly that if the position of a molecule is known more and more exactly ($\varepsilon \to 0$), its velocity will be specified less and less exactly ($\delta \to \infty$). This fact is proved in all its generality in quantum mechanics and is known as *Heisenberg's uncertainty principle*.

Heisenberg's uncertainty principle is intimately connected with the experimentally established dual nature of elementary particles which exhibit corpuscular as well as wavelike characteristics. The latter are brought into evidence by the diffraction patterns produced by streams of elementary particles and normally associated with waves. The motion of elementary particles, unlike that of large macroscopic bodies, is not governed by Newtonian (or classical) mechanics but by *Schrödinger's equation* (or quantum mechanics). Schrödinger's equation, which will be introduced to the reader in Chapter 3, constitutes a theoretical formulation of the laws of quantum mechanics and fully accounts for the wavelike aspects of microscopic particles and naturally leads to Heisenberg's uncertainty principle. Thus the uncertainty principle, contrary to the literal meaning of the word "uncertainty," does not relate to the normal uncertainty or limited precision associated with the use of necessarily imperfect measuring instruments, which are subject to constant improvement; it brings into evidence the more subtle uncertainty in describing the position and momentum of an elementary particle under conditions when its wavelike nature comes to the fore.

We have outlined some of the essential concepts of statistical thermodynamics which can be based on classical or on quantum mechanics, depending on circumstances. This was done with the aid of the model of a gas, pictured as a collection of freely and randomly moving molecules. The models which are suitable for the study of the properties of liquids or solids are evidently different. In fact, no successful microscopic model for a liquid has yet been proposed. In the case of solids, crystalline materials are imagined to consist of regular arrays of atoms arranged in characteristic *lattices* in the physical space. Every crystal lattice is assumed to consist of a large number of identical unit cells, and the resulting symmetries are said to determine the characteristic shapes and properties of macroscopic crystals. The atoms in the lattice are supposed to be linked to all others with elastic bonds. The only motions in which the atoms can participate are characteristic vibrations under the restoring forces which the links exert upon them. In metals, for example, the arrangement of atoms in the lattice is not always perfect, and some of them may become misplaced giving rise to *dislocations* which cause important differences in the strength and, generally, in the behavior of metals as compared with perfect crystals.

The preceding description of a microstate was cast entirely in mechanical terms, and it is not quite clear how the mechanical quantities which characterize such a microstate or a series of microstates associated with one macrostate are related to the properties used in the macroscopic description of the system. It has already been mentioned that measuring instruments are not sensitive to instantaneous and rapid changes in the microstate. It follows, therefore, that macroscopic parameters can only be represented by the common *averages* of the mechanical characteristics of the microstates which correspond to a given macrostate. For example, the pressure exerted on the walls of a vessel by a gas contained in it is interpreted as the average result of the very many impulses which the walls suffer as the molecules impinge on them at high velocities and in large numbers. We shall encounter additional connections between averages of microscopic characteristics and macroscopic parameters later in this book. Here it is sufficient to note that the science of statistical thermodynamics provides a systematic interpretation of macroscopic properties in terms of averages taken over suitable microscopic quantities. It is known from mathematical statistics that only systems containing large numbers of particles will show statistically insignificant departures from averages. In other words, meaningful averages, that is, meaningful and measurable macroscopic properties, can be associated only with statistically large collections of particles. Hence, it follows that small collections of molecules cannot be analyzed by the methods of classical thermodynamics because in them departures from average states are large. Thus, for example, the concept of pressure cannot be associated with a vessel which contains several molecules only, since their impacts occur so rarely that the effect is not equivalent to a uniform constant pressure, as is the case with collections containing enormous numbers of molecules.

2.3. Mechanical Models

The present section, that is, the remainder of this chapter, contains a more detailed discussion of the first three fundamental hypotheses of statistical thermodynamics, namely of the assertion that every macroscopic thermodynamic system, no matter how complex or simple, can be regarded as a collection of discrete, elementary particles. These particles interact with each other through fields of forces and the resulting motion becomes quite complicated. In spite of this complexity, we can master the essential characteristics of such motions in terms of a number of concepts which we are about to review, though many of them may be familiar from mechanics. Furthermore, we shall describe the gross characteristics of the mechanical model of a pure substance, an ideal solid, and a perfect gas, together with those of the intervening forces, and show that they are capable of securing qualitative agreement with certain macroscopic processes, such as thermal expansion, melting, evaporation, and dissociation. It will also be shown that the nature of the force fields postulated in the model is such as to assure the validity of the First law of thermodynamics.

2.3.1. Number and Nature of Elementary Particles

Contrary to superficial observations, it is assumed that all matter possesses a *discrete structure*. The impression of continuity given by macroscopic systems can be reconciled with the assumption of discreteness on a microscopic scale only with the proviso that elementary particles are exceedingly small compared with the macroscopic dimensions, including the smallest, to which we are accustomed in the laboratory.

In statistical thermodynamics it is not always necessary to descend in scale below that of considering molecules and atoms. In other words, it is possible largely to ignore the fact that in a deeper study of the properties of matter, particularly in connection with nuclear reactions, it is necessary to accept the view that atoms in turn possess a complex structure and are composed of a nucleus consisting of protons and neutrons and an outer shell containing a cloud of electrons.

Thus, depending on the macroscopic properties being studied, it is necessary to postulate a more or less complex mechanical model of the system under consideration. The model is most frequently conceived in terms of particles—not waves—the wavelike nature of the system being accounted for mathematically by the methods of quantum mechanics.

2.3.2. Intermolecular Force Potential

The mechanical model of a pure substance is assumed to consist of a large number of atoms which exert forces upon one another. The atoms in a solid are thought of as being packed rather closely, but the mean distance increases enormously in the vapor phase; in the latter case, atoms are grouped in the form of more or less complex molecules. The fact that lumps of a solid do not fly apart or collapse, but persist indefinitely, implies that the interatomic, long-range forces must be attractive. At short range the forces must be repulsive, tending to infinity to account for the very low compressibility of solids and for the fact that even very high, uniform, external pressures can be applied to solids or gases without annihilating them. Thus we come to the conclusion that the force of interaction between two elementary, electrically neutral particles must depend on their distance, varying from strong repulsion at short distance to very weak attraction at very large distance. It follows that the repulsive force must decrease to zero for a particular distance, become attractive, pass through a position of maximum attraction, and decay to zero as the distance is increased. Rather than represent the variation of the force $F(r)$ itself with distance r, it is more convenient to express it with the aid of a *potential* $\phi(r)$ which describes the energy stored by two particles in terms of the distance r between their centers. A sketch of this variation is shown in Figure 2.2.

In assuming the existence of a potential, $\phi(r)$, to describe the forces between two electrically neutral particles, we imply that

> *the internal forces of a mechanical model must be conservative and that the sum-total of all the energies of the particles is equal to the energy E of the thermodynamic system.*

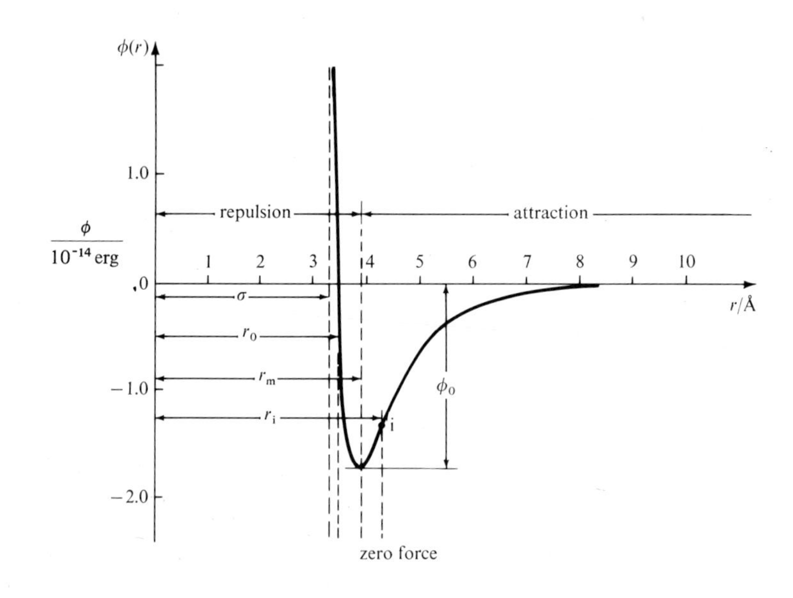

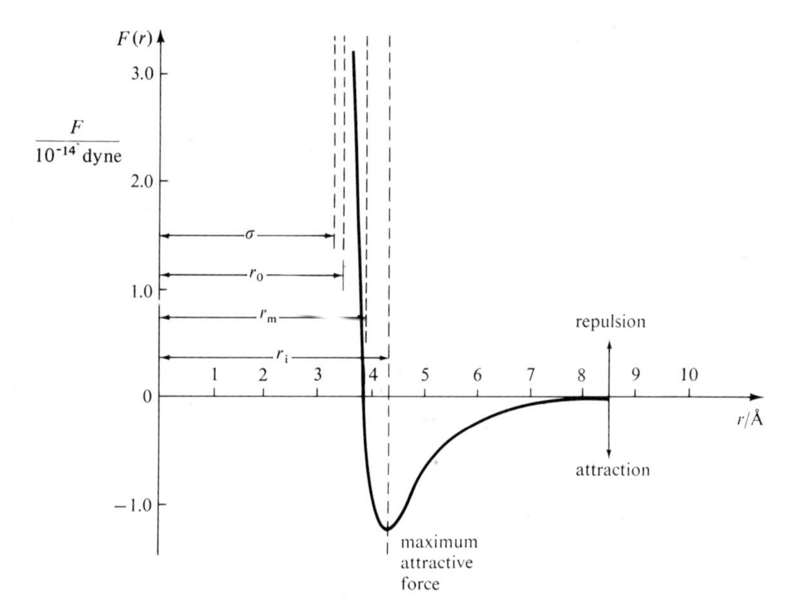

FIGURE 2.2. (Top) Qualitative sketch of interatomic or intermolecular force potential $\phi(r)$.

FIGURE 2.3. (Bottom) Qualitative sketch of interatomic or intermolecular force $F(r) = -\,d\phi(r)/dr$.

Moreover, this energy can depend only on the state of the system. If this were not the case, that is, if nonconservative forces were admitted, an isolated system could not retain its energy or fixed values of the averages which describe its macroscopic properties.

The stipulation that the force potential depends exclusively on the mean distance, r, between particles implies that the field of forces is spherically or centrally symmetrical; this is the case with numerous kinds of molecules. However, not all fields of forces acting between particles are necessarily symmetrical and more complex shapes of fields must be, and have been, assumed in certain mechanical models of more complex substances.

It is known from dynamics that the most general set of forces acting on a body can always be reduced to a force and a couple applied at one point, for example, at its center of mass. For some substances the existence of such a couple in the mechanical model must be taken into account. When this is the case, the substance is said to possess an asymmetric potential. If the moment is due to a separation of electric charges, the molecule is called *polar*. The molecules of H_2O, NH_3, or CO_2 can serve as examples of such polar substances. The substance (or molecule) is called *nonpolar* when the existence of a couple of electrical origin (also referred to as a *dipole moment*) can be neglected. All the noble gases Ar, Ne, Kr, Xe, the molecules N_2, O_2, and many others are nonpolar.

By convention, the repulsive force is assumed positive and the attractive force is assumed negative. Recalling that the force F is usually assumed to be equal to the negative of the derivative of the potential

$$F = -\mathrm{d}\phi/\mathrm{d}r, \tag{2.3}$$

we can verify in Figure 2.2 that the diagram has been drawn in accordance with this convention. Since even solids cannot be compressed to zero volume, the repulsive force must become infinite at some distance, denoted by σ in Figure 2.2. Hence the potential curve $\phi(r)$ must possess a vertical asymptote at $r = \sigma$. At $r = r_0$ the potential $\phi(r) = 0$, and at $r = r_\mathrm{m}$ the negative part of the potential, the so-called potential well, passes through a minimum at which the force $F = 0$. The potential curve must pass through a point of inflection i, say at $r = r_\mathrm{i}$, if the condition $F \to 0$ as $r \to \infty$ is to be satisfied. At this distance, the force of attraction passes through a maximum. When two nonpolar particles are spaced a distance r_m apart, they would remain at rest with respect to each other. It is clear that the values for which $\phi(r) = 0$ can be chosen arbitrarily. The diagram in Figure 2.2 has been drawn according to the convention that $\phi(\infty) = 0$. Hence $\phi(r_0) = \phi(\infty)$, and both are set equal to zero by agreement. The sketch in Figure 2.3 shows that variation of the force $F(r)$ associated with the potential $\phi(r)$ from Figure 2.2.

The area $\pi\sigma^2$ formed with the distance, σ, of minimum possible approach between the centers of two particles is sometimes called their *cross section.* Originally, the term was reserved for the frontal area of a hypothetical rigid molecule of diameter σ which exerts no forces on other molecules except on impact. As shown in Figure 2.4, when such a molecule is in motion, no other center of a molecule can be found within this area at any instant.

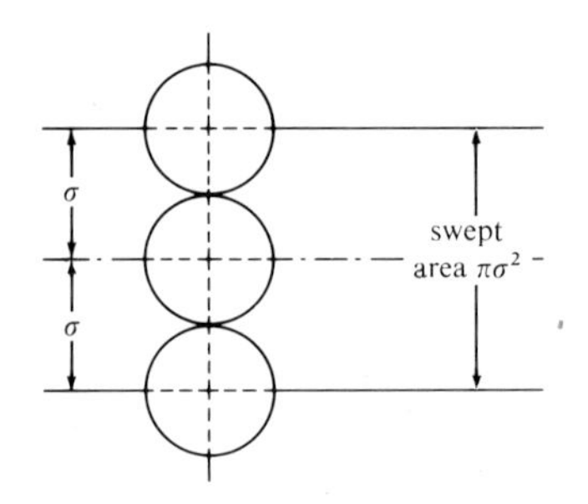

FIGURE 2.4. Cross section of a particle.

The case of electrically charged particles is more complex, since they must exert upon each other forces which obey Coulomb's law. However, in our present, elementary treatment we shall have no occasion to take such *Coulomb forces* into account.

2.3.3. Mechanical Model of a Pure Substance

Bearing in mind the potential $\phi(r)$ sketched in Figure 2.2 as well as the sketch of the force $F(r)$ contained in Figure 2.3, we can now demonstrate that the type of variation with distance postulated in them is in qualitative agreement with the known, gross macroscopic behavior of most substances.

The atoms in a solid are endowed with a relatively small amount of kinetic energy, all of it due to small vibrations about fixed positions. When an external force or pressure is applied to a solid, the mean distances between molecules or atoms decrease. Hence, on the average, the repulsive forces between the particles increase, and thus balance the external force. Even the application of infinite compression does not cause matter to collapse because the repulsive force increases very fast at short range. When tension is applied the average distances between molecules increase and the external force is balanced by the accompanying decrease in the forces of repulsion or increase in the forces of attraction. However, the forces of attraction can only increase as long as the mean distance is less than r_i. An attempt to increase the mean distance beyond this value upsets the balance of forces because the net, average internal force begins to decrease. The particles are torn apart, even if the tensile force remains constant, as observed on a metal specimen subjected to a tensile test.

Adding heat to a solid or liquid or, more precisely, increasing their temperature, enhances the internal motions of atoms and molecules. The forces which act between them are not linear functions of distance, and the increase in repulsion with decreasing distance is much more rapid than the increase in attraction with increasing distance. Consequently, the vibrating particles of a solid move away from each other much more than they move toward each other; the average distances between them increase as observed macroscopically in the form of thermal expansion. During this process, the amplitude as well as the average velocity of the vibrations increase, causing an increase in the total energy of the mechanical model.

The details of this process can be more clearly understood with reference to Figure 2.5. Assuming that the force potential $\phi(r)$ of all the atoms in a solid acting upon a single atom has the same general shape *a* as in Figure 2.2,

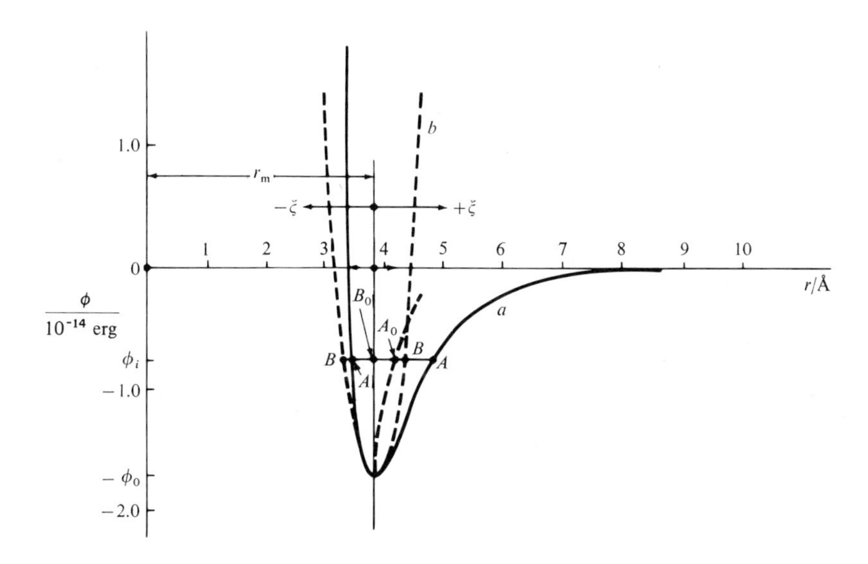

FIGURE 2.5. Thermal expansion and vibration in a solid.

we would conclude that the single atom under consideration will be at rest ($F = -\mathrm{d}\phi/\mathrm{d}r = 0$) at the position $r = r_m$ which corresponds to the minimum value $\phi = -\phi_0$. If the atom is now deflected from its position of rest, it will be attracted to the rest position by a central force. This force is practically linear with distance ξ measured from this position, as seen from Figure 2.3. The atom will therefore perform purely harmonic oscillations with an amplitude X, and its potential energy will be proportional to the square of the instantaneous deflection, ξ^2. The maximum potential energy will exist at $\xi = X$ and will be proportional to X^2. The point at $\xi = X$ where the atom is instantaneously at rest is called the (classical) turning point. If the atom were to perform harmonic oscillations only, its maximum potential energy would remain proportional to X^2, and its equilibrium position would remain stationary. Hence, the potential curve would assume the form of the parabola b. For a given maximum potential energy $\phi_i = KX^2 - \phi_0$ (where K is a constant), the two turning points would correspond to points BB on parabola b, and the position of equilibrium would be at B_0. In fact, the potential curve has the shape a, and the classical turning points are at AA, with the position of equilibrium, A_0, displaced with respect to B_0 toward the right. This is due to the fact that the restoring force is not really linear. For values of potential energy ϕ_i which differ from $-\phi_0$ even by a considerable amount, it is possible to replace the actual potential curve a by the parabola b and to assume that the vibrations are purely harmonic with ξ proportional to $\cos \omega t$, where ω is the characteristic circular frequency.

If the temperature is increased still further, the thermal motion imparts more kinetic energy to the atoms, which eventually sever the bonds in the lattice of the solid; some atoms acquire enough kinetic energy to be able to move from the region of attraction of their nearest neighbors to another region. Macroscopically, the solid begins to melt, but the kinetic energy of its atoms is still insufficient to allow them to escape the combined forces of attraction of all neighboring particles. The acquisition of this kind of mobility by all atoms requires the addition of a certain amount of energy in the macroscopic form of latent heat.

When evaporation begins, the thermal motion has become so vigorous that a substantial number of particles have acquired sufficient kinetic energy to escape the attraction of all their neighbors. A new amount of latent heat must be added in order to impart sufficient kinetic energy to all the particles.

In gases of very low density, the average distances between the molecules become so great that the attractive forces between them become extremely weak. In the limit of extreme dilution ($r \to \infty$), each molecule can be assumed to move unhindered by all the others, and the potential energy of separation can be neglected. In practice, the potential energy $\phi(r)$ becomes very nearly equal to zero at relatively short distances r. Roughly speaking, $\phi(r)$ decreases to a value of $0.01\phi_0$ at $r = 3r_0$. It follows that the mechanical model of a gas at moderate density can be based on the assumption that its *molecules do not interact*, and that their energy is only kinetic. To be sure, upon impact, the distance of approach becomes extremely short and the molecules repel each other with great violence. These are the features of a mechanical model which lead directly to the thermal equation of state of a perfect gas (Section 6.3).

The fact that all substances can be vaporized eventually shows that qualitatively the force potential is the same for all substances.

From the preceding description it is clear that the most essential difference between a solid and its vapor or gas lies in the fact that the kinetic energy of the atoms in the solid is entirely due to vibrations. Consequently, there is a constant transformation of kinetic energy into potential energy and *vice versa*, the maximum values of both being exactly equal if the vibration is harmonic. By contrast, the mobility of the molecules in a gas or vapor imparts to them a large kinetic energy whose value becomes progressively larger compared with the potential energy as the temperature increases.

2.3.4. Empirical Expressions for the Intermolecular Force Potential

The law of forces acting between atoms and molecules can in principle be derived from quantum mechanics. However, owing to mathematical difficulties this has so far been achieved only in very few cases, such as for helium atoms and the diatomic molecule of hydrogen. Consequently, we are forced to introduce empirical approximations in order to render the respective calculations more tractable. The simplest, and crudest, approximation postulates an infinite force (and potential) at $r = \sigma$ and a zero force (and potential) for $r \geq \sigma$. The resulting diagram is shown in Figure 2.6; it is known

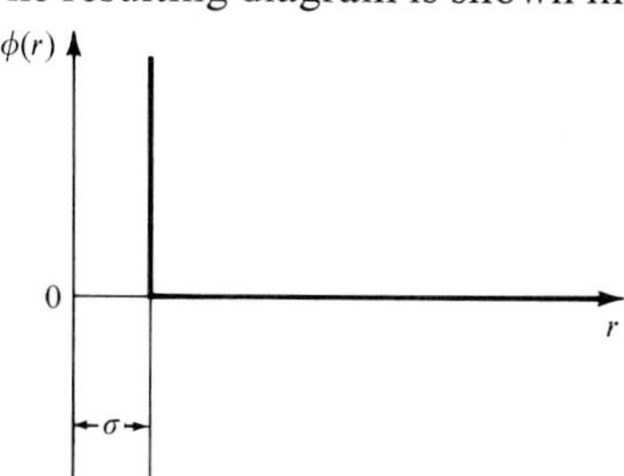

FIGURE 2.6. Potential of hard-sphere (rigid-sphere) model of a molecule.

as the hard-sphere or rigid-sphere model because it represents the law of forces acting between nonattracting, rigid spheres. When their centers are separated by a distance exceeding σ, there are no forces acting between them. Upon impact, which occurs for $r = \sigma$, the force must become infinite because of the perfect rigidity of the molecules. The next approximation leads us to the square-well model whose potential is shown in Figure 2.7. It is given by the specification

$$\phi = \begin{cases} \infty & \text{for} \quad r < \sigma \\ -\phi_0 & \text{for} \quad \sigma \leq r \leq r_0 \\ 0 & \text{for} \quad r > r_0 . \end{cases}$$

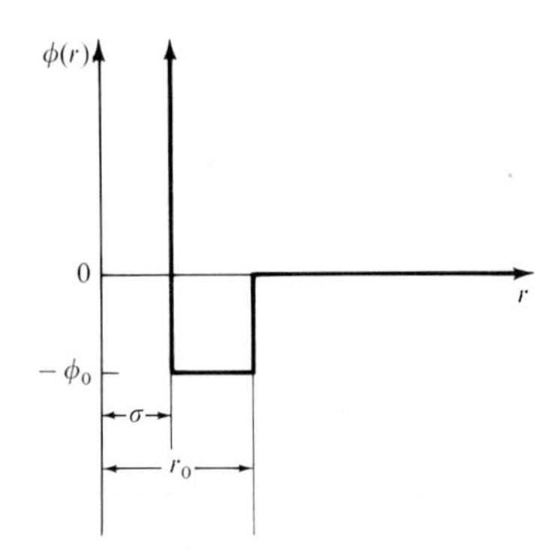

FIGURE 2.7. Potential of square-well model of a molecule.

The Sutherland model corresponding to

$$\phi = \begin{cases} \infty & \text{for} \quad r < \sigma \\ -c/r^m & \text{for} \quad r \geq \sigma, \end{cases}$$

with c and m denoting constants, is shown sketched in Figure 2.8. Frequently it is assumed that $m = 6$.

The most commonly used empirical potential which approximates the shape of Figure 2.2 by a continuous curve was first proposed by J. E. Lennard-Jones; it is seen sketched in Figure 2.9. The part which describes the forces of attraction is assumed proportional to the inverse sixth power, r^{-6}, of distance. This part of the assumption has some justification in quantum theory. The

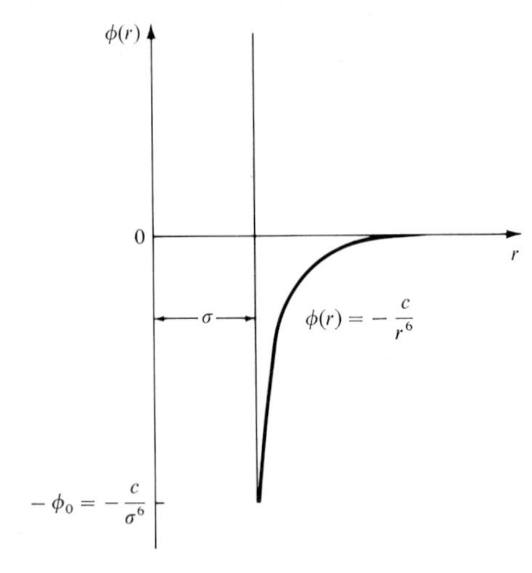

FIGURE 2.8. Sutherland model of a molecule.

term which describes the repulsive forces is assumed proportional to the inverse twelfth power, r^{-12}, of separation. Since this part is purely empirical, alternative exponents (-8, -9, -14, etc.) have also been proposed. Denoting the depth of the potential well by ε and the point where $\phi = 0$ by $r = r_0$, it is easy to derive that a superposition of an attractive and repulsive potential of the above kind leads to the equation

$$\phi(r) = 4\varepsilon\left[\left(\frac{r_0}{r}\right)^{12} - \left(\frac{r_0}{r}\right)^{6}\right]. \tag{2.4}$$

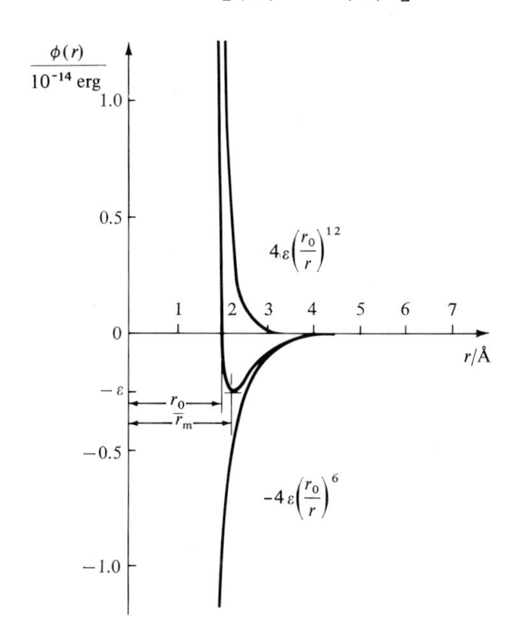

FIGURE 2.9. The Lennard-Jones or six-twelve potential.

The reader can show also that the abscissa for the minimum of potential energy, r_m, is given by

$$r_m = 2^{1/6}\, r_0, \tag{2.5a}$$

and that the form

$$\phi(r) = \varepsilon\left[\left(\frac{r_m}{r}\right)^{12} - 2\left(\frac{r_m}{r}\right)^{6}\right] \tag{2.5b}$$

is equivalent to (2.4). The attractive force is thus taken proportional to r^{-7}, that for repulsion being proportional to r^{-13}. With decreasing distance, the potential increases to positive infinity inversely as the twelfth power of distance. It, therefore, reproduces the essential feature that the forces opposing the interpenetration of molecules increase enormously. However, in the Lennard-Jones model this occurs for $r \to 0$, and not for a finite $r = \sigma$.

Another empirical potential, proposed by R. A. Buckingham, has the form

$$\phi(r) = \frac{\varepsilon}{1 - 6/\alpha}\left\{\frac{6}{\alpha}\exp\left[\alpha\left(1 - \frac{r}{r_m}\right)\right] - \left(\frac{r_m}{r}\right)^6\right\}; \tag{2.6}$$

this is sometimes referred to as the exponential-six potential. Here ε, α, and r_m denote empirical constants. In (2.6), the inverse sixth power of attraction has been retained, but the repulsive part has been made to increase more rapidly. With decreasing distance, the expression in Equation (2.6) increases to a positive, very large maximum and then decreases, tending to negative infinity as the sixth power of the decreasing distance. Hence the potential must be cut off at a distance $r = \sigma$ larger than the abscissa of the maximum and assumed infinite for $r < \sigma$.

2.3.5. The Mechanical Model for a Perfect Gas

A chemically pure perfect gas is pictured as a collection of indistinguishable, independent, and nonlocalized molecules. Naturally, molecules of different chemical species are considered distinguishable. When we picture an array of molecules in random motion in our mind's eye, we are inclined to suppose that the displacement of each one of them can be followed in detail at all times. We tend to assume that a molecule can be identified, at least in the imagination, say by painting a distinctive number on it. Thus, if in a particular microstate we were to replace molecule i by molecule k in such a way that molecule i has acquired the position and momentum of molecule k and *vice versa*, we would conclude that a different microstate has set in. The term *indistinguishable* in this context is meant to imply that this is not the case, and that the two preceding microstates are identical. Similar considerations apply to the atoms within a molecule; they affect the count of microstates in rotation (see Section 3.10).

Molecules are said to be *independent* when the motion of one is unaffected by the motions of all the others, except on impact. Thus, between collisions, forces of interaction are neglected, and the equation of motion of the system consists of N individual equations of motion of the N molecules in the assembly. The molecules are said to be *nonlocalized* because it is postulated that any molecule may occupy any region in the vessel of volume V which contains them.

Molecules of different chemical species differ in the complexity of the arrangement of atoms in them. *Monatomic* molecules consist of single atoms.

Polyatomic molecules contain two (when they are called *diatomic*) or more atoms. A distinction is made between *homonuclear* molecules which contain identical atoms, and *heteronuclear* molecules which contain different atoms.

2.3.6. Number of Degrees of Freedom

The atoms in a complex molecule move as a unit, their center of mass following a definite trajectory. In addition, the atoms perform rotational as well as vibrational motions with respect to the center of mass. The forces which act between any two atoms in a molecule are also conservative and obey a potential whose shape is generally the same as that sketched in Figure 2.2. This shows that a vibrating molecule can only accumulate an amount of energy which corresponds to the height of the potential well, compared with the position of rest at $r = r_m$, for otherwise the distance of mutual separation becomes infinite. When an amount of vibrational energy exceeding this quantity is imparted to the molecule, the atoms in the molecule must move apart and the molecule *dissociates* at the expense of a dissociation energy (see also Section 3.11). Strictly speaking, as will be shown in Figure 3.16, the dissociation energy ε_d, is somewhat smaller than ϕ_0. This is due to the fact that two atoms in a molecule can never be completely at rest with respect to each other. Thus at the *ground state*, the lowest possible energy is somewhat larger than $-\phi_0$, or, numerically, somewhat smaller than ϕ_0.

The internal energy of a gas in a specified state is equal to the mechanical energy of the assembly of molecules in a microstate. Monatomic molecules are assumed to have small moments of inertia in rotation, and only the kinetic energy of translation is taken into account. The *number of degrees of freedom*, f, of a molecule is defined as the number of independent coordinates in the physical space which are required to fix its position (but not its momentum). It is known from mechanics that this number, f, is independent of the nature of the system of coordinates used for the description.

The position of a monatomic molecule in the physical space is described by three independent coordinates, and for this reason the monatomic molecule is said to possess $f = 3$ degrees of freedom. Polyatomic molecules are assumed to move in a more complex fashion, as already mentioned. First, their center of mass performs a translatory motion with

$$f_{tr} = 3 \tag{2.7}$$

degrees of freedom in the same way as a monatomic molecule. Secondly, the molecules perform rotations about axes passing through the center of mass. In this connection it is necessary to distinguish *linear* from *nonlinear* molecules. In a linear molecule, as shown in Figure 2.10, the atoms are imagined arranged along a single axis *a–a* about which the moment of inertia is negligible. Hence independent rotations can be performed about two mutually perpendicular axes, *b–b* and *c–c*, passing through the center of mass, M, and the orientation of such a molecule with respect to its center of mass is described by two independent coordinates. These may be taken as the angles θ and ϕ which describe the orientation of axis *a–a* with respect to the x axis. The

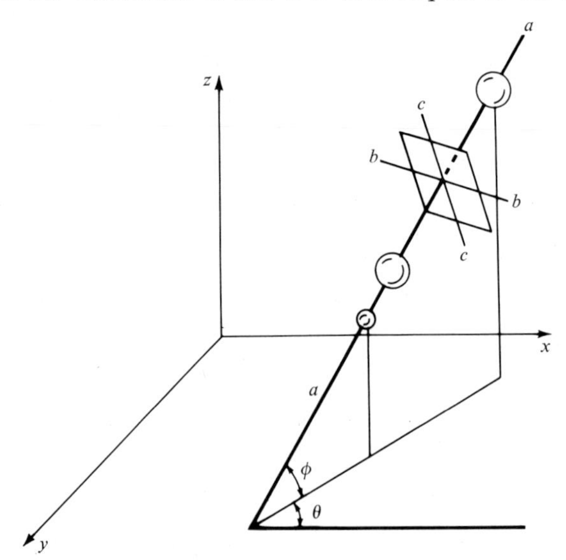

FIGURE 2.10. Linear molecule.

third angle which describes the angular deflection about the *a–a* axis is not counted because motion around it does not contribute to the kinetic energy of rotation. Such a molecule is said to possess

$$f_r = 2 \qquad \text{(linear)} \tag{2.8a}$$

degrees of freedom in rotation. By their nature, all diatomic molecules are linear. In a nonlinear molecule, as shown in Figure 2.11, there exist

$$f_r = 3 \qquad \text{(nonlinear)} \tag{2.8b}$$

degrees of freedom. The three independent coordinates can be chosen as the angles of deflection α_1, α_2, α_3 with respect to the axes *a–a*, *b–b*, and *c–c*, and measured in relation to three fixed directions. Only the angle α_1 has been suggested in the sketch. Finally, the molecules are assumed to perform *vibratory* motions with respect to each other. This possibility releases additional *internal degrees of freedom* f_{int}. The term internal degree of freedom

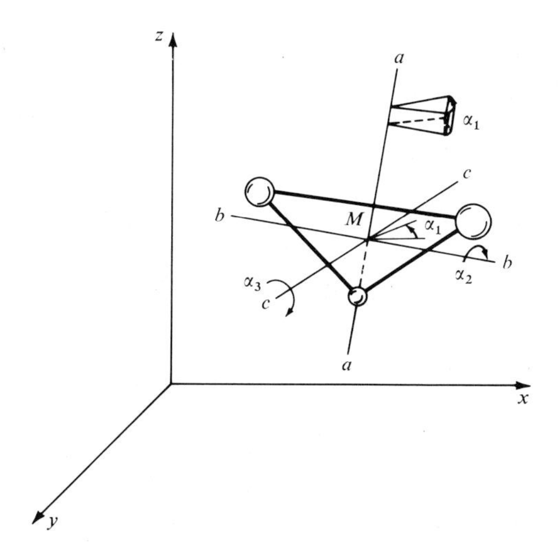

FIGURE 2.11. Nonlinear molecule.

refers to all motions which a molecule can perform about its center of mass. Hence the total number of degrees of freedom is

$$f = f_{tr} + f_{int}, \tag{2.8c}$$

and

$$f_{int} = f_r + f_v. \tag{2.9}$$

In spite of the obvious complexity of vibrational motions, it is possible to simplify drastically their description by a suitable choice of coordinates. When the vibrations can be assumed to be small—as is nearly always done in models of molecules—it is possible to associate *a natural mode of vibration* with each vibrational degree of freedom. This corresponds to the case when all the atoms in the molecule vibrate with the same frequency, the *natural frequency* of the mode, and with the same phase. This is one of the f_v resonant frequencies of the mechanical model, undoubtedly familiar to the reader from his studies of mechanics. The crucial theorem of mechanics which relates to such small-amplitude motions states that every vibrational motion of the system can be represented as a *sum of harmonic oscillations*, each corresponding to one natural (that is normal) mode of vibration.

The number of degrees of freedom in vibration is best computed from Equations (2.8) and (2.9), since the total number of degrees of freedom, f, the translational number, f_{tr}, and the rotational number, f_r, can be computed easily. This is due to the fact that there are now no rigid constraints assumed in the model and the position of every atom in the molecule is specified by three coordinates. Given a polyatomic molecule consisting of n atoms, we must discern two cases. When the molecule is *nonlinear*, we have

$$f = 3n; \qquad f_{tr} + f_r = 3 + 3 = 6,$$

and

$$f_v = 3n - 6 = 3(n - 2) \qquad \text{(nonlinear)}. \tag{2.10}$$

For *linear* molecules,

$$f = 3n; \qquad f_{tr} + f_r = 3 + 2 = 5,$$

and

$$f_v = 3n - 5 \qquad \text{(linear)}. \tag{2.11}$$

When the system contains N noninteracting molecules, the total number of degrees of freedom is simply equal to the above numbers multiplied by N.

2.3.7. Energy of a Molecule

In general, the motion of the mechanical model of a polyatomic molecule is quite complex because the vibration produces a fluctuating change in the moments of inertia with respect to the axes of rotation, and rotations provide centrifugal forces which affect vibration. However, in the model for a perfect gas such interactions between the internal modes of motion are neglected, and each particular motion is analyzed independently of the others. Consequently, the energy of a molecule can be written in the form of the sum

$$E = \mathscr{E}_{tr} + \mathscr{E}_r + \mathscr{E}_v, \tag{2.12}$$

where

$$\mathscr{E}_{tr} = \frac{1}{2m}(p_x^2 + p_y^2 + p_z^2) \tag{2.13a}$$

$$\mathscr{E}_r = \frac{1}{2}\left(\frac{p_1^2}{I_1} + \frac{p_2^2}{I_2} + \frac{p_3^2}{I_3}\right) \tag{2.13b}$$

and

$$\mathscr{E}_{v(i)} = \frac{p_{(i)}^2}{2\mu} + \frac{1}{2}(2\pi\nu_{(i)})^2\mu\xi_{(i)}^2. \tag{2.13c}$$

Here p_x, p_y, p_z denote the linear momenta in translation and m is the mass of the molecule. These momenta are conjugate to the coordinates x, y, z of the center of mass M. The translational energy $\mathscr{E}_{tr}$ is kinetic and consists of

$f_{tr} = 3$ quadratic terms in the momenta. The symbols p_1, p_2, p_3 denote the angular momenta $I_1\dot{\alpha}_1, I_2\dot{\alpha}_2$, and $I_3\dot{\alpha}_3$, where, in turn, I_1, I_2, I_3 represent the moments of inertia about the axes *a–a*, *b–b*, and *c–c*, assumed to be the principal axes in Figure 2.11; and $\dot{\alpha}_1, \dot{\alpha}_2, \dot{\alpha}_3$ denote the angular velocities about the same axes. The momenta p_1, p_2, p_3 are conjugate to the coordinates $\alpha_1, \alpha_2, \alpha_3$. The energy of rotation is also kinetic, and consists of two quadratic terms ($I_1 = 0; f_r = 2$) for a linear molecule and three quadratic terms ($f_r = 3$) for a nonlinear molecule. The expression for the vibrational energy has been written for one normal mode, hence the subscript (i), there being as many terms of the type of Equation (2.13c) as there are vibrational modes f_v. The first term in the equation expresses the instantaneous kinetic energy, with μ denoting the *reduced mass* of the molecule. For example, in a diatomic heteronuclear molecule

$$\mu = \frac{m_1 m_2}{m_1 + m_2}. \tag{2.14}$$

The second term in the equation represents the potential energy, with $\nu_{(i)}$ denoting the *frequency*, which is related to the circular frequency ω by the general equation

$$\nu = \frac{\omega}{2\pi}. \tag{2.15}$$

The symbol $\xi_{(i)}$ denotes the instantaneous linear amplitude and $p_{(i)}$ is the conjugate linear momentum.† In general

$$\mathscr{E}_v = \sum_{1}^{f_v} \mathscr{E}_{v(i)}. \tag{2.16}$$

the sum containing f_v terms of identical form.

It is worth noting that each translational and rotational degree of freedom contributes one quadratic term in the equation for energy, whereas each vibrational degree of freedom contributes two such terms.

2.3.8. The Mechanical Model for a Perfect Crystal

A perfect crystal is pictured as an assembly of vibrating atoms, each attached to a definite point in a regular *lattice*. Atoms of a given chemical species are considered indistinguishable from each other. For many purposes it is necessary to admit the possibility that atoms (and electrons) perform slow,

† Readers not familiar with these expressions may consult, for example, A. Sommerfeld, *Mechanics* (M. O. Stern, transl.), p. 106, Vol. I of "Lectures on Theoretical Physics," Academic Press, New York, 1952.

diffusive motions throughout the lattice. Nevertheless, in the case of a crystalline solid, the atoms are considered to be *localized*, that is, distinguished by their lattice sites. This means that in a particular microstate the atoms vibrate about mean positions which are fixed in space, and that there is only one atom in the neighborhood of each such mean position. Thus the attribute of being localized is associated with the lattice and not with the atom. As far as the interactions between the atoms are concerned, we shall make use of two models. The simplest model assumes that the vibratory motions of the atoms are *independent* of each other. This means that in spite of the fact that the restoring force on each atom is supplied by all the remaining atoms, we shall assume that the vibratory motion of all the other atoms has no effect on the character of the motion of each particular atom. Consequently, it will be stipulated that every atom performs a harmonic oscillation with a particular frequency and amplitude in each of three mutually perpendicular directions. The more elaborate model will assume that the atoms in a solid move like a very complex unit and interact with each other. Thus atoms may perform oscillations of different amplitude and frequency. However, the problem will be simplified to the extent that the amplitudes of all atoms will be assumed very small. This will allow us to regard the total, complex motion as the result of a superposition of as many normal harmonic oscillations as there are degrees of freedom, that is $3N-6$, where N is the (very large) number of atoms in the solid.

Finally, in line with our remarks in Section 2.3.1, we shall stipulate that the internal energy of the system is equal to the sum of the kinetic and potential energies of all the vibrating atoms in a particular microstate, in general accordance with Equation (2.13c).

2.3.9. Mechanical Models for Liquids

Our exposition will not be confined to the two limiting models described in the preceding two sections, but will leave out of account the model of a liquid which cannot be assumed to conform more closely to one than to the other of these extremes. The omission is not, however, unintentional and results from the fact that a well-tested, statistical theory of liquids does not, as yet, exist. The available, partial theories of liquids treat them either as quasi crystals in which mobility is enhanced by the arbitrary addition of "holes" or as very dense gases.†

† Modern accounts of the theory of liquids are given in: G. H. A. Cole, *An Introduction to the Statistical Theory of Simple Dense Fluids*, Pergamon Press, New York, 1967; J. M. H. Levelt and E. G. D. Cohen, A critical study of some theories of the liquid state including a comparison with experiment, in *Studies in Statistical Mechanics*, Volume II, (J. DeBoer and G. E. Uhlenbeck, eds.), North-Holland Publ., Amsterdam, 1964; S. A. Rice and P. Gray, *The Statistical Mechanics of Simple Liquids*, Wiley, New York, 1965.

PROBLEMS FOR CHAPTER 2†

2.1. The kinetic energy of a system of two particles is $E = (m_1 v_1^2 + m_2 v_2^2)/2$. Express the energy in terms of the velocity of the center of mass $\mathbf{V}$, and the relative velocity $\mathbf{v}$, where $\mathbf{V} = (m_1 \mathbf{v}_1 + m_2 \mathbf{v}_2)/(m_1 + m_2)$ and $\mathbf{v} = \mathbf{v}_1 - \mathbf{v}_2$. [*Hint:* Solve for $\mathbf{v}_1$ and $\mathbf{v}_2$ in terms of $\mathbf{V}$ and $\mathbf{v}$ and substitute into the expression for the kinetic energy.] (This result will be needed in Chapter 13.)

2.2. Consider a two-dimensional harmonic oscillator whose energy is

$$E = (p_x^2 + p_y^2)/2m + \alpha(x^2 + y^2)/2.$$

Introduce the polar coordinates $x = r \cos \phi$, $y = r \sin \phi$ and show that the energy can be expressed in the form $E = (p_r^2 + p_\phi^2/r^2)/2m + \alpha r^2/2$. What is the physical interpretation of each term in this expression?

2.3. A point-particle is placed in a one-dimensional potential

$$\phi(x) = ax^2 + bx^4 \qquad \text{for} \quad -\infty < x < +\infty,$$

where b is positive and a may assume positive or negative values. Sketch this potential and locate the equilibrium points of the particle in the case when (a) $a/b > 0$, (b) $a/b < 0$, (c) $a = 0$. Describe the motion of a classical particle which is acted upon by this potential field; include all three cases.

2.4. Describe the structure of the Γ-space. Contrast this with the μ-space for a gas without internal degrees of freedom. How does the geometrical structure of the μ-space change for a triatomic, nonlinear molecule?

2.5. Why is the μ-space unsuitable for the description of systems whose mechanical models consist of interacting particles?

2.6. The μ-space of a particle constrained to move in one dimension may be represented in a space of two dimensions. Construct the appropriate μ-space and sketch the path of a particle with a given energy if the particle is (a) a harmonic oscillator, (b) a bead constrained to move without friction on a circular ring.

SOLUTION: (a) The equation $E = p^2/2m + \alpha x^2/2$ defines an ellipse. (b) The result of Problem 2.2 may be used to express the energy for this case. We put $\alpha = 0$ and $p_r = 0$, so that $E = p_\phi^2/2ma^2$ where a is the radius of the ring. This defines two straight lines in the μ-space.

2.7. A ball, starting from rest, falls a distance d under the influence of gravity, impinges on an elastic floor and rebounds to its original position. Sketch the orbit of this particle in μ-space and denote equal time intervals on the trajectory.

2.8. A particle moves in a two-dimensional rectangular box with dimensions a and b. At some instant, the momentum $\mathbf{p}$ of the particle has components p_x and p_y.

† Problems numbered in boldface type are considered relatively easy; they are recommended for assigning in an elementary first course.

The particle makes specular, elastic collisions with the sides of the box. Sketch the projection of the phase-space trajectory onto the planes x, p_x and p_x, p_y. What modifications should be made to the above sketches if, upon collision with the wall, the particle is reflected elastically, but rebounds in a random direction from the wall? (Sketch the trajectories for several such particles.) [*Hint:* Sketch the region which can be occupied by all possible states after reflection.]

2.9. Two molecules interact according to the Lennard-Jones six-twelve potential, Equation (2.4). Calculate the force, F, of the interaction when the molecules are separated by a distance (a) $r = 0.9r_0$, (b) $r = r_0$, (c) $r = 2^{1/6}r_0$, (d) $r = 2r_0$. How is this force directed in each of the above cases?

2.10. Prove Equations (2.5a) and (2.5b) in the text.

2.11. Assume that two argon atoms interact according to the Lennard-Jones six-twelve potential, Equation (2.4), and compute (a) the magnitude of the maximum force, F, of attraction; (b) the separation between the atoms at the point where the attractive force is a maximum. The Lennard-Jones parameter for argon can be taken to be $\varepsilon/\mathbf{k} = 120$ K, $r_0 = 3.42$ Å.

2.12. The Morse potential for intermolecular forces is

$$\phi(r) = \varepsilon[e^{-2(c/r_0)(r-d)} - 2\,e^{-(c/r_0)(r-d)}].$$

(a) Derive an expression for the force, and show that $d = r_m$. (b) Show that $\varepsilon = -\phi(r_m) = \phi_0$ is the so-called potential-well depth. (c) Show that the relation between the constants r_m, r_0, and c is

$$r_m = r_0\left(1 + \frac{\ln 2}{c}\right).$$

2.13. The Stockmeyer potential,

$$\phi(r, \theta_a, \theta_b, \phi_b - \phi_a) = 4\phi_0\left[\left(\frac{r_0}{r}\right)^{12} - \left(\frac{r_0}{r}\right)^{6}\right] - \frac{\mu_a\mu_b}{r^3}\,g(\theta_a, \theta_b, \phi_b - \phi_a),$$

where μ_a and μ_b are dipole moments and

$$g(\theta_a, \theta_b, \phi_b - \phi_a) = 2\cos\theta_a\cos\theta_b - \sin\theta_a\sin\theta_b\cos(\phi_b - \phi_a),$$

represents the superposition of the Lennard-Jones potential and the interaction of two point dipoles whose orientations with respect to their connecting axis are given by the angles θ and ϕ. The potential is meant to describe polar molecules for which only dipole–dipole interactions are important.

Derive an expression for the intermolecular force from this potential. Does the r_0 here continue to have the same meaning as the r_0 in the Lennard-Jones potential?

2.14. The force F for the single-mode oscillation of a diatomic molecule whose potential $\phi(r)$ has a minimum at $r = r_m$ can be written $F = -k(r - r_m)$ (under what assumptions?).

(a) By expanding $\phi(r)$ in a Taylor series about its minimum, show that the force constant k is

$$k = \left(\frac{d^2\phi}{dr^2}\right)_{r=r_m}$$

(b) Derive an expression for k for the Morse potential (Problem 2.12). Hence, determine the natural frequency for this vibrational mode.
(c) Do the same for the Lennard-Jones potential, Equation (2.4).

2.15. Compute the number, N, of molecules of H_2O contained in $V = 1$ cm³ at atmospheric pressure and at a temperature of $T_1 = 150°C$; $T_2 = 4°C$; $T_3 = -4°C$. The respective specific volumes are: $v_1 = 1.975$ m³/kg; $v_2 = 1.0002 \times 10^{-3}$ m³/kg; $v_3 = 1.0905 \times 10^{-3}$ m³/kg.

2.16. Determine the total number of degrees of freedom

$$f = f_{tr} + f_r + f_v$$

for the following idealized molecular configurations, which are shown in the accompanying sketches, assuming that the connecting springs are both flexible and extensible: (a) homonuclear diatomic molecule; (b) linear, homonuclear triatomic molecule; (c) nonlinear, homonuclear triatomic molecule. Would the number of degrees of freedom be different if the molecules were heteronuclear?

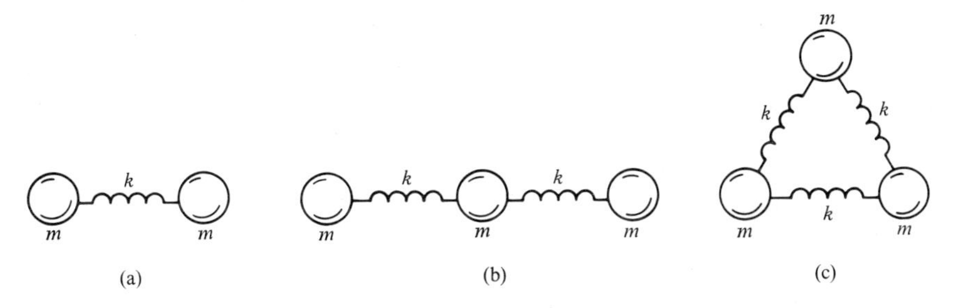

2.17. Determine the number f_v of vibrational and f_r of rotational degrees of freedom for the following molecules: (a) H_2O, water; (b) C_6H_6, benzene; (c) $C_{10}H_8$, naphthalene; (d) C_6H_{12}, cyclohexane; (e) C_2H_2, acetylene (linear).

2.18. The average velocity of a molecule at room temperature is $u \approx 480$ m/sec. What is the kinetic energy, E, of an oxygen molecule with this velocity? Calculate the number, n, of such molecules that would make up a kinetic energy of $E_1 = 1$ J; $E_2 = 10$ eV.

2.19. The cube root of the volume per molecule may be used as a simple estimate for the distance between neighboring atoms in a substance. (a) The density of solid argon at $T = -233°$C is $\rho = 1.65$ gr/cm³, and the mass of an atom is $m = 43 \times 10^{-24}$ gr. What is the spacing between atoms provided by the above estimate? (b) The intermolecular potential of a hard-sphere molecule is illustrated in Figure 2.6. If we assumed that a collection of such molecules were as closely packed as possible in a box, we would find that the volume per molecule is $\sigma^3/\sqrt{2}$, where σ is the diameter of one molecule. What is the effective hard-sphere diameter of an argon atom in solid argon at $T = -233°C$?

Make a mental note of the closeness of the accurate result in (b) and its crude estimate in (a).

2.20. The density of liquid argon at $T = -189.4°C$ is $\rho = 1.419$ gr/cm³. The mass of an argon atom is $m = 43 \times 10^{-24}$ gr. Estimate the average distance, s, between neighboring atoms in liquid argon. Using the result of Problem 2.19, estimate the separation, s_1, of argon atoms in terms of the number of atomic diameters between two atoms. Perform the same calculation for gaseous argon at $T_1 = 0°C$ and $P_1 = 1$ atm if $\rho_1 = 1.78$ kg/m³.

2.21. The density of solid argon at $T = -233°C$ is $\rho = 1.65$ gr/cm³. Suppose that argon atoms interact according to the Lennard-Jones six-twelve potential. Estimate the average potential energy of interaction between a pair of neighboring atoms in the solid. Perform the same calculation for atoms in liquid argon at a density of $\rho = 1.42$ gr/cm³. The parameters ε and r_0 for argon are $\varepsilon/\mathbf{k} = 120$ K, $r_0 = 3.42$ Å.

LIST OF SYMBOLS FOR CHAPTER 2

Latin letters

c	Constant in Sutherland's potential
$\mathbf{c}$	Velocity vector with components u, v, w
E	Energy
$\mathcal{E}$	Energy associated with a particular type of motion
F	Force
f	Number of degrees of freedom
$\mathbf{h}$	Planck's constant
$\hbar$	$\mathbf{h}/2\pi$
I	Moment of inertia
K	Spring constant
$\mathbf{L}$	Loschmidt number
m	Constant in Sutherland's potential; mass
N	Number of molecules
$\mathbf{N}$	Avogadro's number
n	Number of atoms in a polyatomic molecule
p_x, p_y, p_z	Linear momenta
p_1, p_2, p_3	Angular momenta
r	Molecular separation
r_i	Molecular separation at maximum value of attractive force
r_m	Molecular separation at minimum value of the potential, ϕ_0
r_0	Molecular separation at zero value of the potential
V	Volume
$\mathcal{V}$	Volume in six-dimensional phase space
X	Amplitude of harmonic oscillator

Greek letters

α	Constant in Buckingham's exponential-six potential
$\alpha_1, \alpha_2, \alpha_3$	Angular coordinate
Γ	$6N$-dimensional phase space
δ	Uncertainty in momentum
ε	Energy parameter in Lennard-Jones potential; Uncertainty in position
ε_d	Dissociation energy
θ	Angle
μ	Six-dimensional phase space; reduced mass
ν	Frequency of harmonic oscillator
ξ	Displacement of harmonic oscillator from its equilibrium position
σ	Distance of closest approach
ϕ	Intermolecular potential energy
ϕ_0	Depth of potential well
ω	Circular frequency $(=2\pi\nu)$

Subscripts

int	Internal degrees of freedom
m	Minimum of potential
r	Rotational degrees of freedom
tr	Translational degrees of freedom
v	Vibrational degrees of freedom
0	Zero value of potential

Special symbol

·	Time rate of change

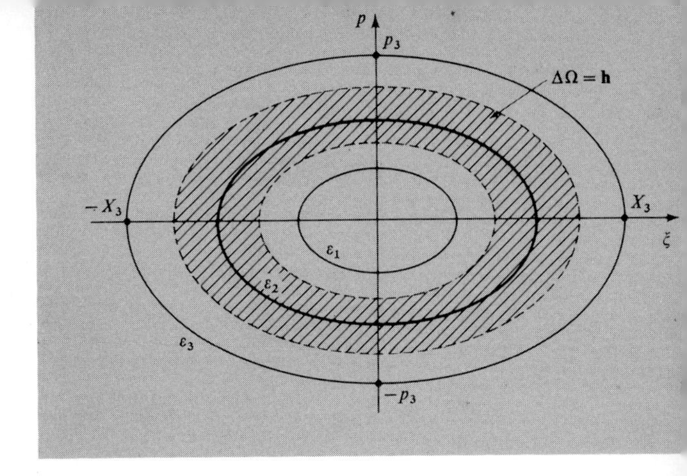

CHAPTER 3

QUANTUM MECHANICS

3.1. Description of the Motion

Disregarding for the time being the difficulties arising from the complexity of the motion of the elements of the mechanical model of a thermodynamic system as well as from their enormously large number, it is necessary to investigate the laws of mechanics which must be used to describe the system. At first this may appear to be a surprising question and the "obvious" answer suggests itself immediately. All our experience with the motions performed by macroscopic systems, from the largest planets to the smallest specks of dust points to the conclusion that the motion of the mechanical model should be governed by Newton's laws, that is, by *classical mechanics*. This is, indeed, the hypothesis which was adopted by the early pioneers of statistical thermodynamics, L. Boltzmann, J. C. Maxwell and their predecessors. During several decades of intensive research it was discovered that the preceding hypothesis leads to resounding successes in the application of the theory to statistical thermodynamics. However, it was also discovered that in many cases conclusions are reached which are in glaring contradiction with experimental evidence. The state of affairs which results from the hypothesis that elementary particles perform motions in accordance with the laws of classical mechanics was brilliantly summarized by Lord Kelvin in 1884 in his famous Baltimore lectures, when he referred to these failures as the "Nineteenth Century Clouds over the Dynamical Theory of Heat," clouds, that is, which obscure the theory. The detailed history of how these clouds were dispersed makes fascinating reading;† we cannot, however, relate it to the reader for lack of space and for fear that his outlook, too, may become obscured by a number of false starts. Today we know that all contradictions disappear if it is recognized that the motion of elementary particles *cannot* be described by classical mechanics—the correct description is provided by *quantum mechanics.*‡ Naturally enough, quantum mechanics does not supersede classical mechanics but supplements it; it provides a description of motion which agrees with experimental facts regardless of the size of the system. And since Newtonian mechanics has succeeded with regard to large bodies, such as the planets, projectiles, etc., it is only to be expected that the description of quantum mechanics applied to them must yield results which are indistinguishable from those of classical mechanics. This fact was first clearly expressed by N. Bohr in the form of his *correspondence principle.* We can formulate it somewhat loosely by asserting that

> *though classical mechanics does not apply to microscopic atomic and molecular motions, it must be expected to hold asymptotically for large bodies.*

When a more general theory is formulated to supplant a less general theory which constitutes a limiting case of the former, it is usually possible to proceed without reference to the more restricted theory. This is the case, for example, with the special theory of relativity which can be formulated without reference to nonrelativistic, Newtonian mechanics. In the case of quantum mechanics this is impossible. Rather, the equations of classical mechanics must be "translated" into quantum "language" or, as it is said, "quantized." The rules for quantization require the development of a new mathematical symbolism which we shall describe in Section 3.4.

There exist several equivalent ways of performing this quantization, but their mention would exceed the scope of this course. We shall, therefore, confine ourselves to a brief discussion of Schrödinger's equation which constitutes one of the several alternative formulations of quantum mechanics.§

† See, for example, A. Sommerfeld, *Thermodynamics and Statistical Mechanics*, (J. Kestin, transl.), p. 233, Vol. V of "Lectures on Theoretical Physics," Academic Press, New York, 1956.

‡ Correct in the sense that no motions are *now* known to exist for which this would not be true.

§ An admirable, introductory (though not elementary) account can be found in P. T. Matthews, *Introduction to Quantum Mechanics*, McGraw-Hill, New York, 1963. The account which follows was consciously modeled on that given in the preceding reference. See also, L. I. Schiff, *Quantum Mechanics*, 3rd ed., McGraw-Hill, New York, 1968; J. L. Powell and B. Crasemann, *Quantum Mechanics*, Addison-Wesley, Reading, Massachusetts, 1961;

Note: Readers who find the ensuing arguments too complex may accept the summary contained in Section 3.16 and proceed from it directly to the statistical theory of Chapter 5. In order to understand the latter it is not necessary to master quantum mechanics in all its details but only to grasp the concept of a stationary quantum state and to convince oneself of the existence of discrete energy levels.

3.2. The Physical Basis of Quantum Mechanics

The need to replace classical with quantum mechanics in the description of the motion of elementary particles arises from the dual nature of matter and radiation alluded to before.

The particle-like characteristics of radiation have been revealed by M. Planck when he discovered the correct expression for the spectral energy distribution of black-body radiation and by A. Einstein in his equation of the photoelectric effect. Both sets of considerations lead to the conclusion that energy can be exchanged between elementary particles only in amounts which consist of discrete quantum jumps. For example, light can exchange energy only in multiples of a *quantum of energy*

$$\varepsilon = \mathbf{h}\nu = \hbar\omega\,, \tag{3.1}$$

where $\mathbf{h} = 2\pi\hbar$ is Planck's constant (Section 2.2). With each wave of wavelength λ there must be associated a momentum p given by

$$p = \frac{\mathbf{h}}{\lambda}. \tag{3.1a}$$

Experiments with streams of particles have revealed that they are capable of forming diffraction patterns, which are also observed when waves are made to interact with each other. Consequently, it becomes necessary to ascribe a wavelike nature to elementary particles, and to accept the fact that there is no essential difference between a particle and a wave. The physical characteristics of a particle, its energy, ε, and its momentum, p, are, therefore, associated with the physical characteristics of a wave, its wavelength, λ, and its circular frequency, ω, (or frequency, ν). This association is expressed in Equations (3.1) and (3.1a). The wavelength λ of a particle of momentum p is called its *de Broglie wavelength.*

A. Messiah, *Quantum Mechanics*, Vols. 1 and 2, North-Holland Publ., Amsterdam, 1961, 1962; and K. Gottfried, *Quantum Mechanics*, Benjamin, New York, 1966. An imaginative, historical presentation can be found in S. I. Tomonaga, *Quantum Mechanics*, (Koshiba, transl.), Vol. I, "Old Quantum Theory," North-Holland Publ., Amsterdam, 1962.

In order to account for the preceding facts it is necessary to recognize two general principles. First, it must be recognized that elementary particles provide an absolute measure of smallness. A system is large when quantum effects are negligible for the description of its properties, otherwise it is small. In a sense, Planck's constant provides a measure of smallness. For example, a change in the energy of radiation which is comparable with $\hbar\omega$ is small. Similarly, a particle is small if its dimensions are comparable with its de Broglie wavelength.

According to this remark, a spherical droplet of water (density $\rho = 1$ gr/cm) of radius $r = 0.1$ cm moving at a speed of $v = 1$ cm/sec is "large," because

$$\lambda = \frac{\mathbf{h}}{mv} = \frac{6.62 \times 10^{-27} \text{ gr cm}^2/\text{sec}}{\frac{4}{3}\pi(0.1)^3 \text{ cm}^3 \times 1 \text{ gr/cm}^3 \times 1 \text{ cm/sec}} \approx 1.56 \times 10^{-24} \text{ cm}.$$

Secondly, it must be recognized that measurements performed on a system necessarily change its state, and this circumstance puts a natural limit on our powers of observation. In normal circumstances this limitation is negligible, but as far as small particles are concerned, the limitation is essential. For example, in order to look at a small object it is necessary to illuminate it with light of a very small wavelength λ. Equation (3.1a) implies that the momentum of the associated particle (the photon) will be relatively large and comparable with the momentum of the observed particle itself. Thus the very act of shining light on an elementary particle deflects its trajectory.

Recognition of all of the preceding circumstances forces us to reformulate our equations of motion and to modify them in a fundamental way.

3.3. The Wave Function

When the motion of a stream of N particles is described with the aid of the equations of Newtonian mechanics, it is possible, at least in principle, to write a differential equation of motion for each degree of freedom. Thus there are Nf differential equations of motion, where f is the numbeı of degrees of freedom of one particle (Section 2.3.6). Given the positions and momenta for translation, rotation, and vibration of all particles at time $t = t_0$, it should be possible to solve the differential equations and so to determine as many trajectories as there are particles, as well as to describe their internal motions. In this manner the microstate at $t = t_0$ together with the classical equations of motion would determine uniquely all the future microstates at times $t > t_0$ (or even the past microstates for $t < t_0$). The sheer impossibility of actually solving the fN simultaneous equations of motion of a mechanical model is beside the point. The essential consequence of the result, even though we can only visualize it by a bold stroke of imagination, is that such a stream would be incapable of producing diffraction patterns under any circumstances.

In order to account for the dual nature of particles which results in the appearance of diffraction patterns and related phenomena, it is necessary to accept the hypothesis that particles do not move along *deterministic trajectories*. Instead, it is assumed that only the *probability* of finding particular values for the coordinates and momenta of elementary particles can be indicated. This probability is assumed to be related to a purely conceptual function. In turn, this function is assumed to satisfy an equation whose solution is wavelike in character. For this reason, the probability function is called the *wave function* of the system. In this manner, the wave–particle duality can be introduced into the theory in a natural way and at its roots.

The wave function for the coordinates (which we denote by the symbol q to conform with usage in quantum mechanics and to stress the fact that the chosen system of coordinates need not be Cartesian) is denoted by the symbol Ψ whereas that for the associated momenta p is denoted by Φ. Since these functions are purely theoretical and do not correspond to anything physically real, it is convenient (and indeed necessary) to assume that they are both complex quantities. The complex conjugates of these functions are denoted by Ψ^* and Φ^*, respectively, and it is noted that the products $\Psi\Psi^*$ and $\Phi\Phi^*$ are real. Both functions depend on time, and

$$\Psi = \Psi(q_1, \ldots, q_{fN}, t) \qquad \text{with} \qquad \Phi = \Phi(p_1, \ldots, p_{fN}, t). \tag{3.2}$$

Given a microstate, we postulate that the probability density W_q of finding the particles at a position between q_i and $q_i + \mathrm{d}q_i$ is

$$W_q = \Psi\Psi^* \qquad (= |\Psi|^2), \tag{3.2a}$$

whereas the probability density W_p of simultaneously finding values of momentum contained between p_i and $p_i + \mathrm{d}p_i$ is

$$W_p = \Phi\Phi^* \qquad (= |\Phi|^2). \tag{3.2b}$$

Since it must be absolutely certain that the coordinates have values confined within the volume of physical space accessible to the system (for example, the vessel containing a gas) and that the momenta have a value ranging from $-\infty$ to $+\infty$, the integrals of the probability density functions W_q and W_p taken over all possible values must be equal to unity. Thus

$$\int \cdots \int W_q(q, t)\, \mathrm{d}q_1 \cdots \mathrm{d}q_{fN} = \int \cdots \int \Psi(q, t)\Psi^*(q, t)\, \mathrm{d}q_1 \cdots \mathrm{d}q_{fN} = 1 \tag{3.3a}$$

and

$$\int \cdots \int W_p(p, t)\, \mathrm{d}p_1 \cdots \mathrm{d}p_{fN} = \int \cdots \int \Phi(p, t)\Phi^*(p, t)\, \mathrm{d}p_1 \cdots \mathrm{d}p_{fN} = 1, \tag{3.3b}$$

at any instant of time t. Here q, t and p, t in the parentheses stand for all fN variables and time. The quantities W_q and W_p are known as *probability densities* by analogy with the integral for the mass of a continuous system.

The fact that the range of values of the momenta p_i extends from $-\infty$ to $+\infty$ implies that some velocities will approach the speed of light. Strictly speaking, relativistic effects may become important, at least for the high-velocity particles, and the range of velocities must be chosen to be in harmony with the theory of relativity. However, this is not normally necessary, and the approximation implied above is sufficient for most purposes.

The two probability densities, and hence the two wave functions Ψ and Φ, are not independent of each other, in analogy with the fact that the momentum p in classical mechanics is related to the time-derivative of its conjugate coordinate. Given one, it is possible to determine the other. The relation which is postulated for them is too complicated mathematically to be described here,† but we shall have no occasion to use it. For our purposes, it is sufficient to recognize that the fundamental equation of quantum mechanics can be written in one of two "languages," the Ψ-language or the Φ-language. We shall use the Ψ-language.

It is recognized that a specification of the two probability densities W_q, W_p replaces the detailed list of values of the fN p_i's and the fN q_i's of classical mechanics. Hence it is seen that the wave function Ψ or Φ represents the microstate of the system.

In accordance with the correspondence principle, the probability densities W_p, W_q for a classical system (quantum system in the limit of large size) tend to Dirac delta functions.‡ In this manner the trajectory of the system

† It involves the Fourier transform integrals

$$\Phi(p,t) = \mathbf{h}^{-Nf/2} \int \Psi(q,t) \exp\left(-\frac{i}{\mathbf{h}} \Sigma\, p_i q_i\right) dq_1 \cdots dq_{fN},$$

$$\Psi(q,t) = \mathbf{h}^{-Nf/2} \int \Phi(p,t) \exp\left(+\frac{i}{\mathbf{h}} \Sigma\, p_i q_i\right) dp_1 \cdots dp_{fN},$$

the sum extending from $i = 1$ to $i = fN$.

‡ The Dirac delta function $\delta(x - x_0)$ is defined as

$$\delta(x - x_0) = \begin{cases} 0 & \text{for } a \leq x \leq x_0 \text{ and } x_0 < x \leq b \\ \infty & \text{for } x = x_0 \end{cases}$$

but with

$$\int_a^b \delta(x - x_0)\, dx = 1.$$

It is shown that

$$\int_a^b f(x)\, \delta(x - x_0)\, dx = f(x_0).$$

in Γ-space, that is, the diagram of its motion, becomes deterministic: to a given value of momentum there corresponds a definite position and *vice versa.*

In order to produce physically meaningful values of probability densities, the wave functions Ψ and Φ must be continuous, single-valued, and finite, in addition to satisfying the *normalization conditions* (3.3a) and (3.3b).

3.4. The Mathematical Basis of Quantum Mechanics. Schrödinger's Equation

The state of a mechanical system is described by the two wave functions Ψ and Φ, that is, by the probabilities W_{q_i} and W_{p_i} instead of the values of the coordinate q_i and momentum p_i of classical mechanics. As already stated, for every particular system it is possible to write a partial differential equation for the wave function Ψ or Φ, as desired, but the most general form of this equation cannot be written as a partial differential equation. In other words, it is impossible to write a general differential equation and to specialize it to a form applicable to each system in turn. Instead, as already mentioned, the characteristic, macroscopic equation for a system must be quantized to yield the differential equation of the problem. In order to achieve this it was necessary to develop a novel mathematical symbolism and theory. This can take the form of W. Heisenberg's and M. Born's *matrix algebra* or of E. Schrödinger's and P. A. M. Dirac's *operator calculus.* In accordance with an earlier remark, we shall discuss briefly only the latter.

Speaking heuristically, an *operator* **A** is defined as any set of rules which transform a given function into another function. For example,

$$\mathbf{A} \equiv x$$

denotes that the function f(x) on which **A** "operates" is simply multiplied by the independent variable x. Hence

$$\mathbf{A}\mathrm{f}(x) \qquad \text{means} \qquad x\mathrm{f}(x)$$

in this case. Similarly, if

$$\mathbf{A} \equiv \frac{\partial}{\partial x}, \qquad \text{then} \quad \mathbf{A}\mathrm{f}(x) \equiv \frac{\partial \mathrm{f}}{\partial x}.$$

Applying the same operator again, we obtain

$$\mathbf{A}[\mathbf{A}\mathrm{f}(x)] = \frac{\partial}{\partial x}\frac{\partial \mathrm{f}}{\partial x} = \frac{\partial^2 \mathrm{f}}{\partial x^2}.$$

In such cases we may put

$$\mathbf{B} \equiv \frac{\partial^2}{\partial x^2} \equiv \mathbf{A}^2,$$

and the convention is adopted that repeated application of an operator is indicated by an exponent.

The formation of the Laplacian of a scalar function $g(x, y, z)$, that is, the expression

$$\nabla^2 g \equiv \frac{\partial^2 g}{\partial x^2} + \frac{\partial^2 g}{\partial y^2} + \frac{\partial^2 g}{\partial z^2}$$

can be interpreted as the result of modifying the function g by the operator

$$\nabla^2 \equiv \frac{\partial^2}{\partial x^2} + \frac{\partial^2}{\partial y^2} + \frac{\partial^2}{\partial z^2}. \tag{3.4}$$

In quantum mechanics, a quantity which can be observed—an *observable*—is always represented by an operator "operating" on the wave function Ψ or Φ. The operation of observation which modifies the state of the system is thus represented by an operator which modifies the wave function. Consequently, the result of two successive applications of two operators **A** and **B** may depend on the order in which they are applied because each of the observations (say that of the momentum followed by that of the position and *vice versa*) may modify the state of the system in different ways. Thus

$$\mathbf{AB} \neq \mathbf{BA},$$

and the difference

$$[\mathbf{A}, \mathbf{B}] \equiv \mathbf{AB} - \mathbf{BA} \tag{3.5}$$

is defined as the *commutator* of the operators **A**, **B** applied in the order indicated in the defining equation (3.5). When

$$[\mathbf{A}, \mathbf{B}] = 0,$$

the operators are said to commute; otherwise we say that they do not commute.

If the result of operating on a function f is the function f itself except for a constant of proportionality a, the function f is called a *characteristic* or *eigen*function† of the operator. Hence if

$$\mathbf{A}\mathrm{f} = a\mathrm{f}, \tag{3.6}$$

f is an eigenfunction of the operator **A** and a is its associated *eigenvalue*. The eigenfunctions of an operator constitute a special set of functions whose form remains unaltered when the operator is applied to each of them, except, as stated, for a possible constant factor. Naturally, the solution of Equation (3.6) need not be unique, and for this reason we speak of a set of eigenfunctions and not of a single eigenfunction.

† From the German *eigen*, meaning intrinsic, belonging to.

As an example we consider

$$\mathbf{A} \equiv -\mathrm{i}\frac{\partial}{\partial x}. \tag{3.7}$$

The eigenfunctions f_n must be determined from the differential equation

$$-\mathrm{i}\frac{\partial \mathrm{f}_n}{\partial x} = a_n \mathrm{f}_n$$

whose general solution is

$$\mathrm{f}_n(x) = \mathrm{e}^{\mathrm{i}a_n x},$$

as is easily verified by substitution. Explicitly,

$$\mathrm{f}_n(x) = \cos(a_n x) + \mathrm{i}\sin(a_n x).$$

If we impose the boundary condition that the imaginary part of the function $\mathrm{f}_n(x)$ must vanish at $x = 0$ and $x = L$ (a constant):

$$\mathrm{Im}[\mathrm{f}_n(x)] = 0 \quad \text{at} \quad x = 0 \quad \text{and} \quad x = L, \tag{3.8}$$

we find that

$$a_n L = \pi n \qquad (n = 0, 1, \ldots),$$

that is, that

$$a_n = \frac{\pi n}{L} \qquad (n = 0, 1, \ldots). \tag{3.8a}$$

Here n denotes an integer, and it is seen that there exists a *discrete*, though infinite, set of eigenvalues of the operator (3.7), subject to the imposed boundary condition (3.8).

In quantum mechanics it is accepted that each type of observation may yield a set of numbers, namely the possible results of observation. *These are identical with the eigenvalues of the corresponding operator applied to the wave function.* Hence, if the system is in a definite state described by a wave function f_n, the result of the observation which corresponds to operator $\mathbf{A}$ is necessarily a_n. Furthermore, it is accepted that the result of repeated observations $\mathbf{A}$ on a set of systems, each one in a state described by the wave function f is the *expectation value*

$$\langle a \rangle = \int_{-\infty}^{\infty} \mathrm{f}^*(x)\mathbf{A}\mathrm{f}(x)\,\mathrm{d}x, \tag{3.9}$$

provided that the normalization condition

$$\int_{-\infty}^{\infty} f^*(x)f(x)\,dx = 1 \tag{3.9a}$$

is satisfied.

The operators which are required for the *Schrödinger equation* related to the wave function Ψ can be listed as follows:

observation of position x:

$$\mathbf{q}_x \equiv x \tag{3.10a}$$

observation of momentum p conjugate to q:†

$$\mathbf{p}_x \equiv -i\hbar \frac{\partial}{\partial x} \equiv \frac{\hbar}{i}\frac{\partial}{\partial x} \tag{3.10b}$$

observation of energy E:

$$\mathbf{H}.$$

The operator $\mathbf{H}$ is obtained by writing down the total energy of the system and by introducing the substitutions (3.10a) and (3.10b) into the resulting expression. The total energy of a system

$$\mathscr{H} = \mathscr{E}_{\text{kin}} + \phi \tag{3.11}$$

is known as the *Hamiltonian* of the system in classical mechanics.‡ Evidently, with the great variety of possible systems, no general expression for the Hamiltonian operator can be indicated, but several examples of the procedure to be adopted will be given in Sections 3.7–3.11. As a preliminary example, we consider a single, independent monatomic molecule of mass m; we have $\phi = 0$, and

$$\mathscr{H} = \mathscr{E}_{\text{kin}} = \frac{p_x^{\,2} + p_y^{\,2} + p_z^{\,2}}{2m}.$$

Hence, by Equations (3.10a) and (3.10b)

$$\mathbf{H} = -\frac{\hbar^2}{2m}\left(\frac{\partial^2}{\partial x^2} + \frac{\partial^2}{\partial y^2} + \frac{\partial^2}{\partial z^2}\right) = -\frac{\hbar^2}{2m}\nabla^2$$

for this particular case.

† It is remembered that a momentum p is said to be conjugate to a position coordinate q when the velocity appearing in the former constitutes a time-derivative of the latter.

‡ Although the Hamiltonian is equal to the total energy for the systems under consideration, the statement is not true for all possible systems. See A. Sommerfeld, *Mechanics*, (M. O. Stern, transl.), p. 181, Vol. I of "Lectures on Theoretical Physics," Academic Press, New York, 1952.

In operator notation the Schrödinger equation has the form

$$\mathbf{H}\Psi + \frac{\hbar}{i}\frac{\partial \Psi}{\partial t} = 0. \tag{3.12}$$

Although it is possible to provide plausibility arguments and to write the Schrödinger equation (3.12) as a result, it must be realized that the Schrödinger equation cannot be *derived*; its validity, like that of Newton's laws of motion or the laws of thermodynamics, is *postulated*; in the present case the postulates include the equation itself as well as the set of principles which provide physical interpretations for the symbols appearing in it.

It may be interesting to note that the form of the Schrödinger equation in momentum (Φ) language is identical with that in the coordinate language(Ψ):

$$\mathbf{H}\Phi + \frac{\hbar}{i}\frac{\partial \Phi}{\partial t} = 0. \tag{3.12a}$$

The Hamiltonian operator $\mathbf{H}$ is formed by the substitutions

$$\mathbf{x} \equiv -\frac{\hbar}{i}\frac{\partial}{\partial p_x}, \qquad \mathbf{p}_x \equiv p_x .$$

3.5. Stationary States. Schrödinger's Time-Independent Equation

When the Hamiltonian $\mathscr{H}$ of the system contains the time, t, explicitly, Schrödinger's equation (3.12) will yield a time-dependent wave function of the system, and the probability density W_q will turn out to be time-dependent at every point in physical space. Thus the solution of the equation will describe a process. Our interest will be centered on the calculation of the values of properties in particular states. Consequently, we must concentrate on examples in which the total energy $\mathscr{H}$ remains constant with time, and this induces us to consider the special class of solutions of Equation (3.12) in which the Hamiltonian operator $\mathbf{H}$ does not contain time explicitly. It is said that such solutions refer to *stationary states*. In fact, the range of solutions required by us is even narrower, as the reader may remember from Section 2.3 on mechanical models of systems. We need only include translatory motion, rotational motion, and oscillatory motion, for which the respective Hamiltonians were displayed in Equations (2.13a)–(2.13c).

We attempt to find a general solution of Equation (3.12) by separation of variables, which means that we suppose that the wave function Ψ can be represented as the product

$$\Psi = \psi(q_1, \ldots, q_{fN}) \cdot \mathrm{f}(t) \tag{3.13}$$

of a function of position $\psi(q_1, \ldots, q_{fN})$ and a function of time $\mathrm{f}(t)$. Substitution into Equation (3.12) leads to the result that

$$\mathrm{f}(t)\mathbf{H}\psi + \psi \frac{\hbar}{\mathrm{i}} \frac{\mathrm{df}}{\mathrm{d}t} = 0,$$

that is, that

$$-\frac{\hbar}{\mathrm{i}} \frac{1}{\mathrm{f}} \frac{\mathrm{df}}{\mathrm{d}t} = \frac{1}{\psi} \mathbf{H}\psi. \tag{3.14}$$

The left-hand side of this equation is independent of the position coordinates, whereas the right-hand side does not contain the time, as assumed. Consequently, equality can prevail only if both sides are equal to a constant—the so-called separation constant, and we may denote it by ε. In this manner Equation (3.14) becomes equivalent to *two* equations:

$$\frac{\hbar}{\mathrm{i}} \frac{1}{\mathrm{f}} \frac{\mathrm{df}}{\mathrm{d}t} = -\varepsilon \tag{3.15a}$$

$$\mathbf{H}\psi = \varepsilon\psi. \tag{3.15b}$$

The general solution of Equation (3.15a) is

$$\mathrm{f}(t) = \exp(-\mathrm{i}\varepsilon t/\hbar) \tag{3.16}$$

or, equivalently

$$\mathrm{f}(t) = \cos(\varepsilon t/\hbar) - \mathrm{i}\,\sin(\varepsilon t/\hbar)$$

with

$$\mathrm{f}^*(t) = \exp(\mathrm{i}\varepsilon t/\hbar) = \cos(\varepsilon t/\hbar) + \mathrm{i}\,\sin(\varepsilon t/\hbar).$$

Equation (3.15b) is identical with Equation (3.6); it defines a set of eigenvalues of energy ε_n together with a set of associated eigenfunctions $\psi_n(q_1, \ldots, q_{fN})$. This set constitutes a characteristic of the energy operator $\mathbf{H}$. Equation (3.15b) must be solved for each case separately, but the rigorous theory of quantum mechanics leads to the conclusion that the eigenvalues ε_n are real, since they constitute observables. We shall confirm this with the reader in relation to our three cases of interest in Sections 3.7–3.11.

The general solution can be formed by summing over all eigenvalues, ε_n, because the Schrödinger equation (3.12) is linear. Hence

$$\Psi = \sum_n \psi_n(q_1, \ldots, q_{fN})\exp(-\mathrm{i}\varepsilon_n t/\hbar),$$

and the probability density is

$$\Psi\Psi^* = \left[\sum_n \psi_n(q_1, \ldots, q_{fN})\exp(-i\varepsilon_n t/\hbar)\right]\left[\sum_m \psi_m^*(q_1, \ldots, q_{fN})\exp(+i\varepsilon_m t/\hbar)\right].$$

This probability density will contain terms of the type

$$|\psi_n|^2 \qquad \text{and} \qquad \psi_n\psi_m^* \exp[(-it/\hbar)(\varepsilon_n - \varepsilon_m)].$$

The terms of the first type are independent of time, and those of the second type contain time unless $\varepsilon_n = \varepsilon_m$. This means that the *energy of the system must be equal to an eigenvalue* ε_n *of the Hamiltonian operator* $\mathbf{H}$ if the state is to be a stationary one.

Thus our solution for one wave function, or stationary state, is

$$\Psi_n = \psi_n(q_1, \ldots, q_{fN})\exp(-i\varepsilon_n t/\hbar). \tag{3.17}$$

The exponential factor represents a periodic function of time, and Ψ_n can be thought of as a *standing wave* whose amplitude ψ_n varies from point to point in the physical space.

The probability density W_q for such a wave can be computed from the relation $W_q = \Psi\Psi^* = |\psi_n|^2 ff^* = |\psi_n|^2$ because $ff^* = 1$, as seen from the earlier relations, and on condition that ε_n is a real number. Thus the probability density turns out to be time-independent, as must be expected for a stationary state. For this reason, the amplitude equation (3.15b) is known as Schrödinger's time-independent equation. Its mathematical properties prove quite generally that the only result of a measurement of energy on a system in a stationary state can be an eigenvalue ε_n of the $\mathbf{H}$ for the system. This can be verified with reference to Equation (3.9), which shows that the expectation value of a set of measurements of the energy in a stationary state is

$$\begin{aligned}\langle E\rangle &= \int_{-\infty}^{\infty} \Psi_n^*(q_1, \ldots, q_{fN})\mathbf{H}\Psi_n(q_1, \ldots, q_{fN})\, dq_1 \cdots dq_{fN} \\ &= \int_{-\infty}^{\infty} \psi_n^*(q_1, \ldots, q_{fN})\varepsilon_n\psi_n(q_1, \ldots, q_{fN})\, dq_1 \cdots dq_{fN} = \varepsilon_n,\end{aligned}$$

since $\mathbf{H}\psi_n = \varepsilon_n\psi_n$ and

$$\int_{-\infty}^{\infty} |\psi_n(q_1, \ldots, q_{fN})|^2\, dq_1 \cdots dq_{fN} = 1.$$

We shall see later that in all three cases of interest, the set of energy eigenvalues ε_n will be discrete, though continuous sets are not excluded *a priori* by any means.

3.6. Complementarity and Heisenberg's Uncertainty Principle

Recalling the definition in Equation (3.5) and the rules for operators in Equations (3.10a) and (3.10b), we can easily verify that for a coordinate and its conjugate momentum, say x and p_x, we have

$$[\mathbf{x}, \mathbf{p}_x]\psi = \frac{\hbar}{i}\left[x\frac{\partial\psi}{\partial x} - \frac{\partial(x\psi)}{\partial x}\right],$$

so that

$$[\mathbf{x}, \mathbf{p}_x] = \hbar i. \tag{3.18}$$

On the other hand, for a coordinate, say x, and a nonconjugate momentum, say p_y, the operators commute, as is easily verified from

$$[\mathbf{x}, \mathbf{p}_y]\psi = \frac{\hbar}{i}\left[x\frac{\partial\psi}{\partial y} - \frac{\partial(x\psi)}{\partial y}\right] = 0,$$

so that

$$[\mathbf{x}, \mathbf{p}_y] = 0. \tag{3.18a}$$

The preceding commutation relations show that there exists a mutual interference between the measurement of a coordinate and its conjugate momentum, whereas no such interference exists between a coordinate and another momentum. Noncommuting operators are said to be *complementary*, and the existence of a relation between the operators of interfering measurements is known as Bohr's *complementarity principle*.

The complementarity principle makes it plausible to infer that the *simultaneous* determination of two observables whose operators are complementary (do not commute) cannot occur with absolute, that is, arbitrarily prescribed, precision. No such limitation operates with respect to noncomplementary observables whose operators commute. The preceding principle can be derived rigorously from the commutation relation and is reflected in the relation between the wave functions ψ and ϕ of a conjugate pair of coordinate and momentum. The resulting quantitative relation is known as *Heisenberg's uncertainty* principle. We are not in a position to display the required mathematical argument; instead, we shall illustrate the principle with the aid of a particular example.

The diagram in Figure 3.1a represents the normalized wave function for a so-called rectangular wave packet at $t = 0$. This indicates that the probability of finding the particle under consideration at $x = \pm\frac{1}{2}\Delta x$ is $(1/\sqrt{\Delta x})^2 = 1/\Delta x$. The distance Δx serves as a measure of the accuracy with which the particle can be "localized" at $x = 0$, and $\Delta x \to 0$ signifies that the particle is at $x = 0$ *precisely*; when Δx is finite, the particle can be said to be

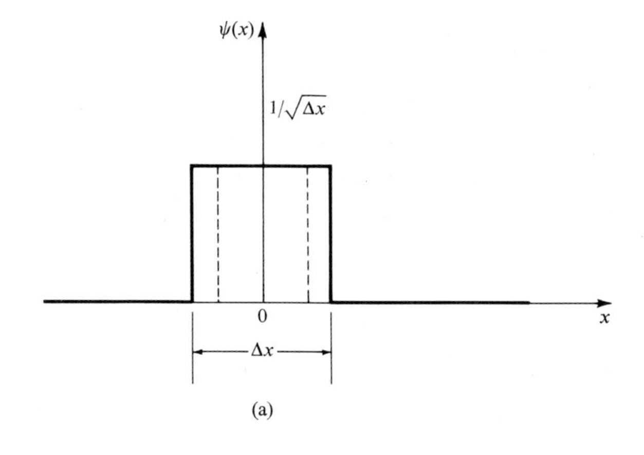

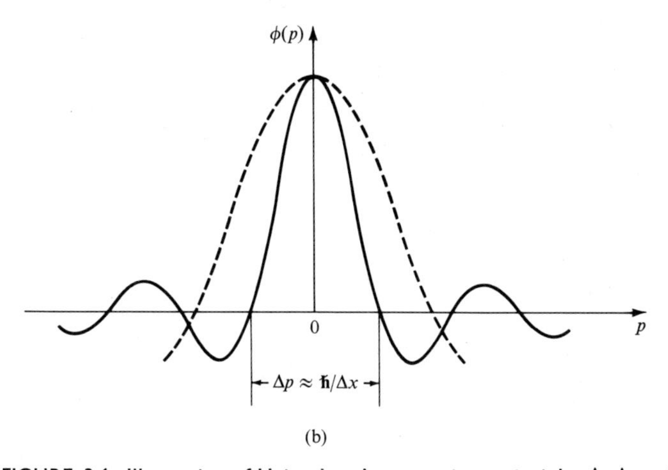

FIGURE 3.1. Illustration of Heisenberg's uncertainty principle, $\Delta p \Delta x \approx \hbar$.

located at $x = 0$ with an uncertainty of $\pm\frac{1}{2}\Delta x$. The corresponding momentum amplitude is shown sketched in Figure 3.1b. This demonstrates that a measurement of the conjugate momentum p_x can yield a value in the range $\pm\infty$ with the exception of the points where $\phi = 0$, and that the overwhelming probability of measuring a value of p is confined to a range $\pm\frac{1}{2}\Delta p$ which can be conveniently taken as being contained between the two zeros of ϕ nearest to $p = 0$. It is proved in quantum mechanics that

$$\Delta p \approx \frac{\hbar}{\Delta x},$$

or that

$$\Delta p \, \Delta x \approx \hbar.$$

This is one example of Heisenberg's uncertainty principle. It shows that attempts to localize the particle more and more precisely ($\Delta x \to 0$) results in a less and less certain value ($\Delta p \to \infty$) of its momentum, as indicated by the broken lines in the diagrams. This is due to the mutual interference of these two measurements and is a direct consequence of the duality of waves and particles. The exact derivation would yield

$$\Delta x \, \Delta p_x \geq \tfrac{1}{2}\hbar . \tag{3.19}$$

It is interesting to note that

$$[\mathbf{p}, \mathbf{H}] = 0 \tag{3.20}$$

if the Hamiltonian does not contain a potential energy term. Furthermore,

$$\Delta E \, \Delta t \geq \tfrac{1}{2}\hbar \tag{3.20b}$$

which shows that the simultaneous measurement of the energy of a particle and of the instant of its passage through a given position suffers from mutual interference. In general, the energy of a system and the time taken to change it cannot be known with an arbitrary accuracy. Thus the energy, E, of a particle can be known only with an error $\Delta E \approx \hbar/2 \, \Delta t$ if the measurement extends over a time interval Δt.

3.7. Translational Motion of a Single, Independent Molecule

The description of the mechanical model adopted for a gas at low density included the assumptions that there was no interaction between its translational, rotational, and vibrational modes of energy storage, and that the interaction between molecules was confined to collisions between them. This set of assumptions makes it possible to analyze each particle and each mode of motion separately, and to describe the motion in μ-space (Section 2.2). Similarly, the assumptions regarding the mechanical model of a solid enable us to confine our study to that of the motion of single harmonic oscillators; this motion is identical with the internal vibration in a polyatomic molecule.

Accordingly, we begin with the case of translational motion and consider a single molecule whose total energy (or Hamiltonian) consists of three quadratic kinetic-energy terms and a single potential-energy term

$$\mathscr{H} = \frac{{p_x}^2 + {p_y}^2 + {p_z}^2}{2m} + \phi(x, y, z) . \tag{3.21}$$

giving

$$\mathbf{H} = -\frac{\hbar^2}{2m} \nabla^2 + \boldsymbol{\phi}(\mathbf{x}, \mathbf{y}, \mathbf{z}).$$

We assume that the molecule moves in a cubic container of volume $V = a^3$ whose one corner is placed at the origin of a Cartesian system of coordinates as shown in Figure 3.2. The time-independent Schrödinger equation for a particular wave function for this system is

$$\frac{\partial^2 \psi}{\partial x^2} + \frac{\partial^2 \psi}{\partial y^2} + \frac{\partial^2 \psi}{\partial z^2} = -\frac{2m}{\hbar^2}(\varepsilon - \phi)\psi . \tag{3.22}$$

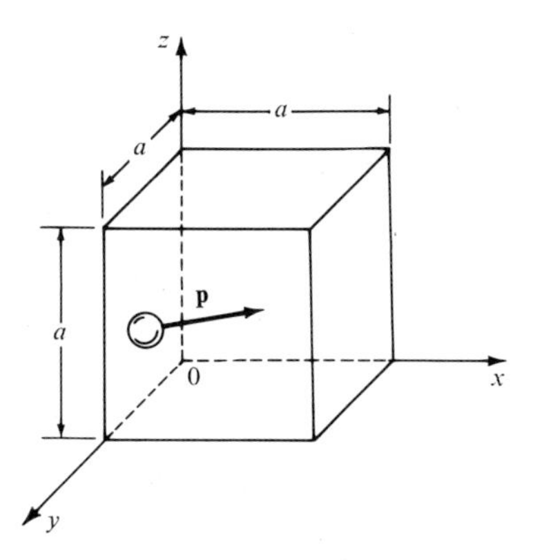

FIGURE 3.2. Molecule in a rectangular container.

Since the potential energy is a pure function of position, the operator $\boldsymbol{\phi}(\mathbf{x}, \mathbf{y}, \mathbf{z})$ for potential energy reverts to $\phi(x, y, z)$ in view of Equation (3.10a) because

$$\boldsymbol{\phi}(\mathbf{x}, \mathbf{y}, \mathbf{z}) = \boldsymbol{\phi}(x, y, z) = \phi(x, y, z) .$$

In our particular problem we put $\phi = 0$, and assume that

$$\psi = 0 \qquad \text{at} \quad x \leq 0, \quad y \leq 0, \quad z \leq 0,$$

and

$$x \geq a; \quad y \geq a; \quad z \geq a . \tag{3.22a}$$

This means that we prescribe a zero probability of finding the particle outside the container.

In classical mechanics the particle would move between the walls and would be reflected from them. Since a constant total energy is postulated to conform with the First law of thermodynamics, the velocity would change direction, but not magnitude. When all possible stationary states are considered we would find that any value of velocity would be admissible and that the kinetic energy could be changed in a continuous manner on going from one stationary state to another.

The corresponding quantum solution is implied in Equation (3.22) together with the boundary conditions (3.22a). Instead of tackling the full problem at once, it will be convenient to assume first that the molecule is constrained to move parallel to the x-axis only. With this stipulation (and with $\phi = 0$), the Schrödinger equation (3.22) assumes the simplified form

$$\frac{d^2\psi}{dx^2} + \frac{2m}{\hbar^2}\,\varepsilon\psi = 0 \qquad \text{with} \quad \psi = 0 \quad \text{at} \quad x = 0 \quad \text{and} \quad x = a\,. \tag{3.23}$$

The general solution to this equation, as is well known, can be written in the form

$$\psi = A \sin \frac{(2m\varepsilon)^{1/2}}{\hbar}\, x + B \cos \frac{(2m\varepsilon)^{1/2}}{\hbar}\, x\,. \tag{3.23a}$$

In order to satisfy the boundary condition at $x = 0$, we must put $B = 0$, and it is easy to see that the value to be assigned to A cannot be adjusted to satisfy the condition that $\psi = 0$ at $x = a$. This can be secured only if the argument $(2m\varepsilon)^{1/2}/\hbar$ satisfies the condition that

$$\frac{2\pi(2m\varepsilon)^{1/2}a}{\mathbf{h}} = s\pi \qquad (s = 0, 1, \ldots)\,.$$

The single quantity in this argument which does not constitute a fixed constant is the separation variable, that is, the energy ε, and the required boundary conditions cannot be satisfied unless the energy is restricted to the discrete class of eigenvalues

$$\varepsilon_s = \frac{\mathbf{h}^2}{8ma^2}\, s^2 \qquad (s = 1, 2, \ldots)\,. \tag{3.23b}$$

Thus the reader can see in detail how a discrete set of eigenvalues (or levels of energy) arises in a particular problem. It may be worth noting that this is due to the fact that ψ is required to vanish at a *finite* distance a from the origin. In view of this condition, the amplitude of the wave function assumes the form

$$\psi = A \sin \frac{s\pi x}{a} \qquad (s = 1, 2, \ldots), \tag{3.23c}$$

and it is seen that the value $s = 0$ must be excluded, as was done in anticipation in Equation (3.23b) because of the normalization condition

$$A^2 \int_0^a \sin^2 \frac{s\pi x}{a}\, dx = 1 \qquad \text{giving} \quad A = \left(\frac{2}{a}\right)^{1/2}.$$

A particular wave function ψ represents one stationary state of the molecule and plays the same part as the equation of a constant-energy trajectory in classical mechanics. For a given value of a, the wave function is completely determined by the indication of a single integer—the *quantum number s*.

In contrast with classical mechanics, the admissible energy levels ε_s form the quadratic series (3.23b), and two neighboring energy levels ε_{s+1} and ε_s can differ only by an amount

$$\varepsilon_{s+1} - \varepsilon_s = \frac{\mathbf{h}^2}{8ma^2}(2s+1), \tag{3.24}$$

the difference being scaled in magnitude with the aid of Planck's constant $\mathbf{h}$. This energy difference constitutes a *quantum jump*, that is, the smallest change in energy consistent with the Schrödinger equation in the circumstances; for $s \gg 1$ the quantum jump is proportional to the quantum number. The minimum energy, the so-called zero-point energy, occurs for $s = 1$; it is

$$\varepsilon_1 = \frac{\mathbf{h}^2}{8ma^2} \tag{3.24a}$$

and differs from zero; the particle can never be found at rest. This is in accord with the uncertainty principle, for otherwise the momentum ($p = 0$) would be known exactly at any exactly specified position.

The eigenfunctions ψ from Equation (3.23a) together with the probability densities ψ^2 have been drawn to scale in Figure 3.3, separately for even and odd values of s. Odd values of the quantum number s lead to symmetric wave

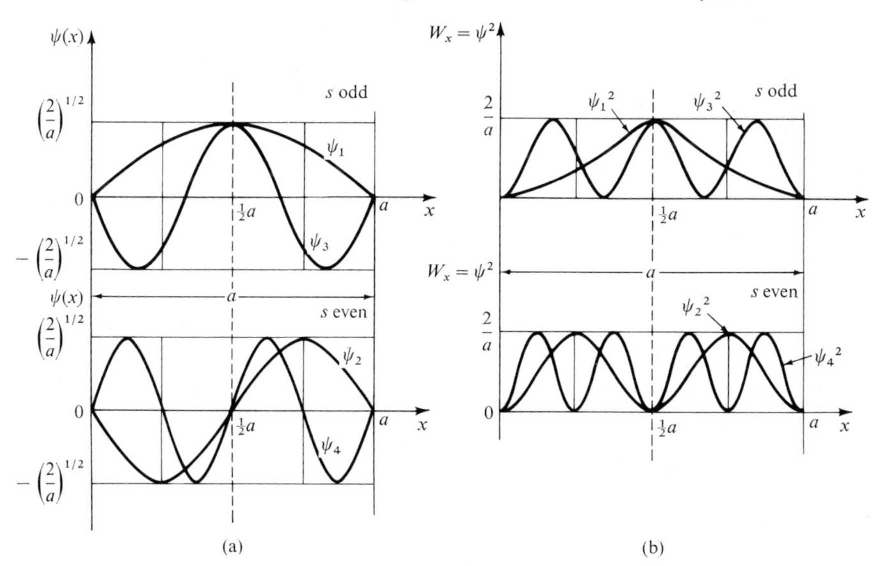

FIGURE 3.3. Wave function for a particle in a container: a, wave function; b, probability density. Note: Area under each curve in b is equal to unity.

functions with respect to $x = \frac{1}{2}a$, whereas even values lead to antisymmetric wave functions with respect to $x = \frac{1}{2}a$. It is easy to verify that negative values of the quantum number s result in wave functions which also satisfy the Schrödinger time-independent equation; they have been omitted from the list because they lead to identical probability densities and do not represent physically distinct solutions. It should be noted that the probability density, $W_x = \psi^2$, becomes less and less distinguishable from a uniform value as s increases; as a consequence, W_p tends to a Dirac delta function. Thus, as we see in this case, but will accept quite generally without proof, the behavior of a quantum system becomes less and less distinct from that of a classical system as the quantum number is made larger and larger.

When the quantum number is prescribed, the energy level ε_s is a function of a, that is, of the dimensions of the vessel confining the particle. This dimension can be changed externally as a result of the quasi-static performance of negative or positive work. We therefore note for future reference (Section 5.14) that

the reversible performance of work changes the values of the energy levels available to the particle.

We further note that positive work (a increasing) decreases the values of the available energy levels, whereas negative work (a decreasing) causes them to increase. The diagram in Figure 3.4 provides a pictorial impression of the effect of external work on the available energy levels.

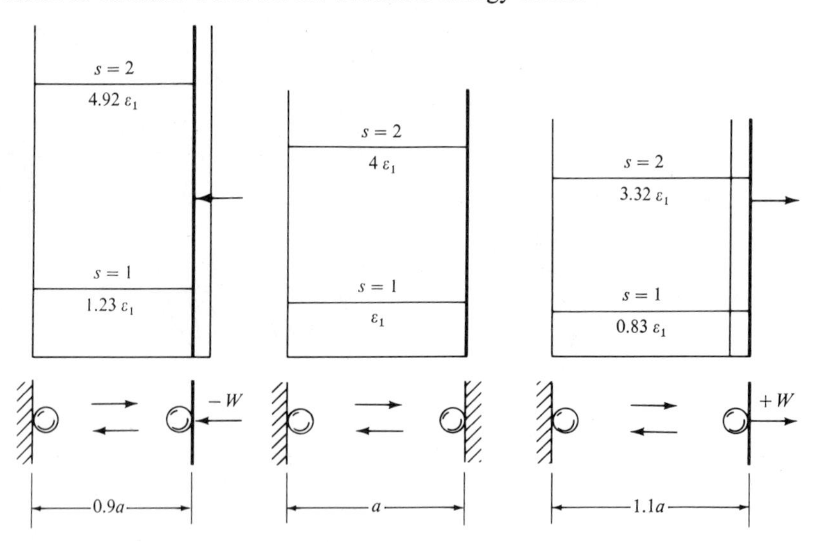

FIGURE 3.4. Effect of external work on the energy levels of a particle constrained to move in one direction.

To form an idea about the order of magnitude of the numerical values which appear in the preceding equations, we consider a molecule of helium constrained to move along a segment of length $a = 1$ cm. The mass of a helium molecule can be computed from the molecular mass $M \approx 4$ gr/gmol and the Avogadro number $\mathbf{N} \approx 6.02 \times 10^{23}$/gmol. Hence $m = M/\mathbf{N} \approx 6.65 \times 10^{-24}$ gr. In Section 6.5, we shall discover that the average energy of translation of a molecule is equal to $\varepsilon_s = \frac{1}{2}\mathbf{k}T$ per degree of freedom in translation. Thus at $T = 300$ K we have $\varepsilon_s = 0.5 \times 1.38 \times 10^{-16}$ (gr cm²/sec² K) $\times$ 300 K $\approx 2.07 \times 10^{-14}$ gr cm²/sec². The zero-point energy is $\varepsilon_1 = \mathbf{h}^2/8ma^2 \approx [6.63 \times 10^{-27}$ (gr cm²/sec)]²/8 $\times$ 6.65 $\times 10^{-24}$ gr $\times$ 1 cm² $\approx 0.83 \times 10^{-30}$ gr cm²/sec². Hence Equation (3.23b) yields $s = (\varepsilon_s/\varepsilon_1)^{1/2} \approx 1.58 \times 10^8$, and it is seen that the quantum number is extremely large, and that energy levels are spaced very closely. The closeness of the spacing of energy levels can be assessed by computing the quantum jump from Equation (3.24), according to which $\Delta\varepsilon_s \approx 2s\varepsilon_1$. In the present case $\Delta\varepsilon_s \approx 2 \times 1.58 \times 10^8 \times 0.83 \times 10^{-30}$ gr cm²/sec² $\approx 2.62 \times 10^{-22}$ gr cm²/sec², so that $\Delta\varepsilon_s/\varepsilon_s \approx 1.2 \times 10^{-8}$.

A change in temperature by 1°C produces a change of $\Delta\varepsilon_s' = \frac{1}{2}\mathbf{k}$ or $\Delta\varepsilon_s'/\varepsilon_s' = 1/300$. This corresponds to a change of $\Delta s \approx \frac{1}{2}s\Delta\varepsilon_s'/\varepsilon_s \approx 1.58 \times 10^8/600 \approx 2.6 \times 10^5$ in the quantum number.

A change in a from 1 cm to 1 Å $= 10^{-8}$ cm produces a change in the zero-point energy by a factor of 10^{16} and by an equal factor in all remaining energy levels. We note that the latter is now equal to $\varepsilon_1 = 0.83 \times 10^{-14}$ gr cm²/sec² and conclude that the quantum jump becomes comparable with the magnitude of the energy of the particle when the quantum number is small and when the size of the vessel is comparable to an atomic dimension.

The probability density diagrams contained in Figure 3.3b demonstrate that there exist locations where the probability of finding the particle is zero However, the enormity of the quantum numbers with which we shall deal habitually causes the peaks in the diagram to crowd so closely together that in practice there exists an equal probability of finding the particle at any position from $x = 0$ to $x = a$. The calculation shows further that at energy levels which we are likely to encounter, its discrete nature will not really come into play owing to the extreme smallness of energy quanta. For our purposes, the allowable set of energy levels may well be regarded as being continuous.

It is instructive to represent the motion of the particle in μ-space. In the present case this phase space consists of a single coordinate x and of a single momentum p_x which may assume the values

$$p_s = \pm\sqrt{2m\varepsilon_s} = \pm\frac{\mathbf{h}s}{2a}. \tag{3.25}$$

The phase space is shown sketched in Figure 3.5. The portion of the μ-space outside the interval between $x = 0$ and $x = a$ is completely inaccessible to the molecule, and the allowable momenta form an arithmetic progression. A particular state is represented by two levels, one at p_s and one at $-p_s$, because the molecule is as likely to move in one as in the opposite direction, as seen from Equation (3.25); in other words, the probability density W_p depends on p_s^2. We can imagine that the phase space has been subdivided into cells as shown by the two cross-hatched areas in the diagram, with the boundaries

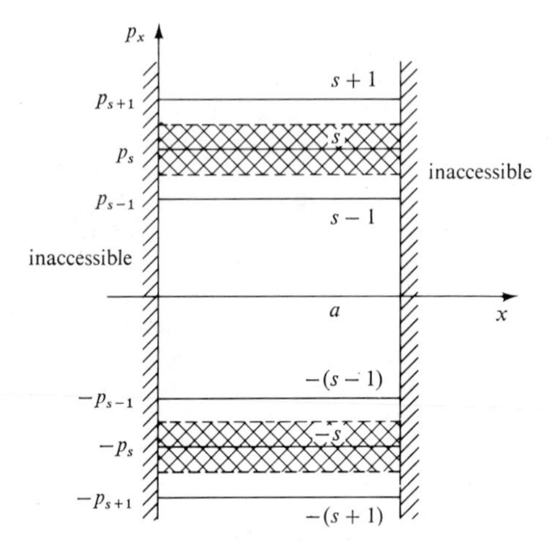

FIGURE 3.5. Phase space (μ-space) for translation in one direction.

drawn midway between two successive levels. The "volume" $\Delta\mathscr{V}$ of such a cell is

$$\Delta\mathscr{V} = 2 \times a \times \frac{\mathbf{h}}{2a} = \mathbf{h}. \tag{3.26}$$

The image of a molecule in a given quantum state can be placed anywhere within this cell, and it is clear that any further subdivision of the μ-space cannot improve the details of the description of the state of a molecule. Equation (3.26) is a particular case of the operation of Heisenberg's uncertainty principle which prescribes cells in phase space of order $\mathbf{h}^f$ where f is the number of degrees of freedom. Interpreting this statement in another way, we can say that the uncertainty of the determination of the position of the particle is $\Delta x = a$, and this results in an uncertainty in the determination of the complementary momentum $\Delta p_x = \mathbf{h}/a$ (if the uncertainty in the direction of motion is included), giving a product $\Delta x \, \Delta p_x$ which is directly related to Planck's constant $\mathbf{h}$.

3.8. Particle in a Container

We now revert to our original problem and assume that the particle is contained in a cube of volume $V = a^3$. The wave function ψ is determined by Equation (3.22) with $\phi = 0$ together with the boundary conditions (3.22a). The reader can verify that this equation is satisfied by a wave amplitude which is a product of three amplitudes of the form indicated in Equation (3.23c). We therefore postulate that

$$\psi(x, y, z) = \psi_1(x)\psi_2(y)\psi_3(z),$$

substitute this expression into the original differential equation, and deduce that

$$\psi = \left(\frac{8}{V}\right)^{1/2} \sin\frac{s_x \pi x}{V^{1/3}} \sin\frac{s_y \pi y}{V^{1/3}} \sin\frac{s_z \pi z}{V^{1/3}} \qquad (s_x, s_y, s_z = 1, 2, \ldots), \tag{3.27}$$

where $V^{1/3}$ has been substitued for a.† The corresponding eigenvalue of energy is given by

$$\varepsilon(s_x, s_y, s_z) = \varepsilon_1 s^2 \tag{3.27a}$$

with

$$\varepsilon_1 = \frac{\mathbf{h}^2}{8mV^{2/3}} \qquad \text{and} \quad s^2 = s_x^{\ 2} + s_y^{\ 2} + s_z^{\ 2}. \tag{3.27b, c}$$

Furthermore,

$$p_x = \pm \frac{\mathbf{h}}{2V^{1/3}} s_x, \tag{3.27d}$$

with analogous equations for p_y, p_z.

The state of motion, as expressed by the wave amplitude (3.27) is now described by *three* quantum numbers s_x, s_y, s_z, since a change in the value of one changes the wave function. In the one-dimensional case considered in the preceding section as well as in the present case, a given state (described by one quantum number, s, or a set of quantum numbers, s_x, s_y, s_z) is characterized by a definite energy level. However, in contrast with the preceding case, now a prescribed energy level ε does not necessarily correspond to a *single* quantum state. In order to see this clearly, we present Table 3.1‡ which lists systematically the energy levels reached by varying the quantum numbers. The number of *quantum states* which corresponds to a given *energy level* is called the *degeneracy* g of the specified energy level. By continuing the entries in Table 3.1 the reader may satisfy himself that very high energy levels are highly degenerate, meaning that g is large for large values of s_x, s_y, s_z.

It is instructive to draw a representation of the μ-space of our "particle in a box." Since the accessible base along the axes x, y, z is the same and equal to a, it is sufficient to sketch the momentum space p_x, p_y, p_z, that is, a

† In a more detailed derivation we would either assume a rectangular box of sides a_x, a_y, a_z or even a container of arbitrary shape. (Compare, for example, R. C. Tolman, *The Principles of Statistical Mechanics*, p. 287, Oxford Univ. Press, London and New York.) This would result in a more complex analysis, but the physical conclusions would remain unaltered.

‡ See J. D. Fast, *Entropy*, p. 183, McGraw-Hill, New York, 1962.

TABLE 3.1

Degeneracy of Energy Levels for a Particle in a Cubic Container

Quantum numbers s_x	s_y	s_z	Energy level in units of $\varepsilon_1 = \mathbf{h}^2/8mV^{2/3}$	Degeneracy g
1	1	1	3	1
2	1	1	6	
1	2	1	6	3
1	1	2	6	
2	2	1	9	
2	1	2	9	3
1	2	2	9	
3	1	1	11	
1	3	1	11	3
1	1	3	11	
2	2	2	12	1

projection of the six-dimensional μ-space into the three-dimensional momentum space. This procedure (though impossible to visualize) is analogous to projecting a three-dimensional object (space of three dimensions) on a plane (space of two dimensions). The momentum space for the particle is shown sketched in Figure 3.6, in which it has been assumed that the momenta are measured in units of $\mathbf{h}/2V^{1/3}$; as seen from Equations (3.27d), the momenta are proportional to $\pm s_x$, $\pm s_y$, and $\pm s_z$. A quantum state is represented by *integral* quantum numbers, and for this reason the set of quantum states is discrete in the momentum space, each state corresponding to a nodal point in a three-dimensional grid drawn through the integral values of s_x, s_y, s_z, as suggested by the sketch. Each quantum state arises from a triplet of integers s_x, s_y, s_z, but to each integer there corresponds a positive and a negative value of momentum. Hence to each nodal point in the first octant we must add seven "satellite" nodal points located in the seven remaining octants of momentum space obtained by mirror reflection with respect to the three coordinate planes. These "satellites" have not been shown in the diagram. Thus a single quantum state is represented by a total of eight points in the momentum space.

With every quantum state we can associate a cubic cell; for example, state (1, 1, 1) can be associated with the cell shown shaded in the diagram with vertices at (0, 0, 0), (1, 0, 0), (0, 1, 0), (0, 0, 1), etc. together with its seven "satellites" (not shown). In this manner, a given state is associated with an element in the momentum space whose volume is

$$\Delta p_x \, \Delta p_y \, \Delta p_z = 8(\mathbf{h}^3/8a^3) = \mathbf{h}^3/V,$$

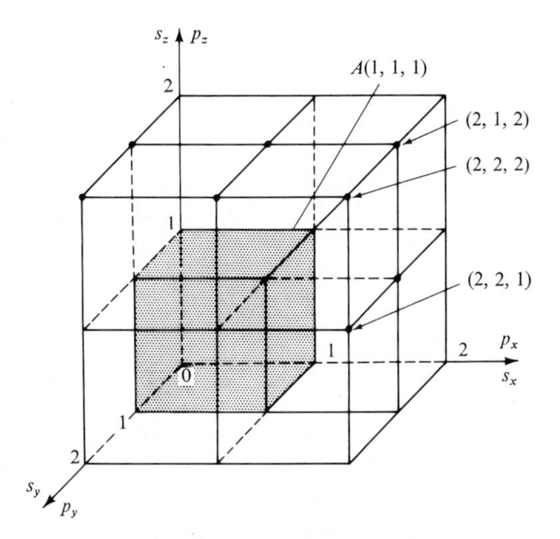

FIGURE 3.6. Quantum states in first octant of momentum space as nodal points in a grid.

The values at $s_x = 0$, $s_y = 0$, and $s_z = 0$ do not correspond to quantum states. The quantum state $A(1,1,1)$ is associated with the shaded cubic cell.

To each quantum state there correspond seven "satellite" grid points which are obtained by considering all possible combinations of $\pm p_x$, $\pm p_y$, $\pm p_z$ for the same quantum numbers s_x, s_y, s_z.

because the volume of a cell in the first octant is but one-eighth of the volume of the whole cell. This is analogous to the two strips per cell in Figure 3.5. The corresponding volume in the μ-space is obtained by multiplication with $\Delta x \, \Delta y \, \Delta z = a^3 = V$. Hence the volume of a cell in phase space is

$$\Delta \mathscr{V} = \mathbf{h}^3 \qquad (= \mathbf{h}^f; \ f = 3). \tag{3.28}$$

This, again, is a particular instance of the operation of the uncertainty principle.

An energy level $\varepsilon_s = \varepsilon(s_x, s_y, s_z)$ is represented by the surface of a sphere; the radius of this sphere is determined by Equation (3.27a). It is now clear that the same energy level ε_s can be obtained with different choices of the quantum numbers and that not every spherical surface represents an admissible energy level. In order to reach an admissible surface, we must choose three steps p_x, p_y, p_z each of which is an integral multiple of $\mathbf{h}/2V^{1/3}$. Spherical surfaces which cannot be reached in this way are not admissible according to quantum mechanics.

Readers familiar with problems of vibrations may have noticed that the preceding argument does not carry over to the case of a rectangular box of sides a_x, a_y, a_z whose lengths are not commensurable. This detail is of no great consequence to us, except to suggest that energy levels will cease to be degenerate. The main purpose of the illustration provided above is to familiarize the student with the *concept* of degeneracy because it plays an important part in statistical thermodynamics.

Reverting to the example of the helium molecule from page 79, we would now find that its average energy at $T = 300$ K is $\varepsilon_s = \frac{3}{2}\mathbf{k}T = 6.21 \times 10^{-14}$ gr cm²/sec² owing to the existence of three degrees of freedom. The value of ε_1 remains unaltered, and $s \approx 2.73 \times 10^8$. This shows that at least one of the three quantum numbers must be very large. Equally, the number of triplets of values of s_x, s_y, s_z whose sum of squares is equal to s^2 becomes enormous.

The degeneracy of an energy level is difficult to calculate if an exact result is wanted, since its change from one energy surface to the next is not smooth, as can be seen from Table 3.1, and complications arise in cases when shapes other than cubes are studied. However, a precise result is not required, and a modified measure of degeneracy will be discussed in Section 3.14. The estimate will lead to the conclusion that a very large number of the available quantum states, or quantum cells in μ-space, of the model of a macroscopic system must necessarily remain unoccupied at every instant of time.

3.9. Two Identical Particles in a Container

When two identical, independent, unlocalized, and indistinguishable particles occupy the same container, the Hamiltonian for the system is the sum of the Hamiltonians of the individual particles. Thus the time-independent Schrödinger equation for the system is

$$\frac{\partial^2\psi}{\partial x_1{}^2} + \frac{\partial^2\psi}{\partial y_1{}^2} + \frac{\partial^2\psi}{\partial z_1{}^2} + \frac{\partial^2\psi}{\partial x_2{}^2} + \frac{\partial^2\psi}{\partial y_2{}^2} + \frac{\partial^2\psi}{\partial z_2{}^2} + \frac{2m}{\hbar^2}\,\varepsilon\psi = 0. \tag{3.29}$$

The solution to this equation can be reduced to the solution for one particle, on condition that the product

$$\psi(x_1, y_1, z_1, x_2, y_2, z_2) = \psi_1(x_1, y_1, z_1)\psi_2(x_2, y_2, z_2) \tag{3.29a}$$

is taken. The validity of this supposition can be tested by substitution, when we obtain

$$\psi_2\left[\nabla_1{}^2\psi_1 + \frac{2m}{\hbar^2}\,\varepsilon_1\psi_1\right] + \psi_1\left[\nabla_2{}^2\psi_2 + \frac{2m}{\hbar^2}\,\varepsilon_2\psi_2\right] = 0.$$

Here the operator $\nabla_1{}^2$ refers to the variables x_1, y_1, z_1, and $\nabla_2{}^2$ refers to x_2, y_2, z_2. By solving the equations

$$\nabla_1{}^2\psi_1 + \frac{2m}{\hbar^2}\,\varepsilon_1\psi_1 = 0 \qquad \text{and} \qquad \nabla_2{}^2\psi_2 + \frac{2m}{\hbar^2}\,\varepsilon_2\psi_2 = 0 \tag{3.30}$$

separately, we shall obtain two solutions, each with a distinct eigenvalue, the two satisfying the condition that

$$\varepsilon_1 + \varepsilon_2 = \varepsilon, \tag{3.31}$$

where ε represents the total energy of the system of two particles. More precisely, ε_1 is associated with three quantum numbers s_x', s_y', s_z', whereas ε_2 is associated with the, generally different, quantum numbers s_x'', s_y'', s_z''.

If we imagine that particle 1 has been interchanged with particle 2, we would find that

$$\psi(x_1, y_1, z_1, x_2, y_2, z_2) = \psi_1(x_2, y_2, z_2)\psi_2(x_1, y_1, z_1), \tag{3.29b}$$

that is, a different wave function. According to the remarks made in Section 2.3.5 concerning the precise meaning of the statement that the particles are indistinguishable, the preceding interchange should *not* lead to a different wave amplitude, that is, it should *not* lead to a different quantum state. It follows that neither of the two wave functions, (3.29a) or (3.29b), represents a physically complete solution to our problem. The solutions must be modified in such a way as to remain invariant when the triplets of variables (x_1, y_1, z_1) and (x_2, y_2, z_2) are interchanged.

In order to achieve our aim we notice that Equation (3.29) is linear and homogeneous. Consequently, if Equations (3.29a) and (3.29b) represent *mathematically* correct solutions, any linear combination of them will also represent a mathematically correct solution. Our task is to find all possible solutions which are also *physically* correct. Accordingly, we consider the two combinations

$$\psi_a(x_1, \ldots, x_2, \ldots) = \frac{1}{\sqrt{2}}[\psi_1(x_1, \ldots)\psi_2(x_2, \ldots) - \psi_1(x_2, \ldots)\psi_2(x_1, \ldots)] \tag{3.32a}$$

and

$$\psi_s(x_1, \ldots, x_2, \ldots) = \frac{1}{\sqrt{2}}[\psi_1(x_1, \ldots)\psi_2(x_2, \ldots) + \psi_1(x_2, \ldots)\psi_2(x_1, \ldots)]; \tag{3.32b}$$

the factor $2^{-1/2}$ arises from the normalization condition for ψ. The wave amplitude ψ_a has the property that an interchange of particles results in a change in sign but not in value, so that

$$\psi_a(x_1, \ldots, x_2, \ldots) = -\psi_a(x_2, \ldots, x_1, \ldots), \tag{3.32c}$$

whereas the wave amplitude ψ_s has the property that

$$\psi_s(x_1, \ldots, x_2, \ldots) = \psi_s(x_2, \ldots, x_1, \ldots). \tag{3.32d}$$

For this reason, the wave function ψ_a is called *antisymmetric* whereas the wave function ψ_s is called *symmetric*. In either case an interchange of particles leaves the probability densities ψ_a^2 and ψ_s^2 unchanged, and both linear combinations now possess the property that an interchange of particles leaves the quantum state unaltered. In the very special case when $s_x' = s_x''$, $s_y' = s_y''$, and $s_z' = s_z''$, we obtain

$$\psi_a \equiv 0 \qquad (s_i' = s_i''). \tag{3.33}$$

This means that in the case of an antisymmetric wave function the two particles *must* be in different quantum states (though we cannot specify which particle is in what state). We express this colloquially by saying that the two particles *cannot occupy the same quantum state*; this statement is known as *Pauli's exclusion principle* (in particular, when applied to electrons).

There exist no general principles which determine the conditions when the wave function for a single particle must be *symmetrized* or *antisymmetrized* to obtain a physically meaningful solution for two particles. This matter is settled by having recourse to observation. In this manner the following lists can be established:

Antisymmetric wave functions—the fundamental constituents of matter: electrons, protons, neutrons. Such particles are known collectively as *fermions.*

Symmetric wave functions—the fundamental particles which do not occur as constituents of matter: photons, mesons. These particles are known as *bosons.*

As far as atoms or molecules as a whole are concerned, it is necessary to count the total number of protons, neutrons, and electrons in their structure; when this number is even, the wave function is symmetric, and when this number is odd, the function is antisymmetric. This rule was first established by P. Ehrenfest and J. R. Oppenheimer. Thus H_2, D_2, N_2, O_2, and He^4 have *symmetric*, but HD, NH_3, NO, and He^3 have *antisymmetric* wave functions.

Analogous considerations apply to systems of many particles, and though many more symmetries become possible, comparison with experiment shows that only wave functions which are either symmetric or antisymmetric with respect to an exchange of *two* particles need to be considered. The latter functions imply the operation of the exclusion principle.

The consequences of the indistinguishability of two particles can be illustrated in phase space with the aid of the sketch in Figure 3.7. In order to simplify the relations, we consider two molecules confined to a "linear box" of "volume" a so that the position of each of them is described by a single space coordinate x. The diagram represents a projection of phase space of four coordinates x_1, x_2, p_{x_1}, p_{x_2} on the plane x_1, x_2. In the diagram, point A cor-

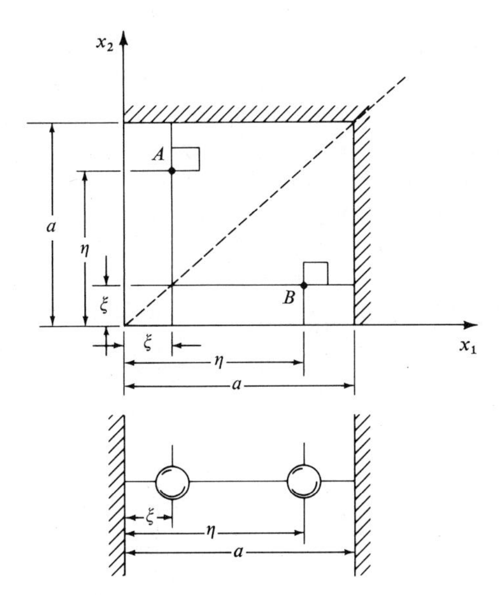

FIGURE 3.7. Indistinguishability of molecules in phase space.

responds to the case when $x_1 = \xi$ and $x_2 = \eta$, whereas at point B the positions have been reversed to $x_1 = \eta$ and $x_2 = \xi$. If the molecules are distinguishable, the two states are considered distinct and different. With indistinguishable particles the two points must be treated as belonging to the same state. This can be achieved by eliminating a portion of the phase space, or by treating the corresponding quantum cells as one.

3.10. Quantization of Rotation

The quantum-mechanical description of translatory motion contained in the preceding sections has served to explain to the reader the essential, qualitative differences between the motion of small particles and that of large bodies. The quantum-mechanical description of the remaining two modes of motion of interest to us—rotation and vibration—differs from translation merely in the greater mathematical complexities involved in the writing and in the solution of the appropriate Schrödinger equations. In our largely heuristic exposition of the subject, it is impossible to enter into a full discussion of the mathematical considerations required for the solution of such problems without exceeding the scope of this course. Consequently, we shall be satisfied with a statement and interpretation of the respective results.

In accordance with the details of our model of a perfect gas, we must now consider the rotational motion of a molecule about its center of gravity realizing that its stationary state corresponds to a motion with no external forces or *moments* acting on its center of mass. Even the classical description of such a motion is complex, notwithstanding the fact that the molecule may be treated as a rigid body owing to the assumption that the interaction between rotation and vibration may be disregarded. It is recalled from mechanics that a general rotation of a rigid body with respect to its center of mass, M, can be studied in relation to its *ellipsoid of inertia* drawn with respect to the center of mass. In general the semiaxes of the ellipsoid differ from each other, and the length of each one of them is inversely proportional to the square root of one of the three *principal moments of inertia*, I_1, I_2, I_3, of the rigid body. Such an ellipsoid of inertia is seen sketched in Figure 3.8. Since no external torques act

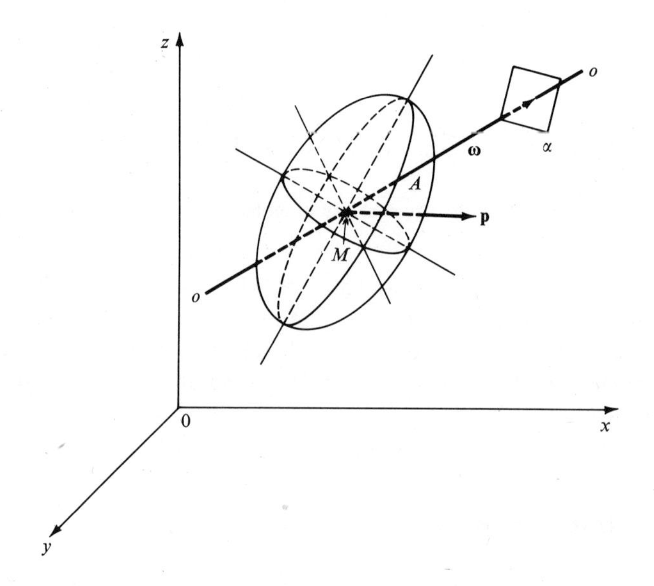

FIGURE 3.8. General rotation.

on the body, the motion must be such as to preserve the magnitude as well as the direction of the vector **p** of angular momentum. The most general, instantaneous rotation can be reduced to a single rotation with an angular velocity **ω** about an axis *o–o* with respect to which the moment of inertia of the body is I_ω. Hence, the kinetic energy of the body is

$$\mathscr{E}_r = \tfrac{1}{2} I_\omega \omega^2 , \tag{3.34}$$

and must be conserved during the motion. Whereas the direction and magnitude of the vector **p** remain fixed, the axis of rotation *o–o*, and hence the moment of inertia I_ω as well as the vector $\boldsymbol{\omega}$ change in a continuous way,† but subject to energy conservation. The vectors $\mathbf{p}$ and $\boldsymbol{\omega}$ coincide only if the direction of $\mathbf{p}$ is that of one of the principal axes of inertia.

The quantum-mechanical counterpart of this general problem has not yet been solved explicitly. However, a solution can be found when two of the principal moments of inertia, I_1, I_2, I_3, are equal (say $I_2 = I_3 = I$). The ellipsoid of inertia is then rotationally symmetrical, and there exists an infinity of axes about which the moment of inertia is equal to I. A general solution of this type will not be needed in our further development because it will turn out that for most polyatomic gases the rotational energy need not be quantized.

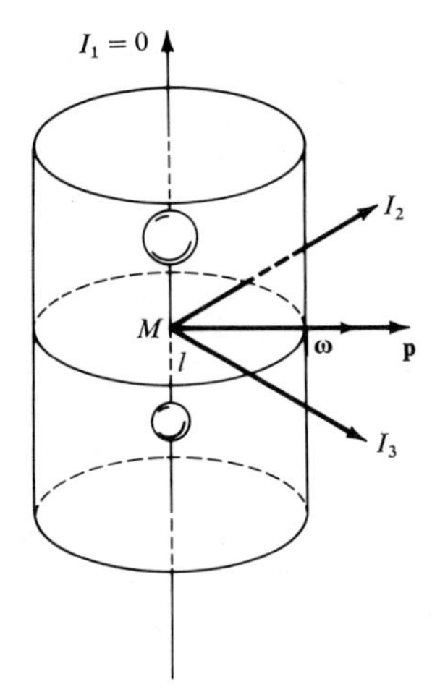

FIGURE 3.9. Linear molecule and its ellipsoid of inertia. (The sketch suggests a heteronuclear, diatomic molecule.)

The case of a linear molecule is even simpler because then $I_2 = I_3$ and $I_1 = 0$ which causes the ellipsoid of inertia to degenerate into a circular cylinder. The relation of this cylinder to a heteronuclear, diatomic molecule is suggested by Figure 3.9. If we imagine that the vector of angular velocity $\boldsymbol{\omega}$ has been expressed by three components $\omega_1, \omega_2, \omega_3$, each along one of the axes I_1, I_2, I_3, we can appreciate that the component ω_1 may be assumed to vanish.

† The reader may recall that the vector of angular momentum $\mathbf{p}$ is normal to a plane α which is, in turn, parallel to the plane tangent to the ellipsoid of inertia at point A. This is the point at which the vector $\boldsymbol{\omega}$ emerges from the ellipsoid of inertia (theorem due to Poinsot). See, for example, S. Timoshenko and D. H. Young, *Advanced Dynamics*, p. 326, McGraw-Hill, New York, 1948.

The equation of motion with respect to axis I_1 is

$$I_1 \frac{d\omega_1}{dt} = 0,$$

since $I_2 = I_3$. Hence

$$I_1\omega_1 = \text{const},$$

and for $I_1 \to 0$, we may assume $\omega_1 \to 0$ if it is not to become infinite. Thus the vectors $\boldsymbol{\omega}$ and $\mathbf{p}$ must lie in the plane I_2, I_3, and both must be collinear and constant as shown in Figure 3.9. This can also be appreciated physically if it is imagined that $\boldsymbol{\omega}$ is not normal to the link l of the molecule, Figure 3.10. Under

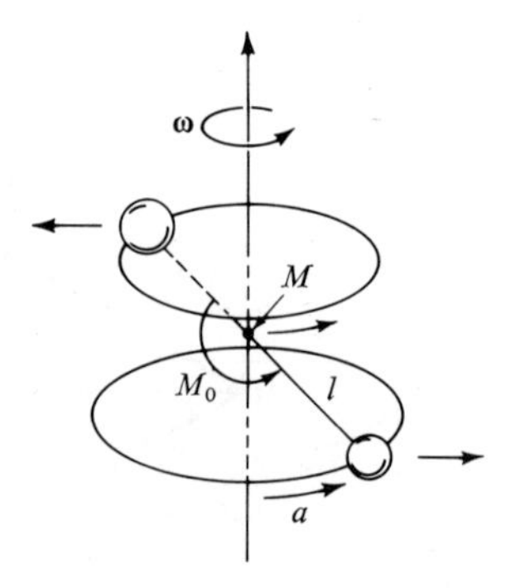

FIGURE 3.10. Moment about the axis of a linear molecule.

such conditions, the link l would be subjected to a torque M_0 which could vanish only in the preceding configuration. Hence the energy becomes

$$\varepsilon = \frac{\mathbf{p}^2}{2I}. \tag{3.35}$$

The Schrödinger equation for this simplified problem has the form

$$\frac{\partial^2\psi}{\partial\theta^2} + \frac{1}{\sin^2\theta}\frac{\partial^2\psi}{\partial\phi^2} + \frac{2I}{\hbar^2}\varepsilon\psi = 0, \tag{3.36}$$

when written in a polar system of coordinates with $r = \text{const}$, since we are discussing rotation with respect to a fixed center of mass. The system of coordinates is seen sketched in Figure 3.11. The solution of this equation subject to the appropriate boundary conditions consists of a set of eigenfunctions $\psi_{jm}(\theta, \phi)$† which involve two integers—the quantum numbers j and m. In the

† $\psi_{jm}(\theta, \phi) = [P_j^{|m|}(x)]e^{im\phi}$, where $P_j^{|m|}(x)$ denotes the so-called associated Legendre function of the argument $x = \cos\theta$. See, for example, E. Jahnke, F. Emde, F. Lösch, *Tables of Higher Functions*, 6th ed., p. 114, McGraw-Hill, New York, 1960.

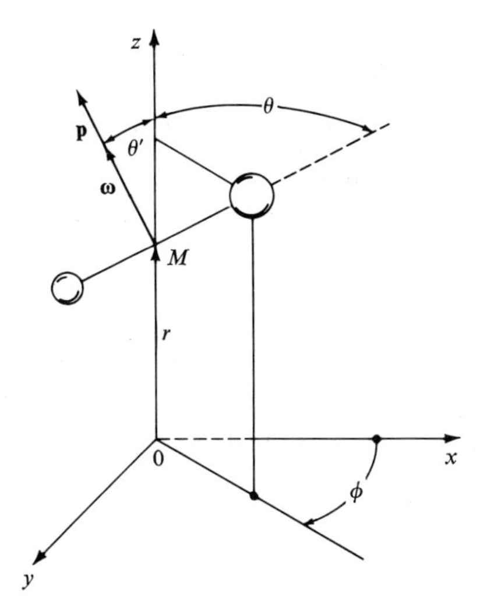

FIGURE 3.11. System of coordinates.

course of finding the solution it is shown that

$$j = 0, 1, 2, \ldots \tag{3.36a}$$

and that for every value of j, the second quantum number

$$m = 0, \pm 1, \pm 2, \ldots, \tag{3.36b}$$

subject to the condition that $|m| \leq j$. In terms of these quantum numbers, the energy levels are

$$\varepsilon_j = \frac{j(j+1)}{2I} \hbar^2 ; \tag{3.37}$$

and by Equation (3.35),

$$\mathbf{p}_j^2 = j(j+1)\hbar^2 . \tag{3.38}$$

The rigorous derivation shows further that

$$\mathbf{p}_z = m\hbar . \tag{3.38a}$$

The preceding results demonstrate that every energy level ε_j can be achieved by as many quantum states as there are allowable values of m. This number, the degeneracy g, is easily counted with the aid of Equation (3.36b); it is

$$g = 2j + 1 . \tag{3.39}$$

For large values of j, Equation (3.38), the total angular momentum $|\mathbf{p}_j| \sim j\hbar$, which shows that $\hbar$ is the "natural" unit for it. The component $p_z = m\hbar$ is exactly quantized in terms of this unit. The degeneracy (3.39) is usually represented pictorially in the suggestive diagram of Figure 3.12 which has been drawn for $j = 4$. The diagram shows the vector of total momentum turning with different angles θ; those angles θ are allowed that cause the projection of the vector on the z axis to assume the quantized values (3.38a).

When the rotation of a large (Newtonian) system is described with the aid of classical mechanics, it is found that the angle θ' between the z axis and the vector $\mathbf{p}$ of angular momentum in Figure 3.11 can change continuously. In quantum mechanics only $g = 2j + 1$ discrete values are admissible. Furthermore, in classical mechanics the energy of rotation $\mathbf{p}^2/2I$ may vary continuously. By contrast, in quantum mechanics only discrete values ε_j may occur.

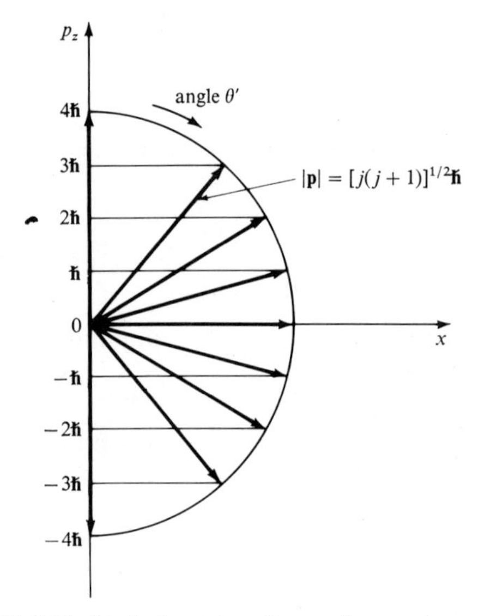

FIGURE 3.12. Symbolic vector diagram for angular momentum.

Axis y is assumed normal to the plane of the drawing. The diagram has been drawn for $j = 4$ and shows $g = 2j + 1 = 9$ allowable quantum states which differ from each other by the angle θ. Here $\mathbf{p} = 20^{1/2}\hbar \approx 4.47\,\hbar$.

Considerations of indistinguishability, which were discussed in Section 3.9 in relation to translation, apply to rotation too. If a diatomic molecule is homonuclear, the wave function must be symmetrized or antisymmetrized in order to erase the distinction between the two atoms. Thus the angle ϕ in Figure 3.11 may now vary only within the interval $0 \leq \phi \leq \pi$ instead of the interval $0 \leq \phi \leq 2\pi$ for a heteronuclear molecule. Whether the wave function should be symmetrized or antisymmetrized depends on a new property of the atoms—the intrinsic momentum of their nuclei called *spin*, which we shall discuss in Section 3.13. In one case only odd values of j are permitted, whereas in the other only even values of j may occur.

However, regardless of the precise details of the form of the wave function, the number of quantum states available to the system becomes affected. In the present case, the number of quantum states is reduced by a factor $\sigma = 2$ when the molecule is homonuclear in comparison with a heteronuclear molecule whose moment of inertia is the same. In the case of more complex, nonlinear molecules the *symmetry number*, σ, may assume different values depending on the details of the spatial distribution of the atoms in it.

The situation is analogous to that illustrated in Figure 3.7, and a similar illustration is given in Figure 3.13. Here, the angular momentum $\mathbf{p} = I\dot{\phi}$ and

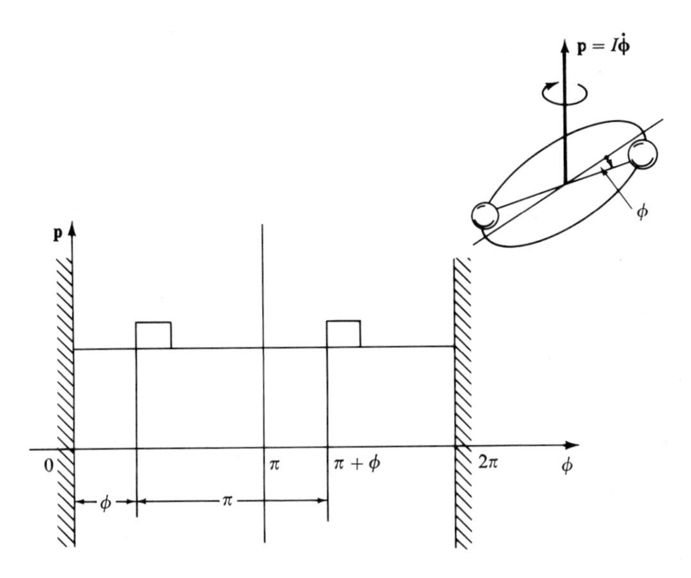

FIGURE 3.13. Indistinguishability of atoms of a homonuclear molecule.

the angle ϕ have been chosen as the coordinates. A change in angle from ϕ to $\pi + \phi$ restores the state, and the number of available states must be halved. Alternatively, the extent of a cell may be doubled.

The reader may wonder why the vector of angular momentum is exactly quantized with respect to what is, after all, an *arbitrarily* chosen axis, z. The complete theory of the phenomenon shows that a different choice of the axes of the coordinate system would lead to wave functions which are mere linear combinations of the $2j + 1$ wave functions obtained with any other system of coordinates. This is a general property of the Schrödinger equation and can be expressed by the statement that the wave functions of an energy level of degeneracy g can be formed arbitrarily by selecting g independent linear combinations of another set of g wave functions.

3.11. Quantization of Vibration

It was stated in Section 2.3.4 that the vibrational motion of a mechanical system can be represented as a sum of harmonic oscillations, each corresponding to one natural mode of vibration which proceeds with a characteristic natural frequency ν. This motion is mathematically identical with that of a mass m moving under the action of a force

$$F = -m(2\pi\nu)^2\xi, \tag{3.40}$$

the trajectory being described by the equation

$$\xi = X\cos 2\pi\nu t. \tag{3.41}$$

The discussion which was given in connection with Figure 2.5 convinces us that this is an acceptable approximation when the vibrational energy is not very large.

Recalling the expression for the Hamiltonian function of a single oscillator from Equation (2.13c), we can see that the Schrödinger equation for this system assumes the form

$$\frac{d^2\psi}{d\xi^2} + \frac{2m}{\hbar^2}\left[\varepsilon - \frac{1}{2}(2\pi\nu)^2 m\xi^2\right]\psi = 0. \tag{3.42}$$

The wave function which satisfies this equation is expressed in terms of so-called Hermite polynomials† and is

$$\psi_v = A_v e^{-x^2/2} H_v(x), \tag{3.43}$$

with

$$x = \left(\frac{2\pi m\nu}{\hbar}\right)^{1/2}\xi; \qquad A_v = (2^v v!\pi^{1/2})^{-1/2}. \tag{3.43a}$$

The corresponding energy levels are

$$\varepsilon_v = (\tfrac{1}{2} + v)\mathbf{h}\nu = (\tfrac{1}{2} + v)\hbar\omega \quad (\omega = 2\pi\nu) \quad (v = 0, 1, 2 \ldots). \tag{3.43b}$$

† Hermite polynomials of orders 0, 1, and 2 are defined as

$$H_0(x) = 1, \qquad H_1(x) = 2x, \qquad H_2(x) = 4x^2 - 2;$$

and higher-order polynomials are formed with the aid of the recurrence formula $H_{n+1} = 2xH_n - H_n'$. See also, E. Jahnke, F. Emde, F. Lösch, *Tables of Higher Functions*, 6th ed., p. 101, McGraw-Hill, New York, 1960.

The quantum state of a single, linear, harmonic oscillator is described by a *single* quantum number, v, and the energy levels are *nondegenerate*.

The wave functions ψ_0, ψ_1, ψ_2, ψ_3, and the corresponding probability densities have been shown sketched in Figure 3.14. Each wave function has v zeros, and there exists a finite, though very small, probability that the particle may be found outside the classical amplitude $\pm X$ (the classical turning points). As v increases, the probability amplitude ψ_v^2 approaches more and

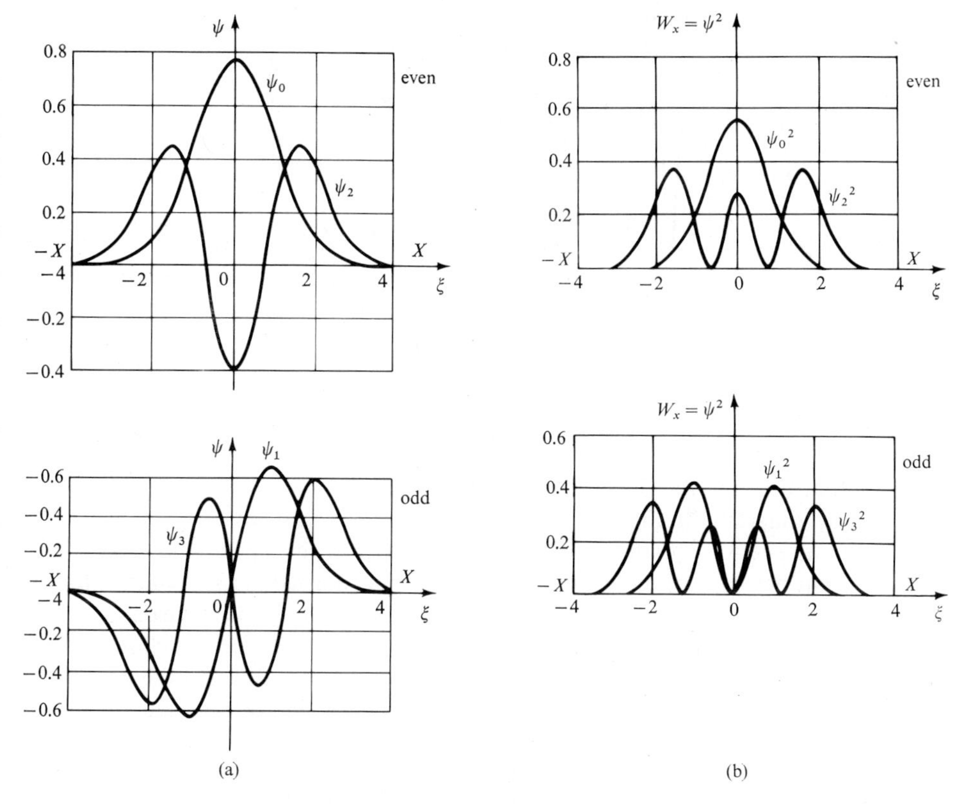

FIGURE 3.14. Wave functions and probability densities for linear harmonic oscillator. (See J. B. Russell, *J. Math. Phys.* **12** (1933), 291.) a. Wave functions, b. Probability densities.

more closely the broken, parabola-like curve shown in all diagrams; the way in which this happens is already very noticeable in Figure 3.15 which represents the probability amplitude for $v = 11$. The parabola-like curve represents the fraction dt/T ($T = 2\pi/\omega$) of total time which a particle spends in an interval $d\xi$ when its motion is governed by the classical equation (3.41). This quantity represents the probability that a random observation will discover the particle in a predetermined position ξ, and the diagram illustrates how the correspondence principle operates in this case.

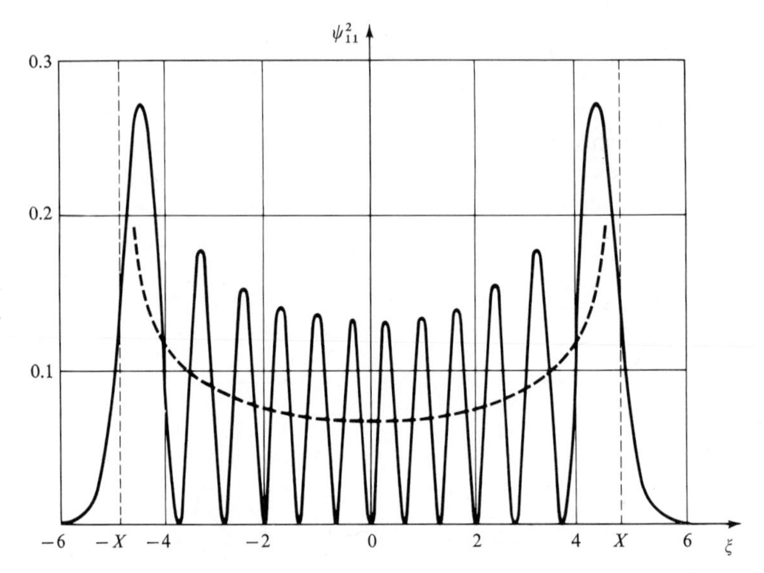

FIGURE 3.15. Probability density of linear oscillator for $v = 11$.

It is noted that for $v = 0$ the particle possesses the ground energy

$$\varepsilon_0 = \tfrac{1}{2}\mathbf{h}\nu = \tfrac{1}{2}\hbar\omega ,$$

and that it can never be completely at rest.

The diagram in Figure 3.16 reproduces the approximate parabola of potential energy $\phi(\xi)$ from Figure 2.5 together with the equally spaced, quantum-mechanical energy levels, ε_v, for a single normal mode. In fact, if the proper form of the potential-energy diagram $\phi(\xi)$ were taken into account (broken line), it would be found that the energy levels were placed closer and closer to each other as the quantum number v increased, and that the oscillation would become anharmonic. The energy levels are represented by the broken lines, the difference between them and the approximate ones being negligible for the lowest quantum numbers. The energy ε_d required to increase the separation between the atoms in a molecule to infinity corresponds to the dissociation energy; this energy is measured with respect to the ground state for which $v = 0$.

The vibrational energy of a polyatomic molecule which is characterized by f_v normal modes is described by f_v quantum numbers v_{1k}, v_{2k}, ... As long as the small-amplitude approximation is acceptable, the total energy is, evidently,

$$\varepsilon = \sum_{k=1}^{f_v} (\tfrac{1}{2} + v_{ik})\mathbf{h}\nu_k \qquad (v_{ik} = 0, 1, \ldots). \tag{3.43c}$$

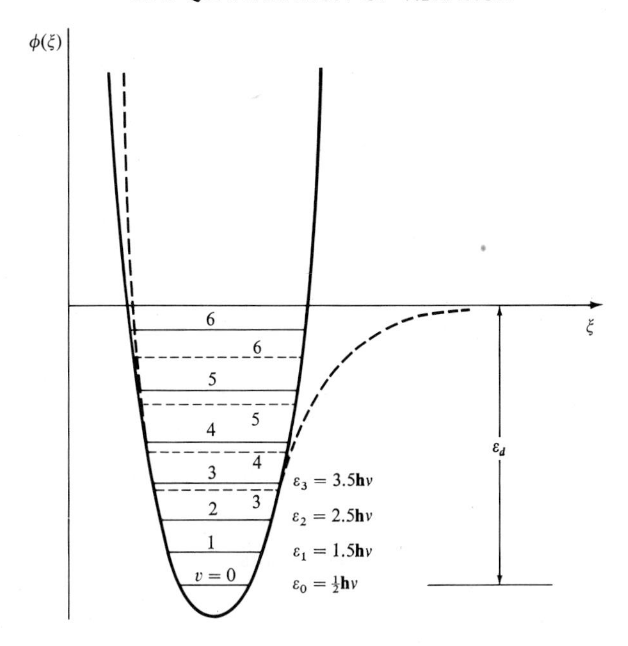

FIGURE 3.16. Quantum-mechanical energy levels of a linear oscillator in relation to the curve of potential energy $\phi(\xi)$. *Note*: The lines $\varepsilon_v =$ const represent total energy, whereas the parabola represents potential energy.

The phase space of a linear oscillator consists of a single displacement coordinate ξ and a single momentum coordinate p. The expression in Equation (2.13c) demonstrates that the surface of constant energy—a line for the linear oscillator—is represented by an ellipse with the semiaxes equal to

$$p_v = (2m\varepsilon_v)^{1/2} \qquad \text{and} \qquad X_v = [2\varepsilon_v/(2\pi\nu)^2 m]^{1/2}, \qquad (3.44\text{a, b})$$

where $\varepsilon_v =$ const, since the sum of the potential and kinetic energy terms remains constant in a stationary state. Such a phase space is shown sketched in Figure 3.17. The consecutive differences between adjoining quantum levels (quantum jumps) are

$$\varepsilon_{v+1} - \varepsilon_v = \mathbf{h}\nu = \hbar\omega .$$

Making use of Equations (3.44a, b), it is easy to calculate the area encompassed by two consecutive midpoint ellipses (each corresponding to $v = 1$, $2, \ldots$), that is, the "volume" of a cell in phase space. Thus we find that

$$\Delta\mathscr{V} = \mathbf{h}, \qquad (3.45)$$

in conformity with Heisenberg's uncertainty principle. With f_v vibrational normal modes, we would have

$$\Delta\mathscr{V} = \mathbf{h}^{f_v}. \qquad (3.45\text{a})$$

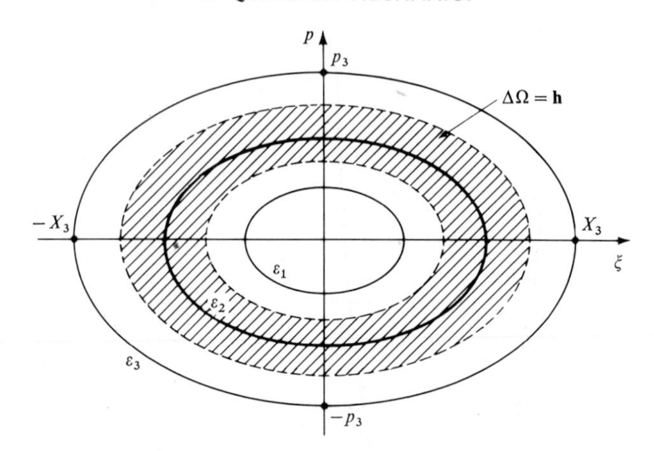

FIGURE 3.17. Phase space (μ-space) for linear, harmonic oscillator.

3.12. Collection of Independent Particles

When the model of the system consists of N independent, that is noninteracting, particles, the Hamiltonian function $\mathcal{H}$ for the whole system consists of N identical terms $\mathcal{H}_i$, each of the latter depending on the coordinates and momenta of one of the N particles,

$$\mathcal{H} = \sum_{i=1}^{N} \mathcal{H}_i . \tag{3.46}$$

Furthermore, the energy of a molecule of a gas may be taken to be equal to the sum of its translational, rotational, and vibrational energies. The energy of a single particle, particle i, is thus expressed in terms of a number of coordinates q_{ki} and their conjugate momenta p_{ki}. The number of such pairs is evidently equal to the number of degrees of freedom $f = 3n$ of a molecule, as discussed in Section 2.3.4, where n denoted the number of atoms in the molecule. Consequently, the Hamiltonian function $\mathcal{H}_i$ of a molecule consists of the following terms:

translational energy; 3 terms: $p_k^2/2m, \quad p_k = p_x, p_y, p_z$,
rotational energy; 2 or 3 terms: $p_k^2/2I, \quad p_k = p_1, p_2, p_3$,
vibrational energy; $3n - 5$ or $3(n - 2)$ terms: $p_k^2/2\mu + \frac{1}{2}(2\pi\nu_k)^2\mu\xi^2$,

there being one term each per normal vibrational mode. Thus the Hamiltonian function for a molecule, in turn, becomes equal to a sum of terms

$$\mathcal{H}_i = \sum_1^3 \mathcal{H}_{\text{tr}} + \sum_1^{f_{\text{r}}} \mathcal{H}_{\text{r}} + \sum_1^{f_{\text{v}}} \mathcal{H}_{\text{v}} . \tag{3.47}$$

The function (3.46) contains $f = 3nN$ independent coordinates and $f = 3nN$ conjugate momenta, it being immaterial that the independent coordinates for translation and rotation do not appear explicitly. The essential characteristic of the Hamiltonian function $\mathscr{H}$ is that each additive term in it, as expressed by Equation (3.47), contains one and only one set of variables. Thus the translational coordinates and momenta do not appear in the expressions for rotational or vibrational energy, and conversely. It follows as a consequence of our assumption of independence that the variables have become *separated.* The same is true for our model of a crystalline solid, because its Hamiltonian function consists of as many terms $\mathscr{H}_v$ as there are normal modes in an assembly of N localized oscillators.

We shall record this fact symbolically by writing

$$\mathscr{H}(p, q) = \mathscr{H}(1) + \mathscr{H}(2) + \cdots + \mathscr{H}(f), \tag{3.48}$$

where p, q stands for all variables, and where the numerals $1, 2, \ldots, f(=3nN)$ serve to remind us that the respective energy function $\mathscr{H}$ depends only on the coordinates and momenta of one degree of freedom of one molecule as far as a gas is concerned or of one normal mode in a solid.

The present situation is analogous to that encountered earlier in Section 3.9 when two particles in a container were discussed. The additional, and inessential, complication consists in the fact that the Hamiltonian *operator* for the system now contains operators for rotation and vibration as well as for translation. The Schrödinger equation

$$\mathbf{H}\psi = \varepsilon\psi \tag{3.49}$$

is now written

$$[\mathbf{H}(1) + \mathbf{H}(2) + \cdots + \mathbf{H}(f)]\psi = \varepsilon\psi,$$

and the reader can easily verify by substitution, or by separation of variables in the same manner as in Section 3.9, that the wave function ψ is equal to the product

$$\psi = \psi(1)\psi(2)\cdots\psi(f), \tag{3.50}$$

where each term in the product satisfies its own Schrödinger equation

$$[\mathbf{H}(1)]\psi = \varepsilon(1)\psi, \quad \text{etc.} \qquad \text{for} \quad 2, 3, \ldots, f.$$

The total energy of the system is simply the sum

$$\varepsilon = \varepsilon(1) + \varepsilon(2) + \cdots + \varepsilon(f). \tag{3.51}$$

In other words, the quantum state of the whole system is described by $3N$ quantum numbers s_x, s_y, s_z for translation, by $2Nf_r$ quantum numbers j, m for rotation in the case of a linear molecule, and by Nf_v quantum numbers v for vibration.

None of the functions (3.50) describes an actual state owing to the general requirement of indistinguishability, and the situation is handled in the same way as in the case of translation. First, we notice that the number of functions ψ is far larger than f,† the actual number P being unimportant. Making use of all such solutions, we form the symmetric wave function

$$\psi_s = \sum_1^P \psi = \sum_1^P \psi(1)\psi(2)\cdots\psi(f),$$

and the antisymmetric function ψ_a. A general method for the construction of the antisymmetric function consists in forming a determinant with the aid of the P functions ψ. The details of this procedure are also unimportant to us, and the only fact that matters is that either the symmetric or the antisymmetric wave function, ψ_s or ψ_a, represents the quantum state of the whole system.

When the state of the system is represented by an antisymmetric wave function, no two particles in it can occupy the same quantum state, as explained in Section 3.9, because then the corresponding wave function is identically zero (exclusion principle). There is no corresponding limitation on symmetric wave functions.

The preceding general result makes it possible to abstain from discussing the wave functions of large collections of independent particles (more precisely, of systems with separable coordinates). Instead, we may confine our attention to the discussion of the quantum numbers which describe the modes of storing energy per degree of freedom and per single molecule. The total energy of the system is simply equal to the sum of all such partial energies, and overall energy levels possess a high degree of degeneracy g.

3.13. Spin

Theoretical considerations, fully supported by experimental evidence, led G. E. Uhlenbeck and S. Goudsmit to the discovery that the quantum state of an elementary particle as well as a molecule cannot be fully described by the quantum numbers s_x, s_y, s_z, j, m, and v considered so far. In addition, it is necessary to ascribe to them a property which has no classical counterpart and which is described as *spin*. Roughly speaking, the wave function for spin resembles one which might be thought of as describing an intrinsic rotation about an axis embedded in the particle. The spin parallel to any axis, somewhat like the component p_z in rotation, can assume only multiples of $\pm \hbar$

† This is due to the fact that a given energy ε of the system can be achieved when the quantum numbers of one particle are interchanged with that of any other thus leading to a large number of permutations.

corresponding to spin quantum numbers which are multiples of $s = \pm \frac{1}{2}$. The spin of particles plays an important part when the interaction of matter with magnetic fields is studied and affects the energy of such particles. Thus the energy levels are determined by the spin quantum number. In the theory of specific heats, spin quantum numbers play a part only in the case of hydrogen, H_2, heavy hydrogen, D_2, and hydrogen deuteride, HD. Here, the energy is virtually independent of the spin, which affects only the degeneracy of the system. We shall mention this effect in Section 6.9.2. The same idea will recur in Chapter 11 on magnetic systems.

There exist principles for the determination of the spin quantum numbers of composite systems when those of their elements are known. For example, the spin quantum number of a nucleus, s_n, can be an integer, or half an odd integer depending on the atomic weight. It is the former for nuclei of even atomic weight, and the latter if the atomic weight is odd. If the energy levels are independent of the spin, the only effect is that the degeneracy becomes multiplied by a factor $2s_n + 1$.

3.14. Density of Quantum Cells in Phase Space

It was stated in Section 3.8 that it would be useful to obtain a modified measure of degeneracy owing to the difficulties of calculating it directly, and owing to the fact that the degeneracy, g, is, strictly speaking, an irregular function of the energy, ε, of the class of quantum states under discussion. In finding such an alternative measure, we shall be guided by two considerations. First, we shall exploit the fact that the degeneracy is represented by a very large number in normal circumstances. Consequently, a relatively crude estimate will yield a sufficient number of decimal places. The large degeneracy is intimately connected with the closeness of the spacing between energy levels. This allows us to approximate the series of discrete numbers, g, which results when the degeneracy is computed in terms of the discrete values of energy, ε, by a continuous function $g(\varepsilon)$.†

The second set of considerations is of a physical nature and relates to the uncertainty principle. At the end of Section 3.6 it was stated that the energy E of a complex system can be known precisely only if the system is left unobserved for an infinite time. Thus the precise quantum energy levels, ε, of elementary systems discussed in the preceding sections must be regarded as constituting mathematical idealizations; in practice, the energy of a system can be determined only with a margin of uncertainty. The margin of uncertainty imposed by quantum mechanics is exceedingly small in relation to macroscopic systems, being of the order of a quantum jump. In practice, the

† See also Section 4.3.

uncertainty of an actual determination of energy is far greater even for the most precise measurements. For example, in the case of a single helium molecule it was found (p. 79), that the quantum jump constituted merely one ten-millionth of its average energy. Consequently, it is necessary to realize that whenever the energy E of a large system is specified, it is implied that its value lies between E and $E + \delta E$, where δE is very small on the one hand, and yet much larger than a quantum jump. Apart from uncertainties in measurement, the fact that the energy of a system can never be specified exactly is also traceable to the imperfection of its isolation from the surroundings. The inevitable remaining interactions, small as they are, cause the energy to fluctuate between the above limits in a haphazard way.

The preceding considerations convince us that a knowledge of the degeneracy of an energy level does not provide a realistic measure of the number of states available to the system. Instead, it is useful to replace it by a count of states—or cells in Γ-space–which are confined between the mathematically precise surfaces of constant energy chosen a distance δE apart.

We shall perform this calculation in some detail for the μ-space of a single molecule in translatory motion and assert, heuristically, that the qualitative conclusions are the same for the Γ-space of large systems. Accordingly, we imagine two energy levels, ε and $\varepsilon + \delta\varepsilon$, in the μ-space of six dimensions: x, y, z, p_x, p_y, p_z. The surface of constant energy is represented by the equation

$$p_x^2 + p_y^2 + p_z^2 = \text{const} \qquad (= p^2), \tag{3.52}$$

and the surface corresponding to energy level $\varepsilon + \delta\varepsilon$ results when the momenta are given the small increments δp_x, δp_y, δp_z. In the physical space, the region of μ-space confined between ε and $\varepsilon + \delta\varepsilon$ is circumscribed by the volume V accessible to the molecule. Equation (3.52) shows that the region under consideration projects itself into the momentum space in the form of a spherical shell confined between the spherical surface $p = \text{const}$ and $p + \delta p = \text{const}$. Thus the surface in μ-space has the attributes of a sphere and a cylinder.

In order to see this more clearly, we imagine that we have introduced spherical coordinates in the momentum space, and introduced the radius p defined in Equation (3.52). As far as the physical space is concerned, we project in the z direction retaining only the coordinates x and y. The projection of μ-space into the space p, x, y is shown sketched in Figure 3.18. Here, the spherical shell projects itself in the form of the two planes $p = \text{const}$ and $p + \delta p = \text{const}$. The portion of μ-space available to the system arises from the intersection of this " spherical shell " with the cylinder whose " base " in physical space is labeled V in the diagram. In fact, we only see the *projection* of this volume on the plane x, y.

The volume $\delta\mathscr{V}$ of the μ-space circumscribed by the two surfaces of constant energy and the cylinder of base V is obtained by integrating the element

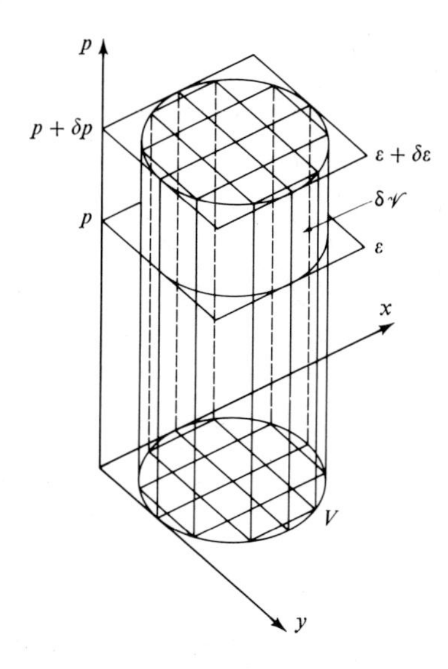

FIGURE 3.18. Portion of μ-space available to a molecule in a container; the portion of space is confined between the surfaces $\varepsilon =$ const and $\varepsilon + \delta\varepsilon =$ const; it has been divided into quantum cells.

$$\mathrm{d}\mathscr{V} = \mathrm{d}x\,\mathrm{d}y\,\mathrm{d}z\,\mathrm{d}p_x\,\mathrm{d}p_y\,\mathrm{d}p_z \tag{3.53}$$

subject to the boundaries discussed earlier. The integration with respect to $\mathrm{d}x\,\mathrm{d}y\,\mathrm{d}z$ yields a factor V in the same way as that which gave the factor a in Section 3.8 and Figure 3.5 because Equation (3.52) does not contain the coordinates x, y, z explicitly. To determine the limits of integration for $\mathrm{d}p_x\,\mathrm{d}p_y\,\mathrm{d}p_y$, we first notice that

$$\iiint \mathrm{d}p_x\,\mathrm{d}p_y\,\mathrm{d}p_z = 4\pi p^2\,\delta p, \tag{3.53a}$$

which corresponds to the spherical shell in momentum space, one octant of which is shown sketched in Figure 3.19. Hence

$$\delta\mathscr{V} = 4\pi p^2 V\delta p\,. \tag{3.53b}$$

The values of p and δp must now be related to ε and $\delta\varepsilon$. Since $\varepsilon = p^2/2m$, we have

$$p = \sqrt{2m\varepsilon} \qquad \text{and} \qquad \delta p \approx \frac{\mathrm{d}p}{\mathrm{d}\varepsilon}\,\delta\varepsilon = \frac{1}{2}\left(\frac{2m}{\varepsilon}\right)^{1/2} \delta\varepsilon\,,$$

and finally

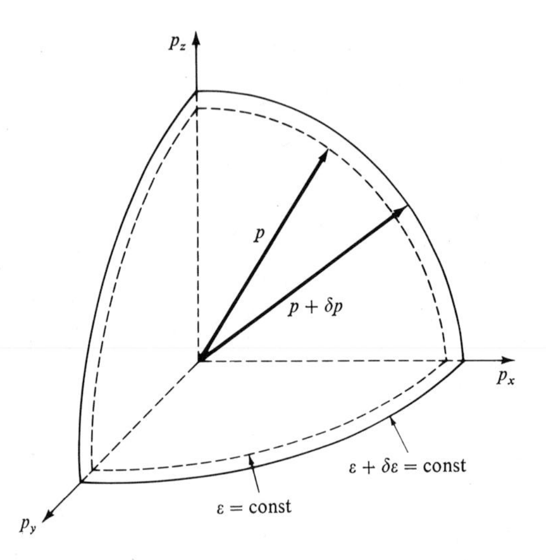

FIGURE 3.19. Spherical shell in momentum space for a single particle in a container.

$$\delta\mathscr{V} = 2\pi V(2m\varepsilon)^{3/2}\frac{\delta\varepsilon}{\varepsilon}. \tag{3.53c}$$

The number of quantum states, Ω, is equal to one-eighth of the number of unit cubes ($s_x = s_y = s_z = 1$) which can be accommodated in a spherical shell of radius $s = (\varepsilon/\varepsilon_1)^{1/2}$ and radius $s + \delta s$, where

$$\delta s \approx \frac{\mathrm{d}s}{\mathrm{d}\varepsilon}\,\delta\varepsilon = \frac{1}{2}\frac{\delta\varepsilon}{\sqrt{\varepsilon_1\varepsilon}}, \tag{3.54}$$

that is, to

$$\Omega \approx \tfrac{1}{8}(4\pi s^2)\delta s\,, \tag{3.54a}$$

as seen from Equation (3.27a). The factor 1/8 arises from the fact that only positive values of the quantum numbers s_x, s_y, s_z possess physical meaning, that is, from the existence of the "satellites" mentioned in Section 3.8. Hence

$$\Omega = \frac{\pi}{4}\left(\frac{\varepsilon}{\varepsilon_1}\right)^{3/2}\frac{\delta\varepsilon}{\varepsilon}. \tag{3.54b}$$

On substituting ε_1 from Equation (3.27b), we can verify that

$$\Omega = \frac{\delta\mathscr{V}}{\mathbf{h}^3}. \tag{3.55}$$

We inform the reader that in a μ-space or in a Γ-space of f dimensions, we would have

$$\Omega = \frac{\delta \mathscr{V}}{\mathbf{h}^f}. \tag{3.55a}$$

This result was already announced in the preliminary discussion of Section 2.2.

In Chapter 5 we shall base our computation of thermodynamic properties of isolated systems on an averaging procedure which assigns equal probability to any one of the Ω quantum states "available" to the system. Such an averaging procedure is thus seen to be equivalent to averaging over the volume $\mathscr{V}$ in Γ-space, provided that the factor $\mathbf{h}^f$ in the denominator is inserted separately. We shall also be informed that an initially uniform distribution of systems over cells in Γ-space or of independent molecules in μ-space must remain uniform in a system in equilibrium (Liouville's theorem in classical and quantum mechanics).

We revert to the example of a single molecule of helium in a container of volume $V = 1\ \text{cm}^3$ and assume that $\delta\varepsilon/\varepsilon = 10^{-6}$, that is that its average energy is known to one part in a million. Thus the number of available quantum states in μ-space is

$$\Omega_\mu = \frac{\pi}{4}\left[\frac{6.21 \times 10^{-14}\ \text{erg}}{0.83 \times 10^{-30}\ \text{erg}}\right]^{3/2} \times 10^{-6} \approx 1.6 \times 10^{19}.$$

At 300 K, 1 cm^3 of helium contains $N \approx 3 \times 10^{19}$ molecules. If these molecules were distributed uniformly over the available quantum states, one state would contain on the average about two molecules only. Since some states are likely to contain many more than two molecules, a large number of states would remain unoccupied. In an assembly of molecules of this kind, one molecule is as likely to be in one quantum state as in another. Now, the number of ways in which N distinguishable objects can be distributed among Ω compartments (Section 4.1.2) is Ω^N. Each such distribution constitutes a microstate, and the total number of microstates available to 1 cm^3 of helium reaches the staggering total of

$$[1.6 \times 10^{19}]^{3 \times 10^{19}} \approx 10^{10^{20}};$$

it is worth noting that this quantity by far exceeds the present estimate of the total number of particles in the universe, which is "only" 10^{80}.

The preceding considerations allow us to recognize that the number of quantum states, Ω, available to a macroscopic system is a function

$$\Omega = \omega(E, V, N)\,\delta E \tag{3.56}$$

of the energy level, E, the volume, V, and the number, N, of elementary particles—atoms or molecules—contained in it; this number is proportional to the uncertainty, δE, in our knowledge of the energy, E. The factor of proportionality, $\omega = \Omega/\delta E$, is known as the *number density* or *density of states* by analogy with the fact that the ratio $\rho = m/\delta V$ represents the density of a volume δV which contains the mass m.

The number density in Γ-space is a very rapidly increasing function of the energy E. In order to see this we consider, by way of a representative example, a system of N molecules which are free to perform translatory motion in a volume V. The Γ-space of such a system will consist of $f = 3N$ coordinates and an equal number of momenta. Retracing the steps of our previous argument, we recognize that integration with respect to the $3N$ spatial coordinates in physical space yields a factor V^N. Integration in the momentum space is difficult to perform, and in order to simplify the task we shall merely write down orders of magnitude, refraining from evaluating numerical factors. Denoting the momenta by $p_1, p_2, \ldots, p_{3N}$, we observe that the projection into the momentum space has a radius

$$p = (p_1{}^2 + \cdots + p_{3N}^2)^{1/2} = (2mE)^{1/2}.$$

The volume $\mathscr{V}$ of the hypersphere–hypercylinder contains the factor V^N from physical space, and a factor p^{3N} from momentum space. Hence

$$\mathscr{V} \sim [(2mE)^{3/2} V]^N. \tag{3.57}$$

On increasing the energy by δE, we add a volume

$$\delta\mathscr{V} \sim \left(\frac{\partial \mathscr{V}}{\partial E}\right)_{V,N} \delta E,$$

or

$$\delta\mathscr{V} \sim \frac{3N}{2} V^N 2m(2mE)^{(3N/2)-1}\,\delta E,$$

so that

$$\frac{\delta\mathscr{V}}{\mathscr{V}} \sim \frac{3N}{2}\frac{\delta E}{E}. \tag{3.58}$$

The fractional change in the volume is $3N/2 \approx N$ times as large as the fractional change in energy. Since N is of the order of the Avogadro number ($\sim 10^{23}$) the ratio is enormous. This means that a small change in energy, δE, introduces a new volume $\delta\mathscr{V}$ (the volume of the shell), and hence a new number of quantum cells in Γ-space, $\Omega = \delta\mathscr{V}/\mathbf{h}^f$, which by far exceeds all those that lie within the level E itself. This is a general property of a hypersphere in multidimensional space, or of any set of similar hypervolumes, for that matter.

We inform the reader that the preceding argument carries over to any macroscopic system and shows that a small change in energy, δE, adds to it roughly $N(\delta E/E)\Omega$ new quantum states. This is another way of saying that the number density $\omega(E, V, N)$ is a rapidly increasing function of energy, the rate of increase being so enormous that a very small change in the argument multiplies the value of the function itself by a very large factor. Thus, as we shall note later, it makes little difference whether averages are taken over available quantum cells which lie within the range of energies E to $E + \delta E$ or over all possible quantum states for all energy levels below $E + \delta E$.

Put more succinctly, we notice that if

$$\mathscr{V} \sim V^N p^{fN},$$

then

$$\frac{\delta \mathscr{V}}{\mathscr{V}} \sim fN \frac{\delta p}{p}.$$

3.15. Spectroscopy

The molecular constants which occur in the equations of the preceding sections are usually determined by spectroscopy. For this reason, the reader may be interested in a very brief, albeit superficial account of the principles which make this possible.† Our account will be restricted to the treatment of the internal motions of gaseous molecules because, as will become clear in Section 6.3, the quantization of translational motion turns out to be unimportant for the calculation of specific heats. At temperatures of the order of 1500 K or more (but only 130–180 K for nitric oxide, NO), it may become necessary to include the energy stored by the excitation of electrons to higher states than the ground state. However, we shall leave this possibility out of account in the present section.

A spectrograph provides a spatial resolution of radiation according to wavelength λ. Since $\mathbf{c} = \lambda \nu$, where $\mathbf{c}$ is the speed of light *in vacuo*, a resolution with respect to frequency is provided simultaneously. When interaction with matter is studied, a beam of light can be regarded as a stream of photons of zero rest-mass, the particles moving with the speed of light $\mathbf{c}$. According to de Broglie's equation (3.1a), a photon of wavelength λ (or frequency ν) has a momentum

† Full details and copious numerical results are contained in three standard works by G. Herzberg: 1. *Atomic Spectra and Atomic Structure*, Prentice Hall, Englewood Cliffs, New Jersey, 1937; reprinted by Dover, New York, 1944; 2. *Molecular Spectra and Molecular Structure, I. Spectra of Diatomic Molecules*, 2nd ed., Van Nostrand, Princeton, New Jersey, 1950; 3. *Infrared and Raman Spectra of Polyatomic Molecules*, Van Nostrand, Princeton, New Jersey, 1945.

$$p = \frac{\mathbf{h}}{\lambda} = \frac{\mathbf{h}\nu}{\mathbf{c}} \tag{3.59}$$

and an energy

$$\varepsilon_{\mathrm{ph}} = p\mathbf{c} = \mathbf{h}\nu \qquad (=\mathbf{hc}/\lambda). \tag{3.59a}$$

Thus there exists a unique relation between the energy $\varepsilon_{\mathrm{ph}}$ of a photon and its frequency ν or wavelength λ. This means that the spectrograph produces a unique ordering of photons according to their energy.

In *absorption spectroscopy* the stream of photons is first allowed to pass through the gas and is resolved afterwards; if necessary, a filter may be used to produce photons in a predetermined frequency range. In this manner some of the photons are made to give up their energy to molecules. As a result, the rotational or vibrational energy of the molecules is increased. A single photon can change the energy of a single molecule only, and can yield to it a single quantum of energy, $\varepsilon_{\mathrm{ph}}$, precisely. On the other hand, a molecule can absorb only a characteristic amount of energy which is determined by its allowable energy levels. Since the two amounts must be equal, there exists a strict correlation between the frequency, ν, of the photon and the quantum jump, $\Delta\varepsilon$, of the molecule. Thus the gas absorbs photons selectively, annihilating those whose energies (and thus also wavelengths, momenta, and frequencies) match the energies of allowable quantum jumps. Such photons will be absent in the spectrum of the transmitted light, and their absence will register as a black line in the spectrogram. To be sure, owing to the fact that the absorption is governed by a probability density, the dark line constitutes a region of continuously varying intensity, but one which possesses a sharp peak of blackness. Heisenberg's uncertainty principle shows that the simultaneous measurement of the position of this peak and of the associated momentum p (and hence frequency ν) is subject to an uncertainty which cannot be less than $\frac{1}{2}\hbar$.

Not all imaginable quantum jumps really do occur in nature. The expected quantum jumps are determined with the aid of *selection rules* which can be deduced theoretically with the aid of quantum mechanics or experimentally, when the attendant mathematical difficulties are too great.

In *emission spectroscopy* energy is supplied to the gas by heating; this, in turn, radiates photons thus losing energy. Transitions to lower energy levels register in the form of bright lines in the spectrum. These must also match the permitted quantum jumps and must also extend over a very narrow region, fading away in both directions.

The allowable energy levels of a rigid rotor have been given in Equation (3.37). Thus a quantum jump from j' to j'' produces an energy difference of

$$\Delta\varepsilon = \frac{\hbar^2}{2I}\,[j'(j'+1) - j''(j''+1)]\,.$$

The selection rule in this case dictates that

$$\Delta j = j' - j'' = \pm 1,$$

so that for an emission line ($j' = j$ and $j'' = j - 1$) we have

$$\Delta \varepsilon = j \frac{\hbar^2}{I} \qquad (j = 1, 2, 3, \ldots). \tag{3.60}$$

A quantum jump of this type produces a line in the spectrum for which

$$\Delta \varepsilon = \mathbf{h}\nu \qquad \text{with} \quad \nu = j \frac{\mathbf{h}}{4\pi^2 I}.$$

In spectroscopy it is usual to introduce the *rotational constant* B defined as

$$B = \frac{\mathbf{h}^2}{8\pi^2 I \mathbf{c}}. \tag{3.61}$$

Hence it follows from Equation (3.60) that the rotational spectra must satisfy the equation

$$k = \frac{\nu}{\mathbf{c}} = 2Bj, \tag{3.62}$$

where $k = 1/\lambda$ is the wave number.† This shows that the wave numbers of the spectral lines are equally spaced. A series of such lines can be used to determine the rotational constant B and hence the single moment of inertia I of a linear molecule. The rotational spectra of most molecules are found in the region where $k = 200\ \mu^{-1}$ or more, that is, in the far infrared.

The quantum jump of a pure vibrational mode was given in Equation (3.43b). Here too the selection rule allows a change of

$$\Delta v = \pm 1 \tag{3.63}$$

only. Denoting the frequency of the photon by ν_{ph} and that of a particular normal mode of the molecule by ν_i, it is easy to see that

$$\nu_{\text{ph}} = \nu_i. \tag{3.64}$$

Thus the spectral lines directly measure the frequencies of the normal modes of the molecule. Vibrational frequencies are not confined to a particular range in the spectrum.

A real spectrum is much more complex than the preceding simple account would lead us to believe. The complications arise from the following circumstances. First, frequently a single photon causes a simultaneous change

† Denoted frequently by ν in books on spectroscopy.

in the rotational and vibrational energy of a molecule, giving rise to *vibration–rotation* spectra. Secondly, the rotational spectra of nonlinear molecules are more complex because their quantum jumps are described by more complex relations. Thirdly, vibrations are anharmonic, and the rotors are nonrigid, as we know from previous remarks. Finally, electronic spectra also appear, and must be separated from the others.†

3.16. Summary of Results from Quantum Mechanics

The present chapter attempted to impart to the reader an appreciation for quantum mechanics. A thorough grasp of this subject requires a good deal more study and a deeper preparation in mathematical analysis. However, the principal results which will be required in the succeeding two chapters on the statistical analysis of very large collections of elementary particles can be summarized rather briefly. If the reader is prepared to accept the validity of this summary, he may postpone his studies of quantum mechanics to a later date and yet be in a position to understand the statistical principles which enable us to deduce the macroscopic properties of systems from a knowledge of the general features of the motion of their mechanical models.

Stationary or Quantum State. When a system performs a more or less complex motion, conserves its total energy, and moves so that the probability amplitude $|\psi|^2$ is independent of time, it is said to be in a stationary state.

Discrete Energy Levels. If we consider translational motion in a confined space of volume V, rotational motion with a constant angular momentum, or a harmonic oscillation in a normal mode, we find that the total energy of the corresponding system cannot be prescribed at will and cannot be altered by arbitrarily small amounts. The motion can occur only at a selected, though infinitely large set of discrete energy levels. The particular energy levels are different for the three types of motion enumerated above; two admissible energy levels are said to be separated by a quantum jump. The magnitude of a quantum jump is different for each of the three types of motion. The energy levels and quantum jumps are scaled with Planck's constant $\mathbf{h}$ (or $\hbar = \mathbf{h}/2\pi$). Since this absolute scaling quantity has an exceedingly small value, quantum jumps are exceedingly small when compared with the energy changes in which we are normally interested as far as macroscopic systems are concerned.

Wave Function. The motion of a system is described by a purely conceptual wave function. The wave function is expressed in terms of generalized

† See for example, S. Glasstone, *Theoretical Chemistry*, Van Nostrand, Princeton, New Jersey, 1944. In particular, see Chapters IV and V on molecular spectra, pp. 141 and 203.

coordinates (Cartesian coordinates, angular coordinates, and so on, as required), and its square represents the probability density of measuring a particular value of the coordinate in question. The wave function is a solution of the appropriate Schrödinger equation subject to given boundary conditions and certain formal requirements: continuity, single-valuedness, boundedness (and therefore normality). A wave function represents a particular state of motion (classically, a trajectory in phase space). Every wave function corresponds to a particular energy level, but a prescribed energy level may be achieved in different motions, each described by a different wave function.

Degeneracy. The number of wave functions, that is, the number of stationary states through which a particular energy level may be realized, is called the degeneracy g of that energy level.

Quantum Numbers. The energy level of a particular stationary state is described by a set of integers known as quantum numbers. A particular set of quantum numbers describes a single stationary state and defines a single wave function. When one of the quantum numbers changes, the wave function, and hence the stationary state, also change. Compensating changes in two or more quantum numbers may produce the same energy level, but the state is different in each case, thus leading to degeneracy.

Translational Motion. Translational motion is described by three quantum numbers: s_x, s_y, s_z, and a particular energy level has a degeneracy g.

Rotational Motion. The rotational motion of a linear molecule is described by two quantum numbers: j and m. The degeneracy of every allowable energy level is $g = 2j + 1$. No general, quantum-mechanical solutions were given for more complex molecules.

Vibrational Motion. A harmonic oscillation with one degree of freedom is characterized by a single quantum number, v, and the energy levels are nondegenerate ($g = 1$).

Spin. In some cases it is necessary to include a set of quantum numbers which describe the intrinsic momentum, or spin, of a particle. This is a quantum number which has no counterpart in classical mechanics.

Collection of Independent Particles. The wave function of a collection of N independent, noninteracting particles is equal to the product of the wave functions of each of the N particles. The wave function of a single molecule which performs translation, rotation, and vibration is equal to the product of the wave functions associated with each degree of freedom of the motion. The energy level of the system is equal to the sum of the energy levels of each particle which, in turn, is equal to the sum of the energy levels of each degree of freedom. Thus the wave function, and hence the stationary state of the whole assembly, is described by N sets of quantum numbers.

Symmetry and Indistinguishability. The square of the total wave function of N identical particles must be invariant when a pair of particles is interchanged, thus ensuring their indistinguishability. It follows that a wave function must be either symmetrical or antisymmetrical with respect to the coordinates of any two particles. Solutions of the Schrödinger equation must be suitably symmetrized or antisymmetrized. A given molecule belongs either to the class of symmetrical (bosons) or antisymmetrical particles (fermions). If any two of the wave functions of a particle are identical, the antisymmetric function is identically zero. It follows that two fermions in an assembly cannot be in the same quantum state (Pauli's exclusion principle). By contrast, there is no limitation on the number of bosons in an assembly which can be in the same quantum state.

Heisenberg's Uncertainty Principle. Heisenberg's uncertainty principle states that the product of the uncertainties in our knowledge of the precise values of two complementary observables, in particular of a coordinate (Δx) and its conjugate momentum (Δp_x) and of energy (ΔE) and time (Δt), must be larger than a quantity related to Planck's constant ($\Delta x \; \Delta p_x \geq \frac{1}{2}\mathbf{\hbar}$; $\Delta E \; \Delta t \geq \frac{1}{2}\mathbf{\hbar}$). It follows that cells in phase space, in particular cells in μ-space, must be ascribed a finite dimension. This is $\Delta \mathscr{V} = \mathbf{h}^3$ for translation and $\Delta \mathscr{V} = \mathbf{h}$ for a single harmonic oscillator.

Correspondence Principle. At very high quantum numbers the quantum-mechanical description of any motion becomes asymptotically identical with that provided by Newton's laws of motion.

Systems Containing Many Particles. In ordinary circumstances the energy levels of systems containing a large number of particles (10^{20}–10^{25}) are extremely closely spaced and their degeneracy is orders of magnitude larger than the number of particles; a small relative increase in the energy level causes the number of available quantum states to increase proportionately, the factor of proportionality being of the order of the number of particles.

PROBLEMS FOR CHAPTER 3†

3.1. Use the uncertainty principle to estimate the minimum energy, E_0, of a particle of mass m in a one-dimensional "box" of side a. Compare your answer with the known energy of the ground state, Equation (3.24a).

† Problems numbered in boldface type are considered relatively easy; they are recommended for assigning in an elementary first course.

SOLUTION: If we set $\Delta x = a$, then the uncertainty in the momentum assumes the value $\Delta p \approx \mathbf{h}/a$. This represents the smallest value of the root mean square of the momentum. The corresponding energy is $E_0 \approx \mathbf{h}^2/2ma^2$.

3.2. The smallest particle that can be seen in an ordinary microscope is about $d \approx 0.1\ \mu$ in diameter. Assuming a density of $\rho = 2$ gr/cm^3, calculate the minimum uncertainty in the velocity of such a particle if the uncertainty in its position is $\Delta d = 0.01\ d$.

3.3. Determine the minimum energy, E_0, of the electrons in an electron microscope to secure a resolution $R = 0.1$ Å. Determine the corresponding energy E_0' of a photon.

3.4. The uncertainty in the position of an electron is given as $\Delta x = 0.5$ Å. Determine the uncertainty, Δp, in the linear momentum of the electron. An electron is placed in a cubical box of side $a = 0.5$ Å. Estimate the lowest energy, E_0, available to the electron.

3.5. Show that the simultaneous measurements of position and energy are not subject to interference.

3.6. An electron of mass $\mathbf{m}_e = 9.11 \times 10^{-31}$ kg and a bullet of mass $m = 10$ gr travel with identical speeds $v = 1000$ m/sec. If the relative precision of the measurement of v is $\Delta v = 0.01\%$, compute the minimum uncertainty which must be ascribed to the position measurement of each mass. Considering today's limit of resolution of a length measurement, does the minimum uncertainty computed for the macroscopic system of the bullet impose any restriction on the procedures used to measure its position?

3.7. Calculate the de Broglie wavelength, λ, for (a) an electron moving at a speed $v = 3 \times 10^4$ m/sec (mass of electron $\mathbf{m}_e = 9.11 \times 10^{-31}$ kg); (b) a projectile of mass $m = 12$ gr moving at a speed $v = 600$ m/sec. Why does the wave nature of the projectile not manifest itself through diffraction effects?

3.8. Write an essay on the physical significance of Heisenberg's uncertainty principle. What is the effect of the uncertainty principle on the structure of the Γ-space and of the μ-space?

3.9. Write an essay on the reasons which compel us to use quantum mechanics rather than classical mechanics in the study of the motion of small particles. Illustrate your reasoning with descriptions of several classical experiments known to you from your study of elementary physics.

3.10. Give a clear physical interpretation of the wave function in Schrödinger's equation.

3.11. One plausibility argument for the Schrödinger equation may be provided with the aid of the results of the classical problem of the vibrating string. To this end, consider a flexible, elastic string which is stretched along the horizontal x axis between two fixed supports located at $x = 0$ and $x = l$. Letting $u(x, t)$ denote small vertical displacements of the string from the x axis, it can be shown† that the differential equation for free vibrations of the string is the wave equation

$$\frac{\partial^2 u}{\partial x^2} = \frac{1}{a^2}\frac{\partial^2 u}{\partial t^2}.$$

The resulting motion can be described as a superposition of two waves propagating with a constant speed a.

(a) Show that $A \sin (2\pi/l) \times \cos (2\pi at/l)$ is a solution (not the most general) to the above wave equation which satisfies the boundary conditions $u(0, t) = u(l, t) = 0$ and the initial condition $u_t(x, 0) = 0$. (b) By analogy, let $\lambda = l$, $v = a$, and $\Psi(x, t) = C\, e^{i(2\pi/\lambda)(x - vt)}$. Substitute the de Broglie wavelength for λ, and express the speed v of a particle of mass m in terms of its momentum p. Finally, show that the one-dimensional time-dependent Schrödinger equation results when this expression for Ψ is substituted into the wave equation. (c) Examine carefully the assumptions made for this illustration, and decide whether the form of the Schrödinger equation given in Equation (3.12) is subject to the same limitations.

3.12. Compute the average value of the kinetic energy, $\langle E \rangle$, for a particle whose wave function is $\psi(x) = A \exp[-(x - x_0)^2/2a^2 + ip_0 x/\hbar]$.

3.13. A one-dimensional harmonic oscillator has a Hamiltonian given by

$$\mathbf{H} = -\left(\frac{h^2}{8m\pi^2}\right)\frac{\partial^2}{\partial x^2} + \tfrac{1}{2}kx^2.$$

The particle is in a stationary state in which its wave function is of the form $\psi(x) = A \exp(-x^2/2a)$. Find

(a) The values of A and a for which $\psi(x)$ is a normalized wave function for the harmonic oscillator; (b) the probability density of observing the particle at point x; (c) the average displacement of the particle, $\langle x \rangle$; (d) the average mean square displacement, $\langle x^2 \rangle$; (e) the average kinetic energy, $\langle p^2/2m \rangle$; (f) the average total energy, $\langle p^2/2m + kx^2/2 \rangle$.

3.14. Show by direct integration that the wave functions given by Equation (3.27) for a particle in a container are orthogonal and normalized, that is, that

$$\int_0^a dx \int_0^a dy \int_0^a dz\, \psi^*_{s_x, s_y, s_z}\, \psi_{s_x', s_y', s_z'} = \begin{cases} 1 & \text{if } s_x = s_x',\ s_y = s_y',\ s_z = s_z' \\ 0, & \text{otherwise.} \end{cases}$$

What is the physical significance of the orthonormality of the wave functions? Do you expect that a similar set of integral formulas will hold for wave functions corresponding to quantum states for more general systems. (*Hint:* The term "orthonormal" refers to the fact that the integral of the square of the modulus of a wave function is equal to unity, signifying that the wave function has been normalized,

† See, for example, I. S. Sokolnikoff and R. M. Redheffer, *Mathematics of Physics and Modern Engineering*, p. 431, McGraw-Hill, New York, 1958.

and that the integral of one wave function multiplied by the complex conjugate of another wave function vanishes. In other words, two wave functions are "orthogonal." [For more details, see L. J. Schiff, *Quantum Mechanics*, 3rd ed., Chapters 2 and 3, McGraw-Hill, New York, 1968.])

3.15. Using physical terms, explain the concept of degeneracy. Give a parallel discussion in mathematical terms.

3.16. Are the energy levels of a three-dimensional oscillator degenerate or not?

3.17. Estimate the values of the quantum numbers, s_n, for a hydrogen atom in a box whose side $a = 1$ cm, if the energy of the hydrogen atom is $E = 1.5\,\mathbf{k}T$, and the temperature is $T = 300$ K.

3.18. What is the lowest energy state available (a) to two identical bosons and (b) to two identical fermions, assuming that they have been placed in a cubical container of side a. Write down the corresponding wave functions, ψ_b and ψ_f. Determine the energy, E_1, of the first excited state, in each case.

SOLUTION: Two bosons may each be in the lowest single-particle state, but if one fermion is in the lowest single-particle state, the other must be in the next single-particle state. Therefore, the lowest energy state for two bosons is $6\mathbf{h}^2/8mV^{2/3}$, while that for two fermions is $9\mathbf{h}^2/8mV^{2/3}$. The energy of the first excited boson state is $9\mathbf{h}^2/8mV^{2/3}$, while that for the fermions is $12\mathbf{h}^2/8mV^{2/3}$.

3.19. The pendulum shown in the figure consists of a mass m at the end of an inextensible weightless string of length R. The pendulum swings through a small arc θ_0 on either side of equilibrium. Since θ_0 is small, the pendulum is assumed to perform simple harmonic motion.

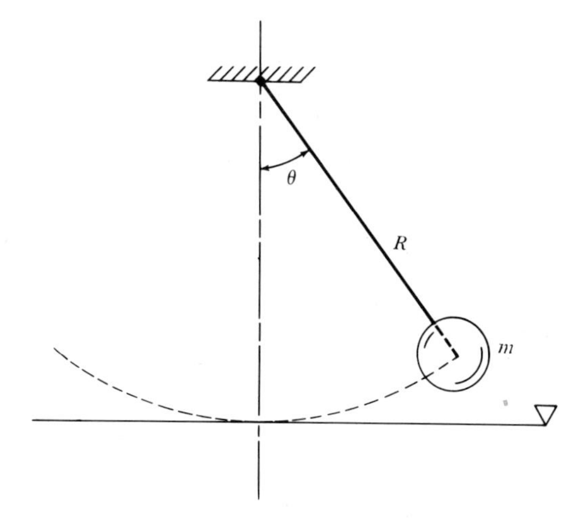

(a) Show that the time-independent Schrödinger equation for the wave function ψ

can be written

$$\frac{d^2\psi}{d\theta^2} + \frac{2mR^2}{\hbar^2}\left(\varepsilon - mgR\frac{\theta^2}{2}\right)\psi(\theta) = 0$$

for this motion. (In deriving this result, it will be helpful to choose the datum for potential energy indicated in the figure.)
(b) By suitably transforming the independent variable θ to a new independent variable x, show that the equation can be transformed to read

$$\frac{d^2\psi}{dx^2} + (\lambda - x^2)\psi(x) = 0,$$

with

$$\lambda = \frac{2R\varepsilon}{\hbar(gR)^{1/2}}.$$

(c) Perform the substitution $\psi(x) = Au(x)\,e^{-x^2/2}$, and find the differential equation for $u(x)$:

$$\frac{d^2u}{dx^2} - 2x\frac{du}{dx} + (\lambda - 1)u(x) = 0.$$

Then demonstrate that the solutions to this equation are

$$u(x) = H_n(x) \qquad (n = 0, 1, 2, \ldots),$$

provided that $\lambda = 2n + 1$ $(n = 0, 1, 2, \ldots)$. Here, $H_n(x)$ are the Hermite polynomials (Section 3.11) which satisfy the relations†

$$H_n'(x) = 2nH_{n-1}(x),$$
$$H_{n+1}(x) = 2xH_n(x) - H_n'(x).$$

As an exercise, write out $H_4(x)$ and verify that it satisfies the differential equation when $\lambda = 9$.

(d) If the pendulum is started from rest at $\theta = \theta_0 = 4°$ and has a mass $m = 0.1$ kg and a length $R = 0.5$ m, calculate its total classical energy ε_c. Compute the vibrational quantum number n which corresponds to the quantum-mechanical energy level $\varepsilon_n = \varepsilon_c$, and then consider the question: As a pendulum slows down, why does its energy appear to decrease continuously? Hence, substantiate the claim that the application of quantum mechanics to macroscopic systems yields results which are in accord with the classical (Newtonian) description.

3.20. The wave function for an electron in a hydrogen atom in its ground state is given by $\psi = (\pi a^3)^{-1/2}\exp(-r/a)$, where

$$a = \varepsilon_0 \mathbf{h}^2/\pi \mathbf{m}_e \mathbf{e}^2 \approx 0.5 \text{ Å}.$$

† It can be shown that Hermite polynomials are the coefficients in the series expansion for $\exp(2x\xi - \xi^2) = \sum_{n=0}^{\infty} H_n(x)\xi^n/n!$. Differentiation of this with respect to x yields the first of these relations; differentiation with respect to ξ yields the second.

(a) Show that the above form of ψ has been normalized, i.e., show that

$$\int_0^\infty \int_0^{2\pi} \int_0^{\pi} r^2 \sin\theta \psi^2(r)\, d\theta\, d\phi\, dr = 1.$$

(b) Determine the probability, W, of finding the electron within a small shell of radius Δr centered at $r=0$.

(c) Determine the probability, W, that the electron will be found within a distance a from the point $r=0$.

3.21. The energy levels for the electron in a hydrogen atom are given by $E_n = -\mathbf{m}_e\mathbf{e}^4/8\varepsilon_0{}^2\mathbf{h}^2 n^2 = (-2.17\times 10^{-18}/n^2)\,J = (-13.6/n^2)$ eV, with $n = 1, 2, 3, \ldots$. Here $\varepsilon_0 = (4\pi\times 9\times 10^9)^{-1}$ Nm^2/C^2 is the permittivity constant. Suppose that a neutron with a kinetic energy $E = 6.0$ eV collides with a hydrogen atom at rest in its ground state ($n=1$). Using the laws of conservation of energy and momentum, show that the collision cannot produce an excited state of the hydrogen atom.

3.22. Refer to Problem 3.21 for the energy levels of the electron in a hydrogen atom. Suppose that an atom makes a transition from a state with $n=3$ to a state where $n=2$. Calculate the frequency, ν, and the wavelength, λ, of the emitted photon.

3.23. A very crude model of the energy levels of an electron in a solid may be obtained by considering the energy levels of a particle moving in a one-dimensional periodic potential which satisfies the following Schrödinger equation

$$-\frac{\hbar^2}{2m}\frac{d^2\psi}{dx^2} - \frac{\hbar^2\pi^2\psi}{20ma^2}\cos\frac{2\pi x}{a} = E\psi.$$

The constants have been chosen for later convenience. Suppose that ψ may be expressed as

$$\psi(x) = \sum_{n=-\infty}^{\infty} C_n \exp i\left(k + \frac{2\pi n}{a}\right)x.$$

Using the identities

$$\cos\frac{2\pi x}{a} = \frac{1}{2}\left[\exp\left(i\frac{2\pi x}{a}\right) + \exp\left(-i\frac{2\pi x}{a}\right)\right]$$

and

$$\int_0^a dx \exp\left[i\frac{2\pi}{a}(n-n')x\right] = \begin{cases} a & \text{if } n = n' \\ 0 & \text{otherwise}, \end{cases}$$

find a relation between the coefficients C_n.

SOLUTION: Inserting the expression for ψ in Schrödinger's equation, we obtain

$$(\hbar^2/2m)\sum_n C_n(k + 2\pi n/a)^2 \exp[i(k+2\pi n/a)x]$$
$$-(\hbar^2\pi^2/40ma^2)\sum_n C_n[\exp i(k+2\pi(n+1)/a)x + \exp i(k+2\pi(n-1)/a)x]$$
$$= E\sum C_n \exp i(k+2\pi n/a)x.$$

Multiply by $\exp[-i(k + 2\pi l/a)x]$ and integrate the resulting expression with respect to x from $x = 0$ to $x = a$. This leads to the relation between the coefficients:

$$(\hbar^2/2m)C_l(k + 2\pi l/a)^2 - (\hbar^2\pi^2/40ma^2)(C_{l+1} + C_{l-1}) = EC_l.$$

3.24. In Problem 3.23 make the approximation that all the C_ns are zero except C_0 and C_{-1}, and determine the resulting energy levels as a function of k. Sketch the wave function, ψ, in this case.

SOLUTION: Assuming that only C_0 and C_{-1} differ from zero, we are led to the two simultaneous equations

$$[(\hbar^2/2m)(k - 2\pi/a)^2 - E]C_{-1} - (\hbar^2\pi^2/40ma^2)C_0 = 0$$

$$[(\hbar^2/2m)k^2 - E]C_0 - (\hbar^2\pi^2/40ma^2)C_{-1} = 0.$$

These two equations have a solution if and only if the determinant of their coefficients vanishes; that is, if

$$[(\hbar^2/2m)(k - 2\pi/a)^2 - E][(\hbar^2/2m)k^2 - E] - (\hbar^2\pi^2/40ma^2)^2 = 0.$$

This is a quadratic equation for E.

3.25. The energy difference between the rotational ground state and the first excited rotational state for the diatomic molecule HCl is $\Delta E = 3.8 \times 10^{-15}$ erg. (a) Calculate the moment of inertia, I, of this molecule. (b) Determine the internuclear distance, r, which corresponds to this moment of inertia. The nuclei of the atoms may be regarded as point masses.

3.26. The energy difference between the vibrational ground state and the first excited vibrational state for HCl is $\Delta\varepsilon = 5.51 \times 10^{-13}$ erg. Calculate the classical vibrational frequency of this molecule.

LIST OF SYMBOLS FOR CHAPTER 3

Latin letters

A	Constant
$\mathbf{A}$	Arbitrary operator
a	Constant of proportionality or eigenvalue; length
B	Constant; rotational constant, see Equation (3.61)
$\mathbf{B}$	Arbitrary operator
c	Speed of light *in vacuo*
E	Energy
$\mathcal{E}$	Energy
F	Force
f	Number of degrees of freedom
f	Arbitrary function; characteristic or eigenfunction; function of time, see Equation (3.13)
g	Degeneracy; density of energy levels
g	Arbitrary function

H	Hermite polynomial
$\mathbf{H}$	Hamiltonian operator
$\mathscr{H}$	Hamiltonian
$\mathbf{h}$	Planck's constant
$\hbar$	$=\mathbf{h}/2\pi$
I	Moment of inertia
j	Quantum number in rotation
k	Wave number
$\mathbf{k}$	Boltzmann's constant
L	Constant
M	Molecular mass
M_0	Torque
m	Mass; quantum number in rotation
N	Number of particles
$\mathbf{N}$	Avogadro's number
n	Integer
p	Momentum; generalized momentum
$\mathbf{p}$	Angular momentum
$\mathbf{p}$	Momentum operator
q	Generalized coordinate
$\mathbf{r}$	Position operator
r	Radius
s	Quantum number in translation
T	Time
t	Time
t_0	Reference time
V	Volume
$\mathscr{V}$	Volume in μ-space or Γ-space
v	Quantum number in vibration
W	Probability density
X	Amplitude
x, y, z	Coordinates
$\mathbf{x}, \mathbf{y}, \mathbf{z}$	Coordinate operators

Greek letters

Γ	Denotes 6N-dimensional phase space
ε	Quantum energy level
ε_1	Zero-point energy
η	Coordinate
θ	Polar coordinate
λ	Wavelength
μ	Reduced mass; denotes six-dimensional phase space
ν	Frequency
ξ	Coordinate; displacement of oscillator from equilibrium position
ρ	Density

σ	Symmetry number
Φ	Wave function for generalized momenta p
ϕ	Polar coordinate; potential energy
Ψ	Wave function for generalized coordinates q
ψ	Function of position, see Equation (3.13); amplitude of standing wave
ψ_a	Antisymmetric wave function
ψ_s	Symmetric wave function
Ω	Number of quantum states
ω	Circular frequency; number density or density of states
$\boldsymbol{\omega}$	Angular velocity

Superscripts

*	Complex conjugate

Subscripts

kin	Kinetic
p	Refers to momentum space
ph	Photon
q	Refers to position space
r	Rotational degrees of freedom
tr	Translational degrees of freedom
v	Vibrational degrees of freedom

Special Symbols

∇^2	Laplacian operator, see Equation (3.4)
$\langle\ \rangle$	Expectation value, see Equation (3.9)

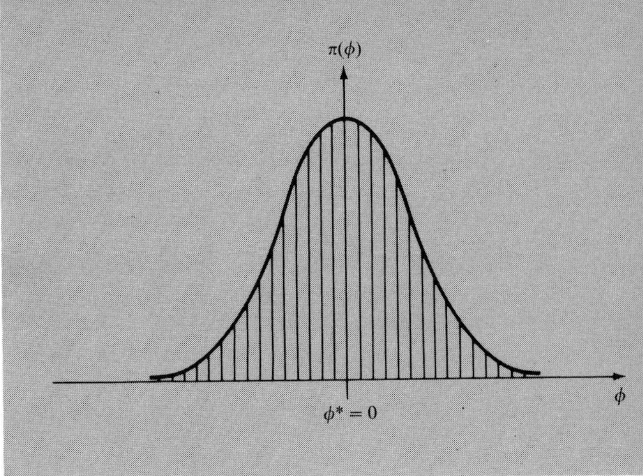

CHAPTER 4

TOPICS IN MATHEMATICS

In the succeeding chapters it will be necessary to make use of certain mathematical formulas and approximations. In this chapter, we provide a short resumé reinforced with some heuristic derivations, in case they should present difficulties to the reader. *Readers who have sufficient training in mathematical methods will find this chapter redundant.*

4.1. Combinatorial Formulas

4.1.1. Ordered Sequences

In statistical mechanics it is often necessary to calculate the number of ways, Ω_v, in which elements can be arranged in groups ("containers"). In order to do this it is necessary to provide a systematic way of counting. Conceptually the simplest way to count is to exhibit explicitly all individual members of a set to be counted. However, in our case this is impracticable because the numbers of elements in those sets are very large. Instead, we must evolve a more systematic way of counting, and this is contained in essentially two combinatorial formulas which we discuss below.

In all counting it is convenient first to assume that we are given N *distinguishable* elements. We shall label them $a_1, \ldots, a_i, \ldots, a_N$. The first problem consists in determining the number of ordered *N-tuples* or ordered sequences. With $N = 3$, we can form the six ordered sequences displayed in Figure 4.1.

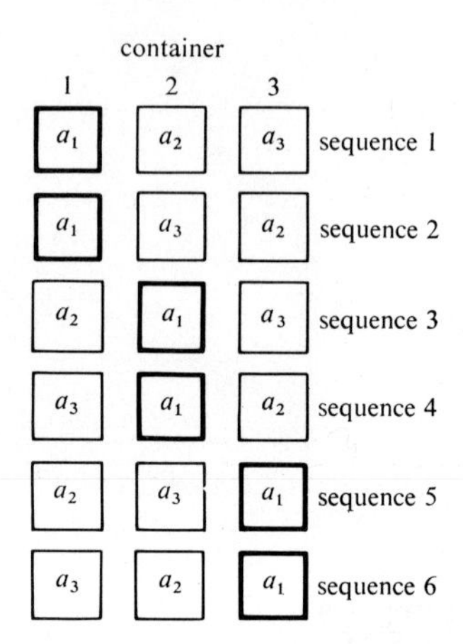

FIGURE 4.1. Ordered sequences of $N = 3$ distinguishable objects.

The system used here consists in imagining three containers and three distinguishable objects. First, we attempt to place a_1: this can be done in three ways, corresponding to the three framed columns in the array. Next we consider a_2 and find that it can be placed only in one of two ways. The third element, a_3, must fit in the remaining space. Thus, we can form a total of $3 \cdot 2 \cdot 1$ ordered 3-tuples. Generalizing to N elements and N containers, we find that a_1 can be placed in N different ways, a_2 in $(N-1)$ different ways, and so on. Hence the number of ordered N-tuples is†

$$\Omega = N(N-1)(N-2)\cdots 1 = N!. \tag{4.1}$$

The preceding method constitutes an example of *sequential counting*.‡ It is noteworthy that even a modest increase in N causes the number Ω to increase enormously owing to the properties of factorials.§

The $\Omega = N!$ permutations are distinguishable, since the elements have to be numbered. If we obliterate the labeling in our numerical example, we obtain a single 3-tuple. The number unity is obtained by dividing Ω by the number of distinguishable N-tuples. Thus

$$\Omega' = \frac{N!}{N!} = 1,$$

† The reader will recall that $0! = 1$.

‡ See T. M. Apostol, *Calculus*, Vol. II, p. 19, Blaisdell, Waltham, Massachusetts, 1962.

§ Tables of factorials are given in E. and F. Spon, *Barlow's Tables*, London, 1947. For larger numbers, the reader may consult D. B. Owen and C. M. Williams, *Logarithms of factorials from 1 to 2000*, Sandia Corporation Monograph, Albuquerque, New Mexico, 1959.

and we establish the rule that distinguishablity is obliterated by dividing the result for distinguishable elements by the number of distinguishable permutations (N-tuples) of the elements under consideration.

4.1.2. Distributions of N Elements over k Containers

The next example concerns the arrangement of N distinguishable elements a_i into k distinct containers without imposing restrictions upon the number of elements in each container. We start again with a numerical example and choose $N = 2$ with $k = 3$. All of the possible arrangements for this case are displayed in Figure 4.2. First, we draw element a_1 and notice that we can

arrangement	container			distribution
	1	2	3	
1	a_1 a_2			2, 0, 0
2	a_1	a_2		1, 1, 0
3	a_1		a_2	1, 0, 1
4		a_1 a_2		0, 2, 0
5		a_1	a_2	0, 1, 1
6	a_2	a_1		1, 1, 0
7			a_1 a_2	0, 0, 2
8	a_2		a_1	1, 0, 1
9		a_2	a_1	0, 1, 1

summary

distribution	ν	Ω_ν
2, 0, 0	1	1
0, 2, 0	2	1
0, 0, 2	3	1
1, 1, 0	4	2
1, 0, 1	5	2
0, 1, 1	6	2

FIGURE 4.2. Arrangement of $N = 2$ elements in $k = 3$ containers.

place it in $k = 3$ different ways. Next, we draw element a_2 and observe that we can still place it in $k = 3$ different ways. Hence, the number of arrangements is $3 \cdot 3 = 9$. In general, the first element can be placed in k ways, the second element in k ways, and so on as far as the Nth element. This leads us to the formula

$$\Omega = k^N. \tag{4.2}$$

In this count, the *order* in which elements have been placed in a particular container is considered irrelevant. Thus, $a_1 a_2$ in container 1 is indistinguishable from $a_2\, a_1$ in container 1, and has not been counted twice in the scheme. Here again, the number Ω is very large even for modest values of k and N.

It is now useful to solve the same problem in a different way. In order to do this, we begin by specifying k numbers n_i. This set of numbers is called a distribution. In Figure 4.2, arrangement 1 corresponds to the distribution $n_1{}^1 = 2$, $n_2{}^1 = 0$, $n_3{}^1 = 0$, abbreviated to 2, 0, 0. There are a total of 6 distributions and the number of arrangements is different for each distribution. However, as expected, the total number of arrangements is the same, namely $3^2 = 9$; it is merely the method of counting them which has changed.

We now suppose that for distribution ν we have specified $n_1{}^\nu, \ldots, n_i{}^\nu, \ldots, n_k{}^\nu$ and ask for the number, Ω_ν, of arrangements which correspond to this distribution. To count this number, we first imagine all possible $N!$ ordered N-tuples. In the numerical examples these are simply

$$a_1 a_2 \qquad a_2 a_1 .$$

Next, we cut off $n_1{}^\nu$ elements. Thus, for the distribution 1, 0, 1, we cut off one element, and proceed in this manner until the n_{k-1}^ν elements have been cut off. The last compartment automatically contains $n_k{}^\nu$ elements, since

$$\sum_{\nu=1}^{k} n_k{}^\nu = N .$$

This allows us to imagine the display:

$$\underbrace{a_1, a_2, a_3,}_{n_1{}^\nu} \underbrace{\ldots\ldots\ldots,}_{n_2{}^\nu} \ldots a_N \qquad N! \text{ times.}$$

and so on

We notice that in all $N!$ displays the $n_1{}^\nu$ elements in container 1 occur in an ordered manner, and that they appear so arranged $n_1{}^\nu!$ times. Since in our count we wish to make no distinction between the order in which the $n_1{}^\nu$ elements have been placed in the first compartment, we must divide $N!$ by $n_1{}^\nu!$ to wipe out this distinction. Performing the same for the k $n_1{}^\nu$s, we are led to the formula

$$\Omega_\nu = \frac{N!}{n_1{}^\nu! \cdots n_k{}^\nu!} = \frac{N!}{\prod_{i=1}^{k} n_i{}^\nu!} . \tag{4.3}$$

This is a very important formula in statistical mechanics; it represents the number of ways in which N distinguishable elements can be placed in k quantum states, the order of placing the elements in one quantum cell being immaterial. However, the transposition of elements a_i and a_j between *different* quantum states is counted as leading to a distinct microstate. The same transposition of elements in a single quantum state is not counted as a distinct microstate.

It is clear that†

$$\sum_{\nu=1}^{\kappa} \frac{N!}{\prod_{i=1}^{k} n_i^{\nu}!} = k^N. \tag{4.4}$$

The determination of the number, κ, of possible distributions is of no importance in statistical mechanics.‡

4.1.3. Distributions of N Indistinguishable Elements over k Containers

In the preceding section we have counted the number of arrangements of N elements over k containers assuming that the elements in two different containers can be distinguished from each other even though the elements in one container could not; in other words, the order of placing elements in one container was irrelevant whereas that of placing them in different containers led to a different arrangement. We now wish to discuss two problems in which no distinction between elements is made under any circumstances: in the first problem we shall allow a container to house at most one element, whereas in the second, it will be stipulated that any number of elements may be placed in any single container. Anticipating the notation of the section in which the succeeding formulas will be called upon, we shall now use the symbol n for the number of elements, denoting the number of containers by g.

The first combinatorial problem consists in filling g containers with $n < g$ objects, one at a time. To facilitate the count, we assume, provisionally, that the elements have been numbered $a_1, a_2, \ldots$. The first element can be placed

† The numbers $\Omega_\nu = N!/\prod_{i=1}^{k} n_i^{\nu}!$ constitute the coefficients of the terms $\prod_{i=1}^{k} x_i^{n_i^{\nu}}$ in the expansion of

$$(x_1 + x_2 + \cdots + \cdots + x_k)^N.$$

That this must be so can be seen by writing

$$\underbrace{(x_1 + \cdots + x_k)(x_1 + \cdots + x_k)\cdots(x_1 + \cdots + x_k)}_{N \text{ times}}.$$

The sum consists of terms $\prod_{i=1}^{k} x_i^{n_i^{\nu}}$ for which $\sum_{i=1}^{k} n_i^{\nu} = N$, and the terms are formed by combining every term in the first set of parentheses with one term chosen from each of the remaining factors.

‡ This is

$$\kappa = \frac{(N + k - 1)!}{N!(k - 1)!},$$

which is the number of ways of placing N indistinguishable objects in k containers without restricting the number of objects per container in any manner. See also Section 4.1.3.

in g distinct ways, the second element in $(g-1)$ distinct ways, and so on until all elements have been exhausted, giving a product of n terms:

$$g(g-1)\cdots[g-(n-1)].$$

In order to erase the temporary identification, we must divide by $n!$ and note that†

$$\Omega = \frac{g(g-1)\cdots(g-n+1)}{n!} = \frac{g!}{(g-n)!n!}. \tag{4.5}$$

The second problem presents a little more difficulty owing to the fact that a particular box can be populated with an arbitrary number of elements. The count is greatly facilitated by reducing the problem to that of forming a linear $(g+n)$-tuple, as we shall explain with the aid of Figure 4.3. In the part-

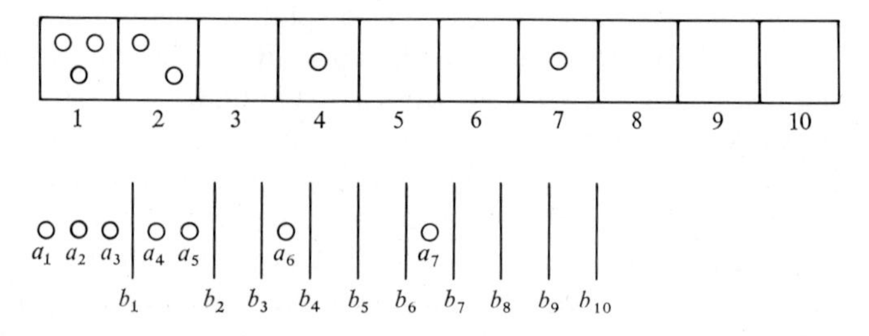

FIGURE 4.3. Arrangements of indistinguishable elements $n=7$, $g=10$.

icular example, we consider $n=7$ and $g=10$. In the upper diagram, the elements are distributed among the $g=10$ boxes with the occupation numbers 3, 2, 0, 1, 0, 0, 1, 0, 0, 0. In order to form a linear chain, we imagine that the first three elements a_1, a_2, a_3, have been placed alongside, as shown in the lower figure. The "divider" b_1 is now placed next to symbolize the end of the first box; two further elements, a_4, a_5 together with divider b_2, represent the second box. The third box is symbolized by a divider, b_3, only, because it is unoccupied. In this way the problem of counting the number of arrangements has been reduced to that of forming linear chains with $g+n$ elements, or of forming $(g+n)$-tuples.

As usual, we assume provisionally that the elements as well as the dividers have been numbered, noting, however, that the last place must always be

† The following equation can be written at once if it is realized that the problem consists in finding the number of ways of dividing g into two groups: a group n (containers filled with one element each) and a group $g-n$ (empty containers). For this reason, Equation (4.5) is identical with Equation (4.3) with the substitution of g for N, 2 for k, n for n_i^{v}. In fact, $n_1^{\mathrm{v}}=n$ and $n_2^{\mathrm{v}}=g-n$.

occupied by a divider, since a divider marks the *end* of a box. For this reason, the number of permissible permutations is *not* $(g+n)!$ We begin the count by placing a divider in the last place; this can be performed in g distinct ways. The remaining $g+n-1$ elements can be arranged in $(g+n-1)!$ ways. In order to erase the ordering of elements $a_1, a_2, \ldots$, we must divide by $n!$ and in order to erase the ordering of the dividers $b_1, b_2, \ldots$, we must divide by $g!$ Thus, we are led to the formula

$$\Omega = \frac{g(g+n-1)!}{g!n!} = \frac{(g+n-1)!}{n!(g-1)!}. \tag{4.6}$$

4.1.4. Very Large Numbers

The number of arrangements in a distribution varies from distribution to distribution. When the number of objects becomes very large, one particular distribution, containing Ω^* arrangements, dominates over the others. Moreover, as $N \to \infty$, the ratio $r = \ln \Omega^*/\ln \Omega \to 1$. A general proof of this statement requires the application of advanced mathematical methods which we prefer to avoid.† Instead, we shall be satisfied with a suggestive argument based on a set of numerical calculations, and will agree to accept this particular result generally.

Accordingly, we simplify the problem drastically and consider N distinguishable elements distributed among $k = 2$ containers in such a way that an interchange of two elements in the same container does not lead to a different distribution. For this simplified problem, a single distribution is specified by the two occupation numbers $n_1^{\nu} = n$ and $n_2^{\nu} = N - n$. The total number of arrangements with $k = 2$ is $\Omega = 2^N$. On the other hand, for a given distribution ν,

$$\Omega_\nu = \frac{N!}{n!(N-n)!} \tag{4.7}$$

as shown earlier in Equation (4.4). Evidently,

$$\Omega = \sum_{n=0}^{N} \frac{N!}{n!(N-n)!} = 2^N.$$

Owing to symmetry, the values of Ω_ν must be the same for $n_1^{\nu} = n$ and $n_1^{\nu} = N-n$. Consequently, the largest value of Ω_ν or of $\ln \Omega_\nu$ occurs for $n^* = \frac{1}{2}N$‡, so that

† This is proved by the application of the central limit theorem. See A. I. Khinchin, *Mathematical Foundations of Statistical Mechanics*, Dover, New York, 1949.

‡ This restricts N to be even, but the gist of the illustration remains unaffected.

$$\Omega^* = \frac{N!}{[(\frac{1}{2}N)!]^2}, \tag{4.7a}$$

and we wish to compute the ratio

$$\frac{\ln N!/[(\frac{1}{2}N)!]^2}{\ln 2^N},$$

when N increases.

The numerical values of Ω, Ω^*, Ω^*/Ω, and $r = \ln \Omega^*/\ln \Omega$ have been computed for $N = 2$, 10, 100, 1000, and 10,000, and the results have been displayed in Table 4.1. In addition, Table 4.2 shows the terms $\Omega_v = N!/n!\ (N-n)!$

TABLE 4.1[a]

The Values of Ω, Ω^*, Ω^*/Ω, and $r = \ln \Omega^*/\ln \Omega$ for $N = 2$, 10, 100, 1000, 10,000, when $\Omega_v = N!/n!(N-n)!$ and $\Omega^* = N!/(\frac{1}{2}N!)^2$

N	Ω^*	Ω	Ω^*/Ω	$r = \ln \Omega^*/\ln \Omega$
2	2	4	5.00×10^{-1}	0.500
10	2.520×10^{2}	1.024×10^{3}	2.46×10^{-1}	0.798
100	1.012×10^{29}	1.268×10^{30}	7.98×10^{-2}	0.964
1000	2.704×10^{299}	1.072×10^{301}	2.52×10^{-2}	0.995
10000	1.592×10^{3008}	1.995×10^{3010}	7.98×10^{-3}	0.999

[a] From J. D. Fast, *Entropy*, p. 54, McGraw-Hill, New York, 1962.

for $N = 10$ and 1000. The ratio of the number of arrangements for a particular distribution, Ω_v, to that of the most probable distribution, Ω^*, is seen plotted in Figure 4.4 against the ratio n/N for different values of N. When n and N are prescribed, the ratio n/N must be a rational fraction, and only discrete values of the independent variable occur. Thus only discrete values of Ω_v/Ω^* can be plotted, as shown for the case when $N = 2$, 6, 10, 20. As the number, N, of elements increases, the discrete values of Ω_v/Ω^* crowd together, and the successions of points can be replaced by continuous curves.

The tables and the diagram allow us to draw a number of qualitative conclusions. First, we notice that as N increases the number of possible arrangements increases exponentially, and that distributions which differ appreciably from the most frequent ones become exceedingly rare. In other words, the probability of finding large deviations from the most probable distribution becomes exceedingly low. For large values of N, the number of arrangements, Ω^*, which correspond to the most probable distribution, $n = n^* = \frac{1}{2}N$, decreases drastically compared with the total number of arrangements, Ω, as seen from the fourth column of Table 4.1. Thus, the probability

TABLE 4.2

Frequency Ω_ν of a Prescribed Distribution[a]

$N = 10$

n	0	1	2	3	4	5	6	7	8	9	10	
$N-n$	10	9	8	7	6	5	4	3	2	1	0	
Ω_ν	1	10	45	120	210	252	210	120	45	10	1	$(\Sigma = 1024)$
Ω_ν/Ω^*	0.0040	0.0397	0.1786	0.476	0.833	1.00	0.833	0.476	0.1786	0.0397	0.0040	

$N = 1000$

n	400	450	460	475	490	500
$N-n$	600	550	540	525	510	500
Ω_ν	6.25×10^{290}	1.82×10^{297}	1.1×10^{298}	7.7×10^{298}	2.21×10^{299}	2.70×10^{299}
Ω_ν/Ω^*	2.31×10^{-9}	6.72×10^{-3}	0.0408	0.287	0.819	1.000

[a] From E. Schmidt, *Thermodynamics* (J. Kestin, transl.), p. 147, Clarendon Press, 1949; reprinted by Dover, New York, 1966.

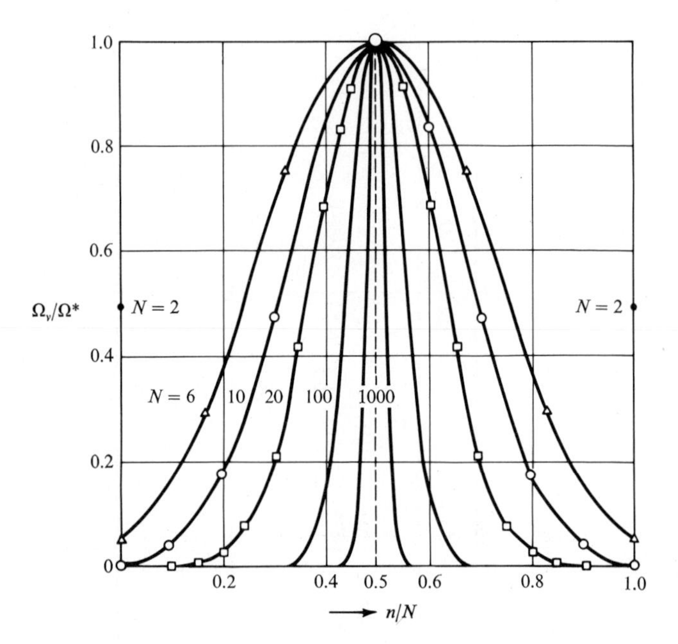

FIGURE 4.4. The ratio Ω_v/Ω^* as a function of n/N for the case when $k = 2$ with $N = 2, 6, 10, 20, 100, 1{,}000$; no restriction on total energy imposed. (From J. D. Fast, *Entropy*, p. 55, McGraw-Hill, New York, 1962).

of obtaining the most probable distribution *exactly* becomes smaller and smaller, but arrangements which differ *appreciably* from the most probable one become very rare, as seen from the diagram of Figure 4.4. This diagram also demonstrates that for N increasing, the number of arrangements which are confined in a narrow band $n^* \pm \Delta n$ with Δn prescribed, increases enormously as N increases. Similarly, the number of arrangements which lie *outside* this band decreases very quickly to zero. In the limit when $N \to \infty$, Ω_v converges to a delta function peaked at $n = n^*$.

The numerical data in the last column of Table 4.1 demonstrate that in spite of the decrease in the ratio Ω^*/Ω, the ratio of their logarithms, $r = \ln \Omega^*/\ln \Omega$, tends to unity very rapidly. In our context, $N = 10^4$ is still a very small number, and yet the error committed by the substitution of $\ln \Omega^*$ for $\ln \Omega$ is only one part in one thousand. When numbers of the order of $N = 10^{23}$ are encountered, the error is insignificant, and for $N \to \infty$, the ratio r converges to unity.

The argument given in this section allows us to infer (but does not prove) that for $n_i^v \to \infty$, the number Ω_v of arrangements tends to a Dirac delta function with respect to each variable, the peaks occurring at $n_i^v = n_i^*$.

4.2. Most Probable Distribution Subject to a Constraint

In statistical mechanics we do not encounter such simple distributions as the one discussed in the preceding section. The most important problem arises when the distribution is constrained to satisfy an additional condition. In turn, the one of greatest importance associates the value of a function, E, with each container and stipulates that each object placed in it acquires this value. As the symbol suggests, in applications the container will represent an energy level and an element placed in it simply acquires that energy. The constraint will demand that only such distributions should be counted as correspond to a prescribed value of *total* energy.

Accordingly, we number the containers $1, 2, \ldots, k$ and associate with each an energy $E_1, E_2, \ldots, E_k$, demanding that

$$\sum_{i=1}^{k} n_i^{\nu} E_i = E \quad (E = \text{const}).$$

A simple, contrived example may help the reader to grasp what is needed and so to recognize how the same end is achieved without displaying the distributions in detail. We shall assume $k = 3$ energy levels, and assign to them the values $E_1 = 1$, $E_2 = 3$, $E_3 = 5$ in some arbitrary units. In order to keep the example simple, we choose $N = 3$ elements a_1, a_2, a_3. All possible distributions of the $k^N = 3^3 = 27$ microstates have been displayed in Table 4.3. Since the total energy E is the same for all arrangements of a distribution, we have displayed in detail only the first four arrangements. The table shows clearly that the total energy differs from distribution to distribution, and that the most frequent distribution—leading to 6 arrangements—is the one for which $n_1{}^* = n_2{}^* = n_3{}^* = 3/3 = 1$.

Imposing a prescribed value of E, we find that the count changes. For example, if we were to stipulate that $E = 9$, the number of arrangements would be reduced from 27 to $6 + 1 = 7$, whereas for $E = 1$ there would be only $3 + 3 = 6$ admissible arrangements. Evidently, the most probable distribution becomes modified, but the details of this modification cannot be exhibited adequately with only three elements. It is, however, noteworthy that specifying low values of the total energy has the effect of rendering the low-energy containers more densely populated. The existence of a single arrangement at the highest energy $E = 15$ is fortuitous because in normal systems the sequence of energy levels is "open-ended;" it has been artificially reduced to $k = 3$ in this example in order to simplify the table.

The Method of Lagrange's Undetermined Multipliers

The example considered in the previous section is still too simple for future applications which will impose the two constraints

TABLE 4.3. Effect of Total Energy on Admissible Distributions.

Distribution number	$n_1^{\nu} = m$	$n_2^{\nu} = n$	$n_3^{\nu} = 3 - (m + n)$
1	0	0	3
2	0	1	2
3	0	2	1
4	0	3	0
5	1	0	2
6	1	1	1
7	1	2	0
8	2	0	1
9	2	1	0
10	3	0	0
$\kappa = 10$			

[a] The most frequent distribution is 1, 1, 1 if the total energy E is disregarded. For $E = 9$, the from 27 to 6. As E decreases, the low-energy quantum states become more densely populated.

$N = 3$, $k = 3$, $\Omega = 3^3 = 27$. Arbitrary Units of Energy[a]

Ω_ν	$E_1 = 1$	$E_2 = 3$	$E_3 = 5$	$E = \sum_{i=1}^{3} n_i{}^\nu E_i$
$\frac{3!}{0!0!3!} = 1$	0×1 empty	0×3 empty	3×5 $a_1 a_2 a_3$	15
$\frac{3!}{0!1!2!} = 3$	0×1 empty empty empty	1×3 a_1 a_2 a_3	2×5 $a_2 a_3$ $a_3 a_1$ $a_1 a_2$	13
$\frac{3!}{0!2!1!} = 3$	0×1	2×3	1×5	11
$\frac{3!}{0!3!0!} = 1$	0×1	3×3	0×5	9
$\frac{3!}{1!0!2!} = 3$	1×1	0×3	2×5	11
$\frac{3!}{1!1!1!} = 6$	1×1	1×3	1×5	9
$\frac{3!}{1!2!0!} = 3$	1×1	2×3	0×5	7
$\frac{3!}{2!0!1!} = 3$	2×1	0×3	1×5	7
$\frac{3!}{2!1!0!} = 3$	2×1	1×3	0×5	5
$\frac{3!}{3!0!0!} = 1$	3×1	0×3	0×5	3
Total $\Omega = 27$				

number of microstates is reduced from 27 to 7. For $E = 11$, the number of microstates is reduced

$$\sum_{i=1}^{k} n_i^{\nu} = N \qquad \text{and} \qquad \sum_{i=1}^{k} n_i^{\nu} E_i = E,$$

rather than one. These will express the fact that the total energy as well as the number of particles has been prescribed, and the problem will be to determine the particular distribution $\Omega_\nu = \Omega^*$ which is a maximum. This is a difficult problem if an attempt is made to solve it exactly, owing to the fact that the independent variables n_i^{ν} must be integers, and cannot, therefore, be varied continuously, as they must if we are to employ the standard methods of the calculus. To achieve our purpose, we will ignore the discrete nature of the variables, and rely on the fact that for our applications they will be enormously large. Taking into account the possible accuracy of measurement, we conclude that nonintegral values of the numbers n_i^* which render Ω a maximum are equally acceptable because the number of significant figures retained will fall far short of the last digit ahead of the decimal point anyway. This will permit us to regard Ω as a continuous function of k continuous variables, n_i, and the same applies to the constraints.

The problem, then, is to determine the maximum of a function of many variables subject to a number of constraints. This problem is solved elegantly by the application of *Lagrange's method of undetermined multipliers*, which can be found explained in any advanced textbook on analysis. For ease of reference, we shall give a brief account of this method.

We consider a simple case first. We suppose that we are given a function $z = F(x, y)$ of two variables x, y and that we wish to determine the maximum value of z for which x and y satisfy the additional relation $f(x, y) = 0$. Geometrically, z can be represented by a surface, such as surface S in Figure 4.5a, or by a set of contour lines, as in Figure 4.5b. The surface possesses a maximum at $A(x_A, y_A)$, where

$$\frac{\partial F}{\partial x} = 0 \qquad \text{and} \qquad \frac{\partial F}{\partial y} = 0.$$

In our problem, however, we do not seek to determine this maximum, but require that maximum value of z which exists at another point, point B, whose coordinates x_B, y_B satisfy the relation $f(x_B, y_B) = 0$. Geometrically, the constraint $f(x, y) = 0$ determines a cylinder whose trace is shown in Figure 4.5b. This cylinder intersects the surface S along curve C in Figure 4.5a, and the problem is to determine the coordinates x_B, y_B at point B on this curve.

The simplest method which comes to mind is to evaluate $y = y(x)$ from the implicit relation $f(x, y) = 0$, and to substitute it into $z = F(x, y)$. Thus

$$z = F[x, y(x)] \tag{4.8}$$

now represents the curve C, and constitutes a function of the single variable x. The condition for a maximum is

$$\frac{dz}{dx} = \frac{\partial F}{\partial x} + \frac{\partial F}{\partial y}\frac{dy}{dx} = 0,$$

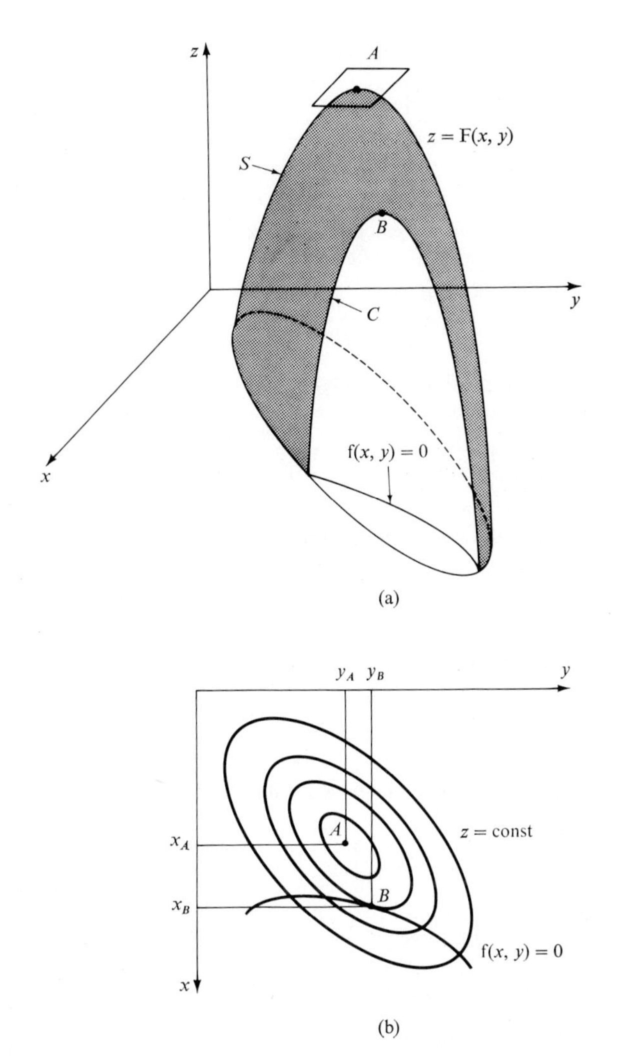

FIGURE 4.5. The surface $z = \mathrm{F}(x, y)$ and its contour lines together with the constraint $\mathrm{f}(x, y) = 0$.

from which x_B can be determined. The value of y_B follows from $y = y(x)$ or $\mathrm{f}(x, y) = 0$, and the value of z_B can be determined as $z_B = \mathrm{F}(x_B, y_B)$.

In the preceding argument, we established merely the necessary conditions for a maximum; they are equally valid for any extremum. In our statistical-mechanical problem, we are assured of the existence of a maximum from physical considerations, and an investigation of the extremum becomes superfluous.

The preceding simple method cannot be employed in complex cases because the evaluation of $y = y(x)$ and the substitution (4.8) turn out to be very cumbersome. Lagrange's method allows us to write out the answers in a symmetric, and correspondingly more lucid, form.

We begin by writing the total differential

$$dz = \frac{\partial F}{\partial x} dx + \frac{\partial F}{\partial y} dy = 0, \tag{4.9}$$

noting that it must vanish at an extremum even if dx and dy are not independent. The relation between dx, dy is given by the differential

$$\frac{\partial f}{\partial x} dx + \frac{\partial f}{\partial y} dy = 0 \tag{4.9a}$$

of $f(x, y) = 0$. We now multiply Equation (4.9a) by an undetermined factor λ, and add it to Equation (4.9). The result obviously must be zero. Hence

$$\left(\frac{\partial F}{\partial x} + \lambda \frac{\partial f}{\partial x}\right) dx + \left(\frac{\partial F}{\partial y} + \lambda \frac{\partial f}{\partial y}\right) dy = 0. \tag{4.10}$$

We can determine a value of λ which satisfies the condition that

$$\frac{\partial F}{\partial y} + \lambda \frac{\partial f}{\partial y} = 0. \tag{4.10a}$$

For this particular value of the *undetermined multiplier* λ, Equation (4.10) reduces to

$$\left(\frac{\partial F}{\partial x} + \lambda \frac{\partial f}{\partial x}\right) dx = 0.$$

Now dx is an independent increment, and we must have

$$\frac{\partial F}{\partial x} + \lambda \frac{\partial f}{\partial x} = 0, \tag{4.10b}$$

which is a symmetric form to Equation (4.10a). The system of three equations (4.10a), (4.10b), together with $f(x, y) = 0$ contains three unknown quantities: x_B, y_B, and λ, which are thus uniquely determined if a maximum is known to exist. It is noted that Equation (4.10) constitutes the total differential of the auxiliary function

$$G(x, y) = F(x, y) + \lambda f(x, y).$$

This remark permits us to write the essential features of the method for a function $F(x_1, \ldots, x_k)$ of k variables, subject to the imposition of l constraints $f_1(x_1, \ldots, x_k) = 0$, $f_2(x_1, \ldots, x_k) = 0$, and so on, as far as $f_l(x_1, \ldots, x_k) = 0$. We introduce the auxiliary function

$$G(x_1, \ldots, x_k) = F(x_1, \ldots, x_k) + \sum_{j=1}^{l} \lambda_j f_j(x_1, \ldots, x_k), \tag{4.11}$$

and form the perfect differential of G. The factors of the dx_is must vanish, and this yields k equations

$$\frac{\partial G}{\partial x_i} = 0 \qquad (i = 1, \ldots, k), \tag{4.12}$$

which together with the l equations of constraint determine the k values x_i at the extremum, and the l "undetermined" multipliers λ_j. More explicitly, the k equations are

$$\begin{gathered}\frac{\partial \mathrm{F}}{\partial x_1}+\sum_{j=1}^{l} \lambda_j \frac{\partial \mathrm{f}_j}{\partial x_1}=0 \\ \vdots \\ \frac{\partial \mathrm{F}}{\partial x_k}+\sum_{j=1}^{l} \lambda_j \frac{\partial \mathrm{f}_j}{\partial x_k}=0 .\end{gathered} \tag{4.13}$$

4.3. On Approximating a Series by an Integral

We are all familiar with the definition of a definite integral of an integrand $y = \mathrm{f}(x)$ between $x = a$ and $x = b$ as a limit of a series. From this it follows that an integral can be *approximated* by a finite series. In statistical thermodynamics we are often faced with the inverse problem of approximating a sum of discrete terms, y_i, which follow one upon another according to a prescribed rule, by a suitably chosen integral. Evidently, we can replace the discrete terms by the continuous variable $y = \mathrm{f}(x)$ and deduce the two equations

$$\sum_{i=0}^{n-1} y_n \approx \frac{1}{\Delta x} \int_a^b \mathrm{f}(x)\,\mathrm{d}x - \tfrac{1}{2} \sum_{i=0}^{n-1} (y_{i+1} - y_i) + \cdots \tag{4.14a}$$

and

$$\sum_{i=1}^{n} y_n \approx \frac{1}{\Delta x} \int_a^b \mathrm{f}(x)\,\mathrm{d}x + \tfrac{1}{2} \sum_{i=1}^{n} (y_i - y_{i-1}) + \cdots . \tag{4.14b}$$

The equations can be understood with reference to Figure 4.6 in which the

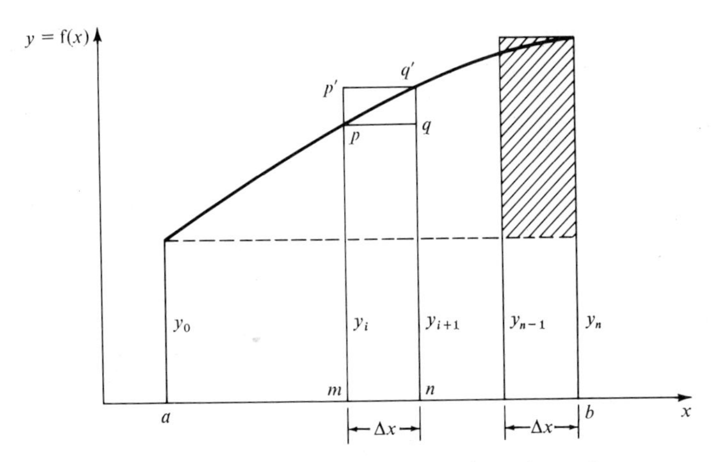

FIGURE 4.6. Approximating a series by an integral.

terms of the two series are represented by areas *mnpq* and *mnp′q′*, respectively. The correction terms are obtained by replacing the curve $y = f(x)$ by straight lines in each interval and by subtracting or adding the triangular areas *pqq′* and *pp′q′*. In either case, we obtain the rule that the approximation of the sum by the integral is satisfactory if *the difference between two consecutive terms is small compared with the terms* themselves. The approximation improves as the number of terms in the series increases and becomes exact when $n \to \infty$. The error is particularly small if the terms in the series are generated by a slowly varying function, and the sign of the error changes when the terms decrease as n increases. It is worth noting that the sum of all the triangular areas is equal to the shaded area, that is, to $\Delta x(y_n - y_0)$.

The approximation can be easily extended to multiple sums. The expression

$$\sum_{i=1}^{n} y_n \approx \frac{1}{\Delta x} \int_a^b f(x)\, dx \tag{4.15}$$

represents the first term in the *Euler–Maclaurin expansion*, or *Euler's summation formula* for a series in terms of integrals.†

4.3.1. The Stirling Approximation

We now apply the preceding approximation to the evaluation of the natural logarithm of the factorial $N!$ of a large number N,

$$\ln N! = \ln 1 + \ln 2 + \cdots + \ln i + \cdots + \ln N. \tag{4.16}$$

The associated function is, evidently, $f(x) = \ln x$, and we must consider the integral

$$\int_1^N \ln x\, dx = (x \ln x - x)\Big|_1^N = N \ln N - N - 1 \approx N \ln N - N, \tag{4.17}$$

which can be easily evaluated by parts: the unity can be neglected compared with N. Since the function $y = \ln x$ satisfies the criteria for a good approximation, we obtain

$$\ln N! = \sum_{i=1}^{N} \ln i \approx N \ln N - N \tag{4.18}$$

or

$$N! \approx \left(\frac{N}{e}\right)^N. \tag{4.18a}$$

The approximation becomes asymptotically exact as $N \to \infty$.

† Compare T. M. Apostol, *Calculus*, Vol. II, p. 449, Blaisdell, Waltham, Massachusetts, 1962.

A more detailed derivation leads to the following, approximate expression for a factorial†

$$N! \approx \sqrt{2\pi N}\left(\frac{N}{\mathrm{e}}\right)^N\left[1+\frac{1}{12N}+\frac{1}{288N^2}+\frac{139}{51{,}840N^3}+\cdots\right]. \tag{4.18b}$$

A comparison with the asymptotic formula (4.18a) reveals that the less exact approximation differs from the first term of the more accurate estimate by a factor $\sqrt{2\pi N}$. This leads to an additive term $\ln\sqrt{2\pi N}$ in $\ln N!$, which however, is utterly negligible in our case. Taking $N \approx 10^5$, we find that $\ln\sqrt{2\pi N} \approx 6.7$ which may be disregarded compared with N, and *a fortiori*, with $N \ln N$. Table 4.4 conveys an idea of the speed with which the two formulas converge to $\ln N!$.

TABLE 4.4

Stirling's Approximation[a]

N	$\ln N!$	$N \ln N - N$	$(N+\frac{1}{2})\ln N - N + \frac{1}{2}\ln 2\pi$
5	4.8	3.0	4.8
10	15.1	13.0	15.1
15	27.9	25.6	27.9
20	42.3	39.9	42.3
25	58.0	55.5	58.0
100	363.7	360.5	363.7

[a] From G. S. Rushbrooke, *Introduction to Statistical Mechanics*, p. 326, Oxford Univ. Press (Clarendon), London and New York, 1949.

In statistical thermodynamics we consistently employ the simplest version of Stirling's approximation, Equation (4.18). Owing to the passage to the limit $N \to \infty$ in the theory, this version leads to exact equations; the use of the more complex expression would not effect an improvement.

4.3.2. Evaluation of a Sum of Exponentials

A very important example where the approximation of a sum by an integral is employed in statistical mechanics involves the series

$$Z = \sum_{i=0}^{k} \mathrm{e}^{-\beta\varepsilon_i}, \tag{4.19}$$

† The derivation of shortened versions of this formula can be found in: R. Courant, *Differential and Integral Calculus*, Vol. 2, p. 361, Blackie, Glasgow and London, 1937; and T. M. Apostol, *Calculus*, Vol. II, p. 450, Blaisdell, Waltham, Massachusetts, 1962.

which occurs when the partition function $\mathscr{Z}$ of a system is computed (Sections 6.3, 6.4, and 6.5). In the theory of a perfect gas, we shall encounter the generating functions

$$\varepsilon_i = ci^2 \qquad \text{or} \qquad \varepsilon_i = ci(i+1)\,,$$

whereas in the study of oscillators

$$\varepsilon_i = c(i + \tfrac{1}{2})\,,$$

with c denoting a constant in each case. The series will include a large number of terms, k, and the fact that the terms decrease as i increases allows us to put $k \to \infty$. Referring to Equation (4.14a), we may use the approximation

$$Z \approx \frac{1}{\Delta x} \int_0^\infty \mathrm{e}^{-\beta\varepsilon(x)}\,\mathrm{d}x\,.$$

As far as the preceding forms of the generating function $\varepsilon(x)$ are concerned, it is found that the evaluation of the integral can proceed with much greater ease than that of the sum. For example, with $\varepsilon_i = cx^2$, we obtain†

$$\sum_{i=0}^{n} \mathrm{e}^{-\beta ci^2} \approx \int_0^\infty \mathrm{e}^{-\beta cx^2}\,\mathrm{d}x = \frac{1}{2}\sqrt{\frac{\pi}{c\beta}}\,. \tag{4.20}$$

In order to use the approximation, it is necessary to ensure that the difference

$$\mathrm{e}^{-\beta\varepsilon_i} - \mathrm{e}^{-\beta\varepsilon_{i+1}}$$

is small compared with the terms themselves, which occurs when

$$\varepsilon_{i+1} - \varepsilon_i \ll 1/\beta\,. \tag{4.20a}$$

In the limiting case when either

$$\varepsilon_{i+1} - \varepsilon_i \to 0 \qquad \text{or} \qquad \beta \to 0\,,$$

the approximation becomes exact.

4.4. The Statistical Method

The laws of statistical mechanics which we shall present in the succeeding chapters take the form of relations among the *average* properties of systems composed of large numbers of particles. The mathematical construction of the required average values and their mathematical properties will turn out to be

† See, for example, H. B. Dwight, *Tables of Integrals and Other Mathematical Data*, p. 200, Macmillan, New York, 1947; or Problem 4.31.

quite similar to those that occur in the theory of games of chance, such as in dice throwing, and so on. The theory of such games of chance is called *mathematical statistics.* In this section we shall discuss an ideal game of chance in order to extract a clear statement of the concepts of the statistical method.† This will enable us to show later that the logical structure of statistical thermodynamics is identical with that employed in mathematical statistics.

Suppose, then, that we discuss the throwing of two ideal, well-balanced dice. Just as in statistical thermodynamics, the attempt to predict the outcome of a single throw presents us with a deterministic problem in mechanics. Given sufficient information regarding the mass, density, and so on of a set of two dice, and given their position at time $t = 0$ together with adequate details about the action of setting them in motion and of stopping them, we could determine the precise outcome of each throw. Owing to the complexity of this deterministic problem, we have recourse to statistical methods and change our questions. Instead of asking for the precise outcome of each single throw, we study the probability that a particular outcome, o, will be observed during the next throw or the *probable frequency* of this outcome in a large number of throws. A single throw is called a *trial,* and a large number of trials forms a *population.* The second line in Table 4.5 represents the outcomes of 432 trials made with two dice; it lists the frequency, f, of each of the eleven possible outcomes:

$$\{o_1 = 2, o_2 = 3, \ldots, o_{11} = 12\} \tag{4.21}$$

in our example. The ratio

$$\pi(o_k) = \frac{f(o_k)}{\omega} \qquad \text{with} \quad \omega = \sum_k f(o_k) \tag{4.22}$$

of the frequency $f(o_k)$ of each outcome, o_k, to the total number of trials is known as its *experimental* or *a posteriori probability.* When the number of trials, ω, is increased ($\omega \to \infty$), we often find that the experimental probabilities, $\pi(o_k)$, tend to definite limits

$$\pi(o_k) = \lim_{\omega \to \infty} \frac{f(o_k)}{\omega}. \tag{4.23}$$

We retain the symbol $\pi(o_k)$ for this limit and agree to redefine the

† We do not intend to provide here even the rudiments of a course in mathematical statistics. There exist many books which do that. Without being comprehensive, we should like to mention three: H. D. Brunk, *An Introduction to Mathematical Statistics*, Ginn, Boston, Massachusetts, 1960; S. Goldberg, *Probability. An Introduction*, Prentice-Hall, Englewood Cliffs, New Jersey, 1960; E. Parzen, *Modern Probability Theory and its Applications*, Wiley, New York, 1960.

TABLE 4.5

Result of $\omega = 432$ Throws with Two Identical Dice Which Are Not "Loaded"[a]

($\Omega = 36$)

Random variable	$\phi =$	2	3	4	5	6	7	8	9	10	11	12
Experimental frequency	$f =$	11	16	38	53	69	75	57	45	27	29	12
Experimental probability	$\pi =$	0.026	0.037	0.088	0.123	0.160	0.174	0.132	0.104	0.063	0.067	0.028
Theoretical frequency in sample space	$F =$	1	2	3	4	5	6	5	4	3	2	1
Classical probability	$\Pi =$	$\frac{1}{36}$ 0.028	$\frac{2}{36}$ 0.056	$\frac{3}{36}$ 0.083	$\frac{4}{36}$ 0.111	$\frac{5}{36}$ 0.139	$\frac{6}{36}$ 0.167	$\frac{5}{36}$ 0.139	$\frac{4}{36}$ 0.111	$\frac{3}{36}$ 0.083	$\frac{2}{36}$ 0.056	$\frac{1}{36}$ 0.028

[a] After E. Schmidt, *Thermodynamics* (J. Kestin, transl.), p. 145, Oxford Univ. Press, London and New York, 1949; reprinted by Dover, New York, 1966.

experimental probability as that limit. Whether the limits displayed in Equation (4.23) do or do not exist can be ascertained only by experiments and by extrapolation. The limits themselves cannot be evaluated except by extrapolation on the basis of progressively larger populations of trials. If the limits exist, we say that we are faced with a *random process*.

Frequently we associate a number $\phi_k = \phi(o_k)$ with each possible outcome. In the present example (Table 4.5), we associate the

$$\text{number} \quad \phi_1 = 2 \qquad \text{with the outcome} \quad o_1 = 2\,,$$

and so on in an obvious (though arbitrary) way. The number ϕ_k is called a *random* or *stochastic variable*. Instead of defining the experimental probability on a set of outcomes o_k (that is, on a set of objects), we may define it with reference to the random variable

$$\pi(\phi_k) = \lim_{\omega \to \infty} \frac{f(\phi_k)}{\omega}, \qquad \omega = \sum_k f(\phi_k)\,. \tag{4.23a}$$

The random variable is used here to count the elements of the set of possible outcomes. In our example, the set $\{\phi_k\}$ contains a finite number of elements. Later we shall encounter denumerable and nondenumerable infinite sets in this context. We note that

$$\sum_k \pi(\phi_k) = 1 \tag{4.23b}$$

and say that the *probability function* $\pi(\phi_k)$ satisfies the normalization condition; the frequencies $f(\phi_k)$ do not satisfy a normalization condition.

The existence of the limits $\pi(o_k)$ or $\pi(\phi_k)$ is usually referred to as the *law of large numbers*. We emphasize that, conceived in this manner, the law of large numbers cannot be proved.† Nevertheless, we remark for the reader's future attention that the assumption of the existence of such definite limits in populations consisting of very large numbers of trials leads to definite laws and regularities in spite of an apparent chaos in the problem. Thus, for our problems, we accept it as a postulate.

In order to formulate a mathematical theory for a *random process*, for example, a game of chance or statistical thermodynamics, it is necessary to construct a suitable mathematical model for it. The set of all possible outcomes, o_j, in a problem is called *a sample space* for the model, and each possible outcome is said to constitute an *event*. In each concrete case there is considerable latitude in the choice of the sample space. In our present case we can choose the eleven elements enumerated in (4.21) as the primary elements of

† The mathematical (and provable) law of large numbers deals with the *probability* associated with the difference between the mean of a population and the mean of a random sample selected from the population.

our sample space. A possible alternative is shown in Figure 4.7, in which the sample space consists of 36 simple outcomes or *simple events* that constitute mutually exclusive possibilities. Here the sample space is so constructed that a distinction is made between die a and die b. In either case, outcomes included are only ones that are consistent with the "rules of the game." In this example, the rules are consistent with the laws of mechanics and the additional restrictions that the dice do not disintegrate during a throw, that they always land with one face up, and so on. An *event* (as distinct from a

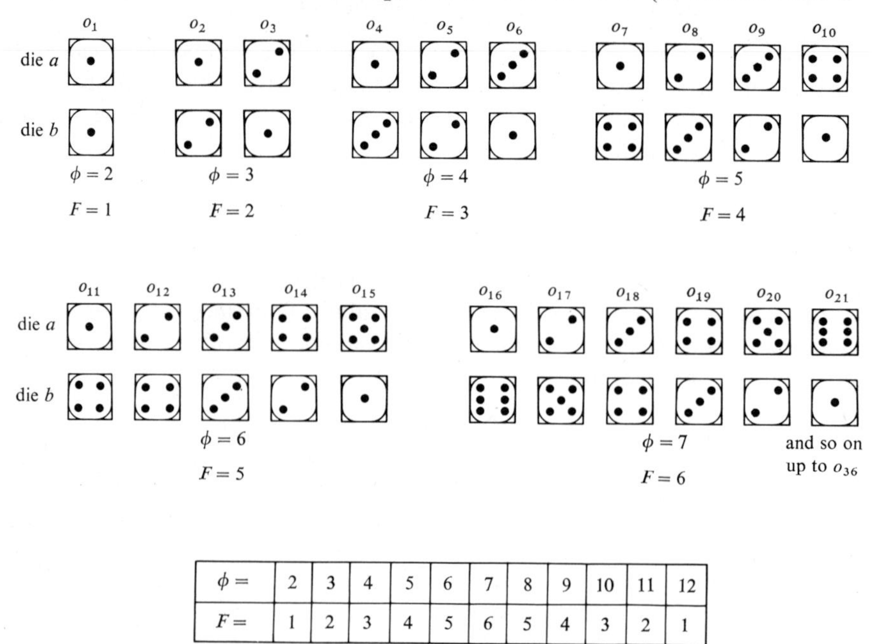

$\phi =$	2	3	4	5	6	7	8	9	10	11	12
$F =$	1	2	3	4	5	6	5	4	3	2	1

$\Omega = 36$

FIGURE 4.7. Sample space for the problem of throwing two dice.

simple event) is any subset of the set of simple events. Thus, throwing a four is an event because it corresponds to the subset.

$$\{ o_4, o_5, o_6 \}.$$

In our game, we wish to distinguish only between such events as those corresponding to different sums of points shown by both dice; it is, therefore, convenient to assign a different number, ϕ_k, to each different event of this class. We choose (arbitrarily) that sum as our value ϕ_k as shown in Figure 4.7. For example, we assign

$$\phi_4 = \phi(o_4) = \phi(o_5) = \phi(o_6) = 4\,.$$

This number ϕ is called a random or stochastic variable. A random variable is a real-valued function defined on the simple events (elements) of a sample space.

To complete the structure of our mathematical model, we must *assign* probabilities to simple as well as compound events. These cannot be deduced mathematically from the data given, and a suitable hypothesis must be made. The result of this mathematically arbitrary assumption† is known as the *a priori* or *classical probability* $\Pi(o_j)$ of the simple event. The probability of any event is *assigned* by our accepting the axiom that, if o_k denotes the union of two or more different simple events o_j, then the probability of o_k is the sum

$$\Pi(o_k) = \sum \Pi(o_j). \tag{4.24}$$

The assignment of probabilities to the simple events may be based on a plausible argument, an inspired guess, and so on, but we must realize that it cannot be based on a derivation. In fact, it represents an essential step in the idealization of a real random process. In the present example, we reason that the two dice are not "loaded" and that therefore the probability of one outcome o_j $(j = 1, \ldots, 36)$ is as great as that of any other. This induces us to assign the probabilities

$$\Pi(o_j) = \cdots = \Pi(o_j) = \cdots = \tfrac{1}{36} \tag{4.25}$$

to the simple events of our present example.

The probability of the compound events characterized by the values $\phi = 2$ to 12 of the random variables defined earlier must also be defined. Accepting the definition (4.24) and assigning equal probabilities to the simple events in accordance with equation (4.25), we conclude that

$$\Pi(\phi_k) = \frac{F(\phi_k)}{\Omega} \qquad \text{with} \quad \Omega = \sum_k F(\phi_k); \tag{4.26}$$

here $F(\phi_k)$ denotes the number of distinct ways a given value ϕ_k can be realized in terms of simple events. Equation (4.26) was first proposed by Laplace. The values F_k for all values of ϕ_k in our problem have been listed in Figure 4.7 and the *probability function* has been defined in Table 4.5 ; it is also sketched in Figure 4.8. The lower part of the latter figure shows that the probability function can also be represented as a set of "masses" on the number line.

We can always assume that the random variable ϕ_k ranges from $-\infty$ to $+\infty$ if we put

$$\Pi(\phi_k) = 0 \qquad \text{for} \quad \phi_k \le 1 \quad \text{and} \quad \phi_k \ge 13.$$

† This is subject to the constraint that each probability must be nonnegative and that their sum over the simple events must be equal to unity.

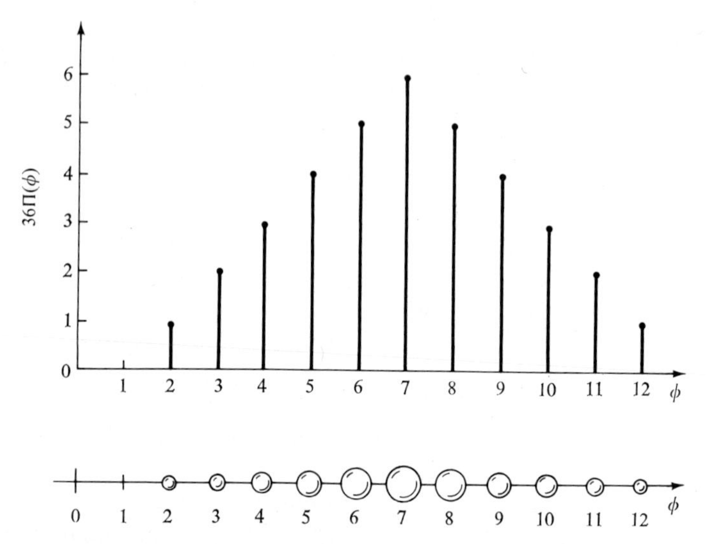

FIGURE 4.8. Geometrical representations of the probability function $\Pi(\phi)$ for the throwing of two indistinguishable dice.

We note that in this problem the *most probable* value of the random variable is $\phi_k = 7$. In statistical thermodynamics the value of the random variable corresponding to the highest value of Π plays an important part (Section 5.11). Frequently, the highest value of Π which occurs for $\phi = \phi^*$, say, persists over a narrow range, the remaining values being very small. This type of probability function† has been sketched in Figure 4.9, in which it is also

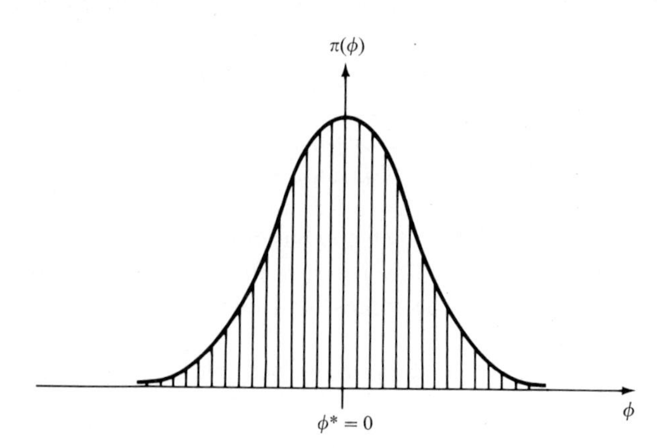

FIGURE 4.9. A typical probability function in statistical thermodynamics (Gaussian).

† It is called "Gaussian" or "normal."

assumed that the set of ϕ_k is represented by a large number of closely spaced points on the number line. For this reason, the discrete end-points have been joined by a continuous curve.

When the experimental probability, π, turns out to be identical with the classical, defined probability, Π,

$$\pi = \Pi \,, \tag{4.27}$$

we *say* that the sample space together with the probability function provide us with a "true" mathematical description of the physical phenomena under consideration, and we reason *by inference* that the assumptions and physical hypotheses adopted in the process of their construction are "correct." The question of whether this is the case (as it is, approximately, in Table 4.5) can be answered only by experiment.

When an experiment is performed to verify the equality in (4.27), it is immaterial whether we execute ω throws with a single set of dice or one throw with ω *identical replicas* of the original dice.

We note once more that the *a priori probability* is defined on the elements or subsets of the sample space, whereas the *a posteriori probability* is defined on the population of data. The equality of the two functions cannot be *proved*; it can only be *verified* by experiment.

4.4.1. Several Random Variables; Functions of Random Variables

In some applications it is convenient to classify events by means of more than one random variable. In our case, we may define a variable $\xi = 1$ through 6 and a variable $\eta = 1$ through 6, each denoting the number of dots on one die, ξ for die a and η for die b. The random process is then characterized by a *joint probability* function $\Pi(\xi, \eta)$, defined on the sample space of the simple events $o_1, \ldots, o_{36}$ from Figure 4.7. From the preceding discussion it is clear that we should define

$$\Pi(\xi, \eta) = \tfrac{1}{36} \qquad (= \text{const}) \,.$$

The diagram in Figure 4.10 represents this function in the form of a field in the plane ξ, η. If

$$\Pi(\xi, \eta) = \Pi_1(\xi)\Pi_2(\eta) \,, \tag{4.27a}$$

that is, if the joint probability function is equal to a product of two probability functions, each of one variable only, the variables ξ, η are called *independent* and the events ξ, η are said to be *disjoint* or *uncorrelated.* We also say that they are *statistically independent.* If this is not the case, the events are termed *statistically correlated.*

The probability for one of the variables, say ξ, to assume a given value

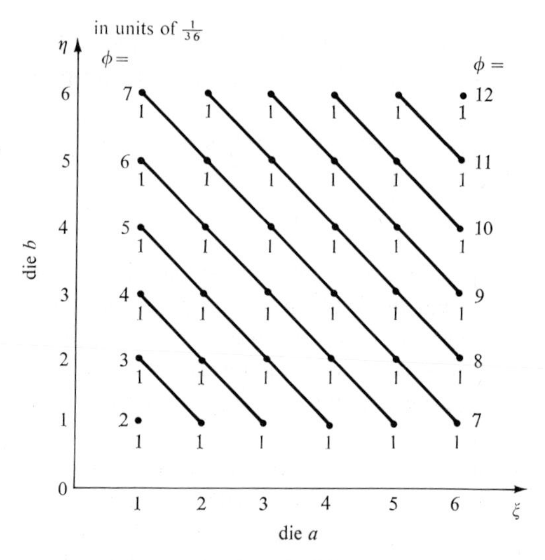

FIGURE 4.10. Joint probability function $\Pi(\xi, \eta)$ for the problem of casting two dice.

regardless of the value of η is called the *marginal probability*. Thus the event under consideration consists of the six simple events

$$(\xi, 1), (\xi, 2), \ldots, (\xi, 6),$$

and its probability is $\frac{1}{6}$ or the sum

$$\Pi_1(\xi) = \sum_\eta \Pi(\xi, \eta);$$

it is called the *marginal probability function* for the variable ξ. Reciprocally,

$$\Pi_2(\eta) = \sum_\xi \Pi(\xi, \eta)$$

is the marginal probability function for η. Evidently, in our case

$$\Pi_1(\xi) = \tfrac{1}{6}, \quad \Pi_2(\eta) = \tfrac{1}{6}, \qquad \text{and} \quad \Pi_1\Pi_2 = \Pi,$$

so that the throwing of one die is statistically independent of the throwing of the other die. This simply means that the outcome of the one has no influence on the outcome of the other.

It is worth noting that the oblique lines in Figure 4.10 join elementary events into events which can be described by the single stochastic variable $\phi = \xi + \eta$, employed earlier.

In general, we may find it useful to introduce n random variables; for example, in kinetic theory (Section 6.4), the number of random variables will be equal to a multiple of the number N of molecules in a volume of gas.

Given a random variable ϕ, defined on a sample space, we are often interested in a value $g(\phi)$ associated with it in accordance with a given rule. This defines a function of a random variable. We can also speak of functions of several random variables.

4.4.2. Formation of Averages

We can characterize a population in a gross way by indicating two measures: (1) its arithmetic mean or average together with (2) a number (variance or standard deviation) that describes the fluctuations of the random variable or of a function of it with respect to the mean. The mean, or *expectation value*, of a function $g(\phi_i)$ of a random variable in a population is

$$\langle g(\phi_i)\rangle = (1/\omega) \sum_i f_i \, g(\phi_i), \tag{4.28}$$

where ω denotes the number of terms in the population. A particular term $g(\phi_i)$ occurs in the sum with a relative frequency f_i/ω equal to the probability function $\pi(\phi_i)$. Owing to the fundamental assumption (4.27,) the probability function $\pi(\phi_i)$ can be deduced from the mathematical model or the adopted sample space. Grouping like terms, we find that the expectation value $\langle g(\phi)\rangle$ is equal to the *weighted average*

$$\langle g(\phi_i)\rangle = \sum_i g(\phi_i)\pi(\phi_i), \tag{4.29}$$

in which each individual value is multiplied by (weighted with) the probability with which it occurs. In the particular case where $g(\phi_i) = \phi_i$, we have

$$\langle \phi_i\rangle = \sum_i \phi_i \, \pi(\phi_i). \tag{4.29a}$$

Bearing in mind the analogy with a distribution of masses illustrated in Figure 4.8, we call the expectation value of the random variable ϕ_i the *first moment* of the probability function $\pi(\phi_i)$. The arithmetic mean of the random variable evaluated with respect to the population is thus seen to be equal to the weighted mean evaluated with respect to the sample space, and the same is true about any function $g(\phi_i)$ of the random variable.

When the probability function $\pi(\phi_i)$ is symmetric with respect to the most probable value of the random variable, ϕ^*, the mean becomes equal to it:

$$\langle \phi_i\rangle = \phi^*, \tag{4.30}$$

because then

$$\pi(\phi^* + \psi) = \pi(\phi^* - \psi) \qquad (= \pi' \text{ say}), \tag{4.30a}$$

where ψ has any suitable value. Indeed, for any pair of symmetric terms, we have

$$(\phi^* + \psi)\pi(\phi^* + \psi) + (\phi^* - \psi)\pi(\phi^* - \psi) = 2\phi^*\pi'$$

and

$$\langle\phi_i\rangle = \tfrac{1}{2}\left(2\phi^* \sum_i \pi'\right) = \phi^* \qquad \text{because} \quad \sum_i \pi' = 1\,.$$

The same is not true about a function $g(\phi_i)$, unless $\pi(\phi_i)$ is symmetric with respect to $g(\phi^*)$.

In statistical thermodynamics we shall encounter probability functions defined over a very large number of values, ϕ_i, which possess sharp peaks near the most probable value, ϕ^*, without necessarily being symmetric. A symmetric example of such a probability function was given in Figure 4.9. In such cases, Equation 4.30, or its equivalent

$$\langle g(\phi_i)\rangle = g(\phi^*) \tag{4.31}$$

for any function $g(\phi_i)$ of a random variable, represents a very good *approximation*. This is due to the fact that in the sum (4.29) we may replace $g(\phi_i)$ by $g(\phi^*)$ for those values of ϕ_i which cluster around ϕ^*, omitting all small terms. Thus the sum is approximated by

$$g(\phi^*) \sum_i \pi(\phi_i) = g(\phi^*)\,,$$

as noted in Equation (4.31).

4.4.3. Variance and Standard Deviation

The fluctuations of any function $g(\phi_i)$ of a random variable about its expectation value, $\langle g(\phi_i)\rangle$, are usually characterized by the expectation value, $\sigma_g{}^2$, of the squares of the differences

$$\sigma_g{}^2 = \langle [g(\phi_i) - \langle g(\phi_i)\rangle]^2\rangle; \tag{4.32}$$

this measure is known as the *variance* of the population. For reasons of dimensional consistency, the positive value of the root $(\sigma^2)^{1/2}$ is often preferred; it is called the *standard deviation* or the root-mean-square of the fluctuation, rms for short.

The variance and the rms can be evaluated in the sample space by the obvious equations

$$\sigma_g{}^2 = \left[\sum_i (g_i - \bar{g})^2 \pi(\phi_i)\right] \tag{4.32a}$$

and

$$\sigma_g = \sqrt{\left[\sum_i (g_i - \bar{g})^2 \pi(\phi_i)\right]} \tag{4.32b}$$

with the abbreviation

$$\bar{g} = \langle g(\phi_i) \rangle . \tag{4.32c}$$

In particular, for the variable itself

$$\sigma_\phi = \sqrt{\left[\sum_i (\phi_i - \bar{\phi})^2 \pi(\phi_i)\right]} . \tag{4.32d}$$

In analogy with a distribution of masses (Figure 4.8), the variance is said to constitute the *second central moment* of the probability function because it represents the second moment of the masses taken with respect to the *central* point $\bar{\phi} = \langle \phi_i \rangle$.

4.4.4. Continuous Probability Functions; Probability Density

We shall see later that the averages which possess physical meaning in statistical thermodynamics are taken over a very large number of closely spaced quantum states. In such cases, as shown in more detail in Section 4.3, it is possible to replace the sums from Equation (4.28) and (4.32d) with integrals (Euler–Maclaurin expansion of a sum). This occurs, in particular, in cases where quantum mechanics can be replaced by classical mechanics in accordance with Bohr's correspondence principle. This approximation is equivalent to replacing the discrete stochastic variable, ϕ_i, with a continuous variable, ϕ. We can always imagine that the latter ranges over the whole number line ($-\infty$ to ∞). Thus, if n values ϕ_i are contained in an interval a, b of ϕ_i, the sum

$$\pi_{ab} = \sum_a^b \pi(\phi_i)$$

is replaced by the integral

$$\pi_{ab} = \int_a^b \rho(\phi)\, d\phi \tag{4.33}$$

of a suitably chosen function $\rho(\phi)$. Reverting to the mass analogy, it is seen that the discrete masses contained in an interval are replaced by a continuous mass distribution. For this reason, the function $\rho(\phi)$ is called a *probability density function.* In this approximation, the probability of finding a *precise* prescribed value ϕ is zero, and the product $\rho(\phi)\, d\phi$ is interpreted as the probability of finding a value contained in the interval from ϕ to $\phi + d\phi$. We encountered such functions in Section 3.3 on wave functions.

The diagram in Figure 4.11 represents the so-called normal or Gaussian probability density function

$$\rho(\phi) = \frac{1}{(2\pi)^{1/2}\sigma_\phi} \exp\left[-\frac{(\phi - \bar{\phi})^2}{2\sigma_\phi^{\,2}}\right] . \tag{4.34}$$

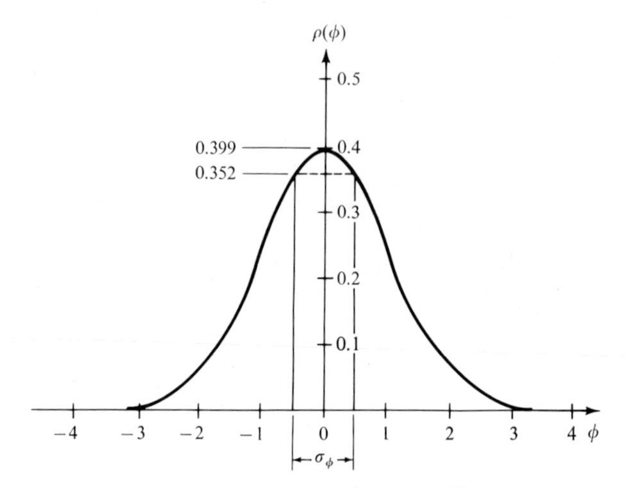

FIGURE 4.11. Normal probability density for $\bar{\phi}=0$ and $\sigma_\phi=1.0$.

In this equation, $\bar{\phi}$ and σ_ϕ denote the mean and the rms which are now computed in the form of integrals. By a simple transposition of Equations (4.28), (4.29), (4.32d), and (4.32e), we may write

$$\langle g(\phi)\rangle = \int_{-\infty}^{\infty} g(\phi)\rho(\phi)\,d\phi = \bar{g}, \tag{4.35a}$$

$$\langle \phi\rangle = \int_{-\infty}^{\infty} \phi\rho(\phi)\,d\phi = \bar{\phi}, \tag{4.35b}$$

$$\sigma_g{}^2 = \int_{-\infty}^{\infty} \{g(\phi) - \bar{g}\}^2\rho(\phi)\,d\phi, \tag{4.35c}$$

$$\sigma_\phi{}^2 = \int_{-\infty}^{\infty} (\phi - \bar{\phi})^2\rho(\phi)\,d\phi. \tag{4.35d}$$

In the case of several random variables, multiple integrals are obtained. The rms σ_ϕ for a Gaussian function is equal to the width of the graph for $\rho(\phi) = 0.352$ as indicated in the diagram.

As the variance becomes smaller and smaller, the probability density function becomes more and more peaked. In the limit, it tends to a Dirac delta function. In such cases, we have

$$\langle \phi\rangle \to \phi^*, \tag{4.36}$$

where ϕ^* is the most probable value. The diagram in Figure 4.12, which is based on the mass analogy, allows us to retain a visual picture of the fact expressed in Equation (4.36). If $\rho(\phi)$ is strictly a delta function, it follows that

$$\langle \phi \rangle = \int_{-\infty}^{\infty} \phi \delta(\phi - \phi^*) \, d\phi = \phi^*$$

exactly.

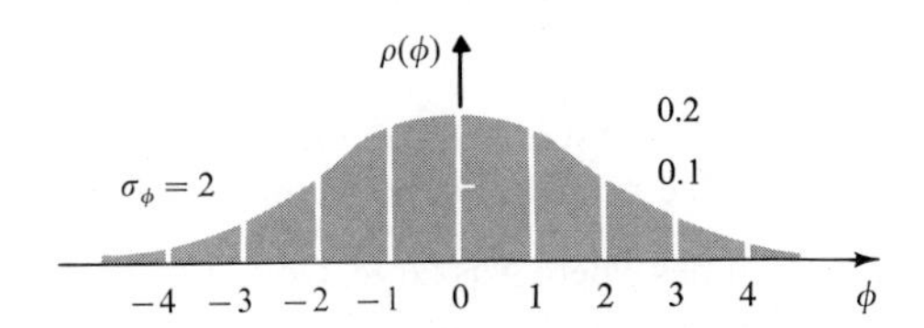

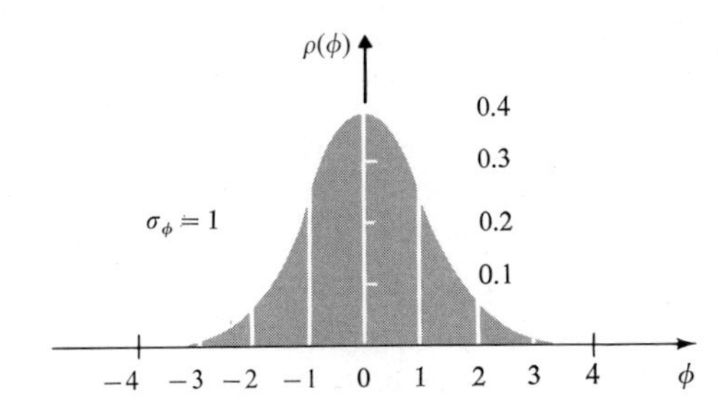

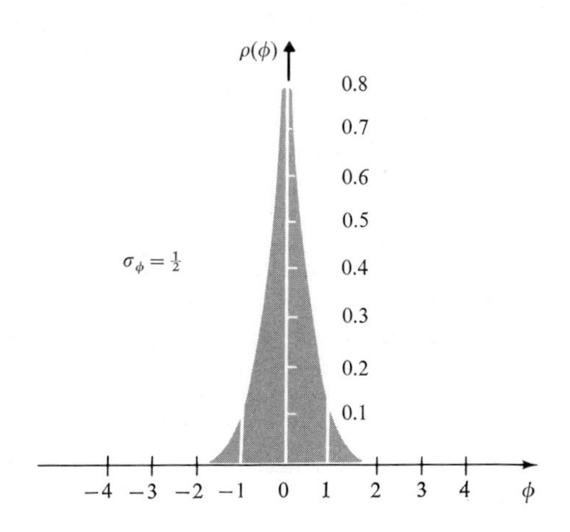

FIGURE 4.12. Effect of changing the rms σ_ϕ on the normal probability density.

PROBLEMS FOR CHAPTER 4†

4.1. A card is drawn from a well-shuffled deck. Compute the probability, π, that it is red *or* a 10.

SOLUTION: There are four 10s and 26 red cards in the deck. Two of these red cards are 10s, so that there are 28 ways a card can be either red or a 10 or both. Therefore, $\pi = 28/52$.

4.2. A coin is tossed six times in succession. Compute the most probable number, n^*, of heads, and determine the probability, π, of this most probable number.

4.3. A coin is tossed N times where N is large. Compute the most probable number, n^*, of heads, and determine the probability, π, of this most probable number.

4.4. (a) A single die is thrown N times. Compute the average value of the number on the top face of the die.

(b) Suppose the die is loaded so that the probability of 4 appearing is $\pi(4) = \frac{1}{2}$, whereas each of the other faces has a probability $\pi(\cdots) = \frac{1}{10}$. Compute then the average value on the top face after N trials.

4.5. A bridge hand consists of thirteen cards taken at random from a group of 52 different cards. Compute the number of different bridge hands. You may use Stirling's formula to estimate this number.

4.6. A box contains n balls of which r are white and the rest are black. Two balls are drawn at random (i) with replacement and (ii) without replacement. Calculate in each case the probability, π_1, that both balls will be black and the probability, π_2, that one will be white and one black.

SOLUTION: If each ball is replaced after being drawn the probabilities of successive drawings are independent, otherwise they are not independent. Therefore, for case (i), the probability of drawing two black balls is $\pi_1 = (n-r)^2/n^2$ and the probability of drawing one white and one black is $\pi_2 = 2r(n-r)/n^2$. The factor 2 accounts for the fact that there are two ways of drawing a white and a black ball. In case (ii), the probability of drawing two black balls is $\pi_1 = (n-r)(n-r-1)/n(n-1)$. The probability of drawing one white and one black ball is $\pi_2 = 2r(n-r)/n(n-1)$.

4.7. (a) Show that the sum 7 is the most probable outcome in a single throw of two dice and compute the probability for its occurrence.

(b) Determine the most probable sum for the single throw of three dice and compute the probability for its occurrence.

4.8. The results of 320 simultaneous tosses of four coins were found to be

Event	No heads	1 head	2 heads	3 heads	4 heads
Number of times observed	18	84	117	79	22

† Problems numbered in boldface type are considered relatively easy; they are recommended for assigning in an elementary first course.

(a) Compute the relative frequency of each event. Compare this with the expected probability of each event and decide whether a bias existed during the tosses. (b) What is the probability that two coins will show heads and two will show tails on the three-hundred-and-twenty-first toss? How should this statement be modified if, after the toss is made, it is known that at least one of the coins shows a head?

4.9. Three dice are tossed. (a) What is the probability that one die shows 3 and another 4? (b) If the total shown on all three dice is 10, what is the probability that one die shows 3 and another 4? (c) If no information is given, what is the probability that the total shown will be 10? (d) If it is known that one die shows 3 and another 4, what is the probability that the total is 10?

4.10. A newspaper reports that the "Odds are two to one against Congress passing a bill, and, if it does pass, the odds are even that the President will veto it." Compute the probability that the bill will become law.

4.11. Compute the probability that two people in a room of 25 people will have the same birthday, irrespective of the year of their birth. You may assume that all years have 365 days.

SOLUTION: Pick one man at random. The probability that another man picked at random does not have the same birthday is $1 - 1/365$. The probability that a third man does not have the same birthday as one of the other two men is $1 - 2/365$. Consequently, the probability that the three men have different birthdays is $(1 - 1/365)(1 - 2/365)$, etc. To compute the final answer you may use the approximation that $\ln(1 - x) \approx -x$ for small x.

4.12. Compute the number of signals that can be made when using four flags in a row if there are available three red, one white, and two black flags.

ANSWER: 38.

4.13. Four objects are distributed at random among six containers. Compute the probability that (a) all objects are in the same container, (b) no two objects are in the same container.

4.14. Show that the maximum value of the sum $-\sum x_i \ln x_i$ subject to the constraint $\sum x_i = 1$ occurs when all the x_i are equal.

4.15. Using Lagrange's multipliers, show that the square has the largest area among all rectangles with a prescribed perimeter.

4.16. In the manufacture of a certain object, the probability of a defect appearing in a certain time δt is equal to $\lambda\, \delta t$ for small δt. Prove that the probability that the object is without defect is equal to $\exp(-\lambda T)$, where T represents the time to manufacture the object.

4.17. Suppose that a quantity $y = \ln x$ is distributed according to the Gaussian or normal distribution function with mean μ and standard deviation σ. Show that the average value of x is given by $\langle x \rangle = \exp(\mu + \sigma^2/2)$.

4.18. Determine the extremum for the curve of intersection of the paraboloid

$$z = ax^2 + 2bxy + cy^2$$

and the cylinder

$$xy = a^2 \qquad (x > 0, \quad y > 0).$$

State the conditions which must be satisfied by the constants a, b, and c to render the surface an elliptic paraboloid and the extremum a minimum.

4.19. Use the method of Lagrange's undetermined multipliers to show that the maximum and minimum distances l_1 and l_2 from the origin to the curve which is the intersection of the ellipsoid

$$\left(\frac{x}{a}\right)^2 + \left(\frac{y}{b}\right)^2 + \left(\frac{z}{c}\right)^2 = 1$$

with the plane

$$Ax + By + Cz = 0$$

are given by the two roots of the quadratic equation

$$\frac{A^2a^2}{a^2 - l} + \frac{B^2b^2}{b^2 - l} + \frac{C^2c^2}{c^2 - l} = 0.$$

4.20. Suppose that we desire to find an extremum of the function $f(x_1, \ldots, x_n)$ whose n variables are subject to the α conditions of constraint

$$\phi_1(x_1, \ldots, x_n) = 0$$

$$\vdots$$

$$\phi_\alpha(x_1, \ldots, x_n) = 0 \qquad (\alpha < n).$$

Demonstrate that the method of Lagrange's undetermined multipliers will fail to give us a solution to the problem if the Jacobian

$$J(x_1, \ldots, x_\alpha) = \frac{\partial(\phi_1, \ldots, \phi_\alpha)}{\partial(x_1, \ldots, x_\alpha)} = 0.$$

When this occurs, does it mean that $f(x_1, \ldots, x_n)$ does not possess an extremum under the given conditions of constraint? [*Reference*: I. S. Sokolnikoff and R. M. Redheffer, *Mathematics of Physics and Modern Engineering*, p. 255, McGraw-Hill, New York, 1958.]

4.21. Starting with the probability function $\pi(\phi_i)$ for the random variable ϕ_i, show that the variance $\sigma_\phi{}^2$ of the distribution is

$$\sigma_\phi{}^2 = \langle\phi^2\rangle - \langle\phi\rangle^2,$$

where $\langle\phi\rangle = \sum_i \phi_i \pi(\phi_i)$. Is this a general result or is it valid only when $\pi(\phi_i)$ is symmetric?

4.22. Show that the maximum value of the integral

$$J = -\int_{-\infty}^{\infty} f(x) \ln f(x)\, dx$$

subject to the constraints

$$\int_{-\infty}^{\infty} \mathrm{f}(x)\,\mathrm{d}x = C_1\,,$$

$$\int_{-\infty}^{\infty} x\,\mathrm{f}(x)\,\mathrm{d}x = C_2\,,$$

$$\int_{-\infty}^{\infty} x^2\,\mathrm{f}(x)\,\mathrm{d}x = C_3\,,$$

where the C_i are constants, occurs when $\mathrm{f}(x)$ has the form

$$\mathrm{f}(x) = A\exp(-Bx^2 + Cx)\,.$$

Here A, B, and C are determined by the C_i's.

4.23. Revert to the reasoning in Section 4.1.4 and *prove* that $\lim \ln \Omega/\ln \Omega^* = 1$ as $N \to \infty$ when the number of compartments is $k = 2$.

SOLUTION: Note that $\Omega = \sum_{n=0}^{\infty} \Omega_v(n)$. From the definition of Ω^* it follows that $\Omega^* \leq \Omega \leq (N+1)\Omega^*$ since $\Omega_v \leq \Omega^*$. Consequently, $\ln \Omega^* \leq \ln \Omega \leq \ln(N+1) + \ln \Omega^*$ or $1 \leq \ln \Omega/\ln \Omega^* \leq 1 + \ln(N+1)/\ln \Omega^*$. Since for large values of N we have $\ln \Omega^* \sim N$, it follows that $\lim \ln \Omega/\ln \Omega^* = 1$ as $N \to \infty$.

4.24. Estimate the sums (a) $\sum_{k=1}^{100} k^{-1/2}$, (b) $\sum_{k=1}^{100} k^{3/2}$.

4.25. The quantity

$$S = \sum_{m=1}^{M} \mathrm{e}^{-m\beta\varepsilon_0}$$

is the sum of a geometric series which can be evaluated explicitly by multiplying each term in the series by the first term to obtain a new series

$$\mathrm{e}^{-\beta\varepsilon_0} S = \sum_{m=1}^{M} \mathrm{e}^{-(m+1)\beta\varepsilon_0}$$

and then subtracting the new series from the original one term by term. (a) Show that

$$S = \mathrm{e}^{-\beta\varepsilon_0}\frac{(1 - \mathrm{e}^{-M\beta\varepsilon_0})}{(1 - \mathrm{e}^{-\beta\varepsilon_0})}\,.$$

(b) Compare $\lim_{M\to\infty} S = S_\infty$ with the integral

$$I = \int_1^{\infty} \mathrm{e}^{-\beta\varepsilon_0 x}\,\mathrm{d}x\,,$$

and discuss the conditions under which $I \approx S_\infty$.

4.26. Evaluate the sum

$$S_1 = \sum_{m=1}^{M} m\,\mathrm{e}^{-\beta\varepsilon_0 m}\,,$$

using the techniques described in Problem 4.25. Compare $\lim_{M\to\infty} S_1 = S_{1\infty}$ with the integral

$$I_1 = \int_1^\infty x\, e^{-\beta\varepsilon_0 x}\, dx,$$

and determine the conditions for which $I_1 \approx S_{1\infty}$.

4.27. Given the probability density (distribution) function

$$\rho(\phi) = A\, e^{-a(\phi - b)^2}$$

for the continuous variable ϕ, determine the constants A, a, and b in terms of the mean $\bar{\phi}$ and variance $\sigma_\phi{}^2$ of the distribution by taking the zeroth, first, and second moments of the distribution with respect to $(\phi - b)$. Having determined the constants, find a relation for the fourth moment of the distribution.

4.28. (a) Prepare a table, analogous to Table 4.3, with $N = 3$ distinguishable particles and $k = 4$ energy levels: 2, 4, 6, 8 (arbitrary units), when there is no restriction on the total energy of the system. Hence, determine the total number of micro-states available to such an imaginary system. (b) Redetermine the number of micro-states when the total energy of the system is 12 (arbitrary units), and find the most probable distribution for this case. (c) How will the count found in part (b) change if the particles are *indistinguishable* and the number of particles in a given energy level is: (i) unlimited (exclusion principle inoperative), (ii) limited to one (exclusion principle operative)?

4.29. The probability density that a particle with mass m has a speed v may be described by

$$g(v) = \begin{cases} \alpha v^2\, e^{-\beta v^2} & \text{for} \quad v \geq 0 \\ 0 & \text{otherwise} \end{cases}$$

with

$$\int_0^\infty g(v)\, dv = 1\,.$$

Compute the mean and rms value of v and v^2.

4.30. Show that the volume of an N-dimensional sphere of radius R, defined by the equation $\sum_{i=1}^N x_i{}^2 = R^2$, has a volume $V_N(R)$ of the form $V_N(R) = R^N V_N(1)$ where $V_N(1)$ is the volume of an N-dimensional sphere of unit radius. Similarly, show that the surface area of such a sphere $S_N(R)$, satisfies the relation $S_N(R) = R^{N-1} S_N(1)$.

4.31. Show that $\int_{-\infty}^\infty e^{-t^2}\, dx = \sqrt{\pi}$.

SOLUTION: Consider the square of the above integral

$$\int_{-\infty}^\infty dx \int_{-\infty}^\infty dy\, e^{-(x^2+y^2)}\,.$$

The region of integration extends over the entire x, y plane. Therefore, we may change to polar coordinates defined by $x = r\cos\theta$, $y = r\sin\theta$ and write

$$\int_{-\infty}^{\infty} dx \int_{-\infty}^{\infty} dy\, e^{-(x^2+y^2)} = \int_0^{\infty} dr \int_0^{2\pi} d\theta\, r\, e^{-r^2} = 2\pi \int_0^{\infty} r\, e^{-r^2}\, dr.$$

The integral $\int_0^\infty r\, e^{-r^2}\, dr$ may be evaluated by the substitution $z = r^2$. Finally,

$$\int_{-\infty}^{\infty} e^{-x^2}\, dx = \left[2\pi \tfrac{1}{2} \int_0^{\infty} e^{-z}\, dz\right]^{1/2} = \sqrt{\pi}.$$

4.32. The Γ-function is defined by the integral

$$\Gamma(N) = \int_0^{\infty} e^{-t} t^{N-1}\, dt.$$

Show that unless $N - 1 \leq 0$, then $\Gamma(N) = (N-1)\Gamma(N-1)$; hence show that if N is a positive integer, $\Gamma(N) = (N-1)!$.

4.33. Stirling's approximation for the factorial $N!$ of a large number N can also be derived from an asymptotic approximation of the integral which defines the gamma function (Γ). By definition,

$$N! = \Gamma(N+1) = \int_0^{\infty} e^{-x} x^N\, dx = \int_0^{\infty} e^{N \ln x - x}\, dx.$$

Letting $g(x) = \ln x - x/N$, we can see that the expression for $N!$ transforms to

$$N! = \int_0^{\infty} e^{Ng(x)}\, dx.$$

If $g(x)$ possesses a maximum at some point, say $x = x_0$, in the range $[0, \infty]$, then for N large the integrand $e^{Ng(x)}$ will exhibit a very peaked behavior, similar to that of a delta function, in the neighborhood of $x = x_0$. When this is the case, the contribution of the integrand to the integral at points outside the neighborhood of x_0 can be neglected and the integration becomes straightforward.

(a) Show that the function $g(x)$ defined above does possess a maximum at $x_0 = N$.

(b) Obtain a Taylor series approximation for $g(x)$ in the neighborhood of $x = x_0$, neglecting all terms in $(x - x_0)$ above the second order. (Is this valid?) Consequently, show that

$$e^{Ng(x)} \simeq e^{Ng(N)}\, e^{-(x-N)^2/2N}.$$

(c) By a suitable transformation of variables, paying careful attention to how the limits of integration transform, evaluate the resulting integral for $N!$; hence, prove that the asymptotic approximation yields the result

$$N! \simeq N^N e^{-N} \sqrt{2\pi N}.$$

4.34. Show that Stirling's formula also holds for the Γ function, that is, that $\ln \Gamma(N) \approx N \ln N - N$ for large N, regardless of whether N is an integer.

4.35. Show that $\Gamma(\frac{1}{2}) = \sqrt{\pi}$ and compute $\Gamma(\frac{7}{2})$.

4.36. It may be shown that the volume of an N-dimensional unit sphere $V_N(1)$ is given by

$$V_N(1) = \pi^{N/2}/\Gamma\left(\frac{N}{2} + 1\right).$$

Using this result, compute the volume of 1, 2, 3, 4, and 5-dimensional spheres of radius R. Obtain an expression for $V_N(R)$ for large N, using the property of the Γ function that $\ln \Gamma(N) \approx N \ln N - N$, for large N.

LIST OF SYMBOLS FOR CHAPTER 4

Latin letters

C	Constant
E	Energy
F	Given function
f	Function of constraint
f	Experimental frequency of an outcome
G	Function defined in Equation (4.11)
g	Number of containers or compartments
k	Number of compartments in a count; number of variables
l	Number of equations of constraint
N	Number of elements to be counted
n	Number of elements in a compartment
t	Time
x	Independent variable
y	Independent variable, dependent variable
Z	Series defined in Equation (4.19)
z	Dependent variable

Greek letters

α	Lagrangian multiplier
β	Constant; Lagrangian multiplier
γ	Lagrangian multiplier
δ	Dirac delta function
ε	Term in series
η	Random variable
κ	Number of distributions
λ	Lagrangian multiplier
ξ	Random variable
o_k	Outcome of an event
$\Pi(\xi, \eta)$	Joint probability
$\Pi_1(\xi)$	Marginal probability for ξ

$\Pi_2(\eta)$	Marginal probability for η
$\Pi(o_j)$	Classical probability or *a priori* probability
$\Pi(\phi_k)$	Probability function
$\pi(o_k)$	Experimental probability or *a posteriori* probabiity
π	Probability in sample space
π'	Probability defined in Equation (4.30a)
$\rho(\phi)$	Probability density
σ	Standard deviation or root mean square of fluctuation
σ^2	Variance
ϕ	Generalized average quantity; continuous random or stochastic variable
ϕ^*	Value of ϕ for which $\pi(\phi)$ is a maximum
ϕ_k	Discrete random or stochastic variable
ψ	Random variable
Ω	Number of permutations; number of arrangements
Ω_ν	Number of elements in a distribution

Superscripts

ν	Refers to distribution
*	Refers to most probable distribution
$\bar{}$	Refers to mean or expectation value

Subscripts

i	Running index
j	Running index
k	Running index
ν	Refers to distribution

Special symbols

!	Factorial product
$\langle\ \rangle$	Average or expectation value

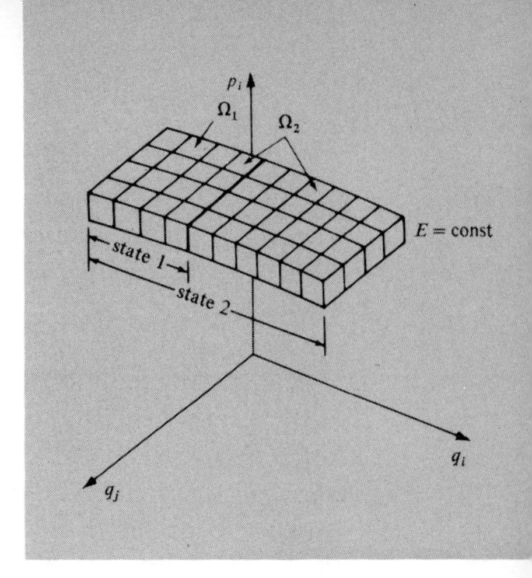

CHAPTER 5

FOUNDATIONS OF STATISTICAL THERMODYNAMICS

This chapter constitutes the most essential part of this course; it introduces the reader to the foundations of the subject and prepares him for the subsequent study of applications. Students who are encountering statistical thermodynamics for the first time should omit Sections 5.4, 5.9, and 5.18. These have been identified later by footnotes.

5.1. Introductory Remarks

The last step which must now be taken to complete the structure of statistical thermodynamics consists in the creation of a basis for relating the macroscopic properties of thermodynamic systems to the microscopic properties of their mechanical models. Classical and quantum mechanics provide methods for the detailed study of individual mechanical systems and are concerned with their individual microstates. Such microstates are described in purely mechanical terms, and such concepts as pressure or entropy do not enter into the description. An extension of the mechanical theory to include thermodynamic quantities can, evidently, be achieved only through the adoption of a set of additional hypotheses, or rules of averaging, and by subsequent appeal to experiment. It is this step which introduces statistics into the subject.

A rigorous exposition of statistical thermodynamics requires the application of methods of analysis whose mastery would demand a large preliminary effort.† In our presentation, this must be replaced by a physical understanding of the problem and by the reader's willingness to accept certain statements on trust or, at best, on the basis of plausibility arguments.

The reader who undertakes a more profound study of statistical thermodynamics will soon notice that the authors of the numerous textbooks differ widely in the way they present the *introduction* to the subject. This will tend to confuse him unless he realizes that the basic mathematical structure is the same in all and that differences consist only in the physical justification advanced for the detailed steps in the derivations. Depending on the path chosen, the justification may appear more or less convincing, but we should observe that the fundamental equation, as given in Section 5.19, for example, is either the same or equivalent in all presentations. Consequently, the working equations are in complete agreement among the various approaches and, since the results of the theory are compared with experiment at this level, the choice between the different methods becomes largely a matter of taste and feeling for rigor.

The method we shall adopt will consist in the creation of a new epistemological basis for thermodynamics in the form of the *postulate of equal a priori probability of quantum states in phase space* (Section 5.5)‡ and in demonstrating that its consequences are consistent with those of classical thermodynamics (Section 5.13). The latter are known to constitute uncontradicted generalizations from experience, and the same can be said to be true about our present theory by inference. A possible alternative method would be to derive the principle of equal *a priori* probability from the first part of the Second law of thermodynamics, whose validity would thus be assumed at the outset. However, the new theory is capable of dealing with a wider range of phenomena

† C. R. Tolman, *The Principles of Statistical Mechanics*, Oxford Univ. Press, New York and London, 1939. The relation between statistical thermodynamics and the whole of contemporary physics has been perceptively analyzed by M. Born in his Waynflete Lectures, published as *Natural Philosophy of Cause and Chance*, Oxford Univ. Press (Clarendon), London and New York, 1949.

The literature on the subject is very large, and we mention only J. E. Mayer and M. G. Mayer, *Statistical Mechanics*, Wiley, New York, 1940; J. A. Fay, *Molecular Thermodynamics*, Addison-Wesley, Reading, Massachusetts, 1965; F. Reif, *Fundamentals of Statistical and Thermal Physics*, containing an extensive bibliography, McGraw-Hill, New York, 1965; G. H. Wannier, *Statistical Physics*, Wiley, New York, 1966; E. L. Knuth, *Statistical Thermodynamics*, McGraw-Hill, New York, 1966; and E. A. Desloge, *Statistical Physics*, Holt, New York, 1966; *see also* R. Kubo, *Statistical Mechanics*, North-Holland Publ., Amsterdam, 1967; and R. Jancel, *Les Fondements de la Mécanique Statistique Classique et Quantique*, Gauthier-Villars, Paris, 1963.

A very concise but rigorous and clear exposition is given in E. Schrödinger, *Statistical Thermodynamics*, Cambridge Univ. Press, London and New York, 1952.

‡ The probability which occurs in this chapter must be carefully distinguished from that represented by the square of the wave amplitude in quantum mechanics.

such as are beyond the grasp of classical thermodynamics. First, it provides means for the computation of equations of state; secondly, it enables us to understand fluctuations; and thirdly, it contains within it the beginning of a theory of nonequilibrium states. For these reasons, the modern tendency is to accept the fact that statistical thermodynamics furnishes us with the more fundamental generalization from experience.

5.2. The Statistical Method

The need to resort to the statistical method arises from complexity and indeterminacy. Complexity results from the very large number of degrees of freedom possessed by even the smallest thermodynamic system we may wish to study in practice. The number of elementary particles is of the order of the Avogadro number (10^{23}), and the number of degrees of freedom of a particle ranges from 3 to about 12, giving a total of 10^{24}–10^{25} variables. Indeterminacy results from the fact that no single microstate of any system can be known from measurement even approximately. If this were not so, the detailed behavior of a system could be determined, at least in principle, with the use of Schrödinger's equation or with the aid of the classical equations of motion in the asymptotic limit of very large quantum numbers. As things are, the initial condition in an individual system (microstate at time $t = 0$) cannot be determined and, besides, a complete description is out of the question, owing to the sheer weight of numbers.†

The statistical method dispenses with a detailed enumeration of the quantum numbers of a particular stationary state and concentrates on the assignment of probabilities, and the formation of averages. In this manner the enormously large number of parameters is reduced to only a few, and to this extent the amount of information about the system is drastically reduced. It might appear that the coupling of a very large number of elementary systems into a macroscopically valid model, together with the attendant loss of detailed knowledge implied in the statistical method, should introduce so much complexity and uncertainty that the model ought to fail. This, however, is not the case. The very presence of extremely large numbers leads to new regularities of behavior on the macroscopic scale, and the realization of this fact provides one of the justifications for the adoption of a statistical method of analysis.

The statistical method is further justified by a clear statement of the aim of our theory. The theory is formulated to provide a theoretical basis for the

† An interesting discussion of the failure of the classical equations to yield deterministic solutions when the initial conditions are imperfectly specified is contained in L. Brillouin, *Scientific Uncertainty and Information*, pp. 85 and 106, Academic Press, New York, 1964.

computation of the properties of a large class of real systems. Each system is thought of as being in thermodynamic equilibrium within a prescribed set of external constraints. The state of equilibrium or *macrostate* of the system can be realized through a very large number of *microstates*† of the mechanical model, but we merely wish to study the properties of a large class of systems which may be in different microstates at different times, on the condition that their macrostates are the same. For example, we may wish to study the properties of helium, realizing that a unit of mass on which measurements are made here and now will display microscopically different properties from one on which measurements have been made at another time or place. The equations of state of a unit mass of helium should merely embrace the common, *average* properties of all possible samples consisting of a unit of mass of helium. On further examination, even the preceding specification is too detailed. Apart from any uncertainties introduced by quantum mechanics, measurements are made within certain limits of precision, however narrow, and systems in equilibrium never maintain a fixed state absolutely precisely. As a result, a system does not persist in a given stationary state but moves from one approximately stationary state to many others in the course of time. Thus, in addition to the act of observation which changes the quantum state, the residual weak interactions produce even larger changes in quantum states, although macroscopically the system is in equilibrium. An appeal to statistical methods is thus seen to be consistent with our physical grasp of the situation.

The preceding view, first clearly formulated by J. W. Gibbs, seems to have become universally accepted now. However, in the course of the historical development of statistical thermodynamics, a different point of view held sway for a long time. The other point of view, present in the writings of L. Boltzmann and J. C. Maxwell, for example, has not been completely discarded by all modern physicists; it deserves to be mentioned explicitly‡ because an awareness of this fact will make it easier for the student to understand the essentials of the subject. According to the second view, the macroscopic properties of a system constitute averages taken over the microstates traversed by a *single system* in the course of time. It is supposed that at statistical equilibrium such *time averages* are independent of the details of the initial conditions (microstate at $t = 0$) and are completely determined by the mathe-

† We recall that a microstate is specified classically by the provision of a detailed list of all the coordinates and momenta or, quantum-mechanically, by the indication of all quantum numbers which describe a stationary state. In some textbooks the term "microstate" is used for what we shall call a distribution in Section 5.11.1.

‡ See, for example, L. D. Landau and E. M. Lifshitz, *Statistical Physics*, (E. Peierls and R. F. Peierls, transl.), p. 4, Vol. V of "Course of Theoretical Physics," Pergamon, New York, and Addison-Wesley, Reading, Massachusetts, 1958 (2nd ed., 1969); G. E. Uhlenbeck and G. W. Ford, *Lectures in Statistical Mechanics*, Am. Math. Soc., Providence, Rhode Island, 1963.

matical properties of the equations of quantum mechanics or classical mechanics, as the case may be. The conjecture known as the *ergodic*† *hypothesis* (Maxwell's *hypothesis of continuity of path*) supposes that systems traverse all possible microstates sufficiently frequently, and thus it attempts to demonstrate that time averages are identical with the averages taken over a large collection of independent systems. If this conjecture could be demonstrated rigorously on the basis of mechanics, there would be no need to introduce any hypotheses of a statistical nature (the hypothesis of equal *a priori* probability), and statistical thermodynamics would turn out to be a necessary consequence of the laws of mechanics.‡

5.3. Gibbsian Ensembles

When a game of chance is played, it is possible to create a population of data; this was done in Table 4.5 for one particular case. However, in statistical thermodynamics, the same cannot be achieved directly.

We explained in Section 1.2.6 that the central problem in thermodynamics consists in the study of a system which performs a process in time starting with an initial equilibrium state and ending in a final equilibrium state, the process itself having been made possible by the withdrawal of a set of internal constraints. If we now center attention on a particular system which performs a particular, well-defined process, we shall notice that it traverses a sequence of nonequilibrium states which we can describe in macroscopic terms (even though such a description would be very complex). In accordance with our introductory remarks in Section 5.2, in statistical thermodynamics we consider an arbitrarily large number, $\mathcal{N}$, of identical replicas of our representative system, all independently performing the same irreversible process, and accept the hypothesis that each instantaneous macroscopic state is related to the average of that quantity taken over the $\mathcal{N}$ systems. Such a collection of noninteracting, independent *identical replicas* of a representative system under consideration is called an *ensemble*, the term having been coined by Gibbs.§

† From the Greek εργον meaning "work" or "energy" and ʻοδος meaning "road"; in other words, "related to a path of constant energy."

‡ P. R. Halmos, *Lectures on Ergodic Theory*, No. 3, Math. Soc. of Japan, Tokyo, 1956; J. E. Farquhar, Ergodic theory in classical statistical mechanics, in *Statistical Mechanics*, p. 1 (T. A. Bak, ed.), Benjamin, New York, 1967; also, *Ergodic Theory in Statistical Mechanics*, Wiley (Interscience), New York, 1964. J. G. Sinai, Ergodicity of Boltzmann's gas model, in *Statistical Mechanics*, p. 559 (T. A. Bak, ed.), Benjamin, New York, 1967.

§ Owing to an unfortunate tendency discernible in certain recent textbooks on the subject, we shall repeatedly emphasize that the systems in an ensemble are independent and do not interact. In these textbooks consideration is given to collections of *interacting* macroscopic systems, as it will also be given in our Section 5.11. The associated argument is unobjectionable, but the term *ensemble* should not be used for its subject.

The corresponding statistical hypothesis can thus be expressed by the following statement.

Statistical hypothesis 1:

The instantaneous macroscopic state of any system is described by parameters which are averages taken over a corresponding ensemble.

Although each system in an ensemble traverses the same sequence of macroscopic states, it traverses a *different* sequence of microscopic states. The number $\mathcal{N}$ is chosen sufficiently large to justify the adoption of the law of large numbers. This means that every possible sequence of microstates is also represented by a large number of systems in the ensemble. Thus a particular sequence of microstates constitutes an outcome in the statistical sense, and an ensemble represents a population of data. In analogy with general statistics, we adopt the following hypothesis.

Statistical hypothesis 2:

At any instant, in a particular microstate the proportion of systems which is present in an ensemble is equal to the probability of finding a randomly selected system in that particular microstate.

Microstates cannot be identified by direct measurements. Consequently, it is impossible to determine the instantaneous probability function by conducting experiments. Thus, the characteristics of ensembles must be *postulated*, and the appropriateness of the adopted set of hypotheses must be ascertained by inference and on the basis of *macroscopic* measurements.

We must now limit our discussion to systems in equilibrium. A study of nonequilibrium states would exceed the scope as far as this chapter is concerned.† Generally speaking, the reader ought to be informed that the statistical theory of nonequilibrium states is still under active development‡ and cannot be explained in simple terms. Since reversible processes are continuous sequences of states of equilibrium, the preceding restriction does not exclude them from our future considerations.

Considered macroscopically, the state of a system in equilibrium does not

† See I. Prigogine, *Non-Equilibrium Statistical Mechanics*, Wiley (Interscience), New York, 1962; R. Jancel, *Les Fondements de la Mécanique Statistique Classique et Quantique*, Gauthier-Villars, Paris, 1963; E. G. D. Cohen (ed.), *Fundamental Problems in Statistical Mechanics*, Vols. I and II, North-Holland Publ., Amsterdam, 1962 and 1968; S. A. Rice and P. Gray, *The Statistical Mechanics of Simple Liquids*, Wiley (Interscience), New York, 1965; M. Kac (with G. E. Uhlenbeck, A. R. Hibbs, and B. van der Pol), *Probability and Related Topics in the Physical Sciences*, Wiley (Interscience), New York, 1958.

‡ An introduction to the study of nonequilibrium states will be given in Chapters 12 and 13.

change with time. Correspondingly, the probability function in an ensemble whose averages are equal to the macroscopic properties of the system must also remain unchanged in time. Such an ensemble is called *stationary*, and we confine our further discussion to stationary ensembles only. In order to make derivations possible, we must now adopt hypotheses which define the elements of the sample space and the form of the probability function, bearing in mind that freedom from contradiction with the laws of mechanics must be assured.

The sample space is defined by the following hypothesis.

Statistical hypothesis 3:

> *At any instant of time, the microstate assumed by any system in a stationary ensemble is a stationary quantum state accessible to the system under its specified external constraints.*

This hypothesis adopts the quantum cells in the Γ-space of the system as the simple elements of the sample space. If the system were isolated perfectly, every microstate would indeed constitute a stationary quantum state. However, as explained earlier, this cannot be achieved under normal circumstances, and residual, small interactions must be reckoned with. Thus, when an isolated system of constant energy E is discussed, we shall think of it as capable of assuming any stationary state which lies in the Γ-space between the energy hypersurfaces $E = \text{const}$ and $E + \delta E = \text{const}$ in the manner outlined in Section 3.14, where δE is interpreted as large compared with a quantum jump but small with respect to the highest precision of a measurement of energy. Naturally, only those cells are admitted that are consistent with the constraints. For example, if the volume V is prescribed, only that portion of the Γ-space is accessible that lies between surfaces $E = \text{const}$ and $E + \delta E = \text{const}$ and within the hypercylinder based on volume V. A pictorial representation of this set is given in Figure 5.1, which is equivalent to Figure 3.18 except that the Γ-space and not the μ-space is now being referred to. We shall retain this picture whenever we suppose that the energy of the system, E, is prescribed.

Before we can state the hypothesis which determines the probability function, we must establish at least sufficient conditions which are imposed on stationary ensembles by the equations of motion. These are provided by Liouville's theorem.

5.4. Liouville's Equation†

Liouville's equation makes use of W. R. Hamilton's form of the equations of motion of classical mechanics and translates them into a statement about the probability function, Equation (5.5) *infra*. In turn, the form of the

† This section may be omitted on first reading.

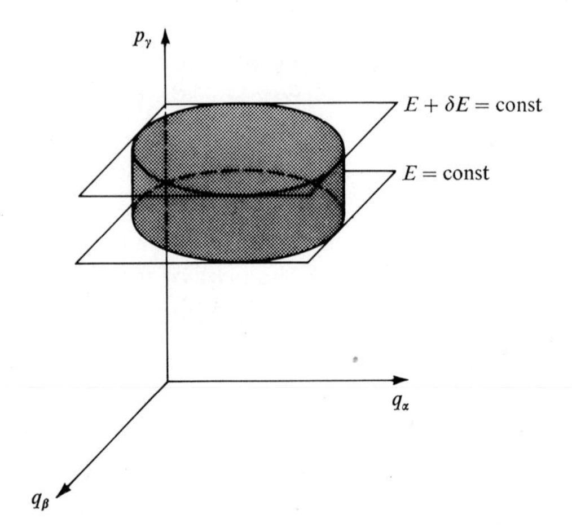

FIGURE 5.1. Portion of Γ-space accessible to isolated system.

differential equation which must be satisfied by the probability function establishes a sufficient condition for the latter to be independent of time regardless of the complexity of the mechanical model of the system and of the motion which it performs. The equation asserts that a probability function which depends only on the energy of each quantum cell in phase space remains constant in time. Consequently, the corresponding ensemble must be stationary.

The remainder of this section establishes this fact in some detail, but the argument in the following sections rests exclusively on the above conclusion.

5.4.1. The Canonical Equations of Motion

We consider a macroscopic system and apply the classical equations of motion to its model, in accordance with N. Bohr's correspondence principle. Amplifying our remarks in Section 5.2, we consider that the motion of the system is fully characterized by fN generalized coordinates q_i and fN generalized, conjugate momenta, p_i, denoting the number of degrees of freedom of each of the N particles by the symbol f. Figure 5.2 shows the projection of the Γ-space into the plane p_i, q_i. The trajectory a, that is, the locus of microstates traversed by the system, passes through the point q_i°, p_i° which corresponds to the initial conditions, and the vector $\mathbf{c}(\dot{q}_i, \dot{p}_i)$ which is equal to the velocity of the representative point A in Γ-space is tangent to the trajectory. Evidently, the values of these two components are related to each other through the equations of motion, and this fact determines a direction at each point. The direction determines an element of the trajectory at that point.

The two components of the velocity vector $\mathbf{c}$ are given by the equations

$$\dot{p}_i = F_i(q_i, p_i) \qquad \text{and} \qquad \dot{q}_i = p_i/m_i, \tag{5.1a, b}$$

where F_i denotes the generalized force—a function of all coordinates and momenta—and m_i denotes the mass, moment of inertia, and so on, depending on the physical nature of the momentum p_i. The right-hand sides of Equations (5.1a, b) can be expressed as derivatives of the Hamiltonian, $\mathscr{H}$, that is, of the total energy of a conservative system. Since

$$\mathscr{H}(q_i, p_i) = \sum_i \frac{p_i^2}{2m_i} + \phi(q_1, \ldots, q_{fN}),$$

where ϕ is the potential energy, we have

$$F_i = -\frac{\partial \phi}{\partial q_i} = -\frac{\partial \mathscr{H}}{\partial q_i} \quad \text{and} \quad \frac{p_i}{m_i} = \frac{\partial \mathscr{H}}{\partial p_i}. \tag{5.2a, b}$$

Hence we obtain†

$$\dot{p}_i = -\frac{\partial \mathscr{H}}{\partial q_i}, \qquad \dot{q}_i = \frac{\partial \mathscr{H}}{\partial p_i}. \tag{5.3a, b}$$

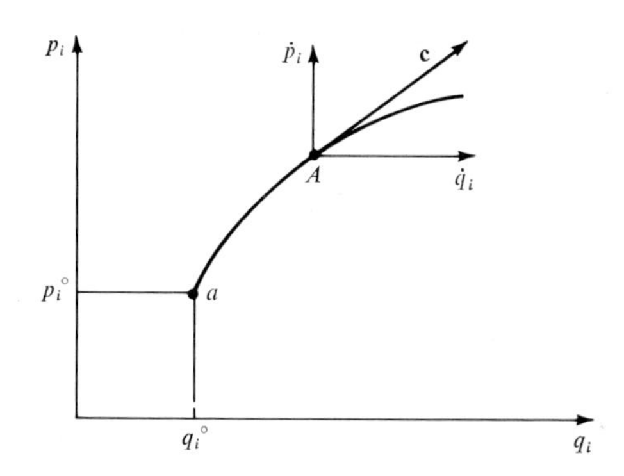

FIGURE 5.2. Projection of Γ-space into the plane q_i, p_i.

Thus a knowledge of the Hamiltonian at every point allows us to determine the element of the trajectory passing through the point. The preceding equations are known as the *canonical equations of motion* or *Hamilton's equations.*

Recalling the uniqueness theorem for systems of ordinary, nonlinear differential equations,‡ we conclude that through every regular point in Γ-space there passes one and only one trajectory. We must, however, realize that the trajectories can become extremely complex and tortuous in shape. Fortunately, for our purposes, the solutions to these equations are not required. The formal simplicity of the fN equations (5.3a, b) allows us to establish a set of sufficient conditions for the probability function of a stationary ensemble.

† The resulting equations are more general than suggested by this derivation, but this is unimportant in statistical thermodynamics.

‡ See, for example, R. Courant, *Differential and Integral Calculus*, (E. J. McShane, transl.), Vol. II, p. 454, Blackie, Glasgow and London, 1937, or E. Goursat, *A Course in Mathematical Analysis*, Vol. II, Part II, pp. 45ff, (E. R. Hedrick and O. Dunkel, transls.), Ginn, Boston, Massachusetts, 1917, reprinted by Dover, New York, 1959.

5.4.2. Probability Density

We now return to an ensemble of $\mathcal{N}$ identical, independent, and noninteracting systems. Since we consider the asymptotic limit when $\mathcal{N} \to \infty$, it is convenient to replace the probability function with a probability density, ρ. If n_i denotes the number of systems whose representative points are confined at instant t to a volume

$$\Delta\mathscr{V} = \Delta q_1 \cdots \Delta q_{fN} \Delta p_1 \cdots \Delta p_{fN}$$

of the Γ-space, then the average density of such points is

$$\rho(q_i, p_i, t) = n_i/\Delta\mathscr{V}.$$

The probability function associated with the element $\Delta\mathscr{V}$ is thus

$$\pi_i(t) = \frac{n_i}{\mathcal{N}} = \frac{\rho\Delta\mathscr{V}}{\mathcal{N}} = \rho'\Delta\mathscr{V},$$

where

$$\rho'(q_i, p_i, t) = \frac{\rho}{\mathcal{N}} \quad \text{with} \quad \mathcal{N} = \int_{\Gamma-\text{space}} \cdots \int \rho(q, p, t)\, d\mathscr{V}.$$

Liouville's theorem relates to the variation in the probability density ρ with time which is imposed on it by the canonical equations of motion.

5.4.3. Liouville's Theorem

As each of the independent systems in the ensemble carries out its complex motion, the swarm of representative points, now imagined to form a continuum, sweeps the Γ-space. At every instant t, a volume $\Delta\mathscr{V}$ centered on point q_i, p_i contains $\rho(q_i, p_i, t)\,\Delta\mathscr{V}$ such points. With the passage of time, representative points enter and leave the element, rather like the particles of a fluid, and this causes the local density to change. This set of circumstances is shown graphically in Figure 5.3, in which the fluxes (numbers of representative

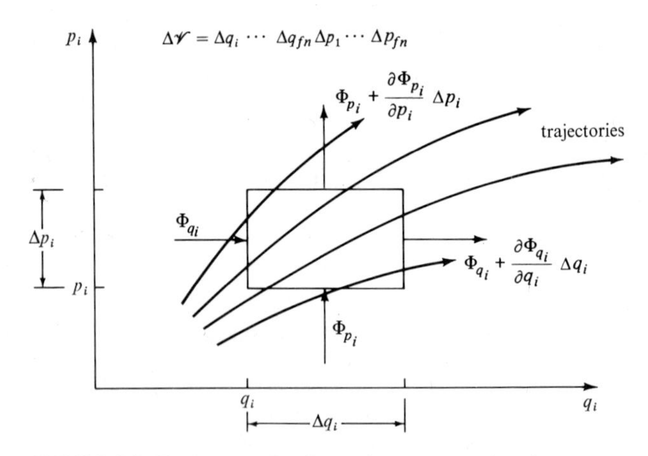

FIGURE 5.3. Projection of volume element into the plane q_i, p_i.

points per unit time) entering and leaving the element have been denoted by $\Phi(\Phi_{qi}, \Phi_{pi})$ and $\Phi + \Delta\Phi$ with components

$$\Phi_{qi} + \frac{\partial \Phi_{qi}}{\partial q_i} \Delta q_i \qquad \text{and} \qquad \Phi_{p_i} + \frac{\partial \Phi_{pi}}{\partial p_i} \Delta p_i .$$

Proceeding in a manner identical with that employed in fluid mechanics when the continuity equation is derived, we compute the *net* change in the number of representative points inside the element. Thus, the flux in the q_i direction is

$$\Phi_{qi} = \rho \dot{q}_i \Delta p_i = \rho \frac{\partial \mathscr{H}}{\partial p_i} \Delta p_i$$

because the flux is equal to the product of density, area and component of velocity normal to the surface. The second form on the right-hand side arises from Equation (5.3b). The *net* flux in the q_i direction and *out* of the element of volume is thus

$$\Delta\Phi_{qi} = \frac{\partial \Phi_{qi}}{\partial q_i} \Delta q_i = \frac{\partial}{\partial q_i}\left(\rho \frac{\partial \mathscr{H}}{\partial p_i}\right) \Delta q_i \, \Delta p_i .$$

Similarly, in the p_i direction, we obtain

$$\Phi_{pi} = \rho \dot{p}_i \, \Delta q_i = -\rho \frac{\partial \mathscr{H}}{\partial q_i} \Delta p_i$$

and

$$\Delta\Phi_{pi} = -\frac{\partial}{\partial p_i}\left(\rho \frac{\partial \mathscr{H}}{\partial q_i}\right) \Delta q_i \, \Delta p_i .$$

Taking into account all the faces on the element $\Delta\mathscr{V}$ and so involving fN directions q_i and p_i, we can see that the net change in representative points is equal to

$$\Delta\mathscr{V} \sum_i \left[\frac{\partial}{\partial q_i}\left(\rho \frac{\partial \mathscr{H}}{\partial p_i}\right) - \frac{\partial}{\partial p_i}\left(\rho \frac{\partial \mathscr{H}}{\partial q_i}\right)\right]$$

because the "area" of each face, say normal to q_k, is equal to a product of $2fN - 1$ terms $\Delta p_i \Delta q_i$, excluding Δq_k, the volume $\Delta\mathscr{V}$ appearing when multiplication with Δq_k is performed. Each term of this sum yields a term

$$\frac{\partial \rho}{\partial q_i} \frac{\partial \mathscr{H}}{\partial p_i} - \frac{\partial \rho}{\partial p_i} \frac{\partial \mathscr{H}}{\partial q_i}$$

and a term

$$\rho\left(\frac{\partial^2 \mathscr{H}}{\partial p_i \, \partial q_i} - \frac{\partial^2 \mathscr{H}}{\partial p_i \, \partial q_i}\right) = 0 .$$

By virtue of the application of Hamilton's equations, the net flux simplifies to

$$\Delta\Phi = \Delta\mathscr{V} \sum_i \left[\frac{\partial \rho}{\partial q_i} \frac{\partial \mathscr{H}}{\partial p_i} - \frac{\partial \rho}{\partial p_i} \frac{\partial \mathscr{H}}{\partial q_i}\right] = \Delta\mathscr{V} \sum_i \left[\dot{q}_i \frac{\partial \rho}{\partial q_i} + \dot{p}_i \frac{\partial \rho}{\partial p_i}\right]$$

The number of points inside the element changes by $(\partial\rho/\partial t)\Delta\mathscr{V}$, and this change must balance the *net* flux $\Delta\Phi$ outward because representative points are neither created nor destroyed in an ensemble. This enables us to establish the continuity equation in two alternative forms

$$\frac{\partial\rho}{\partial t}+\sum_i\left(\frac{\partial\rho}{\partial q_i}\frac{\partial\mathscr{H}}{\partial p_i}-\frac{\partial\rho}{\partial p_i}\frac{\partial\mathscr{H}}{\partial q_i}\right)=0 \tag{5.4}$$

or

$$\frac{\partial\rho}{\partial t}+\sum_i\left[\dot{q}_i\frac{\partial\rho}{\partial q_i}+\dot{p}_i\frac{\partial\rho}{\partial p_i}\right]=0\,. \tag{5.5}$$

Each of the preceding two equations embodies *Liouville's theorem.*

5.4.4. Geometrical Interpretation of Liouville's Theorem

The second form, Equation (5.5), allows us to obtain a remarkable geometric interpretation of Liouville's theorem if we notice that the factors $\dot{q}_i$ and $\dot{p}_i$ in the brackets represent the $2fN$ components of the vector $\mathbf{c}$ from Figure 5.2. We recall from fluid mechanics that the sum of the operators

$$\frac{\partial}{\partial t}+\sum_i c_i\frac{\partial}{\partial x_i}\equiv\frac{\mathrm{d}}{\mathrm{d}t}$$

represents the total, substantive derivative of any scalar quantity such as the density or the probability density in the present case. Hence, Equation 5.5 can be simplified to

$$\frac{\mathrm{d}\rho}{\mathrm{d}t}=0\,. \tag{5.6}$$

The total derivative $\mathrm{d}\rho/\mathrm{d}t$ represents the variation in the number of representative points contained in an element $\Delta\mathscr{V}$ of Γ-space as we follow its motion along the trajectories which have now become analogous to the streamlines in a fluid. The equation asserts that this number remains constant, rather like the density in an incompressible fluid. As we follow the motion of a small volume $\Delta\mathscr{V}$, we discover that its shape distorts in a manner suggested by Figure 5.4. Owing to the large number of degrees of freedom and the complexity of the motion, the distortion is impossible to visualize in detail. The diagram gives but a pale impression of what goes on in reality. Nevertheless, and regardless of all this complexity, the number of representative points contained in this volume (or, by integration, in any other volume) remains constant. In particular, if ρ is uniform throughout the Γ-space, it will remain uniform at all times.

The physical content of Liouville's theorem is easy to grasp if it is remembered that ρ describes the fraction of the systems in an ensemble which exist in it in a prescribed microstate. Determinism demands that each of those systems follow the same evolution, that is, the same trajectory. Thus, some time later, the fraction of systems which have reached another, but accessible, state is still the same. Liouville's theorem shows that this is also true for systems whose microstates have been close at $t=t_0$ and may have diverged violently at $t>t_0$.

5.4.5. Sufficient Conditions for an Ensemble To Be Stationary

In a given ensemble, the energy E and, hence, the Hamiltonian $\mathscr{H}$ of the conservative mechanical systems which form the ensemble need not be the same, although they will be equal for a smaller number of them. Thus we may choose the probability density to be a

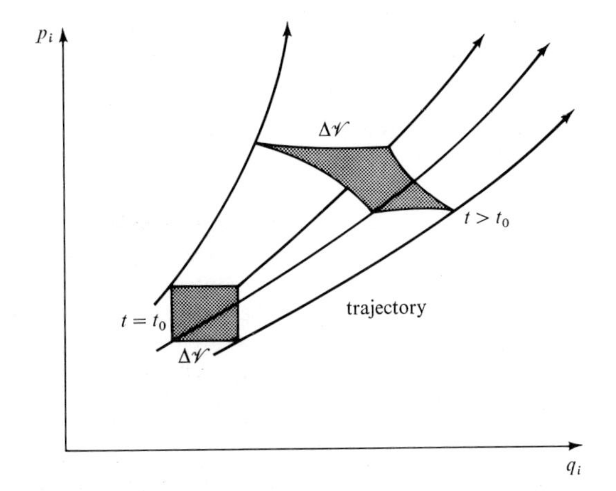

FIGURE 5.4. Geometrical interpretation of Liouville's theorem. A given number of representative points always occupies the same volume.

function of the energy associated with every point in the portion of the Γ-space accessible to the systems:

$$\rho = \rho(E). \tag{5.7}$$

In a conservative system, the energy E remains constant along every trajectory and has a constant value at each point of the Γ-space. Thus for any element of phase space, we have

$$\frac{\partial E}{\partial t} = 0$$

and

$$\frac{\mathrm{d}E}{\mathrm{d}t} = \frac{\partial E}{\partial t} + \sum_i \left[\dot{q}_i \frac{\partial E}{\partial q_i} + \dot{p}_i \frac{\partial E}{\partial p_i}\right] = 0,$$

or

$$\sum_i \left[\dot{q}_i \frac{\partial E}{\partial q_i} + \dot{p}_i \frac{\partial E}{\partial p_i}\right] = 0,$$

Reference to the Liouville equation (5.5) allows us to write for this case

$$\frac{\partial \rho}{\partial t} = -\sum_i \left[\dot{q}_i \frac{\partial E}{\partial q_i} + \dot{p}_i \frac{\partial E}{\partial p_i}\right] \frac{\mathrm{d}\rho}{\mathrm{d}E} = 0,$$

that is, explicitly,

$$\frac{\partial \rho}{\partial t} = 0. \tag{5.8}$$

Equation (5.8) demonstrates that a probability density which is uniquely related to the energy of each system in an ensemble remains constant with respect to time throughout the Γ-space, thus rendering the corresponding ensemble stationary. There is no need to establish the necessary conditions for this to be the case because we can obviously found our statistical theory on a single postulate for the probability density, being able to think of many, ultimately equivalent alternatives. In fact, later we shall explore three possible, equivalent forms of the probability-density function $\rho = \rho(E)$.

5.4.6. Concluding Remarks

To round off the present discussion, we should note that ρ can be chosen to be a function of any *constant of the motion*, that is, of a constant of integration of Hamilton's equations. There are $2fN + 1$ such constants, in general, and the energy E can be chosen as one of them in a conservative system. It is also interesting to be aware of H. Poincaré's theorem, mentioned by L. Brillouin, which states that "For a large class of conservative systems, the canonical equations of classical mechanics do not admit any *analytical* and uniform integrals of motion apart from the energy integral. Thus the choice of energy as the independent variable in $\rho(E)$ turns out to be most convenient."†

The final remark concerns the fact that classical instead of quantum mechanics has been used to derive the sufficient condition (5.7). This is entirely due to our desire for as elementary an exposition as possible under the circumstances. At this point, the reader is asked to accept the assurance that an exact counterpart of Liouville's equation can be obtained with the aid of quantum mechanics.† Furthermore, the appeal to classical mechanics is not restrictive when large systems are discussed.

5.5. Geometrical Structure of the Statistical Sample Space

According to statistical hypothesis 3, the sample space of a system in equilibrium consists of the quantum cells in the accessible portion of the Γ-space. These can now be imagined arranged in shells of given energy E_i in a manner suggested by Figure 5.5. The reader is assured that the combining of many quantum levels, as in Figure 3.18, into one of "thickness" δE has no influence on future developments, owing to the extremely small value of quantum jumps when the quantum numbers are large. In other words, the shells in Figure 5.5 may be imagined formed with the aid of energy surfaces drawn many quantum jumps apart.

In every shell, we imagine that all quantum cells corresponding to the degeneracy of the energy level have been drawn. The probability function for these cells, the simple elements of the sample space, is assigned in accordance with the following hypothesis.

Statistical hypothesis 4:

> *In a state of macroscopic equilibrium, all stationary quantum states of equal energy have equal* a priori *probability.*

† For further details, see R. C. Tolman, *The Principles of Statistical Mechanics*, p. 39 and p. 335, Oxford Univ. Press, London and New York, 1938.

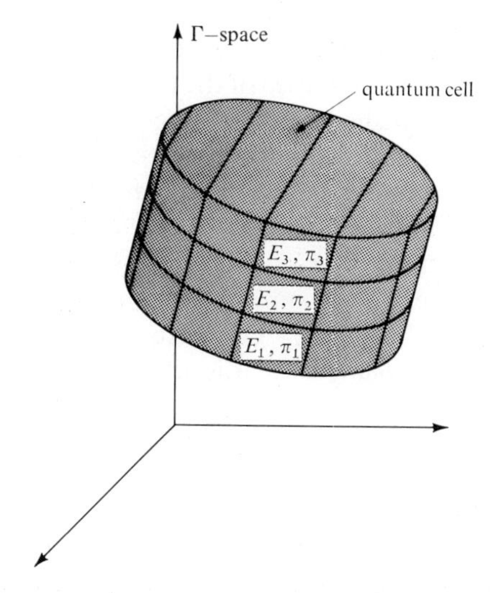

FIGURE 5.5. Geometrical structure of the statistical sample space.

This is the famous hypothesis of equal *a priori* probability of microstates in phase space; it specifically relates to quantum cells in the phase space of the whole system (Γ-space) and not to those in the μ-space. In this manner, interactions between elementary particles within the system are fully accounted for. The acceptance of this hypothesis assures us that the state is indeed one of macroscopic equilibrium.

Passing to the classical limit, we would replace the probability function π by a probability density ρ and note that each surface of constant energy is traced by a continuum of trajectories which do not cross into other surfaces of constant energy. The detailed nature of the probability function to be prescribed depends on the constraints imposed on the system under consideration, each set of constraints defining a particular type of stationary ensemble and so a different convenient sample space. Whatever the form of the probability function, the ensemble average of any quantity g is expressed by Equation (4.29).

5.6. Relation between Theories Based on Different Ensembles

There exist three forms of the function $\pi(E)$, or of its classical analog $\rho(E)$, which play an important part in statistical thermodynamics. They correspond, in turn, to a *microcanonical ensemble*, *a canonical ensemble*, and a *grand canonical ensemble*. In principle, though not in practice, there is no need to study all three of them.

As the theories are developed, it must be expected that each probability function will lead us to a particular form of the fundamental equation, namely that which is appropriate to the corresponding set of external constraints. However, since each development must lead to the same expressions for the *equilibrium* properties and to the same relations between them regardless of *how* the state of equilibrium is enforced by the external constraints, we expect that the fundamental equations derived statistically, like those identified by the methods of classical thermodynamics in Section 1.4.4, must be equivalent and must go over into each other by Legendre transformations. The same cannot be asserted about *fluctuations*, that is, about possible departures from equilibrium. This is due to the fact that fluctuations about an equilibrium state depend on that state as well as on the external constraints which maintain it. We shall base our development on a study of microcanonical ensembles and on combinations of systems which approximate a canonical ensemble.

5.7. Microcanonical Ensemble

First we center our attention on an isolated system in equilibrium. We assume that it consists of N particles constrained to move within a given volume V. This means that the energy E of the system is prescribed.

In the most general case, the external constraints may include a set of extensive variables, z_i, the volume V being merely one of them. To simplify the expression, we shall continue our argument as if only the volume were specified. Whenever necessary, we shall replace the volume by the full set of variables z_i. Similarly, N may consist of a sum $N_1 + \cdots + N_k$ of k different kinds of particles, as in a mixture or an alloy.

A collection of a very large number, $\mathcal{N}$, of such systems, each pursuing its own trajectory in Γ-space, is called a *microcanonical ensemble*. Hypothesis 4 imposes on such an ensemble a probability function of the most striking simplicity: it is a constant! Assuming that there are Ω accessible quantum states, we put

$$\pi = \frac{1}{\Omega} \qquad (E_0 < E < E_0 + \delta E_0), \tag{5.9}$$

noting that

$$\Omega = \Omega(E, V, N), \tag{5.10}$$

and conceive of π as being a function of the quantum numbers.

In view of the very large numbers of quantum states accessible to a localizeable macroscopic system, we may adopt a probability density, ρ, in the Γ-space as we did in Liouville's theorem. Then ρ becomes a function of the generalized coordinates and momenta, and we may put

$$\rho(q_i, p_i) = A\,\delta(E - E_0), \tag{5.11}$$

where δ denotes the Dirac delta function† and q_i, p_i stand for all the fN variables of each kind. The normalization condition is written

$$\int_{\substack{\text{accessible}\\ \Gamma\text{-space}}} \rho(q_i, p_i)\,\mathrm{d}q_i\,\mathrm{d}p_i = 1\,, \tag{5.11a}$$

and this determines the constant A.

5.8. Canonical Ensemble

Another class of systems which will interest us later is an ensemble whose representative system of specified volume V (generally with a specified set of external variables z_i) and a number of identical or diverse particles N is in contact with an infinitely large heat reservoir of constant temperature T. Such an ensemble is called *canonical.* In a microcanonical ensemble, we must expect the temperature to differ from system to system. Similarly, in a canonical ensemble we must expect that the energy will differ from system to system. The form of the probability function, π, in the sample space is again settled by hypothesis.

Statistical hypothesis 5:

> *The probability function in the sample space of a canonical ensemble is a function of the energy E_i of each system, and its form is exponential; specifically,*

$$\pi = A\,\mathrm{e}^{-\beta E_i}\,. \tag{5.12}$$

The constant A is fixed by the normalization condition. The constant β is determined by a suitable physical argument (Section 5.13).

When the number of quantum states in the sample space is large, as it is for localizeable macroscopic systems, we may again prefer to employ a probability density. For a canonical ensemble this would be of the form

$$\rho = B\,\mathrm{e}^{-\beta\mathscr{H}}\,, \tag{5.12a}$$

where $\mathscr{H}$ denotes the Hamiltonian of the system. In our case the Hamiltonian is identical with the total energy. The constant B is also obtained from the normalization condition.

The hypothesis in Equation (5.12) appears quite arbitrary, as indeed it is.

† See footnote on p. 64.

Its only justification is that it allows us to recover the familiar macroscopic equations of classical thermodynamics as they apply to states of equilibrium. This, however, will become apparent to the student much later. At this stage we merely note three points about it. First, we see that, for $E_i = \text{const}$, the value of π becomes constant and assigns an equal *a priori* probability to each quantum state of given energy E_i and thus does not conflict with statistical hypothesis 4. The second point to note is that the exponential factor $\exp(-\beta E_i)$, known as *Boltzmann's factor*, represents a rapidly decreasing function of the energy level E_i. In accordance with our previous remarks, the macroscopic energy at prescribed values of T, V, and N is

$$\langle E \rangle = A \sum_{\substack{\text{quantum} \\ \text{states}}} E_i \, e^{-\beta E_i}, \tag{5.13}$$

because this is the form of the weighted average of energy, Equation (4.29), for the canonical ensemble. The sum must be extended over all quantum states in accordance with statistical hypothesis 3, as indicated, and E_i represents a function of some random variable which need not be specified in detail; it may be a number, $k = 1, \ldots$, which identifies a particular quantum cell. Recalling our considerations in Section 3.14, we realize that the number of accessible quantum states increases enormously as the energy increases. This means that for i increasing, the number of terms E_i in Equation (5.13) increases enormously. On the other hand the sum of products

$$E_i \, e^{-\beta E_i}$$

consists of terms in which very small values of energy are multiplied by Boltzmann factors of order unity, whereas very large values of energy are multiplied by Boltzmann factors which are very small. Consequently there must exist a narrow range in which the terms pass through a very sharp maximum centered about some value, E^*, of energy. This will be the level which contains more quantum states than any lower level, since the contributions from higher levels are sharply curtained by the very small values of the associated Boltzmann factors. It follows that the average $\langle E \rangle$ will be very close to E^*, and E^* will represent the most probable energy level, that is, the energy found in an overwhelming number of members in the ensemble.

The third point to be noted is that the combined probability of finding two systems of energies E_j, and E_k, that is, of total energy $E_j + E_k$, is

$$\pi_{jk} = A' \, e^{-\beta(E_j + E_k)} = A' \, e^{-\beta E_j} \, e^{-\beta E_k}.$$

This is then equal to the product $\pi_j \pi_k$ of finding each system separately; the reader may verify that the normalization constant A' is equal to A^2. Thus the postulate for the probability function in a canonical ensemble satisfies the condition of statistical independence of two systems.

At this stage it is not clear that a stationary ensemble of systems interacting with thermal baths can be assigned a form which has been justified for stationary ensembles of noninteracting systems. We shall explain in Section 5.11.5 that for $\mathcal{N} \to \infty$ the energy distribution tends to a delta function centered on the most probable energy, E^*. Thus in this limit the systems cease to interact with their baths, and their energies become asymptotically equal.

5.9. Grand Canonical Ensemble†

The last class of stationary ensembles commonly studied in statistical thermodynamics consists of $\mathcal{N}$ independent, noninteracting open systems, each of which remains in equilibrium with a large thermal bath of temperature T as well as with a number of receptacles, also maintained at temperature T, each of which communicates with the system through a semipermeable membrane and contains particles of one of the k species present in the main system. Such a collection of systems is known as a *grand canonical ensemble*. The appropriate probability function is given by the following hypothesis.

Statistical hypothesis 6:

The probability function for a grand canonical ensemble is of the form

$$\pi = A \exp\left[- \beta E_i - \beta \sum_k \mu_k N_k \right]. \tag{5.14}$$

Here E_i is the energy level and N_k is the number of elementary particles of kind k. The parameter β has the same significance as in Equation (5.12a) and μ_k is interpreted as the chemical potential. In the limit of $\mathcal{N} \to \infty$, it is found that the distribution of energy as well as that of the number of particles of each species tend to Dirac delta functions, so that asymptotically this case also reduces to the one covered by statistical hypothesis 4.

5.10. Statistical Interpretation of Entropy

The rules for the formation of macroscopic properties, such as entropy and temperature, which have no direct connection to microscopic properties, will emerge naturally in Sections 5.13 and 5.14. We will see that statistical considerations lead to differential forms which have the same structure as those in classical thermodynamics. This will yield explicit expressions for entropy and temperature. For the sake of clarity of exposition, it is advisable to

† This section can be omitted on first reading.

anticipate one of these results and to *postulate* here the connection between entropy and the microscopic characteristics of a thermodynamic system. This postulate, first enunciated by Boltzmann, will subsequently be verified in Section 5.16. Accordingly, we accept the following statement.

> *The entropy $S(E, V, N)$ of an isolated macroscopic system is directly related to the number of microstates or Γ-cells, Ω, which are accessible to the thermodynamic system.*

It should be noted that Ω denotes here the number of microstates of a single isolated system.

Boltzmann's postulate, together with the requirement that entropy must be an extensive property, is sufficient to determine uniquely the relation between $S(E, V, N)$ and $\Omega(E, V, N)$, thus providing a "bridge" between statistical and classical thermodynamics. The remaining functions of interest such as thermodynamic temperature T and the thermodynamic potentials F and G follow naturally from known relations.

In accordance with the preceding hypothesis, we stipulate that

$$S = \mathrm{f}(\Omega) \tag{5.15}$$

and ask for the appropriate form of this function. We consider two systems, 1 and 2, whose entropies are S_1 and S_2, respectively, and the composite system $1 + 2$ whose entropy is denoted by S. The composite system is obtained by simply joining the isolated systems 1 and 2. Owing to the property of additivity for entropy, we must have

$$S = S_1 + S_2 . \tag{5.15a}$$

The rule for combining microstates is different, and reads

$$\Omega = \Omega_1 \Omega_2 \tag{5.16}$$

because any one microstate of system 1 can combine with any microstate of system 2 to yield a microstate of the combined system.

Since the form of the function in Equation (5.15) must reflect the property (5.15a) we obtain the relation

$$\mathrm{f}(\Omega_1 \Omega_2) = \mathrm{f}(\Omega_1) + \mathrm{f}(\Omega_2). \tag{5.17}$$

This is a functional equation, because we seek to determine an unknown function f of the argument Ω which possesses the property, expressed by Equation (5.17), that the value of the function f of a product $\Omega_1 \Omega_2$ must be equal to the sum of the values of the same function f, each taken at Ω_1 and Ω_2, respectively.

It is easy to verify that the logarithmic function

$$S = \mathbf{k} \ln \Omega \tag{5.18}$$

possesses this property. Here $\mathbf{k}$ denotes an arbitrary constant. Later, in Section 5.13, we shall show that this arbitrary constant (as far as the functional equation is concerned) turns out to be equal to Boltzmann's constant $\mathbf{k}$. In anticipation of this result, we use the symbol $\mathbf{k}$ rather than k. Indeed, we see that

$$\mathbf{k} \ln (\Omega_1 \Omega_2) = \mathbf{k} \ln \Omega_1 + \mathbf{k} \ln \Omega_2 .$$

Thus we have shown that relation (5.18) is *sufficient* for our purposes. We can also show that this relation is *necessary*, meaning that the logarithmic function $\mathbf{k} \ln \Omega$ is the only one which possesses the necessary property.

The preceding statement can be proved by taking a partial derivative of the terms in Equation (5.17) first with respect to Ω_1 and then with respect to Ω_2. Denoting the derivatives of the function f with respect to the argument displayed in parentheses by primes, we deduce consecutively that

$$\Omega_2 \, \mathrm{f}'(\Omega_1 \Omega_2) = \mathrm{f}'(\Omega_1)$$

and

$$\Omega_1 \Omega_2 \, \mathrm{f}''(\Omega_1 \Omega_2) + \mathrm{f}'(\Omega_1 \Omega_2) = 0 ,$$

or

$$\Omega \, \mathrm{f}''(\Omega) + \mathrm{f}'(\Omega) = 0 .$$

This transforms the functional equation for f into a differential equation for $\mathrm{f}(\Omega)$. The unique solution for this equation is

$$\mathrm{f}(\Omega) = \mathbf{k} \ln \Omega + C .$$

The constant C is independent of the state of the system because it is independent of the number of microstates, Ω. In spite of this, it is not a universal constant, being arbitrary for each subsystem. If we select the values C_1 and C_2 for the two subsystems, substitution into Equation (5.17) shows that we must select

$$C = C_1 + C_2$$

to satisfy it. This is an example of the familiar problem of normalizing additive constants when larger systems are compounded from a number of subsystems. In order to evade this problem, we find it is best to stipulate

$$C = C_1 = C_2 = 0 , \tag{5.17a}$$

noting that the independence of the constant C on the state of the system, or of C_1 and C_2 on the respective states of the subsystems, must have important physical consequences. We shall discuss these in Section 5.17.2 in connection with the formulation of the Third law of thermodynamics.

Equation (5.18) was first perceived by Boltzmann, though he never wrote it in this form. This was first achieved in 1906 by M. Planck. Following A. Einstein, we now call it *Boltzmann's principle*. In recognition of Boltzmann's contributions to statistical thermodynamics, the formula

$$S = k \log W$$

has been carved on his tombstone in the Central Cemetery in Vienna.†

The preceding remarks demonstrate that the entropy of a macroscopic system does not vary continuously, since Ω can only change by an integer. However, as we already know, Ω is very large, and the property of being discrete cannot be normally detected on a macroscopic scale. Hence, in practice, the entropy S may be assumed to vary continuously. In Section 5.17.2, we shall see that, near the absolute zero of temperature, Ω becomes of order unity. Consequently, at very low temperatures, the discreteness of S may become significant.

Writing Equation (5.18) in the form

$$\Omega = e^{S/\mathbf{k}}, \tag{5.18a}$$

we recognize that an increase in the entropy of an isolated system is associated with an increase in the number of available stationary quantum states in the Γ-space. (The physical reasons for this will be discussed in Section 5.17.1. It now follows that the increase in the entropy of an isolated system after the performance of an irreversible process is paralleled by an exponential increase in the number of Γ-states.

Employing the value $\mathbf{k} = 1.38 \times 10^{-23}$ J/K and the value $S = 29.7$ kcal/K for 1 kmol of helium at 0°C and low pressure‡ we calculate

$$\Omega \approx e^{10^{28}} \approx 10^{10^{27.6}}. \tag{5.19}$$

This gives us an idea of the magnitude of Ω for a single system; we have estimated the same number in a different way in Section 3.14.

Since Ω constitutes a well-defined integer for each macrostate of a system, it is possible to introduce an absolute measure of entropy. This statement is the microscopic forerunner of the Third law of thermodynamics, which we defer to Section 5.17.2. At this stage, we note that $S = 0$ implies that $\Omega = 1$, which means that the entropy of a system vanishes when its macrostate can be

† This is the same as Equation (5.18) except for the notation.

‡ See, for example, Table XVI in J. Kestin, *A Course in Thermodynamics*, Vol. II, Blaisdell, Waltham, Massachusetts, 1969.

realized through a single and, therefore, nondegenerate microstate. When only one microstate is available, we can say that the system exists in a state of perfect order because all elementary particles of the system must assume prescribed positions in μ-space. As the number of possibilities open to the system increases, the system becomes free to display more and more micro-states. We then say that its macrostate has become more "chaotic," because successive microstates present an ever changing appearance, instead of the immutable state where $S = 0$. Thus the value of entropy can be used as a measure of disorder (or order) in the system. For this reason, the second part of the Second law of themodynamics may be interpreted as stating that the system undergoes a transition from a less probable state of order to a more probable state of disorder at constant values of E, V, and N.

5.11. Method of the Most Probable Distribution

We have now established the basic elements of our statistical theory of equilibrium states and proceed to derive the fundamental equation. To this end we consider $\mathcal{N}$ *interacting*, identical, closed systems, each containing N particles in a volume V. The assembly of systems, shown in Figure 5.6, is

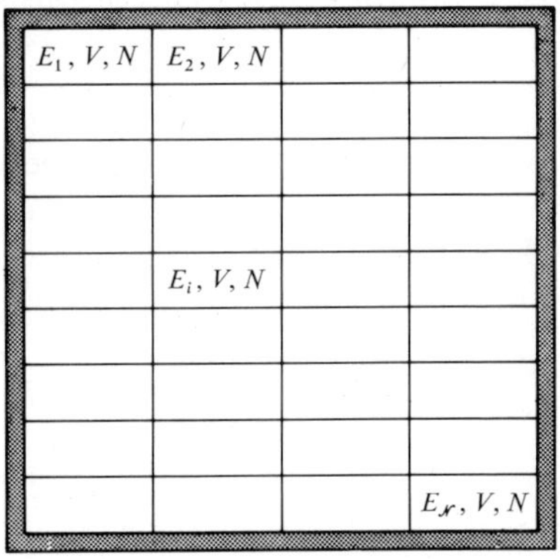

E_1, V, N	E_2, V, N		
	E_i, V, N		
			$E_{\mathcal{N}}, V, N$

FIGURE 5.6. A macrosystem composed of $\mathcal{N}$ interacting macroscopic, localizable systems.

imagined enclosed in an adiabatic envelope; it constitutes what we shall call a *macrosystem*. We further imagine that the macrosystem arose from $\mathcal{N}$ isolated systems, each containing an energy E, so that the total energy of the isolated macrosystem is $\mathcal{N}E$. Owing to interactions, the instantaneous energy E_i differs from system to system and is, generally speaking, different from the original average energy E.

A macrosystem does not constitute an ensemble, because the systems of a macrosystem interact with each other. Thus the systems are not statistically independent, and Liouville's theorem does not apply to them.† However, a collection of $\mathscr{N}$ systems combined into $\mathscr{N}/\mathscr{N}$ isolated macrosystems constitutes a microcanonical ensemble. The problem we now propose to consider is that of determining the distribution of energy, E_i, within a macrosystem in accordance with hypotheses 1, 2, 3, and 4. More generally, we propose to determine the form of the probability function which must be defined in the Γ-space of an individual system, given that it is a constant in the Γ-space of an isolated macrosystem. The latter Γ-space contains $\mathscr{N}$ times more coordinates than the former, and to avoid misunderstanding we shall refer to it as the Γ_m-space.

The available energies in the Γ-space form an "open-ended" sequence governed by the properties of the mechanical model, by the number of particles N, and by the volume V. The accessible energy levels of a single system must now be combined in all possible ways on the condition that $\mathscr{N}$ of them are taken at a time and that the sum of the energies over the macrosystem is equal to $\mathscr{N}E$. This count will yield the number Ω of quantum states which the macrosystem can possibly assume; each of them must be assigned an equal *a priori* probability. Thus we are faced with a combinatorial problem,‡ best solved with the aid of the concept of a *distribution.*§

5.11.1. Distribution

A distribution arises when we indicate a set of numbers n_i^{ν} each of which describes the number of systems existing in a given quantum state i. The number n_i^{ν} is called the *occupation number* for the respective quantum state. Evidently, in a given distribution denoted by the superscript ν we must have

$$\sum_{i=1}^{k} n_i^{\nu} = \mathscr{N} \qquad \text{and} \qquad \sum_{i=1}^{k} n_i^{\nu} E_i = \mathscr{N}E. \tag{5.20a, b}$$

Although the set of microstates in Γ-space is "open-ended," we can assume for our present purposes that it has been terminated at the number k chosen in such a way that $E_k > \mathscr{N}E$. Energy levels higher than this are clearly inaccessible to any one of the systems in the macrosystem.

† If the representative system interacts with a heat reservoir, an ensemble consists of N independent replicas of the system plus reservoir. Unfortunately, some authors do use the term "canonical ensemble" for what we call here a macrosystem. This creates another source of confusion.

‡ At this point the reader may find it useful to review the auxiliary Section 4.1.

§ Many authors replace the systems by molecules and use the term "microstate" for what we call a distribution. The question of how to treat assemblies of molecules will be answered in Sections 5.18.1 and 6.2.

The diagram in Figure 5.7 gives a geometrical interpretation of our counting scheme; it attempts to represent the common Γ-space and the resulting Γ_m-space, albeit very schematically. The cells in the accessible portion of the Γ-space, numbered 1, 2, ..., i, ..., constitute the open sequence mentioned earlier, and a suitable set of $\mathcal{N}$ of them produces one cell in Γ_m-space, all the latter being confined to the accessible portion of a shell in which $\mathcal{N}E$ = const. The problem is to find how many of them can exist, given the structure of the Γ-space and the total energy $\mathcal{N}E$. Thus the elements of our sample space (Γ_m) constitute suitable combinations of the elements of the sample space (Γ) of a single system.

The energy levels which belong to the quantum states in the Γ-space are imagined arranged in an ascending series, $E_1, E_2, \ldots, E_i, \ldots, E_k$, with

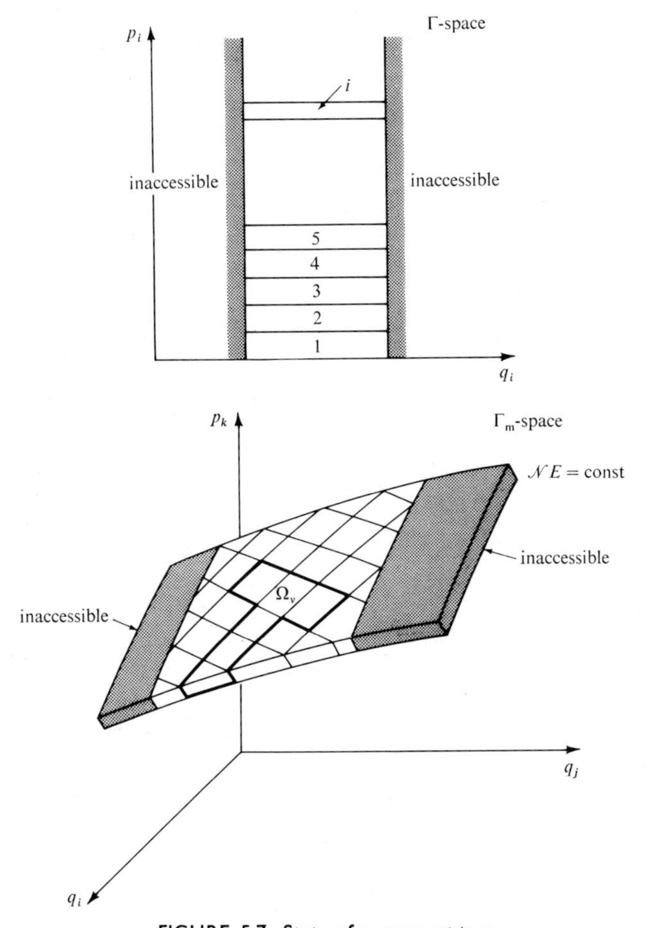

FIGURE 5.7. State of a macrosystem.

$E_{i+1} \geq E_i$. This means that energy levels in the series are repeated precisely as many times as their degeneracy demands. A distribution determines a set of quantum states in the Γ_m-space because interchanging the microstates of two systems creates a new quantum state, owing to the fact that macroscopic systems are localized and distinguishable. We shall denote the number of Γ_m-states in a given distribution v by Ω_v and assume that there exist κ distinct distributions. One such distribution is suggested by the framed area in Figure 5.7.

The total number of Γ_m-states is

$$\Omega = \sum_{v=1}^{\kappa} \Omega_v, \tag{5.21}$$

and within each distribution the constraints (5.20a, b) must be satisfied. We shall see presently that this expression is more suitable for our purposes than the equivalent combinatorial formula†

$$\Omega = k^{\mathcal{N}}. \tag{5.21a}$$

This counting procedure can be recorded in the following table.

νth Distribution of Systems[a]	
Quantum state number (Γ-space)	$1, 2, \ldots, i, \ldots, k$
Energy level of system	$E_1, E_2, \ldots, E_i, \ldots, E_k$
Occupation number in Γ-space	$n_1^v, n_2^v, \ldots, n_i^v, \ldots, n_k^v$

[a] The number of supermicrostates in the distribution is Ω_v.

It is easy to see now that the sum (5.21) exhausts all possible supermicrostates, that is, that each of the v distributions constitutes a subset of the set of simple elements in the sample space, with no two subsets containing common elements.

Reference to Equation (4.3) shows that

$$\Omega_v = \frac{\mathcal{N}!}{\prod_{i=1}^{k} n_i^v!}, \tag{5.22}$$

the number of elements in a distribution being equal to the number of ways in which $\mathcal{N}$ systems can be distributed over k quantum states with *prescribed* occupation numbers n_i^v. Certain of the occupation numbers in a distribution may, evidently, be equal to zero, but this introduces no complication.

† For a derivation and discussion see Equation (4.2) in Section 4.1.2.

5.11.2. Averages and Probability Functions

Later, we shall be interested in a number of averages of quantities which characterize the quantum states of the system in the Γ-space; we denote them generally by the symbol ϕ. For example, ϕ_i may denote the energy E_i. Thus we treat the current index i as the random variable, and

$$\phi_i = \phi(i) \tag{5.23}$$

is an arbitrary function of this random variable. The expectation value must now be taken over our microcanonical ensemble, which we construct so that it consists of an arbitrary multiple of Ω macrosystems, say, of $\mathcal{N}\Omega$ systems for simplicity. If we choose the Γ_m-space as our sample space, statistical hypothesis 4 lays down a statistical probability function

$$\Pi = 1/\Omega = \text{const} \tag{5.24}$$

defined on the quantum states in Γ_m-space. If we adopt the Γ-space itself as our sample space, the probability function will be different, but it must be consistent with Equation (5.24). The required expression can be easily deduced by forming the ensemble average of ϕ_i.

One system placed in Γ-state i contributes ϕ_i to our count. However, since there are n_i^ν systems in state i, they will contribute an amount $n_i^\nu \phi_i$. The value n_i^ν occurs Ω_ν times because it is present in each quantum state of a given distribution. Summing over all distributions,

$$\sum_{\nu=1}^{\kappa} \Omega_\nu \, n_i^\nu \phi_i ,$$

we find the total contribution contained in all distributions. The expectation value is formed when we sum over all microstates in the Γ-space and divide this sum by $\mathcal{N}\Omega$, that is, the total number of quantum states of a single system which can occur in Ω distributions. In every distribution there are as many quantum states as systems, and the total number of distributions is Ω. Hence, we have

$$\langle \phi \rangle = \frac{\sum_{i=1}^{k} \sum_{\nu=1}^{\kappa} \Omega_\nu \, n_i^\nu \phi_i}{\mathcal{N}\Omega} . \tag{5.25}$$

The sum in the numerator can be transformed to read

$$\sum_{i=1}^{k} \sum_{\nu=1}^{\kappa} \Omega_\nu \, n_i^\nu \phi_i = \sum_{i=1}^{k} \phi_i \left[\sum_{\nu=1}^{\kappa} \Omega_\nu \, n_i^\nu \right] .$$

Thus, we obtain

$$\langle \phi \rangle = \sum_{i=1}^{k} \pi_i \phi_i , \tag{5.25a}$$

where

$$\pi_i = \frac{\sum_{\nu=1}^{K} \Omega_\nu \, n_i^\nu}{\mathcal{N}\Omega} \qquad \text{and} \qquad \sum_{i=1}^{k} \pi_i = 1 . \tag{5.25b}$$

This is the weighting factor associated with a value ϕ_i, which characterizes a single quantum state in the Γ-space of a single system. Equation (5.25a), together with the definition contained in Equation (5.25b), encompasses the averaging rule which is a consequence of our statistical hypotheses. The weighting factor for one Γ-state is equal to the ratio of the number of times it occurs in all distributions to the total number of systems and represents the probability of finding it in a random sampling. Thus Equation (5.25b) represents the appropriate probability function defined on the elements of the Γ-space conceived as the sample space.

5.11.3. The Most Probable Distribution

The numerical example discussed in Section 4.4.1 convinces us that the number of arrangements in a distribution Ω_ν regarded as a function of the occupation numbers n_i^ν possesses a very sharp maximum Ω^* for a particular set n_i^* of occupation numbers, tending in the limit to a product of Dirac delta functions. Thus in the sum

$$\sum_{\nu=1}^{K} \Omega_\nu \, n_i^\nu$$

the term $\Omega^* n_i^*$ dominates over all the others. In the limit when $\mathcal{N} \to \infty$ and when, therefore, all the n_i^ν also tend to infinity, this sum can be replaced by the single term $\Omega^* \, n_i^*$. Similarly, the sum

$$\Omega = \sum_{\nu=1}^{K} \Omega_\nu \qquad (= \Omega^* \text{ in the limit})$$

can be replaced by the single term Ω^*. [The values n_i^* as well as Ω^* must, of course, be computed subject to the constraints imposed by Equations (5.20a, b).]

If this is the case, we may put

$$\pi_i^* = \frac{n_i^*}{\mathcal{N}}, \tag{5.26}$$

and the expectation value becomes simply

$$\langle \phi \rangle = \sum_{i=1}^{k} \pi_i^* \phi_i \qquad \text{with} \qquad \sum_{i=1}^{k} \pi_i^* = 1 . \tag{5.27}$$

This signifies that the expectation value may be computed by taking into account only the *most probable occupation numbers*, with the averaging performed over a single distribution. This single distribution is characterized by the fact that it yields the largest number of quantum states in the Γ_m-space and is therefore called the *most probable distribution*. The designation is justified because this particular distribution will be observed so much more frequently than any other that we may ignore the others completely. For this reason, the method which we adopt is known as the *method of the most probable distribution*. It amounts to accepting the probability function (5.26) as an excellent approximation to that derived in Equation (5.25b) from statistical hypothesis 4.

5.11.4. Mathematical Form of the Probability Function

If we were to find that distribution which renders Ω_v in Equation (5.22) a maximum without the constraints imposed by Equation (5.20a, b), the most probable distribution would be uniform with

$$\bar{n}_1^* = \bar{n}_2^* = \cdots = \bar{n}_i^* = \frac{\mathcal{N}}{k} \qquad (= \text{const}).$$

However, the example discussed in Section 4.2, or a moment's reflection, will convince the reader that, in our case, the most probable distribution cannot possibly be uniform. If it were, the average would be

$$\frac{1}{\mathcal{N}}\left[\sum_{i=1}^{k} \bar{n}_i^* E_i\right] = \frac{\sum_{i=1}^{k} E_i}{k} \neq E$$

and could not attain the *prescribed* value E, except accidentally.

To determine the most probable distribution, we consider that the occupation numbers n_i^v constitute continuous variables. This is justified on the grounds that the numbers n_i^v can be made as large as we please, if $\mathcal{N}$ has been chosen large enough. For this reason we omit superscript v in Equation (5.22) and proceed to find the maximum of

$$\Omega(E, V, N) = \frac{\mathcal{N}!}{\prod_{i=1}^{k} n_i!}, \tag{5.28}$$

now treated as a function of the n_i. We shall employ the method of Lagrange's undetermined multipliers. Our interest rests actually on $\ln \Omega$, as might be guessed from Boltzmann's principle, Equation (5.18). This allows us to express the factorials with the aid of Stirling's approximation given in Equation (4.18). Since the n_i may be chosen as large as we please by choosing $\mathcal{N}$ large enough, no essential error is committed.

Referring to Section 4.2, where we explain Lagrange's method, we seek the maximum of the function

$$\ln \Omega(n_i) = \ln \mathcal{N}! - \sum_{i=1}^{k} \ln n_i! . \tag{5.28a}$$

Applying Stirling's formula, $\ln x! = x \ln x - x$, we consider the function

$$\ln \mathcal{F} = \ln \mathcal{N}! - \sum_{i=1}^{k} (n_i \ln n_i - n_i) + \alpha \left[\left(\sum_{i=1}^{k} n_i \right) - \mathcal{N} \right] - \beta \left[\left(\sum_{i=1}^{k} n_i E_i \right) - \mathcal{N} E \right] \tag{5.28b}$$

Here α and $-\beta$ denote the two undetermined multipliers which stem from the auxiliary conditions (5.20a) and (5.20b):

$$\sum_{i=1}^{k} n_i = \mathcal{N} \qquad \text{and} \qquad \sum_{i=1}^{k} n_i E_i = \mathcal{N} E . \tag{5.29a, b}$$

A negative multiplier, $-\beta$, has been chosen for future convenience.

The k occupation numbers n_i^* must satisfy the two Equations (5.29a, b) and k counterparts of Equations (4.12). These are of the general form

$$\frac{\partial \ln \mathcal{F}}{\partial n_i} = -\ln n_i - 1 + 1 + \alpha - \beta E_i = 0 ,$$

and the occupation numbers n_i^* which render $\ln \Omega$ a maximum are determined by the k equations

$$\ln n_i^* = \alpha - \beta E_i \qquad (i = 1, 2, \ldots, k) . \tag{5.30}$$

Hence, we have

$$n_i^* = e^{\alpha} e^{-\beta E_i} , \tag{5.30a}$$

The Lagrangian multipliers should now be evaluated with reference to Equations (5.29a, b), but this is possible only with respect to α. Substituting the result from Equation (5.30a) into (5.29a), we find that

$$e^{\alpha} = \frac{\mathcal{N}}{\sum_{i=1}^{k} e^{-\beta E_i}} . \tag{5.30b}$$

The undetermined multiplier β should be obtained from the relation

$$E = \frac{\sum_{i=1}^{k} n_i^* E_i}{\mathcal{N}} = \frac{\sum_{i=1}^{k} E_i e^{-\beta E_i}}{\sum_{i=1}^{k} e^{-\beta E_i}} , \tag{5.31}$$

which cannot be solved explicitly. However, this is not a serious obstacle for our further development. Instead of solving Equation (5.31), we retain β as a supplementary parameter and carry it along with the equation, merely noting that it will be necessary to discover its physical significance; this will be accomplished in Section 5.13.

The expression

$$\mathscr{Z} = \sum_{i=1}^{k} e^{-\beta E_i} \tag{5.32}$$

is known as the canonical *partition function* of the system; we shall see shortly that it plays the same part in our theory as a fundamental equation does in classical thermodynamics. It will turn out to be closely related to the Massieu function $J = -(E - TS)/T$ (Section 5.13). Here the partition function appears expressed in terms of the k values E_i and of the parameter β, and we may put

$$\mathscr{Z} = \mathscr{Z}(\beta, E_i) \qquad (k + 1 \text{ variables})\,. \tag{5.32a}$$

In our present reasoning, the parameter β plays the part of a constant. However, if we examine a reversible process whereby the energy E and the external parameter V of the system change, β will change also. Therefore, we note that in arguments which involve processes, β will appear as a variable, and the same is true of the energy levels E_i.

We have now determined the most probable occupation numbers n_i^*; the set of them is known as the *Maxwell–Boltzmann* distribution. Combining Equation (5.30b) with Equation (5.30a), we can express them in the form of the probability function

$$\pi_i^* = \frac{n_i^*}{\mathscr{N}} = \frac{e^{-\beta E_i}}{\sum_{i=1}^{k} e^{-\beta E_i}} = \frac{e^{-\beta E_i}}{\mathscr{Z}}. \tag{5.33}$$

The multiplicative term $\exp(-\beta E_i)$ is the *Boltzmann factor*.

The preceding derivation shows clearly how the imposition of the constraint on energy, that is (ultimately), the need to satisfy the First law of thermodynamics, modulates the values of the most probable occupation numbers n_i^* and causes them to depart from equality. When $\beta \to 0$, the Boltzmann factors become equal to unity, and

$$\mathscr{Z} \to k. \tag{5.34}$$

Now, k is the total number of microstates (Γ-cells) of the system, and

$$n_i^* \to \frac{\mathscr{N}}{k}.$$

In this special case, the occupation numbers n_i^* become equal, each being equal to the *average* $\mathscr{N}/k$, as is the case in the absence of the constraint on energy. This fact can also be inferred directly from Equation (5.28b), because putting $\beta = 0$ in it amounts to disregarding the constraint on energy.

Figure 5.8 is a graphical representation of the probability function π_i^* from Equation (5.33) in terms of energy denoted here by E_j. The diagram is only suggestive, because normally the energy levels are closely spaced and

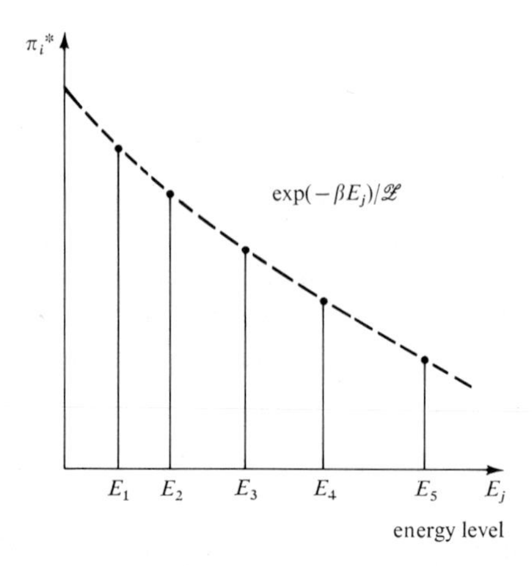

FIGURE 5.8. Probability function π_i^* in terms of the energy level E_j.

their number is enormous. The Boltzmann factor in the numerator has the largest value for the lowest energy, decreasing very sharply as E_j increases. This tendency for the lowest energy levels to acquire the highest population of systems can also be observed in the crude numerical example discussed in Section 4.2. This diagram may be contrasted with Figure 5.9, in which

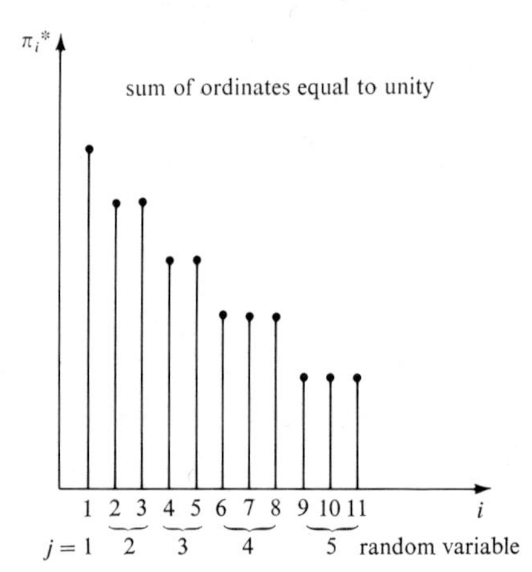

FIGURE 5.9. Probability function π_i^* in terms of the random variable i (counting number for quantum cells in Γ-space).

the probability function has been plotted as a function of the random variable i, on the supposition that energy levels, E_2, E_3 are doubly degenerate, levels E_4 and E_5 having a degeneracy of three. In this diagram, all ordinates must add up to unity. The ratio of the most probable occupation numbers in two arbitrary microstates λ and μ in Γ-space is

$$\frac{n_\lambda^*}{n_\mu^*} = \frac{e^{-\beta E_\lambda}}{e^{-\beta E_\mu}}. \tag{5.35}$$

5.11.5. Canonical Ensemble

The probability function π_i^* from Equation (5.33), which we have adopted for future calculations, is identical with that stipulated in statistical hypothesis 5, Equation (5.12), for a canonical ensemble. This is not accidental. The macrosystem, for which Equation (5.33) represents an excellent approximation in the case where the number of systems, $\mathcal{N}$, is very large, consists of interacting systems. Thus we can regard each system as being in equilibrium with $\mathcal{N} - 1$ systems. As $\mathcal{N} \to \infty$, the $\mathcal{N} - 1$ systems approach an ideal thermal bath more and more closely.

As the number of systems, $\mathcal{N}$, increases, the expectation value

$$\langle E \rangle = \sum_{i=1}^{k} \pi_i^* E_i = \frac{\sum_{i=1}^{k} E_i \, e^{-\beta E_i}}{\mathscr{Z}} \tag{5.36}$$

approaches the average prescribed value, E, more and more closely; it would be exactly equal to it if the (still unevaluated) probability function π_i from Equation (5.25b) had been used. This can be seen if Equation (5.36) is rewritten in terms of energy levels. Since every degenerate microstate appeared in our original list on page 188 as many times as its degeneracy g_j demanded, the partition function may also be computed from the formula

$$\mathscr{Z} = \sum_{j=1}^{l} g_j \, e^{-\beta E_j}. \tag{5.37}$$

We have changed the running index (as we did in Figure 5.8) from i to j to avoid confusion, and we have denoted the number of distinct (though degenerate) energy levels by l. In the same notation, Equation (5.36) becomes

$$\langle E \rangle = \frac{\sum_{j=1}^{l} g_j E_j \, e^{-\beta E_j}}{\mathscr{Z}}; \tag{5.38}$$

it shows that the probability function defined on the energy, E_j, treated as the stochastic variable, is of the form

$$\pi(E_j) = \frac{g_j \, e^{-\beta E_j}}{\mathscr{Z}}. \tag{5.38a}$$

Since the degeneracy in a macroscopic system is very large and increases sharply with E_j, the product of a sharply increasing function g_j and a sharply decreasing function $\exp(-\beta E_j)$ in the numerator develops a maximum and tends to a Dirac delta function.† This maximum occurs closer and closer to the given, average value E as the number $\mathcal{N}$ increases, and the corresponding term in the sum (5.38) dominates over all the others. Thus in the limit when $\mathcal{N} \to \infty$, the systems in an isolated macrosystem become increasingly statistically independent and behave more and more closely like the systems of a canonical (or a microcanonical) ensemble.

5.12. Partition Function

The partition function

$$\mathscr{Z} = \sum_{i=1}^{k} e^{-\beta E_i} = \sum_{j=1}^{l} g_j \, e^{-\beta E_j} \tag{5.39}$$

justifies its name because its terms govern the distributions or "partition" of systems among the quantum states in Γ-space. Since it extends over all quantum states, it is called *Zustandsumme*—"sum-over-the-states"—in the German language.

> *The principal problem in statistical mechanics has now been reduced to that of computing the partition function for various systems on the basis of the mechanical models assumed for them.*

The calculation of the partition function is particularly simple when the mechanical model of the system assumes the existence of independent and distinguishable elements. In such cases the energy E of the system becomes equal to the sum of the energies $\varepsilon_a, \varepsilon_b, \ldots$ of the elements. An example of the existence of such independent, distinguishable elements occurs in an ideal solid crystal. The various degrees of freedom in a gaseous molecule also constitute *independent* elements, but the problem is complicated in that the molecules themselves as well as identical atoms in a molecule are *not distinguishable*. This affects the count of quantum states, as we know from Section 3.9, owing to the need to symmetrize or to antisymmetrize the wave function. We leave this latter problem aside (see Section 6.2) and concentrate on distinguishable elements. Each element obeys its own, *separated* Schrödinger equation, each containing only f coordinates in their Hamiltonian operators instead of the Nf which must be included in the Hamiltonian of N interacting elements.

We may now define a partition function for each element separately. To begin with, we suppose that there are only two such elements and define a random variable a $(=1, 2, \ldots, A)$ for element a and a random variable

† We have made essentially the same observation earlier on p. 180.

b $(=1, 2, \ldots, B)$ for element b. The separate partition functions are defined as formal analogs of Equation (5.32):

$$Z_a = \sum_{a=1}^{A} e^{-\beta\varepsilon_a}; \qquad Z_b = \sum_{b=1}^{B} e^{-\beta\varepsilon_b}. \tag{5.40}$$

Evidently, the sums are taken here over the restricted phase space of each element.† Since the energy E_i of the whole system is equal to the sum $E_i = \varepsilon_a + \varepsilon_b$ at every quantum state, we find that

$$\mathscr{Z} = \sum_{i=1}^{k} e^{-\beta E_i} = \sum_{i=1}^{k} e^{-\beta(\varepsilon_a + \varepsilon_b)}.$$

Every possible combination of states i will be included if, instead of summing over i $(=1, 2, \ldots, k)$, we take the sum over both a $(=1, 2, \ldots, A)$ and b $(= 1, 2, \ldots, B)$. Hence we see that

$$\mathscr{Z} = \sum_{a=1}^{A} \sum_{b=1}^{B} e^{-\beta\varepsilon_a} e^{-\beta\varepsilon_b},$$

or

$$\mathscr{Z} = \sum_{a=1}^{A} \left[e^{-\beta\varepsilon_a} \sum_{b=1}^{B} e^{-\beta\varepsilon_b} \right],$$

because we may sum all the terms with a particular value of ε_a and sum over the index b first. In this manner we prove that

$$\mathscr{Z} = Z_b \sum_{a=1}^{A} e^{-\beta\varepsilon_a},$$

or that

$$\mathscr{Z} = Z_a Z_b. \tag{5.41}$$

The reader who cannot grasp this derivation at once may write out several terms of the sum or notice that the product

$$Z_a Z_b = (e^{-\beta\varepsilon_1{}^a} + e^{-\beta\varepsilon_2{}^a} + \cdots)(e^{-\beta\varepsilon_1{}^b} + e^{-\beta\varepsilon_2{}^b} + \cdots)$$

consists of a sum of exponentials,

$$e^{-\beta(\varepsilon_1{}^a + \varepsilon_1{}^b)} + e^{-\beta(\varepsilon_1{}^a + \varepsilon_2{}^b)} + \cdots,$$

in which all combinations of $\varepsilon_a + \varepsilon_b$ are represented.

In the case of an arbitrary number of independent, distinguishable elements, we derive by induction that

$$\mathscr{Z} = \prod_k Z_k \qquad (= Z_a Z_b Z_c \cdots). \tag{5.42}$$

† It would be the μ-space in the case of N independent molecules in a gas.

When there exist N *identical* elements, we obtain

$$\mathscr{Z} = Z^N. \tag{5.43}$$

We conclude that

the partition function of independent distinguishable elements is equal to the product of the partition functions of the elements.

This means that their logarithms are additive.

5.13. Change in the Partition Function during a Reversible Process

The statistical derivations of the preceding sections can be summarized in the form of four equations.

Statistical function: $$\pi_i^* = \frac{n_i^*}{\mathcal{N}} = \frac{e^{-\beta E_i}}{\mathscr{Z}}; \tag{5.44}$$

Energy: $$E = \sum_{i=1}^{k} \pi_i^* E_i = \frac{\sum_{i=1}^{k} E_i\, e^{-\beta E_i}}{\mathscr{Z}}; \tag{5.45}$$

Any mechanical quantity: $$\phi = \sum_{i=1}^{k} \pi_i^* \phi_i = \frac{\sum_{i=1}^{k} \phi_i\, e^{-\beta E_i}}{\mathscr{Z}}; \tag{5.46}$$

Partition function: $$\mathscr{Z}(\beta, E_i) = \sum_{i=1}^{k} e^{-\beta E_i}. \tag{5.47}$$

The equations describe states of equilibrium and relate the mechanical averages (we have omitted the sign $\langle\ \rangle$ for expectation values because it is now superfluous) to the energy levels E_i and the Lagrangian constant β.

We are now ready to construct the still missing link with classical thermodynamics by considering a most general, elementary reversible process in a closed system ($N = \text{const}$), during which the partition function changes by $d\mathscr{Z}$ as a result of changes in β and E_i by $d\beta$ and dE_i, respectively. The additivity property of $\ln \mathscr{Z}$ suggests that we should give preference to $\ln \mathscr{Z}$ over $\mathscr{Z}$ itself. Our task, then, is to calculate

$$d \ln \mathscr{Z}(\beta, E_i) = \frac{\partial \ln \mathscr{Z}}{\partial \beta} d\beta + \sum_{i=1}^{k} \frac{\partial \ln \mathscr{Z}}{\partial E_i} dE_i. \tag{5.48}$$

To facilitate the computation of the partial derivatives, we observe the following algebraic properties of the partition function which rest on the properties of the exponential function:

$$\frac{\partial \mathscr{Z}}{\partial \beta} = -\sum_{i=1}^{k} E_i\, e^{-\beta E_i}, \qquad \text{so that} \qquad -\frac{\sum_{i=1}^{k} E_i\, e^{-\beta E_i}}{\mathscr{Z}} = \frac{1}{\mathscr{Z}} \frac{\partial \mathscr{Z}}{\partial \beta}. \tag{a}$$

Similarly, we have

$$\frac{\partial \mathscr{Z}}{\partial E_i} = -\beta\, e^{-\beta E_i}, \qquad \text{so that} \quad -\beta \frac{e^{-\beta E_i}}{\mathscr{Z}} = \frac{1}{\mathscr{Z}} \frac{\partial \mathscr{Z}}{\partial E_i}. \tag{b}$$

Equation (a) combined with Equation (5.45) leads to

$$\frac{\partial \ln \mathscr{Z}}{\partial \beta} = -E, \tag{c}$$

and Equation (b) combined with Equation (5.44) yields

$$\frac{\partial \ln \mathscr{Z}}{\partial E_i} = -\beta \frac{n_i^*}{\mathscr{N}}. \tag{d}$$

Substitution of Equations (c) and (d) into Equation (5.48) provides us with the desired total change in d ln $\mathscr{Z}$:

$$d \ln \mathscr{Z} = -E\, d\beta - \frac{\beta}{\mathscr{N}} \sum_{i=1}^{k} n_i^*\, dE_i. \tag{5.48a}$$

Before we draw our final conclusions, we notice that the preceding equation contains E instead of its increment dE which appears in the expression $dQ^\circ = dE + dW^\circ$ of the First law. It is, therefore, convenient to perform a Legendre transformation in Equation (5.48a) and to change it to

$$d(\ln \mathscr{Z} + \beta E) = \beta \left[dE - \frac{1}{\mathscr{N}} \sum_{i=1}^{k} n_i^* dE_i \right]. \tag{5.48b}$$

In this form, the left-hand side of the equation represents a total differential of the function

$$\sigma = \ln \mathscr{Z} + \beta E,$$

which signifies that the right-hand side must also be a perfect differential. If this is the case, then β must play the part of an integrating factor, and we are justified in guessing that Equation (5.48b) constitutes the statistical-mechanical counterpart of the familiar classical equation

$$dS = \frac{dE + \sum_r Y_r\, dz_r}{T}, \tag{5.48c}$$

where the Y_r are the generalized forces and the z_r are the generalized coordinates for the system. This would interpret $1/\beta$ as related to the thermodynamic temperature, T, and σ as related to the entropy, S, of the system.

The preceding results could be accepted quite formally and made the starting point of a complete theory of thermodynamics. We would then *say* that we had *defined* temperature and entropy as related to β^{-1} and σ, respectively, and would proceed to explore their physical significance. In the methodology adopted in this book, it is necessary to satisfy ourselves that the preceding *interpretation* of terms is, indeed, correct. This we shall proceed to do in the next section but must not fail to notice that the *mathematical structure* of the two equations is identical, which proves that, *fundamentally*, the microscopic and macroscopic points of view are *consistent*. This is the remarkable achievement of statistical thermodynamics.

As a final introductory point, it is useful to note at this stage, subject to future detailed verification (Section 6.3), that the parameter β should be made proportional, rather than equal to T^{-1}, the factor of proportionality being related to Boltzmann's constant. Thus we put

$$\beta = \frac{1}{\mathbf{k}T}, \tag{5.48d}$$

the need to do so stemming from our desire to preserve the existing, perfect-gas temperature scale intact. Assuming that $\mathbf{k} \equiv 1$ would define a different unit of temperature, still preserving the properties of conventional temperature scales. By a comparison of terms, it is then possible to verify that making

$$S = \mathbf{k}(\ln \mathscr{Z} + \beta E) = \mathbf{k} \ln \mathscr{Z} + E/T \tag{5.48e}$$

reduces Equation (5.48b) to Equation (5.48c) *exactly*, provided that we also accept the equality

$$dW^\circ = \sum_r Y_r \, dz_r = -\frac{1}{\mathscr{N}} \sum_{i=1}^{k} n_i^* \, dE_i. \tag{5.48f}$$

These substitutions demonstrate that

$$\mathbf{k}T \ln \mathscr{Z} = TS - E,$$

or that the Helmholtz function is given by

$$F = -\mathbf{k}T \ln \mathscr{Z}. \tag{5.49}$$

In other words, except for a constant of proportionality, the logarithm of the partition function turns out to be identical with the Massieu function

$$J = -\frac{E - TS}{T} = \mathbf{k} \ln \mathscr{Z}.$$

5.14. Comparison with Classical Thermodynamics

An understanding of the reasons which cause the statistical equation (5.48b) to be identical in physical content with the macroscopic equation (5.48c) can be found through a discussion of the effect of interactions on the microscopic parameters. This provides a physical parallel to the mathematical discussion of the preceding section. It is known from classical thermodynamics that it is necessary to consider two kinds of interactions with the surroundings when a closed system undergoes a reversible process, namely, those resulting in a transference of heat and work, respectively. On the other hand, we know from quantum mechanics that the energy levels E_i accessible to a system can be influenced only by a change in the external parameters, that is, *by the performance of work*. For example, in our discussion of the quantum states accessible to a particle in a container given in Section 3.8, Equation (3.27b), we found that an increase in volume V reduced the energy of a quantum state described by the quantum numbers s_x, s_y, s_z. Hence a positive displacement ($\mathrm{d}V > 0$) caused the energy levels to decrease ($\mathrm{d}\varepsilon_i < 0$). The same is true of a macroscopic system. The factor $1/\mathcal{N}$ arises from the existence of $\mathcal{N}$ systems in our macrosystem, the sum

$$\sum_{i=1}^{k} n_i^* \mathrm{d}E_i$$

representing the work of changing the energy levels of $\mathcal{N}$ representative systems.

When heat alone is transferred to a system, the energy E of the macrosystem changes with all levels E_i remaining intact. This can occur only by a rearrangement of the occupation numbers n_i^* in agreement with Equation (5.45). Thus, in Equation (5.48b), the term

$$\sum_{i=1}^{k} n_i^* \mathrm{d}E_i$$

vanishes, and the term

$$\frac{\mathrm{d}(\ln \mathscr{Z} + \beta E)}{\beta} = \mathrm{d}E$$

must be equal to the reversible heat, so that

$$\mathrm{d}Q^\circ = T\,\mathrm{d}S = \frac{\mathrm{d}(\ln \mathscr{Z} + \beta E)}{\beta}.$$

The structure of the term is such that we must have

$$T \text{ proportional to } 1/\beta \quad \text{and} \quad S \text{ proportional to } (\ln \mathscr{Z} + \beta E), \tag{5.50}$$

and we choose Boltzmann's constant $\mathbf{k}$ as our coefficient of proportionality. This leads us from Equation (5.48d) to the relation

$$\mathrm{d}S = \mathbf{k}\,\mathrm{d}(\ln \mathscr{Z}) + \mathrm{d}(E/T),$$

and so to Equation (5.49) for the Helmholtz function; simultaneously, the argument shows that β, like T, must be positive for systems with "open-ended" energy levels.

When work alone is transferred reversibly, $dS = 0$, and the left-hand side of Equation (5.48b) vanishes. The change in the external parameters causes dE_i to differ from zero, and the change in the energy of the system is

$$dE = \frac{1}{\mathcal{N}} \sum_{i=1}^{k} n_i^* dE_i \,;$$

it is equal to the sum of the products of the most probable equilibrium occupation numbers and the energy changes per one of the $\mathcal{N}$ systems in the ensemble. This change in energy represents the negative of reversible work, the negative sign in Equation (5.48b) being dictated by convention. Thus the identification of terms in Equation (5.48f) becomes understandable.

The two preceding types of interactions are illustrated in Figures 5.10 and 5.11, which represent the effect of a reversible addition of heat or work,

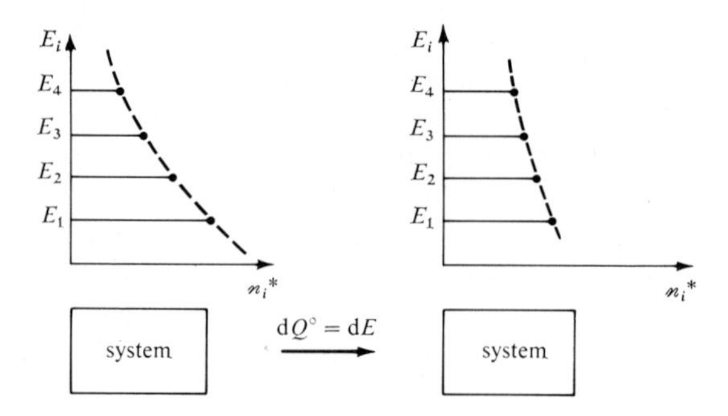

FIGURE 5.10. Effect on the most probable distribution of reversibly adding heat to a system. (Energy levels remain unchanged, but systems are "forced" into higher energy levels.)

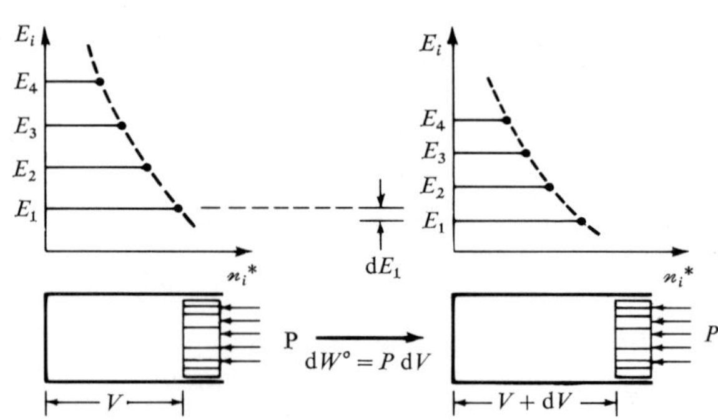

FIGURE 5.11. Effect on the most probable distribution of reversibly performing work. (Energy levels are lowered, but occupation numbers remain unchanged.)

respectively, on the distribution of systems among the energy levels. Degeneracy has been disregarded for the sake of simplicity.

The clear link between statistical and classical thermodynamics, for example, in the form of Equation (5.49) for the Helmholtz function or in the form of (5.48e) for entropy, proves that the logarithm of the partition function expressed in terms of β and the E_i plays the part of a fundamental equation, as we have anticipated. Since

$$\beta = 1/\mathbf{k}T \qquad \text{and} \qquad E_i = E_i(N, z_r)\,,$$

the Helmholtz function in Equation (5.49) contains implicitly the variables

$$T,\ N,\ V \qquad (\text{or} \quad T,\ N,\ z_r)\,.$$

Thus statistical considerations reconfirm that the above constitutes the canonical set of variables for this potential. Performing Legendre transformations, we can retrieve the formulas known to us from previous studies.

5.15. Explicit Formulas; Chemical Potential

The last observation in the preceding section makes it unnecessary to calculate the thermodynamic properties of systems, such as pressure or any of the generalized forces, Y_r, by direct averaging. Instead, we may make use of the numerous thermodynamic relations, being certain that the equations of all such properties are implied in $\ln \mathscr{Z}$.

For example, we may infer that

$$Y_r = \frac{1}{\beta}\left(\frac{\partial \ln \mathscr{Z}}{\partial z_r}\right)_{z_s,\,T,\,N} \tag{5.51}$$

from the fact that

$$Y_r = -\left(\frac{\partial F}{\partial z_r}\right)_{z_s,\,T,\,N},$$

as seen from Equation (1.19) with the obvious substitution of

$$\sum_r Y_r\,\mathrm{d}z_r \qquad \text{for} \quad P\,\mathrm{d}V.$$

Here the subscript z_s reminds us that all external parameters are kept constant, except the one involved in the differentiation. Equation (5.51) implies that explicit expressions for the energy levels E_i have been substituted into $\ln \mathscr{Z}$ in terms of the parameters z_r as a result of having solved the appropriate Schrödinger equation.

As a special case of this equation, we notice that

$$P(V, T) = -\left(\frac{\partial F}{\partial V}\right)_T = \mathbf{k}T\left(\frac{\partial \ln \mathscr{Z}}{\partial V}\right)_T, \tag{5.51a}$$

where the constancy of N is understood. The resulting functional relation is the thermal equation of state of the system. An explicit expression is obtained when the energy levels E_i are known as functions of the volume V.

Similarly, the internal energy U of a system may be calculated directly from Equation (5.45) or from

$$U(T, V) = \left[\frac{\partial}{\partial(1/T)}\left(\frac{F}{T}\right)\right]_V = -T^2\left[\frac{\partial}{\partial T}\left(\frac{F}{T}\right)\right]_V,$$

that is, from

$$U = -\left(\frac{\partial \ln \mathscr{Z}}{\partial \beta}\right)_{E_i}. \tag{5.52}$$

The last equation is a direct consequence of Equation (a) in Section 5.13 and the preceding equation can be derived from the fact that

$$\mathrm{d}\left(\frac{F}{T}\right) = U\,\mathrm{d}\left(\frac{1}{T}\right) - \left(\frac{P}{T}\right)\mathrm{d}V$$

is a perfect differential.

Apart from the external parameters z_r, the energy levels E_i depend on the number of particles N present in a system. The statistical derivations are independent of whether the elementary particles are identical or whether several species of them are present. The only assumption is that the solution to the appropriate Schrödinger equation for the multicomponent system is available and that the series of energy levels $E_i = E_i(z_r, N_k)$ is known. We denote here a representative species by the running subscript k and propose to use the symbol N_l for all species except N_k. The fact that the derivations stipulate a set of closed systems in the macrosystem has not affected the form of the equilibrium equations which represent their correct dependence on all the N_k. The same reasoning which allowed us to discuss a reversible process in Section 5.13 can now be invoked to discuss the effect of reversibly changing the number of particles, N_k, of one kind. Consequently, the chemical potential of species k can be obtained with the aid of Equation (1.33). Hence, we have

$$\mu_k = \left(\frac{\partial F}{\partial N_k}\right)_{T,\, z_r,\, N_l} = -\mathbf{k}T\left(\frac{\partial \ln \mathscr{Z}}{\partial N_k}\right)_{T,\, z_r,\, N_l}. \tag{5.53}$$

It is now useful to summarize the most important equations which link classical with statistical formulas. In writing them, we assume the existence of several external, extensive parameters z_r (with an equal number of conjugate intensive parameters Y_r) and of several species N_k. (It should be noted that N_k is the number of *particles*, not mols.) The equations are

$$\mathscr{Z} = \sum_{i=1}^{k} \exp\left[-E_i(z_r, N_k)/\mathbf{k}T\right]; \tag{5.54}$$

$$F(T, z_r, N_k) = -\mathbf{k}T \ln \mathscr{Z}; \tag{5.55}$$

$$S = -\left(\frac{\partial F}{\partial T}\right)_{z_r, N_k} = \mathbf{k}T\left(\frac{\partial \ln \mathscr{Z}}{\partial T}\right)_{z_r, N_k} + \mathbf{k} \ln \mathscr{Z}; \tag{5.56}$$

$$Y_r = -\left(\frac{\partial F}{\partial z_r}\right)_{T, z_s, N_k} = \mathbf{k}T\left(\frac{\partial \ln \mathscr{Z}}{\partial z_r}\right)_{T, z_s, N_k} \tag{5.57}$$

(z_s denotes all external, extensive parameters except z_r);

$$E = -T^2\left(\frac{\partial(F/T)}{\partial T}\right)_{z_r, N_k} = \mathbf{k}T^2\left(\frac{\partial \ln \mathscr{Z}}{\partial T}\right)_{z_r, N_k}; \tag{5.58}$$

$$\mu_k = \left(\frac{\partial F}{\partial N_k}\right)_{T, z_r, N_l} = -\mathbf{k}T\left(\frac{\partial \ln \mathscr{Z}}{\partial N_k}\right)_{T, z_r, N_l}. \tag{5.59}$$

5.16. Boltzmann's Principle

Boltzmann's principle, which was stated in anticipation in Section 5.11, provides a direct interpretation of the number Ω^* of most probable microstates, and it is useful now to derive it, realizing that Ω^* is interchangeable with the total number of microstates Ω. The derivation will simultaneously allow us to acquire a new interpretation of entropy in terms of the probabilities π_i.

We revert to the macrosystem and consider Equation (5.28a), replace Ω by Ω^*, approximate the factorials by Stirling's formula, and introduce $\pi_i^* = n_i^*/\mathscr{N}$ into it. This leads us to the following succession of formulas:

$$\ln \Omega^* = \mathscr{N} \ln \mathscr{N} - \mathscr{N} - \sum_{i=1}^{k} n_i^* \ln n_i^* + \sum_{i=1}^{k} n_i^*,$$

that is,

$$\ln \Omega^* = \mathscr{N} \ln \mathscr{N} - \mathscr{N} \sum_{i=1}^{k} \pi_i^* \ln (\mathscr{N} \pi_i^*)$$

because

$$\mathscr{N} = \sum_{i=1}^{k} n_i^*.$$

Hence, we have

$$\ln \Omega^* = \mathcal{N} \ln \mathcal{N} - \mathcal{N} \ln \mathcal{N} \left(\sum_{i=1}^{k} \pi_i^* \right) - \mathcal{N} \sum_{i=1}^{k} \pi_i^* \ln \pi_i^* ,$$

and, since

$$\sum_{i=1}^{k} \pi_i^* = 1 ,$$

we obtain finally

$$\ln \Omega^* = - \mathcal{N} \sum_{i=1}^{k} \pi_i^* \ln \pi_i^*. \tag{5.60}$$

This important relation demonstrates that the probabilities π_i^* are chosen to render the function

$$\sum_{i=1}^{k} \pi_i \ln \pi_i$$

a *minimum*, subject to the condition that the energy

$$E = \sum_{i=1}^{k} \pi_i E_i$$

is prescribed, and that

$$\sum_{i=1}^{k} \pi_i = 1 .$$

Next, we turn to Equation (5.48e) and notice that it can also be written

$$S = \mathbf{k} \ln \mathcal{Z} + \mathbf{k}\beta \sum_{i=1}^{k} \pi_i^* E_i .$$

On the other hand, from Equation (5.44), we infer that

$$-\beta E_i = \ln(\pi_i^* \mathcal{Z}) .$$

Hence, we have

$$S = \mathbf{k} \ln \mathcal{Z} + \mathbf{k} \sum_{i=1}^{k} \pi_i^* (\beta E_i) = \mathbf{k} \ln \mathcal{Z} - \mathbf{k} \sum_{i=1}^{k} \pi_i^* \ln \pi_i^* - \mathbf{k} \sum_{i=1}^{k} \pi_i^* \ln \mathcal{Z},$$

or, finally,

$$S = -\mathbf{k} \sum_{i=1}^{k} \pi_i^* \ln \pi_i^* . \tag{5.61}$$

A comparison with Equation (5.60) leads to Boltzmann's principle for an isolated macrosystem in the form

$$\underset{\text{macrosystem}}{S} = \mathbf{k} \ln \Omega \qquad (= \mathscr{N} S), \tag{5.62}$$

where Ω has been substituted for Ω^* because $\ln \Omega \to \ln \Omega^*$ for $\mathscr{N} \to \infty$. This result is identical with that in Equation (5.18a). Applying the combinatorial Equation (5.21), we can now easily understand the relation expressed in Equation (5.18), in which Ω refers to the number of quantum states in the Γ-space of a single system. In a macrosystem,

$$\Omega = \Omega_{\text{sys}}^{\mathscr{N}} \tag{5.63}$$

because every quantum state of a system can be combined with every quantum state of any other to form a cell in Γ_{m}-space.

Equation (5.62) leads to the discovery of a direct link between entropy and probability. It is useful to combine Boltzmann's principle with Equation (5.48e) and to record the relation

$$\Omega = \mathscr{Z}^{\mathscr{N}} \, \mathrm{e}^{\mathscr{N} E/\mathbf{k}T}. \tag{5.64}$$

The preceding exhibits clearly the fact that the sum over the states, or partition function $\mathscr{Z}$, can be used interchangeably with the total number of quantum states of an isolated system, Ω_{sys}, for the development of theory. Either expression constitutes the statistical analog of a fundamental equation, and in given circumstances that function is used which is found to be more convenient, in the firm knowledge that both must lead to exactly the same conclusions. It is easy to guess that for every thermodynamic potential it is possible to construct a statistical analog. This circumstance leads to a great variety of different expositions of the subject in different treatises and proves to be a source of confusion for the beginner, as anticipated in Section 5.1.

At equilibrium, the number of most probable states is a maximum, and this leads to the conclusion that the function

$$-\sum \pi_i \ln \pi_i$$

is a maximum too (it is a minimum when the negative sign is omitted). It follows, in turn, that the entropy must be a maximum for given values of energy and external parameters. Thus, we find a statistical confirmation of the equilibrium principle given in Section 1.4.8. In other words, the most probable distribution corresponds to the equilibrium state.

5.17. The Laws of Thermodynamics

The considerations of Section 5.14 make it possible to assert that the equations of statistical thermodynamics can be used to deepen the meaning of the laws of thermodynamics. The First law of thermodynamics is a direct consequence of the conservative nature of all models, and the first part of the

Second law was dealt with in Sections 5.13 and 5.14. To complete the study, we turn to the second part of the Second law, which was summarized in Chapter 1. Subsequently, we shall formulate the Third law of thermodynamics.

5.17.1. The Second Part of the Second Law of Thermodynamics

When an isolated system is in an equilibrium state, 1, it has accessible to it a number of quantum states in its phase space, all equally probable. Generally speaking, the removal of a constraint extends the range of one or several of the external parameters z_r of the isolated system without changing its energy. In turn, the range of one or several coordinates q_i is thus extended, and a new set of quantum cells becomes available to the system in its phase space. It is true that a change in the range of the parameters z_r affects the structure of the phase space of the elements of the system and the distribution of energy among them, but in Γ-space the effect can be illustrated with the aid of Figure 5.12. In the second equilibrium state, 2, the number of available quantum cells, Ω_2, exceeds the original number, Ω_1, and we see that

$$\Omega_2 > \Omega_1 .$$

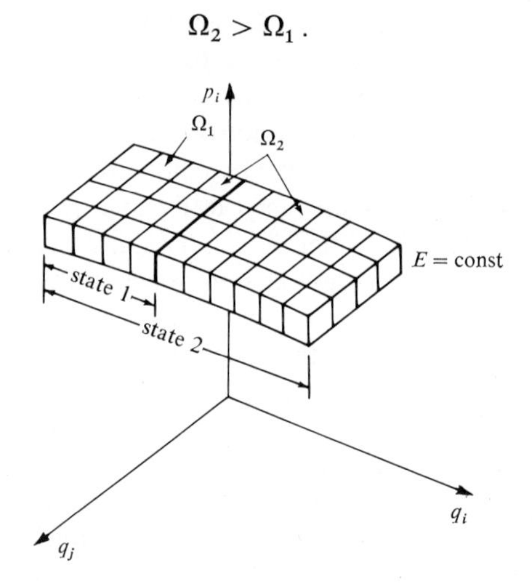

FIGURE 5.12. Second part of the Second law: effect of extending the range of a coordinate q_i in Γ-space.

Statistical hypotheses 1 and 3 assert that microstates are never observed and that only averages over suitable stationary ensembles can be measured. If we now imagine a microcanonical ensemble of isolated systems, we notice that the sample space associated with it expands. Thus the macroscopic properties of the system must change, with Boltzmann's principle providing the simplest connection between the microscopic and the macroscopic characteristics. Equation (5.18) shows immediately that

$$S_2 > S_1,$$

which is the second part of the Second law. It states that, if the energy E of the system does not change, the entropy must achieve a new maximum. In this manner we obtain a statistical interpretation of the second part of the Second law and of the principle of entropy maximum.

These principles do not specify the time required to effect the change. The order of magnitude of this time is known as the *relaxation time* of the system with respect to the process. If this relaxation time is short compared with the time during which measurements are made, we say that a process at a finite rate is observed. If the relaxation time is very large compared with the observation time, the rate is said to be very slow and the process may be ignored altogether. The latter occurs frequently when systems tend to chemical or phase equilibrium.

It is interesting to calculate the ratio of Ω_2 to Ω_1 in a particular example. We consider 1 mol of gas whose volume is allowed to double. In such conditions, we find that

$$S_2 - S_1 = \mathbf{R} \ln 2 \qquad \text{and} \qquad \mathbf{R} \ln 2 = \mathbf{k} \ln \frac{\Omega_2}{\Omega_1}.$$

Since $\mathbf{N} = \mathbf{R}/\mathbf{k}$, we obtain

$$\frac{\Omega_2}{\Omega_1} = 2^{\mathbf{N}}.$$

The calculation shows that there is an enormous increase in the number of states available to the system.

Since the original sample space is a subset of the new sample space, we can calculate the probability that the system will occupy the old sample space when the new space has become fully occupied after the lapse of time equal to the relaxation time. This is

$$\pi = \frac{\Omega_1}{\Omega_2} = 2^{-\mathbf{N}},$$

which is an exceedingly small number; it expresses the probability of observing the properties of state 1 compared with those of state 2.

Strictly speaking, we have not *proved* the second part of the Second law of thermodynamics. However, we made it very plausible to assert that the removal of constraints can never succeed in decreasing the number of states available to an isolated system. This was easily seen in the examples discussed earlier, but it is possible to think of cases where this is more difficult. For example, when a supersaturated liquid or undercooled vapor is turned into a two-phase system by the introduction of nuclei of condensation, the structure of the Γ-space is not easy to perceive. Similarly, it is not easy to see that the number of states does not change during a reversible adiabatic process in an isolated system.

It is instructive to show that the reinsertion of the constraint does not materially alter the number of states.†

5.17.2. Third Law of Thermodynamics

In the introductory section on Boltzmann's principle, Section 5.10, it was mentioned that the constant C in the expression $S = \mathbf{k} \ln \Omega_{\text{sys}} + C$ was independent of the parameters of the system and that we could assume it to be equal to zero for convenience. As the state of the system is changed, the constant C remains unaffected, and only the term

$$S = \mathbf{k} \ln \Omega_{\text{sys}}$$

becomes modified. The Third law of thermodynamics is concerned with the limit of this quantity when $T \to 0$, that is, when $\beta \to \infty$. This limit can be evaluated with reference to Equation (5.48e), in which the explicit expressions from Equations (5.45) and (5.47) are substituted:

$$S = \mathbf{k} \ln \sum_{i=1}^{k} \mathrm{e}^{-E_i/\mathbf{k}T} + \frac{\sum_{i=1}^{k} E_i \, \mathrm{e}^{-E_i/\mathbf{k}T}}{T \sum_{i=1}^{k} \mathrm{e}^{-E_i/\mathbf{k}T}}. \tag{5.65}$$

As T decreases, the exponents $-E_i/\mathbf{k}T$ become larger and larger, causing $\exp(-E_i/\mathbf{k}T)$ to be very small. Thus the terms in the sums which correspond to high values of energy, E_i, may be omitted, and ultimately only the lowest level needs to be retained. Suppose that this lowest, or ground, level is E_0 and that its degeneracy is g_0; since the sums are taken over the quantum states, we obtain‡

$$S \approx \mathbf{k} \ln(g_0 \, \mathrm{e}^{-E_0/\mathbf{k}T}) + \frac{g_0 \, E_0 \, \mathrm{e}^{-E_0/\mathbf{k}T}}{T g_0 \, \mathrm{e}^{-E_0/\mathbf{k}T}}$$

and

$$S = \mathbf{k} \ln g_0 .$$

† The reader may wish to work out this example himself or to consult the following two references: J. E. Mayer and M. G. Mayer, *Statistical Mechanics*, pp. 81f and 90, Wiley, New York, 1940; J. A. Fay, *Molecular Thermodynamics*, pp. 105f, Addison-Wesley, Reading, Massachusetts, 1965.

‡ Note that $\mathbf{k} \ln \mathrm{e}^{-E_0/\mathbf{k}T} = -E_0/T$.

We can derive this expression at once from Boltzmann's principle if we realize that as $T \to 0$ only the ground state needs to be counted.

Experimental evidence seems to be consistent with the statement that

the ground state of any system which is in internal equilibrium is non-degenerate.†

Consequently, we put $g_0 = 1$ and conclude that

$$\lim_{T \to 0} S = 0, \tag{5.66}$$

regardless of the variation of any other parameter of the system. If we did not assume that $C = 0$, we would have

$$\lim_{T \to 0} S = C, \tag{5.66a}$$

with the arbitrary constant C having the same value at any temperature, also regardless of the value of any other parameter. The preceding statement is known as the *Third law of thermodynamics* or as *Nernst's heat theorem.* When $C = 0$ is assumed, the resulting value of S is called *absolute entropy*; the corresponding formulation of the Third law was conceived by Planck.

When applied to a homogeneous system, the Third law provides a convenient, absolute value of entropy, and the assumption that $C = 0$ relieves us of the need to remember that the constant C for a combined system consisting of systems 1 and 2 must be put equal to the sum, $C_1 + C_2$, of the constants chosen for the systems 1 and 2 separately. In addition, the Third law asserts that the difference

$$\Delta S = S(T, x_2) - S(T, x_1) \to 0 \quad \text{as} \quad T \to 0 \tag{5.67}$$

of the entropy of the system at temperature T and at a value x_2 of any other property (such as pressure or volume) and of the entropy at T and x_1 must vanish in the limit when $T \to 0$.

The full significance of the Third law becomes evident when chemical reactions and phase transformations are examined. For example, if we consider an allotropic transition between two crystalline forms of the same substance or the phase transition between solid and liquid helium, we are dealing in each case with two equilibrium states of the same substance which co-exist at some temperature T and have different entropies. The Third law asserts that

† The problem of whether experimental results can yield definite statements about the degeneracy of the ground state is discussed in M. Klein, *Laws of Thermodynamics*, in *Rendiconti S.I.F.*, X Corso, p.122, Academic Press, New York, (1960). See also, H. B. G. Casimir, *Zeitschr. F. Phys.* **171** (1963), 246.

the entropy difference, ΔS, vanishes as $T \to 0$. The same is true for the difference in the entropy of the reactants and products of a chemical reaction.

The foregoing statements remain true even in the case where the system exists in a state of internally constrained equilibrium, on the condition that both states are constrained in an identical manner. For example, at low temperatures there may exist solid solutions which are capable of performing a chemical reaction but in which such a reaction does not occur, owing to its slow rate. Nevertheless, the entropy difference between two such constrained states at different pressures vanishes in the limit $T \to 0$. On the other hand, the entropy difference between one such system which is in the constrained state of internal equilibrium and another which is not may be different. In other words, the removal of the internal constraint causes a change in entropy even in the limit when $T \to 0$. The same applies to changes in isotopic composition, to the alignment of nuclear spins, and so on.

For example, the rotational degrees of freedom of hydrogen differ in their spins, and a normal sample exists in the form of a metastable mixture of parahydrogen with spin number $s = 0$ and of orthohydrogen with spin number $s = 1$. The relative abundance encountered in nature is 1 molecule of parahydrogen to 3 molecules of orthohydrogen. The same is true of deuterium; there is 1 molecule of paradeuterium to 2 molecules of orthodeuterium. Transitions between states with different spins are rare, and the process may be thought of as impeded by an internal constraint. Each species can be said to attain zero entropy at $T = 0$ separately, but the difference between the entropies of the two species is not zero at $T = 0$.

The possibility of defining an absolute value of entropy depends on the degeneracy of the ground state. The same cannot be said about energy which depends on the value E_0 assigned to the ground state. Referring to Equation (5.45), we see that at very low temperatures

$$E \approx \frac{g_0 E_0 \, e^{-E_0/\mathbf{k}T}}{g_0 \, e^{-E_0/\mathbf{k}T}} = E_0 .$$

If the value E_0 assigned to the ground state is changed arbitrarily to $E_0 = E_0'$, the energy will change by an amount E_0', that is, by an arbitrary constant. Thus, the zero-point energy of a system, that is, its energy at absolute zero, is arbitrary. Nevertheless, differences in the energy between any two states of a closed system are not arbitrary, as the reader knows from his earlier studies.

5.17.3. Concluding Remarks

The preceding discussion shows that in the context of statistical thermodynamics the laws of thermodynamics appear as merely probable rather than certain. This is due to the fact that our derivations were based on the properties of the most probable distribution. It is clear, however, that in the large number of microstates assumed by a system there may exist those that deviate from the average by a macroscopically detectable amount and the probability of such occurrence increases as the number of particles in a system becomes smaller. Thus it must be expected that *fluctuations* about a state of equilibrium will sometimes be observed (Chapter 14).

Recognizing the importance of such phenomena, it is nevertheless useful for us to know that they do not play a big part in systems which are our main concern in this book. A feeling for what is involved can be obtained from the calculation of density fluctuations in a subvolume of 1 cm^3 of a gas under standard conditions performed by R. Becker.† The calculation analyzed the average recurrence time of a prescribed deviation from the equilibrium density and yielded the following numerical results:

Relative deviation in density	Average recurrence time
2×10^{-10}	4×10^{-3} sec
3×10^{-10}	1 sec
4×10^{-10}	1.3×10^{3} sec = 21 min
5×10^{-10}	1.3×10^{7} sec = 5 month
6×10^{-10}	10^{12} sec = 3×10^{4} yr
7×10^{-10}	5×10^{17} sec = 2×10^{10} yr

5.18. The Method of the Most Probable Distribution and the Grand Canonical Ensemble‡

In the course of the historical development of statistical mechanics, the method of the most probable distribution was not applied at once to an isolated collection of $\mathcal{N}$ localized, macroscopic systems, as we have done in Section 5.11. Instead, the method was used directly to evaluate the most probable distribution of molecules among the different quantum states in μ-space for an isolated system. This argument led to correct results in the case of a crystalline solid whose elements are localized, but produced erroneous results for gases whose molecules are indistinguishable. Therefore, it is interesting to understand the circumstances in which the method of Section 5.11 can be applied to cases other than those discussed in that section.

† *Theorie der Wärme*, p. 103, Springer, Berlin, 1961.
‡ This section can be omitted on first reading.

5.18.1. Collection of Subsystems

Instead of building a macrosystem in the form of a collection of $\mathcal{N}$ systems, we may mentally subdivide an isolated system into $\mathcal{N}$ macroscopic subsystems. Given that the number of molecules in a gas is of the order of 10^{23}, $\mathcal{N}$ can be chosen as large as 10^8 leaving still 10^{15} molecules in a subsystem. However, in a gas the subsystems cannot be closed because molecules are free to move over the whole volume. We may argue that the number of molecules crossing a boundary is proportional to the surface, whereas that inside it is proportional to the volume. This would lead us to expect that the exchange of molecules between subsystems is of minor importance. With this proviso we may apply the reasoning of Section 5.11 to such a collection of subsystems, realizing that the Γ_m-space is now analogous to the Γ-space, and that each Γ-space of a subsystem (Γ_s-space) contains vastly fewer coordinates than the Γ-space proper. This would lead us to the conclusion that the probability of finding a subsystem in a quantum state i in Γ_s-space is proportional to the Boltzmann factor. In short, the theory would be unchanged, and we would find that in the most probable distribution the subsystems are statistically independent even if they are quite small measured on a macroscopic scale. A similar idea recurs in continuum thermodynamics and is at the back of the principle of local state.

However, there exists one important difference: we may not pass to the limit $\mathcal{N} \to \infty$ in the face of a finite number of molecules, N. If we increase the number of subsystems, the number of molecules in a system will decrease. Eventually the absence of the localizability of molecules will assert itself and the method will break down. This means that a Maxwell–Boltzmann distribution among subsystems will exist approximately, the approximation being very good on the condition that the subsystems are small enough, but sufficiently large to be localizable. No such distribution may be expected among the molecules of a perfect gas and over the quantum states in μ-space, even though the molecules themselves are treated as statistically independent. It is this fact that created seemingly insurmountable difficulties during the nineteenth century (Lord Kelvin's "clouds"). We shall see in Section 6.9 how this defect ought to be corrected. No such difficulties occur in the theory of crystalline solids because their elements *are* localizable.

It is useful to recognize another point at which the theory fails to carry over. When the subsystem consists of one molecule in a gas or one element of a solid, the combinatorial formula (5.22) refers to distributions in Γ-space, and the occupation numbers are those in μ-space. We recall from Section 3.14 that the number of quantum states in μ-space vastly exceeds the number of molecules. Hence an overwhelming number of occupation numbers must assume very small values, many of them being equal to 0 or 1. Thus the use of

Stirling's approximation can no longer be justified. Nevertheless, the result stands, as was shown by C. S. Darwin and R. H. Fowler, who obtained it as well as results for nonlocalizable molecules by the method of mean values (saddle-point method).†

5.18.2. Grand Canonical Distribution

The circumstances that the combinatorial formula (5.22) presupposes closed subsystems is easily remedied. Instead of defining occupation numbers n_i^{ν} for quantum states at a constant number of elements, N, we specify occupation numbers n_{ij}^{ν}, each corresponding to a quantum state i and a specified number of elements N_j. The occupation numbers can be displayed in an array (Figure 5.13). A similar array can be imagined for the energies E_{ij}. To

Quantum state		Number of molecules index j from 1 to l						
		N_1	N_2		N_j		N_l
Quantum states index i from 1 to k	1	n_{11}^{ν}	n_{12}^{ν}	$\cdots$	n_{1j}^{ν}	$\cdots$	n_{1l}^{ν}
	2	n_{21}^{ν}	n_{22}^{ν}	$\cdots$	n_{2j}^{ν}	$\cdots$	n_{2l}^{ν}
		$\vdots$	$\vdots$		$\vdots$		$\vdots$
	i	n_{i1}^{ν}	n_{i2}^{ν}	$\cdots$	n_{ij}^{ν}	$\cdots$	n_{il}^{ν}
		$\vdots$	$\vdots$		$\vdots$		$\vdots$
	k	n_{k1}^{ν}	n_{k2}^{ν}	$\cdots$	n_{kj}^{ν}	$\cdots$	n_{kl}^{ν}

FIGURE 5.13. Array of occupation numbers for grand canonical distribution. (A similar array can be written for the energies E_{ij}.)

each number of molecules N_j there corresponds a different Γ-space, but a double set of numbers n_{ij}^{ν} still specifies a distribution. One column in the array corresponds to a set of closed systems which now forms a subset of the set of n_{ij}^{ν}.

The combinatorial formula remains the same, except that we now have to deal with kl "containers." Here k denotes the highest number of quantum states and l ranges from 1 (or 0) to $\mathcal{N}N$. Thus we obtain

$$\Omega_{\nu} = \frac{\mathcal{N}\,!}{\prod_{i=1,j=1}^{k,l} n_{ij}^{\nu}\,!}\,. \tag{5.68}$$

† For details see, for example, A. Sommerfeld, *Thermodynamics and Statistical Mechanics*, Vol. V of "Lectures on Theoretical Physics," (J. Kestin, transl.), p. 259, Academic Press, New York, 1956 and 1964. Another method of overcoming the difficulties with Stirling's approximation consists in defining occupation numbers for large collections of cells ("macro-cells"), *ibid*, p. 219.

The most probable distribution is found subject to three constraints:

$$\sum_{j=1}^{l}\sum_{i=1}^{k} n_{ij}^{\nu} = \mathcal{N}, \qquad \sum_{j=1}^{l}\sum_{i=1}^{k} n_{ij}^{\nu} E_{ij} = \mathcal{N}E, \qquad \sum_{j=1}^{l}\sum_{i=1}^{k} n_{ij}^{\nu} N_j = \mathcal{N}N. \tag{5.68a}$$

Thus there will appear three Lagrangian multipliers: α, $-\beta$, and $-\gamma$.

We leave the details of the computation to the reader and note only the result. The most probable distribution is

$$n_{ij}^{*} = e^{\alpha}\, e^{-\beta E_{ij}}\, e^{-\gamma N_j}. \tag{5.69}$$

The probability function is

$$\pi_{ij}^{*} = \frac{e^{-\beta E_{ij}}\, e^{-\gamma N_j}}{\Xi}, \tag{5.70}$$

where

$$\Xi = \sum_{j=1}^{l}\sum_{i=1}^{k} e^{-\beta E_{ij}}\, e^{-\gamma N_j} \tag{5.71}$$

is the *grand canonical partition function*. The probability function is identical with that quoted in Section 5.9, except that Equation (5.70) refers to molecules of one kind.

Following the argument given in Section 5.13 for the canonical partition function, we now proceed to establish the analogous link between the grand canonical partition function and the equations of classical thermodynamics. To this end we consider the change in the logarithm of the grand canonical partition function during a reversible process when β changes by $d\beta$, γ changes by $d\gamma$ and the E_{ij} change by dE_{ij}. The change dE_{ij} is brought about by varying the external parameters, z_r, symbolized by the volume, V. Since the number of particles in the macrosystem remains constant, the N_i's remain constant, because they range from 1 (or 0) to $N\mathcal{N}$, as already stated. In other words, the index j in the symbol n_{ij}^{ν} really stands for N_j, and no derivative with respect to it needs to be included. Thus we can write the counterparts of Equations (a)–(d) in Section 5.13,

$$\frac{\partial \ln \Xi}{\partial \beta} = -E; \tag{a$'$}$$

$$\frac{\partial \ln \Xi}{\partial \gamma} = -N; \tag{b$'$}$$

$$\frac{\partial \ln \Xi}{\partial E_{ij}} = -\beta \frac{n_{ij}^{*}}{\mathcal{N}}; \tag{c$'$}$$

$$\frac{\partial \ln \Xi}{\partial N_j} = -\gamma \sum_{i=1}^{k} \frac{n_{ij}^{*}}{\mathcal{N}}. \tag{d$'$}$$

In this manner, the total differential of the partition function becomes

$$\mathrm{d} \ln \Xi = -E \, \mathrm{d}\beta - N \, \mathrm{d}\gamma - \beta \sum_{j=1}^{l} \sum_{i=1}^{k} \frac{n_{ij}^{*}}{\mathcal{N}} \, \mathrm{d}E_{ij} . \tag{5.72}$$

The Legendre transform of Equation (5.72) with respect to the pairs E, β and N, γ yields the form

$$\mathrm{d}(\ln \Xi + \beta E + N\gamma) = \beta \left[\mathrm{d}E + \frac{\gamma}{\beta} \mathrm{d}N - \sum_{j=1}^{l} \sum_{k=1}^{k} \frac{n_{ij}^{*}}{\mathcal{N}} \, \mathrm{d}E_{ij} \right], \tag{5.73}$$

which can be compared directly with Equation (1.31), namely

$$\mathrm{d}S = \frac{\mathrm{d}U}{T} - \frac{\mu \, \mathrm{d}N}{T} + \frac{P \, \mathrm{d}V}{T} .$$

Consequently, the various terms in Equation (5.73) acquire the following interpretations:

$$\beta = \frac{1}{\mathbf{k}T}, \tag{5.74a}$$

$$\gamma = -\beta\mu, \tag{5.74b}$$

$$-\sum_{j=1}^{l} \sum_{i=1}^{k} \frac{n_{ij}^{*}}{\mathcal{N}} \, \mathrm{d}E_{ij} = P \, \mathrm{d}V \qquad \left(\text{or} \sum_{r} Y_r \, \mathrm{d}z_r\right), \tag{5.74c}$$

$$\ln \Xi + \beta E + N\gamma = \frac{S}{\mathbf{k}}, \tag{5.74d}$$

where μ is the chemical potential of the system.

The last of the above equations can be put in the form

$$PV = \mathbf{k}T \ln \Xi(T, V, \mu) . \tag{5.75}$$

We leave it to the reader to derive the succeeding, explicit formulas that allow us to compute the thermodynamic properties of systems from their grand canonical partition function.

$$S = \mathbf{k}T \left(\frac{\partial \ln \Xi}{\partial T} \right)_{z_r, \mu} + \mathbf{k} \ln \Xi, \tag{5.76a}$$

$$N = \mathbf{k}T \left(\frac{\partial \ln \Xi}{\partial \mu} \right)_{z_r, T} . \tag{5.76b}$$

$$Y_r = \mathbf{k}T \left(\frac{\partial \ln \Xi}{\partial z_r} \right)_{\mu, T} = \mathbf{k}T \frac{\ln \Xi}{z_r} . \tag{5.76c}$$

It is, of course, possible to think of a large variety of ensembles, each leading to a characteristic partition function which corresponds to one of the very many possible thermodynamic potentials and each conceived as a function of the appropriate canonical variables.† These are, usually, of marginal importance compared with $\mathscr{Z}$ and Ξ.

5.19. Summary

The present chapter was devoted to the study of the statistical method as it is applied to thermodynamics. The line of reasoning proved to be lengthy as well as intricate, and for these reasons it may be useful to review it quickly with the aid of the block diagram contained in Figure 5.14.

The science of statistical thermodynamics rests on a number of statistical hypotheses. Our exposition of it was based on hypotheses 1, 2, 3, and 4 (Sections 5.3 and 5.4). With their aid, together with the assumption that the mechanical models of thermodynamic systems must be conservative, it was possible to deduce the laws of thermodynamics (Sections 5.15 and 5.17) and the extremum principles (Section 5.16).

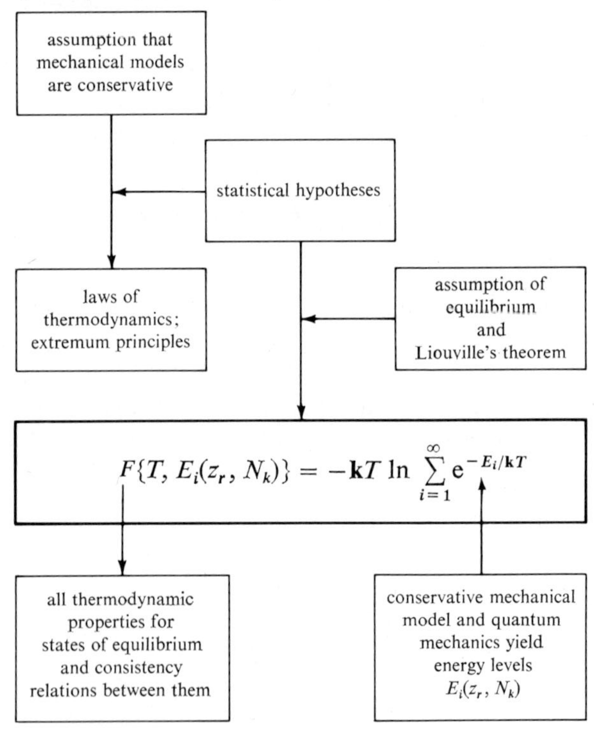

FIGURE 5.14. Block diagram for statistical thermodynamics of equilibrium states.

† See R. Kubo, *Statistical Mechanics*, p. 23, North-Holland Publ., Amsterdam, 1967.

The statistical hypotheses restricted by the assumption of equilibrium and an argument based on Liouville's theorem (Section 5.4) yielded one form of the fundamental equation. In our case it was the equation for the Helmholtz function, displayed in the central block of the diagram. In other presentations of the subject, this may be replaced by one of a number of equivalent forms. Paraphrasing E. Schrödinger, it is possible to state that *this equation contains all of equilibrium thermodynamics in a nutshell.*

The fundamental equation represents the Helmholtz potential in the form of an explicit function of the thermodynamic temperature T and of the set E_i of the quantum energy levels which the system may assume as a whole. The Boltzmann constant **k** appears in it only because we wish to retain our customary scale of temperatures. In contrast with Section 5.11, we now extend the sum over the infinite set of energy levels. This introduces no errors because the Boltzmann factor for the very high energy levels is negligibly small, but simplifies explicit calculations.

The set of quantum energy levels and their degeneracy is obtained when the mechanical properties of the conservative model have been established. This is done on the basis of quantum mechanics, for example, with the aid of Schrödinger's equation (Section 3.4). The energy levels depend on the prescribed external constraints z_r and the number $(N_1, \ldots, N_k)$ and nature of the particles in the model. This step in the procedure presents great mathematical difficulties in practice; it has yet to be taken in a large number of cases of practical importance, for example, in the theory of gases of high density.

The mechanical problem may sometimes be simplified by an appeal to N. Bohr's correspondence principle (Section 3.1) which allows us to use classical instead of quantum mechanics. In such cases the sum in the equation is replaced by an integral in accordance with the Euler–Maclaurin formula (Section 4.3).

A successful completion of the scheme yields numerical values of thermodynamic properties for states of equilibrium as well as the consistency relations between them.

PROBLEMS FOR CHAPTER 5†

5.1. Show that two trajectories in Γ-space cannot cross or combine into one trajectory.

5.2. Estimate the number of distributions for a macrosystem consisting of $N = 1000$ systems, each of which can assume one of $k = 1000$ energy levels. Disregard the constraint on total energy and apply Stirling's formula.

† Problems numbered in boldface type are considered relatively easy; they are recommended for assigning in an elementary first course.

5.3. Give a clear definition of an ensemble and of a macrosystem (as used in this textbook).

5.4. Compute the number of states, $\Omega(E)$, classically for a perfect gas of N atoms in a box of volume V, and using Equation (5.18) determine the equation of state.

SOLUTION: The number of states is

$$\Omega(E) = \frac{V^N}{\mathbf{h}^{3N}} \int \cdots \int d\mathbf{p}_1 \cdots d\mathbf{p}_n \qquad \text{with} \quad \sum p_i^2 \leq 2mE.$$

The integral is simply the volume of a $3N$-dimensional sphere of radius $2mE$. Therefore, using the results of Problems 4.30 and 4.36, we may write

$$\Omega(E) = \frac{V^N(2\pi mE)^{3N/2}}{\mathbf{h}^{3N}\Gamma[(3N/2)+1]}.$$

For large N, we may use Stirling's formula for $\ln \Gamma\left[\left(\frac{3N}{2}\right)+1\right]$. To obtain the pressure, we use the facts that $S = \mathbf{k} \ln \Omega$ and $P = T(\partial S/\partial V)_{E,N}$, which leads to $P = N\mathbf{k}T/V$.

5.5. Consider an isolated system of N independent particles, each of which can be in two energy states, $\varepsilon_1 = 0$ and $\varepsilon_2 > 0$. The total energy of the system is measured to be E. (a) Find the entropy and temperature of the system as a function of E. (b) At what energy will the entropy be a maximum?

5.6. Consider a classical harmonic oscillator with energy E. Find the volume $\mathscr{V}(E)$ in phase space for states with energy $\leq E$. Also determine the number of quantum states, $\Omega(E)$, for this oscillator, and show that for large values of E, $\mathscr{V}(E)/\mathbf{h} \sim \Omega(E)$. (*Hint:* The area of an ellipse with semiaxes a and b is πab. Use this fact to show that $\mathscr{V}(E) = E/\nu$, where ν is the classical frequency of the oscillator.)

5.7. Construct a statistical analogue of the Zeroth law of thermodynamics.

5.8. One form of the Second law of thermodynamics states that in a cyclic process heat cannot be entirely converted to work. Provide a molecular explanation for this statement.

5.9. Suppose that a constant value E_0 is added to all of the energy levels, E_k, for a quantum system, so that $E_k' = E_k + E_0$. Show that this transformation has no effect on the thermodynamic properties of the system.

5.10. Verify that the canonical equations of motion, $\dot{p}_i = -\partial\mathscr{H}/\partial q_i$ and $\dot{q}_i = \partial\mathscr{H}/\partial p_i$, are equivalent to the Newtonian equations when the system consists of (a) a free particle, with $\mathscr{H} = p^2/2m$; (b) a harmonic oscillator, with $\mathscr{H} = p^2/2m + kx^2/2$; (c) a collection of interacting particles for which the Hamiltonian is

$$\mathscr{H} = \sum_i \mathbf{p}_i^2/2m + \sum_{1 \leq i < j \leq N} \phi(|\mathbf{r}_i - \mathbf{r}_j|).$$

5.11. Derive and solve the Liouville equation for an ensemble of systems each of which consists of a single, free particle. Provide a geometrical interpretation for the equation.

5.12. Derive the Liouville equation for an ensemble of systems each of which consists of a single harmonic oscillator.

5.13. Derive Equations (555)–(559) in the text.

5.14. The energy levels for a harmonic oscillator with frequency ν are given by $\frac{1}{2}\mathbf{h}\nu$, $\frac{3}{2}\mathbf{h}\nu$, ..., $(n+\frac{1}{2})\mathbf{h}\nu$, ... Consider a system of N oscillators with total energy $E = \frac{1}{2}N\mathbf{h}\nu + M\mathbf{h}\nu$, and show that this energy state may be realized in

$$\Omega = (M + N - 1)!/M!(N-1)!$$

ways. Compute the entropy, S, and temperature, T. Express E as a function of T.

5.15. Show that the grand canonical partition function may be expressed in the form $\Xi = \sum \mathscr{Z}_N \exp(\mu\beta N)$ where $\mathscr{Z}_N$ is the canonical partition function of a system of N particles, and $\mathscr{Z}_0 \equiv 1$.

5.16. Prove that the property $p = PV$ expressed in terms of the variables T, P, μ_i yields a fundamental equation of state. (This is the appropriate potential for the grand canonical ensemble.)

5.17. Derive Equation (5.14) for the case of one species by considering a macrosystem and by applying an argument modeled on that discussed in Section 5.11. The macrosystem may be constructed by subdividing a system of $\mathscr{N}N$ molecules into $\mathscr{N}$ subsystems, each containing $N_1, N_2, \ldots, N_l$ molecules, respectively. The probability function

$$\pi_{ij} = A \exp[\beta(\mu N_j - E_{ij})]$$

should then be obtained as that for the most probable distribution, Equation (5.70), with

$$\frac{1}{A} = \Xi = \sum_{j=1}^{l} \sum_{i=1}^{k} \exp[\beta(\mu N_j - E_{ij})]$$

representing the grand canonical partition function.

Indicate the generalization to r species.

5.18. (a) Referring to Problems 5.16 and 5.17 and to Section 5.18.2, show that

$$PV = \mathbf{k}T \ln[\Xi(T, V, \mu)].$$

(b) Using first classical formulas, and then statistical formulas, prove the validity of the following relations:

$$P = \left[\frac{\partial(PV)}{\partial V}\right]_{T,\mu} = \frac{\mathbf{k}T}{V} \ln \Xi;$$

$$N = \left[\frac{\partial(PV)}{\partial \mu}\right]_{T,V} = \mathbf{k}T\left[\frac{\partial \ln \Xi}{\partial \mu}\right]_{T,V};$$

$$S = \left[\frac{\partial(PV)}{\partial T}\right]_{V,\mu} = \mathbf{k} \ln \Xi + \mathbf{k}T\left[\frac{\partial \ln \Xi}{\partial T}\right]_{V,\mu}.$$

LIST OF SYMBOLS FOR CHAPTER 5

Latin letters

A, A'	Normalization factor
a	Random variable
B	Normalization factor
b	Random variable
C	Constant
$\mathbf{c}$	Velocity of a point in Γ-space
E	Energy
E_i	Energy level
E^*	Most probable energy
F	Helmholtz function; probability function
F_i	Generalized force
$\mathscr{F}$	Function defined by equation (5.28b)
f	Function appearing in Equation (5.15)
f	Number of degrees of freedom
G	Gibbs function
g	Degeneracy
$\mathscr{H}$	Hamiltonian
J	Massieu function
j	Running index
k	Running index; index for number of quantum cells
$\mathbf{k}$	Boltzmann's constant
l	Number of distinct energy levels
m_i	Generalized mass
N	Number of particles
$\mathbf{N}$	Avogadro number
$\mathscr{N}$	Number of identical replicas of a system; number of systems constituting a microcanonical ensemble
$\mathscr{N}$	Number of systems in microconical ensemble in Section 5.11
n_{ij}^{v}	Number of systems in a quantum state i and a specified number of elements N_j
n_i	Number of systems whose points are confined within volume $\Delta\mathscr{V}$
n_i^{v}	Number of systems in a quantum state i (occupation number in Γ-space)
p_i	Generalized conjugate momenta
Q	Heat
q_i	Generalized coordinates
$\mathbf{R}$	Universal gas constant
S	Entropy
s	Quantum number
T	Temperature
t	Time
U	Internal energy
V	Volume
$\mathscr{V}$	Volume in Γ-space

W	Number of accessible microstates or Γ-cells; work
x	Generalized property
Y_r	Generalized forces
Z	Individual partition function for each element of a system
$\mathscr{Z}$	Canonical partition function for a system
z_i	Set of extensive variables
z_r	Generalized coordinates

Greek letters

α, β, γ	Lagrangian multipliers
β	$(\mathbf{k}T)^{-1}$
Γ	Denotes 6N-dimensional phase space
δ	Dirac delta function
κ	Number of distinct distributions
μ	Chemical potential; 6-dimensional phase space
ν	Number of distributions in Γ-space
Ξ	Grand canonical partition function
Π	Probability function
π	Probability function in sample space
π_i	Probability function associated with volume element $\Delta\mathscr{V}$; or with quantum state i
ρ	Probability density function
ρ'	Average probability density function associated with volume element $\Delta\mathscr{V}$
σ	$(=\ln \mathscr{Z} + \beta E)$ in Equation (5.48b)
Φ	Flux through volume element $\Delta\mathscr{V}$
ϕ	Generalized average quantity; random variable; potential energy
Ω	Number of accessible microstates or Γ-cells
Ω_ν	Number of supermicrostates in distribution ν

Superscripts

—	Mean or expectation value
·	Time rate of change
′	First derivative of function with respect to its argument
″	Second derivative of function with respect to its argument
*	Most probable or maximum value
°	Reversible process; initial value

Subscript

sys	Denotes property of system
0	Lowest level or ground state

Special symbols

{ }	Denotes a set of variables
⟨ ⟩	Denotes mean or expectation value

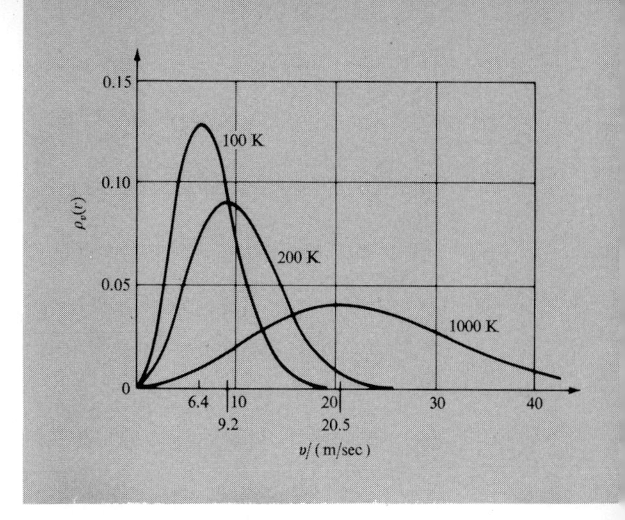

CHAPTER 6

PROPERTIES OF PERFECT GASES

6.1. Method

The statistical considerations presented in Chapter 5 reduced the problem of the evaluation of the equilibrium properties of thermodynamic systems to the determination of a single function: the partition function $\mathscr{Z}$. In general, an explicit expression for the partition function is obtained from the postulated mechanical model of the system. Once this has been made available, all the properties of interest are derived in a straightforward manner from the equations which have been summarized in Section 5.15, Equations (5.54)–(5.59).

The preceding general method has been applied with great success to a large variety of systems, and much work in this direction is currently still in progress. The difficulty in each case consists in postulating a physically satisfactory model which is simple enough to permit us to solve the corresponding Schrödinger equation and so to determine the energy levels. As stated earlier, we propose to devote the present chapter to the study of the simplest possible system, a perfect gas. The mathematical simplicity of this case hinges on the possibility of assuming that the elements of the model are statistically independent. The assumption of statistical independence permits us to represent the partition function in the form of a product of partition functions for the statistically independent elements. The elements themselves are endowed with simple properties, and their energy levels can be extracted from Chapter 3.

Speaking geometrically, the system partition function $\mathscr{Z}$ expresses the structure of quantum cells for the most probable distribution in Γ_m-space in terms of the energy levels of a single system in Γ-space. The assumption of statistical independence for the elements of a macroscopic system permits us to descend one level lower and to express the energy levels in Γ-space in terms of the quantum states of the elements in their respective μ-spaces. The elements considered need not be atoms or molecules. On the contrary, we shall find it convenient to select the various degrees of freedom of atoms or molecules as our statistically independent elements; in the theory of solid crystals and in the theory of vibrating polyatomic gaseous molecules, the independent elements will turn out to be synonymous with their normal modes.

6.2. The Partition Function of a Perfect Gas

We now undertake the task of specializing the general statistical theory to the case of a perfect gas. We consider the gas as an assembly of N statistically independent, identical, but nonlocalizable elements—the molecules. An exact calculation of the partition function which represents the properties of a gas over the whole range of temperatures is quite complex. For this reason we shall restrict ourselves to the derivation of an approximation; the approximation to be derived breaks down at very low temperatures, that is, in the neighborhood of absolute zero.

To achieve our goal, it seems best to start with Equation (5.57), which we repeat here for ease of reference:

$$\mathscr{L} = \sum_{j=1}^{k'} g_j \, e^{-E_j/\mathbf{k}T}, \tag{6.1}$$

and recall that E_j denotes a representative energy level of a single system in the common Γ-space of all systems. The degeneracy of one such level, denoted here by g_j, must now be calculated from the known properties of the elements in μ-space. The systems in a macrosystem were distinguishable, but the molecules in a system are not, and this fact forces us to modify the count as explained in Section 5.18.

We now recall the table given for a macrosystem on page 188 and define a distribution ν by indicating a set of occupation numbers n_i^{ν} in μ-space. The degeneracy of an energy level in Γ-space is simply given by the number Ω_ν of microstates, because a distribution uniquely defines an energy level E_ν of the system. Owing to the assumption of statistical independence, we have

$$E_\nu = \sum_{i=1}^{k} n_i^{\nu} \varepsilon_i, \tag{6.2}$$

where ε_i is the energy of a single molecule in a single particle quantum state i. As long as the occupation numbers n_i^{ν} for molecules in μ-space remain constant, the energy E_ν does not change. A change in any one n_i^ν (actually two owing to the requirement that

$$\sum_{i=1}^{k} n_i^\nu = N \tag{6.3}$$

for every distribution) does not necessarily cause the level E_ν to change, because the sum (6.2) is taken over all quantum states of a single molecule in μ-space and not over quantum levels. However, this is of no consequence when the partition function $\mathscr{Z}$ is evaluated because the sum extends over all microstates in Γ-space. In our present notation, Equation (6.1) must be written

$$\mathscr{Z} = \sum_{\nu=1}^{k} \Omega_\nu \, \mathrm{e}^{-E_\nu/\mathbf{k}T}. \tag{6.4}$$

It must be remembered that Ω_ν refers here to the Γ-space and not to the Γ_{m}-space as it did in Section 5.11.

If the molecules of a gas were localized, we would put

$$\Omega_\nu = \frac{N!}{\prod_{i=1}^{k} n_i^\nu !} \qquad \text{(distinguishable)} \tag{6.5a}$$

and promptly proceed to complete the theory. However, in this count the interchange of two molecules between two quantum states in μ-space produces a new microstate in Γ-space, contrary to our present requirements. With indistinguishable molecules, the mere specification of the occupation numbers n_i^ν for quantum states in μ-space produces a *single microstate*. Hence we obtain

$$\Omega_\nu = 1 \qquad \text{(indistinguishable)} \tag{6.5b}$$

in our present case. If all occupation numbers for the molecules were equal either to unity or to zero, the correction for indistinguishability would be very simple. As we know from Chapter 3, the number of quantum states in μ-space is very large at sufficiently high temperatures and exceeds the number of molecules in a system by orders of magnitude (see the example on page 105). Thus the preceding situation is likely to prevail under most circumstances, since most of the μ-states remain unoccupied.

The desired approximation is obtained by making the provisional assumption that the molecules in a gas *are* distinguishable. When this is done, we speak about *Maxwell–Boltzmann statistics*. The change from *systems* to *molecules* requires no more than a change from script to italic symbols. The corresponding partition function is then

$$\mathscr{Z}_{\mathrm{MB}} = \sum_{\nu=1}^{\kappa} \frac{N!}{n_1^\nu ! \, n_2^\nu ! \cdots} \exp[-\beta(n_1^\nu \varepsilon_1 + n_2^\nu \varepsilon_2 + \cdots)] \, ,$$

where β has been inserted for $(\mathbf{k}T)^{-1}$ to simplify the appearance of the formulas. The exponential in the preceding equation can be written as

$$(\mathrm{e}^{-\beta\varepsilon_1})^{n_1{}^\nu}(\mathrm{e}^{-\beta\varepsilon_2})^{n_2{}^\nu}\cdots,$$

and the observation made in Section 5.12 allows us to write

$$\mathscr{Z}_{\mathrm{MB}} = (\mathrm{e}^{-\beta\varepsilon_1} + \mathrm{e}^{-\beta\varepsilon_2} + \cdots)^N .$$

If we introduce the particle partition function

$$Z = \sum_{i=1}^{k} \mathrm{e}^{-\beta\varepsilon_i}, \tag{6.6}$$

we record that

$$\mathscr{Z}_{\mathrm{MB}} = Z^N \qquad \text{or} \qquad \ln \mathscr{Z}_{\mathrm{MB}} = N \ln Z . \tag{6.7a, b}$$

The fact that the occupation numbers are either zero or unity implies that every one of the N molecules is in a different quantum state in μ-space, and that

$$\prod_{i=1}^{k} n_i{}^\nu! \equiv 1 .$$

Thus the weighting factors Ω_ν used to derive Equations (6.7a, b) are each $N!$ times larger than the ones we should use, as shown in Equation (6.5b). Consequently, we may correct the results in Equations (6.7a, b) by introducing the divisor $N!$ into the first equation or by subtracting the Stirling approximation for $\ln N! = N \ln N - N$ from the second. This leads us to *corrected Maxwell–Boltzmann statistics* (also known as *quantum statistics in the classical limit*) for which

$$\mathscr{Z}_{\mathrm{cMB}} = \frac{Z^N}{N!} \tag{6.8a}$$

or

$$\ln \mathscr{Z}_{\mathrm{cMB}} = N \ln Z - N \ln N + N, \tag{6.8b}$$

with the particle partition function defined in Equation (6.6); it is once again a sum over *states*, but this time a sum over the quantum states of a single molecule in μ-space.

The introduction of the divisor $N!$ into the partition function of a macroscopic quantity of gas is crucial for the whole theory. Without it, as we shall see in Section 6.8.1, the expression for entropy would not yield an extensive property, though the formulas for other properties (except those that contain entropy) would remain unaffected. Before the advent of quantum mechanics, the classical theory assumed that the molecules of a gas were distinct, which led to the expressions in Equations (6.7a, b). The paradox involving entropy was discovered quite early, notably by J. W. Gibbs, and the correction factor was introduced in an artificial, *ad hoc* manner. Quantum mechanics resolved this paradox, and this fact contributed greatly to its acceptance.

The critical reader will object that the factor $(N!)^{-1}$ was introduced here also in a vague, *ad hoc* manner. This, actually, must be conceded with the excuse that a full justification would require more space than we could devote to this detail at this point. In a rigorous derivation it is necessary separately to evaluate the partition function for bosons (*Bose–Einstein statistics*) and for fermions (*Fermi–Dirac statistics*), and to show that for high enough temperatures both converge asymptotically to the form given in Equations (6.8a, b).† The rigorous argument will remove another difficulty which we shall encounter in Section 6.8.2, where the reader will find that the present theory does not lead to agreement with the Third law of thermodynamics. To achieve such agreement it is necessary to recognize that boson and fermion gases differ from perfect gases at very low temperatures in that they exhibit *gas degeneracy* which leads to departures from the perfect-gas law even for statistically independent molecules.‡

Apart from such difficulties, the standard theory of perfect gases is always based on the *corrected* Maxwell–Boltzmann partition function. However, to give the reader a somewhat better intuitive grasp of the point at issue, we advance a contrived, illustrative example in the next section.

Number of Microstates§

We assume that there are $N = 3$ molecules which can occupy equally spaced non-degenerate states ε_i, so that $\varepsilon_i = ih$ where h is an unspecified multiple of Planck's constant **h**. We assume that we are concerned with all distributions which lead to an energy level $E_\nu = 9h$. All such possible distributions have been displayed in Table 6.1, but the reader should attempt to reproduce the table for himself, starting with the highest energy level. Clearly, with three molecules, the highest occupied energy level can only be $\varepsilon_7 = 7h$. With $n_7{}^1 = 1$, we can admit just $n_1{}^1 = 2$ if the constraint on energy is to be satisfied. Moving down to $\nu = 2$, we put $n_6{}^2 = 1$, and this forces us to place one molecule each in ε_1 and ε_2; the remaining occupation numbers (not shown in the table) are obviously $n_2{}^1 = n_3{}^1 = \cdots = n_7{}^2 = 0$. When the highest-energy molecule is placed in level $\varepsilon_5 = 5h$, there are two choices, shown as $\nu = 3$ and $\nu = 4$, respectively. Proceeding in this way, the table can be completed in a systematic manner.

† This calculation is performed in Chapter 8.

‡ To be precise, neither fermions nor bosons are completely independent *statistically* because the state available to one molecule is influenced by that occupied by another; this is particularly clear for fermions which must satisfy the exclusion principle. What is implied here is simply that the Hamiltonian operator for the system is equal to a sum of N identical Hamiltonian operators, each of which is evaluated for an element.

§ This example has been modeled on one given by T. L. Hill, *Introduction to Statistical Thermodynamics*, p. 433, Addison-Wesley, Reading, Massachusetts, 1960.

TABLE 6.1

Count of Microstates for $N=3$, $k=7$, $E=9h$ with Equally Spaced Levels

									Number of microstates		
State number		1	2	3	4	5	6	7		$\Omega_\nu^{\text{MB}} = \dfrac{N!}{\prod_{i=1}^{7} n_i^\nu !}$	
Energy level		h	$2h$	$3h$	$4h$	$5h$	$6h$	$7h$	Ω_ν^{FD}		Ω_ν^{BE}
Occupation	1	2						1	0	3	1
numbers for	2	1	1				1		1	$6=3!$	1
distribution	3	1		1		1			1	$6=3!$	1
	4		2			1			0	3	1
	5	1			2				0	3	1
$\nu=$	6		1	1	1				1	$6=3!$	1
	7			3					0	1	1
Number of microstates in Γ-space,									$\Omega=3$	28	7
Total energy $E_\nu = 9h$ for every distribution.											

As long as the molecules are distinguishable, we have

$$\Omega_\nu^{\text{MB}} = \frac{N!}{\prod_{i=1}^{k} n_i^\nu !},$$

as we know from previous studies. When the molecules are indistinguishable, the count changes and becomes affected by Pauli's principle. When the exclusion principle is inoperative (Bose–Einstein), every set of occupation numbers describes a *single* Γ-state as indicated in the column labeled Ω_ν^{BE}. The same is true when the exclusion principle is operative (Fermi–Dirac), except that distributions with $n_i^\nu \neq 0$ or 1 become inadmissible. This means that distributions $\nu = 1, 4, 5, 7$ are excluded, leaving $\nu = 2, 3, 6$.

The present example shows the following total counts:

$$\Omega^{\text{FD}} = 3, \qquad \Omega^{\text{MB}} = 28, \qquad \Omega^{\text{BE}} = 7;$$

and it is clear from the method of counting that Ω_ν^{MB} is always larger than either Ω_ν^{FD} or Ω_ν^{BE}. With large numbers involved, these differences may become very large indeed. Upon further inspection, it is noticed that each admissible Fermi–Dirac distribution leads to a single state, whereas the same distribution leads to $N!$ Γ-states of distinguishable molecules. Hence we have

$$N!\Omega_\nu^{\text{FD}} \leq \Omega_\nu^{\text{MB}}.$$

Similarly, every distribution leads to a single Bose–Einstein state, and to *at most* $N!$ Γ-states of distinguishable molecules. This means that

$$N!\Omega_\nu^{\text{BE}} \geq \Omega_\nu^{\text{MB}}.$$

In other words,

$$\Omega_v^{FD} \le \frac{\Omega_v^{MB}}{N!} \le \Omega_v^{BE},$$

as can be verified with reference to the sums displayed at the bottom of the table.

When the total energy E_v becomes very large, the number of available μ-states also becomes large, and this happens generally at sufficiently high temperatures. It follows that most of the occupation numbers will be 1 or 0, thus tending to obliterate the distinction between the three numbers Ω_v^{FD}, $\Omega_v^{MB}/N!$, and Ω_v^{BE}. We may record this as

$$\Omega_v^{FD} \rightarrow \frac{\Omega_v^{MB}}{N!} \leftarrow \Omega_v^{BE} \qquad (E_v \text{ or } T \text{ sufficiently large}).$$

If the preceding result is accepted provisionally (an exact proof will be given in Chapter 8), we can immediately write Equations (6.8a, b), as was done in the preceding section.

6.3. Pressure and Thermal Equation of State of a Perfect Gas

The form of the Hamiltonian for a gas molecule given in Equation (3.47) shows that it is possible to associate a sum over states, Z_f, with every one of the independent degrees of freedom of translational and internal motion of a single molecule. The considerations of Sections 3.7 to 3.13 allow us to recognize that the energy levels of the translational degrees of freedom alone depend on volume V, those for the internal motion being independent of it. Therefore, it is convenient first to identify the *translational partition function per one degree of freedom*

$$Z_{tr} = \sum_{s=1}^{k} e^{-\varepsilon_s/\mathbf{k}T}, \tag{6.9}$$

where the series for ε_s is given in Equation (3.23b). We may write it as

$$\varepsilon_s = \frac{\mathbf{h}^2 s^2}{8mV^{2/3}} \qquad (s = 1, 2, \ldots). \tag{6.9a}$$

We shall denote the partition function for all remaining degrees of freedom by Z_{int}, and we may put

$$\mathscr{Z} = \frac{(Z_{tr}^3\, Z_{int})^N}{N!}, \tag{6.10a}$$

or

$$\ln \mathscr{Z} = N[\ln Z_{tr}^3(V) + \ln Z_{int}] - N \ln N + N. \tag{6.10b}$$

This form allows us to perceive that the pressure P computed from Equation (5.77) will be unaffected by the details of the internal structure of the molecules or by the quantum-mechanical correction term $N!$, since the derivatives of the logarithms of these terms with respect to volume vanish. In other words,

> *the pressure exerted on the walls of the vessel is due entirely to the translational motion of the molecules.*

The partition function Z_{tr} can be approximated by an integral, in accordance with the argument in Section 4.3. Such an approximation is valid for

$$\varepsilon_{s+1} - \varepsilon_s \ll \mathbf{k}T,$$

as was proved in Equation (4.20a). With reference to Equations (6.9a), we arrive at the condition that the use of an integral is justified if

$$\frac{\mathbf{h}^2}{8m\mathbf{k}TV^{2/3}} \ll 1. \tag{6.11}$$

Since the exponential is a decreasing function, we may assume that the number of μ-states $k \to \infty$. This allows us to accept the approximation

$$Z_{tr} \approx \int_0^\infty \exp\left(-\frac{\mathbf{h}^2 s^2}{8m\mathbf{k}TV^{2/3}}\right) \mathrm{d}s = \left[\frac{(2\pi m\mathbf{k}T)^{3/2}V}{\mathbf{h}^3}\right]^{1/3}, \tag{6.12}$$

as we know from Equation (4.20), in which $\Delta x = \Delta n = 1$ has been substituted.

The replacement of the sum of states by the integral of Equation (6.12) is essentially equivalent to the application of the correspondence principle of quantum mechanics, as we shall see in more detail in Section 6.4 on the classical partition function. Applying Equation (5.57), we conclude that

$$P = N\mathbf{k}T\left(\frac{\partial \ln Z_{tr}^3}{\partial V}\right)_{T,N} = \frac{N\mathbf{k}T}{V}. \tag{6.13}$$

Not surprisingly, we discover that the thermal equation of state of one mol of a perfect gas containing $N = \mathbf{N}$ molecules is

$$Pv_m = \mathbf{R}T \tag{6.13a}$$

since

$$\mathbf{Nk} = \mathbf{R}, \tag{6.13b}$$

where $\mathbf{N}$ denotes the Avogadro number and $\mathbf{R}$ is the universal gas constant. We have substituted v_m for V to emphasize that one mol of gas is being investigated.

We now revert to the remarks made in Section 5.13, where in Equation (5.48d) we provisionally substituted the Boltzmann constant $\mathbf{k}$ for the arbitrary constant k. If an arbitrary constant k had been retained, the right-hand side of Equation (6.13) would contain the factor

$$\mathbf{N}kT = \mathbf{Nk}\left(\frac{k}{\mathbf{k}}\,T\right) = \mathbf{R}\tilde{T}\,,$$

showing that a different convention would have led us to the adoption of a new temperature scale $\tilde{T}$ instead of the perfect-gas temperature scale T, the new scale being proportional to the old one. Thus we have justified the anticipatory remarks made just after Equation (5.48d).

Statistical considerations allow us to retrieve the perfect-gas law, but the derivation uncovers the existence of a restriction, Equation (6.11), unsuspected in classical thermodynamics. The inequality stipulates that the perfect-gas law must break down when the temperature T becomes too low. This causes the restriction to become related to the restriction which governs gas degeneracy and under which the asymptotic form of the partition function may be used at all. We quote this restriction in anticipation of its derivation in Section 8.5, and state that corrected Maxwell–Boltzmann statistics may be used if

$$\frac{Z_{\mathrm{tr}}^3}{N} = \left(\frac{2\pi m\mathbf{k}T}{\mathbf{h}^2}\right)^{3/2}\left(\frac{v_{\mathrm{m}}}{\mathbf{N}}\right) \gg 1\,. \tag{6.11a}$$

The limitation expressed in inequality (6.11a) results in a higher minimum temperature than that expressed in inequality (6.11), owing to the presence of the factor **N** in the denominator. Hence it is concluded that condition (6.11) is satisfied *a fortiori* if condition (6.11a) is satisfied.

The data collected in Table 6.2 serve to show that Equation (6.12) constitutes a very good approximation under almost all conditions which are likely to be encountered in practice. The term $Z_{\mathrm{tr}}^3/\mathbf{N}$ has been evaluated here at the critical point of several gases, that is, at the lowest value of the molar specific volume v_{m} for a gas. It is not suggested that the perfect-gas law is

TABLE 6.2

Numerical Values of Z_{tr}^3/N for Several Gases at the Critical Point[a]

Gas	T_c (K)	P_c (atm)	$v_{\mathrm{m}}/\mathbf{N}$ (10^{-22} cm^3)	$Z_{\mathrm{tr}}^3/\mathbf{N}$
Helium, He	5.3	2.26	0.96	1.8
Argon, Ar	151	48	1.25	11,200
Hydrogen, H_2	33.3	12.8	1.08	6
Oxygen, O_2	154.4	49.7	1.23	8,000
Water vapor, H_2O	647.4	218	0.93	22,000

[a] From J. A. Fay, *Molecular Thermodynamics*, p. 177, Addison-Wesley, Reading, Massachusetts, 1965.

applicable at the critical point, since this value was chosen merely to obtain a measure of the highest possible lower limit for temperature. The numerical data included in the last column demonstrate that for all gases, with the exception of helium and hydrogen, corrected Maxwell–Boltzmann statistics can be used at low pressures right down to liquefaction. The effect of gas degeneracy at very low temperatures can only be expected to become detectable experimentally in helium and possibly in hydrogen.

A more general expression for pressure can be derived without the use of the approximation (6.12). Since

$$\frac{\partial \varepsilon_s}{\partial V} = -\frac{2}{3}\frac{\mathbf{h}^2 s^2}{8mV^{2/3}V} = -\frac{2}{3}\frac{\varepsilon_s}{V}, \tag{6.13c}$$

as seen from Equation (6.9a), and since

$$P = -\left(\frac{\partial E}{\partial V}\right)_S,$$

as shown in Equation (1.15), we may compute the pressure from the weighted average of the derivatives (6.13c). Thus we obtain

$$P = \frac{1}{V}\frac{2}{3}\frac{\sum_{s=1}^{k} \varepsilon_s \exp(-\mathbf{h}^2 s^2/8m\mathbf{k}TV^{2/3})}{\sum_{s=1}^{k} \exp(-\mathbf{h}^2 s^2/8m\mathbf{k}TV^{2/3})}. \tag{6.13d}$$

The fraction in this expression is equal to the energy. It is convenient to denote this energy by the symbol u_{tr}°, because it represents the portion of the internal energy of v_{m} units of volume of a perfect gas which is stored in the translational degrees of freedom. Equation (6.13d) is equivalent to

$$Pv_{\mathrm{m}} = \tfrac{2}{3}u_{\mathrm{tr}}^{\circ}, \tag{6.14}$$

and the derivation demonstrates that its validity, unlike that of Equation (6.13a), extends to degenerate gases.

6.4. The Classical Partition Function

The correspondence principle of quantum mechanics asserts that for very high quantum numbers the motion is asymptotically described by the equations of classical mechanics. Thus the replacement of a sum by an integral (employed in the preceding section) which becomes possible when the energy levels are very closely spaced—that is, when the quantum numbers are very large—constitutes an example of the operation of this principle. It follows that in many cases it should be possible to adopt a completely classical-mechanical approach to statistical thermodynamics (kinetic theory), provided that the limitations of this approach are understood. In fact, the whole subject developed first along such lines, and the many paradoxes and "clouds," to use Lord Kelvin's words once more, sprung up because the limitations were unappreciated before the logic of the development forced upon us the acceptance of quantum mechanics. The classical approach is useful owing to our inability to solve the Schrödinger equations in many cases when an approximate solution of the classical equations is possible, and because, generally speaking, classical equations of motion are more amenable to solution than the corresponding quantal equations. For these reasons it may be useful to review quickly the classical approach.

A single term in Equation (6.4) represents the sum of $\exp(-\beta E_\nu)$ over the accessible portion of a single shell in Γ-space, the thickness of the shell representing one quantum jump or a band δE_ν, as explained in Section 3.14. Such a shell of approximately constant energy E_ν contains a volume

$$d\mathscr{V} = \int_{E_\nu}^{E_\nu+\delta E_\nu} d\mathbf{q}\, d\mathbf{p}.$$

This form uses a compact notation in which **q** *represents* $q_1, \ldots, q_{Nf}$, where f is the number of degrees of freedom of each of the N molecules, and **p** *represents* $p_1, \ldots, p_{Nf}$. Consequently, $d\mathbf{q}\, d\mathbf{p}$ *stands for* the product $dq_1 \cdots dq_{fN}\, dp_1 \cdots dp_{fN}$. Recalling that one cell has a volume $\mathbf{h}^{fN}$, we find that

$$d\Omega = \frac{1}{\mathbf{h}^{fN}} \int_{E}^{E+\delta E} d\mathbf{q}\, d\mathbf{p}.$$

In the classical limit, the energy at every point in Γ-space is given by the Hamiltonian $\mathscr{H}(\mathbf{q}, \mathbf{p})$, and for this reason the term $\Omega_\nu \exp(-\beta E_\nu)$ is written

$$e^{-\beta E_\nu}\, d\Omega = \frac{e^{-\beta \mathscr{H}(\mathbf{q},\mathbf{p})}}{\mathbf{h}^{fN}} \int_{E}^{E+\delta E} d\mathbf{q}\, d\mathbf{p},$$

and the partition function is obtained by integrating over the accessible portion of the phase space. Thus we have

$$\mathscr{Z}_{\mathrm{MB}} = \frac{1}{\mathbf{h}^{fN}} \int_{\Gamma\text{-space}} e^{-\beta \mathscr{H}(\mathbf{q},\mathbf{p})}\, d\mathbf{q}\, d\mathbf{p}. \qquad (6.15)$$

We have introduced the exponential under the sign of the multiple integral because its mean value over one shell was implied in the earlier differential form. This equation would be obtained directly in *the kinetic theory of gases*. To correct for indistinguishability, we introduce the factor $N!$ and omit the subscript MB as superfluous. Hence we obtain

$$\mathscr{Z} = \frac{1}{\mathbf{h}^{fN} N!} \int_{\Gamma\text{-space}} e^{-\beta \mathscr{H}(\mathbf{q},\mathbf{p})}\, d\mathbf{q}\, d\mathbf{p}. \qquad (6.15\text{a})$$

The preceding derivation suffers from the justified criticism that no account has been taken of the symmetry properties of wave functions for bosons and fermions. Here the reader must accept the assurance that a mathematically rigorous transition from a quantal to a classical description yields the preceding equation *exactly*.

In most practical cases the Hamiltonian is of the form

$$\mathscr{H}(\mathbf{q}, \mathbf{p}) = \sum_{i=1}^{fN} \frac{p_i^2}{2m_i} + \phi(\mathbf{q}), \tag{6.15b}$$

in which the potential energy is a function of the coordinates only. The contemporary theory of dense gases and liquids† is based on Equations (6.15a, b).

The equation becomes simplified to a large extent in the theory of perfect gases when the assumption is made that the molecules are statistically independent. Equation (6.15a) allows us to recognize that the probability density is

$$\rho(\mathbf{r}, \mathbf{p}) = A\, \mathrm{e}^{-\beta \mathscr{H}(\mathbf{r}, \mathbf{p})}, \tag{6.15c}$$

where $\mathbf{r}$ now stands for x, y, z and $\mathbf{p}$ for all the momenta of a single molecule. This can also be seen from the fact that a swarm of independent, distinct molecules of an adiabatically isolated volume of gas behaves statistically exactly like the supersystem treated in Section 5.11. The factor A is determined by the normalization condition

$$A \iint_{\mu\text{-space}} \mathrm{e}^{-\beta \mathscr{H}(\mathbf{r}, \mathbf{p})}\, \mathrm{d}\mathbf{r}\, \mathrm{d}\mathbf{p} = 1\,.$$

We now limit our attention to the translational degrees of freedom when

$$\mathscr{H}(\mathbf{r}, \mathbf{p}) = \frac{p_x^2 + p_y^2 + p_z^2}{2m},$$

and find that

$$\frac{1}{A} = \iiint_{\substack{x,\,y,\,z \\ \text{over } V}} \int_{p_x=-\infty}^{\infty} \int_{p_y=-\infty}^{\infty} \int_{p_z=-\infty}^{\infty} \mathrm{e}^{-\beta(p_x^2+p_y^2+p_z^2)/2m}\, \mathrm{d}x\, \mathrm{d}y\, \mathrm{d}z\, \mathrm{d}p_x\, \mathrm{d}p_y\, \mathrm{d}p_z$$

is given by a sextuple integral. Here the limits of integration for x, y, z extend over the volume V of the vessel. Since the integrand is independent of the coordinates, integration with respect to x, y, z produces a factor V. The limits for p_x, p_y, p_z range from $-\infty$ to $+\infty$ because relativistic effects are ignored.

† Compare Chapter 7, and S. A. Rice and P. Gray, *The Statistical Mechanics of Simple Liquids*, Wiley (Interscience), New York, 1965.

The integral with respect to p_x, p_y, p_z separates into a product of three integrals of the type

$$\int_{-\infty}^{\infty} e^{-\xi^2/2m\mathbf{k}T}\, d\xi = (2\pi m\mathbf{k}T)^{1/2},$$

where ξ is a dummy variable of integration, and the value of the integral follows from Equation (4.20). Hence we have

$$\frac{1}{A} = (2\pi m\mathbf{k}T)^{3/2} V. \tag{6.15d}$$

Apart from the factor $\mathbf{h}^{-3}$, this expression is identical with the partition function Z_{tr}^3 from Equation (6.13). Instead of representing a sum, it represents an *integral* over states. The factor $\mathbf{h}^{-3}$ can be introduced at this stage if it is remembered that integration should not be performed over the volume $d\mathbf{r}\, d\mathbf{p}$ in μ-space but over the number of quantum states

$$d\Omega = \frac{d\mathbf{r}\, d\mathbf{p}}{\mathbf{h}^3},$$

because the number of degrees of freedom is $f = 3$. The same factor is obtained if it is realized that Δx in Equation (4.15) should be replaced by $\mathbf{h}$. Hence the *classical partition function* for a monatomic gas or for the translational degrees of freedom of a polyatomic gas is

$$Z_{tr}^{cl} = Z_{tr}^3 = \frac{(2\pi m\mathbf{k}T)^{3/2} V}{\mathbf{h}^3}, \tag{6.15e}$$

in complete agreement with Equation (6.12).

Substituting the value from Equation (6.15d) into Equation (6.15c), we obtain

$$\rho(\varepsilon, \mathbf{r}) = \frac{e^{-\varepsilon/\mathbf{k}T}}{(2\pi m\mathbf{k}T)^{3/2} V},$$

an expression for probability density in terms of energy. The coordinate $\mathbf{r}$ does not appear explicitly in ε in the absence of a potential energy field. The marginal probability density of finding a molecule with energy ε anywhere in the vessel is deduced by integration over the whole vessel, and this gives

$$\rho'(\varepsilon) = \int_{\mathbf{r}} \frac{e^{-\varepsilon/\mathbf{k}T}\, d\mathbf{r}}{(2\pi m\mathbf{k}T)^{3/2} V} = \frac{e^{-\varepsilon/\mathbf{k}T}}{(2\pi m\mathbf{k}T)^{3/2}}. \tag{6.15f}$$

This *energy probability distribution function* is expressed in terms of the energy ε and the Boltzmann factor $\exp(-\varepsilon/\mathbf{k}T)$. This distribution is shown plotted graphically in Figure 6.1 for the case of a helium molecule at the

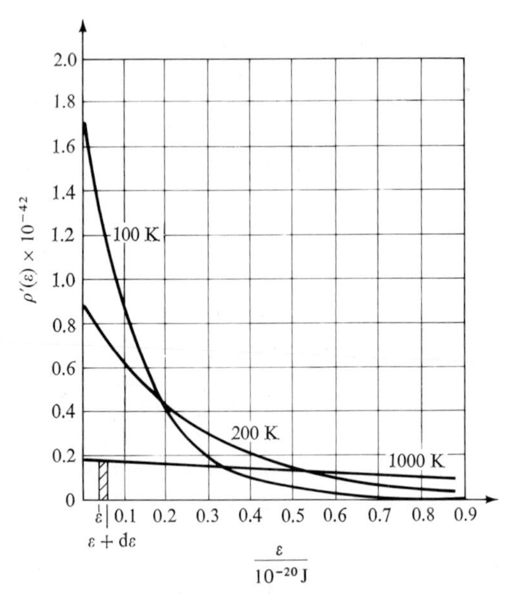

FIGURE 6.1. Probability density as a function of the energy of a molecule (helium with $m = 6.65 \times 10^{-23}$ gr).

three temperatures $T = 100$, 200, 1000 K. The diagram confirms that the lowest energy levels are most densely populated by molecules and that the distribution tends to a uniform one as $T \to \infty$. As the temperature is raised, an increasing number of molecules acquire higher energies. The relative number of molecules whose energies are confined to the interval ε to $\varepsilon + d\varepsilon$ at $T =$ 1000 K is proportional to the shaded strip in the graph.

6.5. Equipartition of Energy in Classical Statistical Mechanics

The expectation value of the translational energy of a molecule can be written by analogy with Equation (5.36) in the form

$$\langle \varepsilon \rangle = \int_{\mathbf{r}} \int_{\mathbf{p}} \varepsilon(\mathbf{r}, \mathbf{p}) \rho(\mathbf{r}, \mathbf{p}) \, d\mathbf{r} \, d\mathbf{p} \,. \tag{6.16}$$

The integration over $\mathbf{r}$ once more produces the factor V, and the remainder can be separated into a sum of three integrals of the form

$$\frac{1}{(2\pi m\mathbf{k}T)^{3/2}} \int_{-\infty}^{\infty} \int_{-\infty}^{\infty} \int_{-\infty}^{\infty} \frac{p_x^{\,2}}{2m} \, e^{-p_x^2/2m\mathbf{k}T} \, e^{-p_y^2/2m\mathbf{k}T} \, e^{-p_z^2/2m\mathbf{k}T} \, dp_x \, dp_y \, dp_z$$

$$= \frac{(2m)^{3/2}}{(2\pi m\mathbf{k}T)^{3/2}} \int_{-\infty}^{\infty} e^{-p_y^2/2m\mathbf{k}T} \, d\left(\frac{p_y}{\sqrt{2m}}\right) \int_{-\infty}^{\infty} e^{-p_z^2/2m\mathbf{k}T} \, d\left(\frac{p_z}{\sqrt{2m}}\right)$$

$$\times \int_{-\infty}^{\infty} \frac{p_x^{\,2}}{2m} \, e^{-p_x^2/2m\mathbf{k}T} \, d\left(\frac{p_x}{\sqrt{2m}}\right).$$

The definite integrals are given in standard tables,† and we obtain,

$$\langle \varepsilon \rangle = \tfrac{3}{2}\mathbf{k}T, \tag{6.16a}$$

or for **N** particles in one mol

$$u_{\mathrm{tr}}^{\circ} = \tfrac{3}{2}\mathbf{R}T. \tag{6.16b}$$

According to this equation, we have

$$c_{v,\,\mathrm{tr}}^{\circ} = \left(\frac{\partial u_{\mathrm{tr}}^{\circ}}{\partial T}\right)_v = \tfrac{3}{2}\mathbf{R},$$

and it is still true that

$$c_{p,\,\mathrm{tr}}^{\circ} = c_{v,\,\mathrm{tr}}^{\circ} + \mathbf{R} = \tfrac{5}{2}\mathbf{R}. \tag{6.16c}$$

The preceding calculation for the translational degrees of freedom can be repeated *verbatim* for the internal degrees of freedom. The two rotational degrees of freedom of a linear molecule and the three rotational degrees of freedom of a nonlinear molecule each contribute a term of the form $a_i x_i^2$ to the Hamiltonian. Here x_i may be interpreted as an angular momentum. Every normal mode in vibration contributes two such terms, one representing the kinetic energy for which x_i denotes the momentum, and one representing the potential energy for which x_i may be taken as the coordinate. A linear molecule with $f_{\mathrm{v}} = 3n - 5$ [see Equation (2.11)] will contribute

$$k = 3 + 2 + 2(3n - 5) = 6n - 5 \qquad \text{(linear)} \tag{6.17a}$$

terms to the Hamiltonian, whereas a nonlinear molecule with $f_{\mathrm{v}} = 3n - 6$ [see Equation (2.10)] will contribute

$$k = 3 + 3 + 2(3n - 6) = 6n - 6 \qquad \text{(nonlinear)} \tag{6.17b}$$

such terms. Equation (6.16) is now replaced by

$$\langle \varepsilon \rangle = \frac{\int_{-\infty}^{\infty} \cdots \int_{-\infty}^{\infty} dx_1\, dx_2 \cdots dx_k (\sum_{i=1}^{k} a_i x_i^2) \exp[(-\sum_{i=1}^{k} a_i x_i^2)/\mathbf{k}T]}{\int_{-\infty}^{\infty} \cdots \int_{-\infty}^{\infty} dx_1\, dx_2 \cdots dx_k \exp[(-\sum_{i=1}^{k} a_i x_i^2)/\mathbf{k}T]}.$$

We have eliminated integration over the position **r** of the center of a molecule in physical space because the volume in the numerator cancels that in the denominator. The limits of integration for each variable may be assumed to range from $-\infty$ to $+\infty$ because terms in which $a_i x_i^2$ in the exponential is much larger than $\mathbf{k}T$ make a negligible contribution to the integral.

The integral for $\langle \varepsilon \rangle$ separates into a sum of k integrals, each of which is of the form

† $\int_{-\infty}^{\infty} e^{-\beta\xi^2}\, d\xi = \pi^{1/2}/\beta^{1/2}$ and $\int_{-\infty}^{\infty} \xi^2 e^{-\beta\xi^2}\, d\xi = \pi^{1/2}/2\beta^{3/2}$.

$$\frac{\int_{-\infty}^{\infty} \cdots \int_{-\infty}^{\infty} \mathrm{d}x_1\,\mathrm{d}x_2 \cdots \mathrm{d}x_k (a_i x_i^2) \exp(-a_i x_i^2/\mathbf{k}T) \exp[(-\sum_{j \neq i} a_j x_j^2)/\mathbf{k}T]}{\int_{-\infty}^{\infty} \cdots \int_{-\infty}^{\infty} \mathrm{d}x_1\,\mathrm{d}x_2 \cdots \mathrm{d}x_k \exp[(-\sum_{i=1}^{k} a_i x_i^2)/\mathbf{k}T]}.$$

Here the symbol $j \neq i$ in the second exponential in the numerator denotes a sum with $a_i x_i^2$ *excluded.* The integration with respect to all values $x_1 \cdots x_k$ except the chosen x_i contributes identical terms in the numerator and the denominator, and we are left with

$$\frac{\int_{-\infty}^{\infty} a_i x_i^2 \exp(-a_i x_i^2/\mathbf{k}T)\,\mathrm{d}x_i}{\int_{-\infty}^{\infty} \exp(-a_i x_i^2/\mathbf{k}T)\,\mathrm{d}x_i}.$$

The definite integrals are given in standard tables,† and we see that every square term in the Hamiltonian produces an *additive* term

$$\varepsilon_i = \tfrac{1}{2}\mathbf{k}T \tag{6.18}$$

in the average energy of a molecule, or

$$u_i^\circ = \tfrac{1}{2}\mathbf{R}T \tag{6.18a}$$

in the molar energy, and a term

$$c_{v,i}^\circ = \tfrac{1}{2}\mathbf{R}$$

in the specific heat at constant volume or pressure.

The preceding results represent the *principle of equipartition* of energy in classical statistical mechanics. The principle of equipartition states that *the energy of a monatomic gas is distributed equally between the degrees of freedom of translational motion, each contributing an amount of* $\frac{1}{2}\mathbf{k}T$ *per molecule or* $\frac{1}{2}\mathbf{R}T$ *per mol.* When applied to the internal degrees of freedom, the principle asserts that *each square term in the Hamiltonian of the molecule contributes to the internal energy an amount* $\frac{1}{2}\mathbf{k}T$ per molecule or $\frac{1}{2}\mathbf{R}T$ per mol.

Depending on whether we regard molecules as rigid or vibrating, we come to the conclusion that the principle of equipartition leads to constant values of the specific heats c_p°, c_v° and their ratio γ°. The corresponding formulas have been summarized in Table 6.3. A comparison between this table and experimental results shows that real agreement exists only for monatomic gases. As far as diatomic gases are concerned there exists a measure of agreement with the assumption of rigidity at low temperatures and with the assumption of an elastic link at high temperatures. However, the experimental results for polyatomic gases as well as diatomic gases decidedly point to the fact that their specific heats *are functions of temperature.*

† From the preceding footnote, it follows that

$$\text{numerator} = \pi^{1/2}(\mathbf{k}T)^{3/2}/2a_i^{1/2},$$
$$\text{denominator} = \pi^{1/2}(\mathbf{k}T)^{1/2}/a_i^{1/2}.$$

TABLE 6.3

The Molar Specific Heats of Perfect Gases According to the Principle of Equipartition of Energy in Classical Statistical Mechanics

Structure	c_v°	$c_p^\circ = c_v^\circ + \mathbf{R}$	$\gamma^\circ = c_p^\circ/c_v^\circ$
Monatomic	$\frac{3}{2}\mathbf{R}$	$\frac{5}{2}\mathbf{R}$	$\frac{5}{3} = 1.67$
Linear, rigid	$\frac{5}{2}\mathbf{R}$	$\frac{7}{2}\mathbf{R}$	$\frac{7}{5} = 1.40$
Linear, vibrating[a]	$\frac{1}{2}(6n - 5)\mathbf{R}$	$\frac{1}{2}(6n - 3)\mathbf{R}$	$\dfrac{6n-3}{6n-5}$
Nonlinear, rigid	$3\mathbf{R}$	$4\mathbf{R}$	$\frac{4}{3} = 1.33$
Nonlinear, vibrating[a]	$3(n - 1)\mathbf{R}$	$(3n - 2)\mathbf{R}$	$\dfrac{3n-2}{3(n-1)}$

[a] n is the number of atoms in a molecule.

The derivation given in this section shows that there is no possibility of reconciling classical statistical mechanics with experiment, because no modification can be introduced which would alter the fact that this theory will always lead to *constant values*. The dependence on temperature, and the evaluation of the temperature at which equipartition occurs in practice, can only be brought about by quantum statistics.

The preceding glaring discrepancy between the classical theory and experiment was brilliantly commented upon by Lord Kelvin (who had no inkling of quantum mechanics) in his Baltimore lectures delivered in 1884. The success of quantum statistics in settling this paradox constituted one of the many compelling reasons for its adoption at the time when it was being evolved.

6.6. The Maxwellian Velocity Distribution

It should be clear to the reader from Section 6.4 that the classical theory naturally leads to the perfect-gas law without, however, recognizing that gases must become degenerate at very low temperatures. As we know, the perfect-gas law results from a consideration of the translational motion alone; the internal modes, being independent of volume, contribute no pressure terms. Therefore, it is instructive to obtain a more detailed representation of the velocity distribution which exists in a perfect gas in equilibrium. This is obtained by considering the *momentum distribution function for one molecule* (marginal probability density). To obtain this, it is necessary to integrate $\rho(\mathbf{r}, \mathbf{p})$ from equations (6.15c, d) over the coordinates in physical space, thus counting all molecules whose momenta are confined within the interval $\mathbf{p}$ to $\mathbf{p} + \mathrm{d}\mathbf{p}$ regardless of position.

We compute

$$\rho_1(\mathbf{p}) = \int_{\mathbf{r}} \rho(\mathbf{r}, \mathbf{p}) \, \mathrm{d}\mathbf{r} = \int_{\mathbf{r}} A \, \mathrm{e}^{-\varepsilon(\mathbf{r}, \mathbf{p})/\mathbf{k}T} \, \mathrm{d}\mathbf{r}.$$

This yields

$$\rho_1(\mathbf{p}) = \frac{\mathrm{e}^{-p^2/2m\mathbf{k}T}}{(2\pi m\mathbf{k}T)^{3/2} V} \int_{\mathbf{r}} \mathrm{d}\mathbf{r};$$

that is,

$$\rho_1(\mathbf{p}) = \frac{\mathrm{e}^{-p^2/2m\mathbf{k}T}}{(2\pi m\mathbf{k}T)^{3/2}}. \tag{6.19}$$

Since $\mathbf{p} = m\mathbf{v}$, we can obtain the *velocity distribution function* (also a marginal probability density) $\rho_2(\mathbf{v})$ by performing a transformation of variables:

$$\rho_1(\mathbf{p}) \, \mathrm{d}\mathbf{p} = \rho_2(\mathbf{v}) \, \mathrm{d}\mathbf{v} \qquad \text{or} \qquad \rho_1(\mathbf{p}) m^3 \, \mathrm{d}\mathbf{v} = \rho_2(\mathbf{v}) \, \mathrm{d}\mathbf{v};$$

here we remember that $\mathrm{d}\mathbf{p}$ stands for $\mathrm{d}p_x \, \mathrm{d}p_y \, \mathrm{d}p_z$, so that

$$\mathrm{d}\mathbf{p} = \mathrm{d}p_x \, \mathrm{d}p_y \, \mathrm{d}p_z = (m \, \mathrm{d}v_x)(m \, \mathrm{d}v_y)(m \, \mathrm{d}v_z) = m^3 \, \mathrm{d}\mathbf{v}.$$

Thus we have

$$\rho_2(\mathbf{v}) = m^3 \rho_1(\mathbf{p}) = \left(\frac{m}{2\pi\mathbf{k}T}\right)^{3/2} \mathrm{e}^{-mv^2/2\mathbf{k}T}. \tag{6.19a}$$

The result expressed in Equation (6.19a) represents a *Gaussian probability* distribution curve when applied to one velocity component; we may then write

$$\rho_2(v_i) = \left(\frac{m}{2\pi\mathbf{k}T}\right)^{1/2} \exp(-mv_i^2/2\mathbf{k}T), \tag{6.19b}$$

where v_i stands for v_x, v_y, or v_z. The factoring can be performed because the three velocity components are statistically independent, and the joint probability for a velocity $\mathbf{v}$ is the product of that for each component.

Figure 6.2 represents this distribution for the molecules of helium ($m = 6.65 \times 10^{-23}$ gr) at the three temperatures $T = 100$, 200, and 1000 K. The area under each curve is equal to unity owing to the normalization condition, and

$$\rho_2(0) = \left(\frac{m}{2\pi\mathbf{k}T}\right)^{1/2}.$$

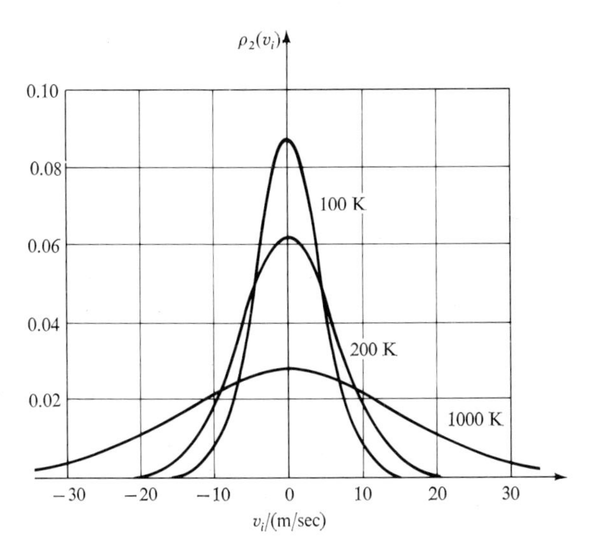

FIGURE 6.2. Gaussian distribution of velocity components in a classical perfect gas (helium with $m = 6.65 \times 10^{-23}$ gr).

The half-width, that is, the value v_i^* for which $\rho_2(v_i^*) = \frac{1}{2}\rho_2(0)$, is given by

$$v_i^* = \pm\left(\frac{2\mathbf{k}T \ln 2}{m}\right)^{1/2} = \pm\left(\frac{1.39\mathbf{k}T}{m}\right)^{1/2}.$$

As the temperature is increased, an increasing number of particles move at high velocities, and the diagram tends to flatten.

A more important representation is obtained when the probability distribution function for the absolute value $v = |\mathbf{v}|$ of the velocity is investigated. We denote this marginal probability density by ρ_v and observe that it is obtained by calculating the volume $4\pi v^2\, dv$ contained in the spherical shell between the radii v and $v + dv$ and multiplying it by the value of ρ_2 from Equation (6.19a). This leads to

$$\rho_v(v) = 4\pi v^2 \left(\frac{m}{2\pi\mathbf{k}T}\right)^{3/2} \mathrm{e}^{-mv^2/2\mathbf{k}T}. \tag{6.20}$$

The distribution ceases to be Gaussian owing to the appearance of the factor v^2; it is known as the *Maxwellian velocity distribution.* Figure 6.3 represents this distribution as a function of the absolute value of velocity v for helium and for the three temperatures 100, 200, and 1000 K employed in Figure 6.1. The diagram reveals the existence of a maximum probability density

$$\rho_v^* = \frac{4m^{1/2}}{\mathrm{e}}(2\pi\mathbf{k}T)^{-1/2} \qquad \text{for} \quad v^* = \left(\frac{2\mathbf{k}T}{m}\right)^{1/2} = \left(\frac{2\mathbf{R}T}{M}\right)^{1/2}, \tag{6.21a}$$

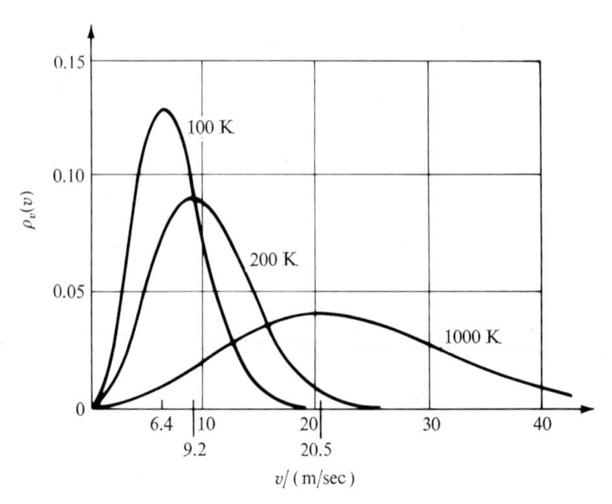

FIGURE 6.3. Maxwellian distribution of absolute velocity in a classical perfect gas (helium with $m = 6.65 \times 10^{-23}$ gr).

where $M = m\mathbf{N}$ is the molar mass. The absolute velocity v^* is the most probable one, and its value increases as the temperature increases. The diagram shows how an increase in temperature causes more and more particles to move at higher speeds, the probability tending to a more uniform shape.

Sometimes it is useful to consider the root-mean-square velocity

$$v_{\text{rms}} = \left[\int_0^\infty v^2 \rho_v(v)\, \mathrm{d}v\right]^{1/2} = \left(\frac{3\mathbf{k}T}{m}\right)^{1/2} = \left(\frac{3\mathbf{R}T}{M}\right)^{1/2}, \tag{6.21b}$$

or the average velocity

$$v_{\text{av}} = \int_0^\infty v \rho_v(v)\, \mathrm{d}v = \left(\frac{8\mathbf{k}T}{\pi m}\right)^{1/2} = \left(\frac{8\mathbf{R}T}{\pi M}\right)^{1/2}. \tag{6.21c}$$

The averages are in the ratio

$$v^* : v_{\text{av}} : v_{\text{rms}} = 1 : 1.13 : 1.22 \qquad (=\sqrt{2} : \sqrt{8/\pi} : \sqrt{3}). \tag{6.21d}$$

It is interesting to note that the temperature T is a measure of the average kinetic *energy* of a molecule. In the kinetic theory of gases, this relation is employed to *define* the absolute temperature of a perfect gas.

A quantitative measure of the Maxwellian distribution can be obtained experimentally in *molecular beams*; a corresponding qualitative measure is provided by the observation of the Gaussian *broadening of spectral lines* due to the *Doppler effect* in a luminous gas.

6.7. Monatomic Gases

Monatomic gases include the so-called noble gases (helium, He; neon, Ne; argon, Ar; krypton, Kr; and xenon, Xe) at normal temperatures and pressures as well as the vapors of the alkali metals (lithium, Li; sodium, Na; potassium, K; rubidium, Ru; and cesium, Cs) at higher temperatures and low pressures. At sufficiently high temperatures many of the diatomic gases, such as hydrogen or nitrogen, dissociate and become monatomic. The halides (such as fluorides or chlorides, for example, hydrogen fluoride, HF; or hydrogen chloride, HCl) may also dissociate into mixtures of monatomic halogens (fluorine, F; chlorine, Cl; bromine, Br; and iodine, I).

Monatomic gases possess no internal degrees of freedom, except those which are contributed by the nuclear and electronic quantum states. The lowest two nuclear energy states are characterized by very large energy differences which are of the order of $\mathbf{k}T$ when T is of the order of 10^{10} K. Under normal conditions the atoms are always found in their nuclear ground state. Since this state does not change in normal processes, its degeneracy may be disregarded. As far as the electronic quantum states are concerned, it is found that in most gases the separation of the first two energy levels become of the order $\mathbf{k}T$ when T is of the order 10,000 K. Therefore, in most cases it may be assumed that the electrons are in their ground state too. However, there exist exceptions which we will discuss in Section 6.9.4. To allow for such exceptions we shall include a degeneracy factor g_0 in the partition function, and write

$$\mathscr{Z} = \frac{(g_0 Z_{\text{tr}}^3)^N}{N!}. \tag{6.22}$$

For the present, the reader may assume that

$$g_0 = 1 \qquad \text{(noble gases)} \tag{6.23}$$

and wait for its justification until he reaches Section 6.9.4.

The theory of a monatomic gas is implied in the equation

$$\ln \mathscr{Z} = N \ln\left[\frac{g_0(2\pi m\mathbf{k}T)^{3/2} V}{\mathbf{h}^3}\right] - N \ln N + N.$$

This equation arises from the substitution of Equation (6.12) into Equation (6.22), with $\ln N!$ replaced by its Stirling approximation. It can also be written

$$\ln \mathscr{Z} = N \ln\left[\frac{g_0(2\pi m\mathbf{k}T)^{3/2}\, eV}{\mathbf{h}^3 N}\right], \tag{6.24}$$

where the full particle translational partition function has the form

$$Z'_{tr} = g_0 Z^3_{tr} = \frac{g_0(2\pi m\mathbf{k}T)^{3/2}V}{\mathbf{h}^3}. \tag{6.24a}$$

With $g_0 = 1$, the equation also contains the theory of the contributions to the properties of polyatomic perfect gases which are due to their translational motion, because under our assumption all the terms which stem from the internal motions are additive to the one shown above.

Application of Equations (5.54) through (5.59) with $N = \mathbf{N}$ leads us to the following succession of working formulas for 1 mol of gas:†

$$f^\circ_{tr} = -\mathbf{N}\mathbf{k}T \ln\left[\frac{g_0(2\pi m\mathbf{k}T)^{3/2}}{\mathbf{h}^3}\frac{ev_m}{\mathbf{N}}\right]$$
$$= -\mathbf{R}T \ln\left[\frac{g_0(2\pi m\mathbf{k}T)^{3/2}}{\mathbf{h}^3}\frac{ev_m}{\mathbf{N}}\right]; \tag{6.25}$$

$$u^\circ_{tr} = \tfrac{3}{2}\mathbf{R}T; \tag{6.25a}$$

$$s^\circ_{tr} = \tfrac{5}{2}\mathbf{R} + \mathbf{R}\ln\left[\frac{g_0(2\pi m\mathbf{k}T)^{3/2}}{\mathbf{h}^3}\frac{v_m}{\mathbf{N}}\right]$$
$$= \tfrac{5}{2}\mathbf{R} + \mathbf{R}\ln\left[\frac{g_0(2\pi m)^{3/2}(\mathbf{k}T)^{5/2}}{\mathbf{h}^3 P}\right] \tag{6.25b}$$

(with $v_m/\mathbf{N} = \mathbf{k}T/P$ substituted from the perfect-gas law);

$$\mu^\circ_{tr} = \left(\frac{\partial f_{tr}}{\partial n}\right)_{T,v_m} = -\mathbf{R}T\ln\left[\frac{g_0(2\pi m\mathbf{k}T)^{3/2}}{\mathbf{h}^3}\frac{v_m}{\mathbf{N}}\right]$$
$$= -\mathbf{R}T\ln\left[\frac{g_0(2\pi m)^{3/2}(\mathbf{k}T)^{5/2}}{\mathbf{h}^3 P}\right] \tag{6.25c}$$

(with the number of mols $n = N/\mathbf{N}$ and $\partial/\partial n = \mathbf{N}\,\partial/\partial N$). In addition we derive

$$c^\circ_{v,tr} = \tfrac{3}{2}\mathbf{R}, \qquad c^\circ_{p,tr} = c^\circ_{v,tr} + \mathbf{R} = \tfrac{5}{2}\mathbf{R}, \qquad \gamma^\circ = \tfrac{5}{3} = 1.667. \tag{6.26a, b, c}$$

Since the translational partition function has been computed by integration in the classical limit, we find that equipartition of energy is present, thus leading to constant values of the specific heats and of their ratio. This is actually observed in practice, demonstrating that the approximation implied in the classical limit is adequate for all purposes as far as the translational degrees of freedom are concerned.

6.8. Entropy and the Sackur–Tetrode Equation

The expression for entropy contained in Equation (6.25b) is of special as well as historical interest. The expression is known as the *Sackur–Tetrode*

† Note that $m\mathbf{N} = M$ and $\mathbf{k}\mathbf{N} = \mathbf{R}$.

equation, having been derived independently by O. Sackur† and H. Tetrode.‡ For the purpose of our present discussion, it is convenient to rewrite the equations in a somewhat simplified notation; we assume $g_0 = 1$, drop the subscript tr, postulate a system of $n = N/\mathbf{N}$ mols, and recall for future reference that for a monatomic gas the equipartition values of the specific heats are

$$c_v^\circ/\mathbf{R} = \tfrac{3}{2} \qquad \text{and} \qquad c_p^\circ/\mathbf{R} = \tfrac{5}{2}.$$

Hence we have the modified form

$$S = \tfrac{5}{2}n\mathbf{R} + n\mathbf{R} \ln\left[\frac{(2\pi mkT)^{3/2}}{\mathbf{h}^3}\frac{V}{N}\right] \tag{6.27a}$$

$$= \tfrac{5}{2}n\mathbf{R} + n\mathbf{R} \ln\left[\frac{(2\pi m)^{3/2}(\mathbf{k}T)^{5/2}}{\mathbf{h}^3 P}\right]. \tag{6.27b}$$

6.8.1. Effect of the Factorial Correction Term

If the equation for entropy had been derived directly from the relation

$$S' = \mathbf{k}\left[\frac{\partial(T \ln \mathscr{Z})}{\partial T}\right]_{V,N} = \mathbf{k}\left[\frac{\partial(T \ln Z^N)}{\partial T}\right]_{V,N},$$

without the factorial correction term $(N!)^{-1}$ for indistinguishability, we would have obtained

$$\ln \mathscr{Z} = N \ln\left[\frac{(2\pi m\mathbf{k}T)^{3/2}}{\mathbf{h}^3} V\right]$$

and

$$S' = \tfrac{3}{2}n\mathbf{R} + n\mathbf{R} \ln\left[\frac{(2\pi m\mathbf{k}T)^{3/2}}{\mathbf{h}^3} V\right] \tag{6.27c}$$

instead of Equation (6.27a). Here Z^N denotes the complete translational partition function for a monatomic gas.

The functional dependence of S' in terms of the volume V and the number of molecules N or mols n is not a homogeneous one, because the simultaneous multiplication of V, N, or n by a factor λ does not produce the required results, namely, $\lambda S'$, but rather

$$\lambda S' + \lambda n\mathbf{R} \ln \lambda.$$

The addition of the correction term

$$-\ln N! = N - N \ln N$$

to $\ln \mathscr{Z}$ for indistinguishability transforms the equation for S' into the equation for S, and it is easy to see that the latter is homogeneous of order one, since the ratio V/N in the logarithmic term of Equation (6.27a) remains unchanged when both are multiplied by the factor λ.

Historically, Equation (6.27c) was derived with the aid of classical,

† *Ann. Physik* **36** (1911), 958; **40** (1913), 67.
‡ *Ann. Physik* **38** (1912), 434.

Maxwell–Boltzmann statistics (without a correction for the quantization of μ-space as outlined in Sections 6.4–6.6), and its failure to yield a homogeneous function of order one was regarded as an unexplained paradox. The development of quantum mechanics removed the paradox, and this fact served as yet another compelling reason for the adoption of quantum statistics. Classical mechanics can still be used owing to the existence of the correspondence principle, but the availability of quantum mechanics allows us to assess whether the passage to the limit has been performed correctly, as we have done in this and the preceding sections.

6.8.2 The Entropy of a Perfect Gas near Absolute Zero

We remember from Section 5.17.2 that the entropy of any system should reduce to zero as the limit of absolute zero is approached. Thus, according to the Third law of thermodynamics, passing to the limit of $T \to 0$ at $P = \text{const}$ in Equation (6.27b) should yield $S=0$. However, the presence of the logarithmic term causes the entropy to become singular with $S \to -\infty$. Similarly, the passage to the limit $T \to 0$ at $V = \text{const}$ in Equation (6.27a) will yield the same singular value.

This irregular behavior is due to our having neglected to account for gas degeneracy and to the fact that the classical limit of the particle partition function has been employed in Equation (6.12). For this reason the specific heats c_v° and c_p° have acquired their equipartition values of $\frac{3}{2}\mathbf{R}$ and $\frac{5}{2}\mathbf{R}$, respectively, as evidenced by the powers of the terms $\mathbf{k}T$ in the equations. In other words, the range of validity of the Sackur–Tetrode equation is restricted by Equation (6.11a) to temperatures which arc much higher than the characteristic temperature

$$T' = \frac{1}{\mathbf{k}}\left[\frac{\mathbf{h}^3 P}{(2\pi m)^{3/2}}\right]^{2/5} = \frac{1}{\mathbf{k}}\left(\frac{\mathbf{N}}{v_\mathrm{m}}\right)^{2/3}\frac{\mathbf{h}^2}{2\pi m}, \tag{6.28}$$

and must be expected to fail near absolute zero. The characteristic temperature T' is that for which $Z_{\mathrm{tr}}^3/\mathbf{N} = 1$ in Equation (6.11a), and the first form follows if $v_\mathrm{m}/\mathbf{N} = \mathbf{k}T/P$ is substituted. This characteristic temperature is different for different isobars or isochores:

$$T' = T'(P) \qquad \text{or} \qquad T' = T'(v_\mathrm{m}).$$

Consequently, degeneracy affects the behavior of a gas differently when the limit $T \to 0$ is approached along different isobars or isochores.

To pass to the limit $T \to 0$ rigorously, it is necessary to develop separately the complete theory of boson and fermion degeneracy which we defer to Chapter 8. Instead, we have prepared the quantitative diagram of Figure 6.4, which represents the variation of the entropy of 1 mol of helium with temperature at the constant pressure of $P = 1$ atm. Curve a represents Equation

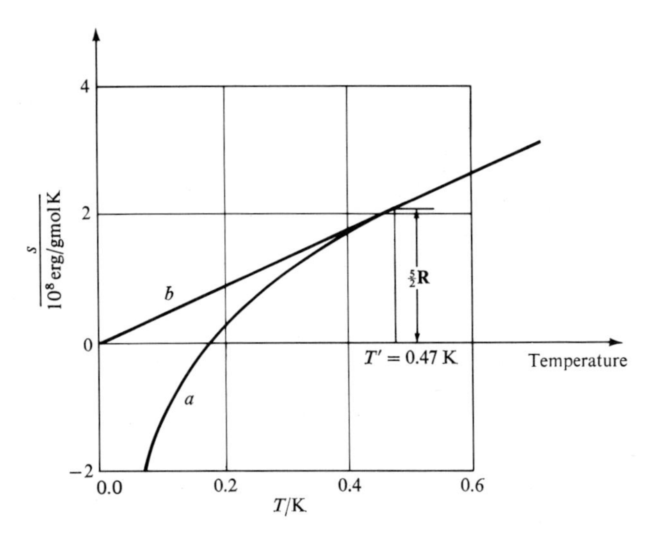

FIGURE 6.4 The entropy of helium at $P = 1$ atm near absolute zero ($m = 6.65 \times 10^{-23}$ gr, $T' = 0.47$ K). Curve *a*, obtained from the Sackur–Tetrode equation, has been drawn to scale; curve *b* represents the behavior of a degenerate gas only qualitatively and has not been drawn to scale.

(6.27b), whereas curve *b* suggests qualitatively the proper functional relationship for a degenerate gas along an isobar. The characteristic temperature T' is also shown, and it is seen that for T larger than T' the asymptotically valid Sackur–Tetrode equation can be used as an excellent approximation. Therefore, we conclude that the Sackur–Tetrode formula gives the absolute value of entropy on the supposition that the entropy at $T = 0$ and the given pressure (or volume) has been normalized at the value

$$S(0, P) = S(0, V) = 0.$$

6.8.3. Comparison with Previous Formulas of Classical Thermodynamics

The Sackur–Tetrode equation must be consistent with the equations which we assume to be familiar to the reader. As long as the gas is unaffected by degeneracy, its equation of state is $PV = n\mathbf{R}T$ and its specific heats are functions of temperature only. However, this is no longer the case for a degenerate gas of bosons or fermions.

To understand the relation between our present and more standard equations, we choose a reference state P^*, T^*, $v_m{}^*$ and write the equations in the forms:

$$S(T, V) - S(T^*, V^*) = n \int_{T^*}^{T} c_v{}^{\circ}(T) \frac{\mathrm{d}T}{T} + n\mathbf{R} \ln \frac{v_m}{v_m{}^*}, \tag{6.29a}$$

and

$$S(T, P) - S(T^*, P^*) = n \int_{T^*}^{T} c_p^\circ(T) \frac{\mathrm{d}T}{T} - n\mathbf{R} \ln \frac{P}{P^*}. \qquad (6.29\text{b})$$

For a monatomic gas, the specific heats have constant values for $T \gg T'$. Hence we obtain

$$\int_{T^*}^{T} c_v^\circ \frac{\mathrm{d}T}{T} = c_v^\circ \ln \frac{T}{T^*} = \frac{3}{2} \mathbf{R} \ln \frac{T}{T^*}$$

or

$$\int_{T^*}^{T} c_p^\circ \frac{\mathrm{d}T}{T} = c_p^\circ \ln \frac{T}{T^*} = \frac{5}{2} \mathbf{R} \ln \frac{T}{T^*}.$$

In the general equations it is not possible to choose T^* arbitrarily small as long as the specific heats are given constant values, because the term $\ln T/T^*$ would be singular and because the specific heats of degenerate gases are no longer pure functions of temperature. Thus without the Sackur–Tetrode equation it would not be possible to standardize the entropy of a gas at the value $S = 0$ for $T = 0$ and any pressure or volume, as demanded by the Third law. By contrast, the statistical-mechanical equations achieve this result automatically.

The detailed way in which this has been brought about can be recognized by transforming Equations (6.27a, b) with the aid of Equation (6.28). In this manner we obtain

$$S(T, V) = \tfrac{5}{2} n\mathbf{R} + \tfrac{3}{2} n\mathbf{R} \ln \frac{T}{T'} \qquad (v_\mathrm{m} = v_\mathrm{m}^* = \text{const}),$$

or

$$S(T, P) = \tfrac{5}{2} n\mathbf{R} + \tfrac{5}{2} n\mathbf{R} \ln \frac{T}{T'} \qquad (P = P^* = \text{const}).$$

These forms may be compared with Equations (6.29a,b) in which we put $T^* = 0$ and $S(T^*, V^*) = S(T^*, P^*) = 0$. This shows that

$$\begin{aligned} n \int_0^{T} c_v^\circ(T) \frac{\mathrm{d}T}{T} &= n \int_0^{T'(v_\mathrm{m}^*)} c_v(T, v_\mathrm{m}^*) \frac{\mathrm{d}T}{T} + n \int_{T'(v_\mathrm{m}^*)}^{T} \tfrac{3}{2}\mathbf{R} \frac{\mathrm{d}T}{T} \\ &= n \int_0^{T'} c_v(T, v_\mathrm{m}^*) \frac{\mathrm{d}T}{T} + \tfrac{3}{2} n\mathbf{R} \ln \frac{T}{T'}, \end{aligned}$$

and similarly that

$$n \int_0^{T} c_p^\circ(T) \frac{\mathrm{d}T}{T} = n \int_0^{T'(P^*)} c_p(T, P^*) \frac{\mathrm{d}T}{T} + \tfrac{5}{2} n\mathbf{R} \ln \frac{T}{T'}.$$

The integrals from 0 to T' along $v_m{}^*$ or P^* constant would diverge if we continued to suppose that the specific heats are constant. Statistical thermodynamics shows that these integrals are *not* divergent and supplies their numerical values, namely,

$$\int_0^{T'(v_m^*)} c_v(T, v_m{}^*)\frac{dT}{T} = \int_0^{T'(P^*)} c_p(T, P^*)\frac{dT}{T} = \tfrac{5}{2}\mathbf{R}. \tag{6.30}$$

Since T' is a function of volume or pressure, the term $\ln T/T'$ in the remainder of the equation for entropy depends on the choice of the reference isochore or isobar. Consequently, it is impossible to provide a tabulation of the entropy functions $s_v{}^\circ(T)$ and $s_p{}^\circ(T)$ without specifying the reference pressure P^* or reference volume $v_m{}^*$, if the reference temperature is to be placed at $T^* = 0$. This explains the use of the term *standard entropy function* in tables of thermodynamic properties.

Figure 6.5 shows a comparison between the specific heat $c_v{}^\circ$ of a classical

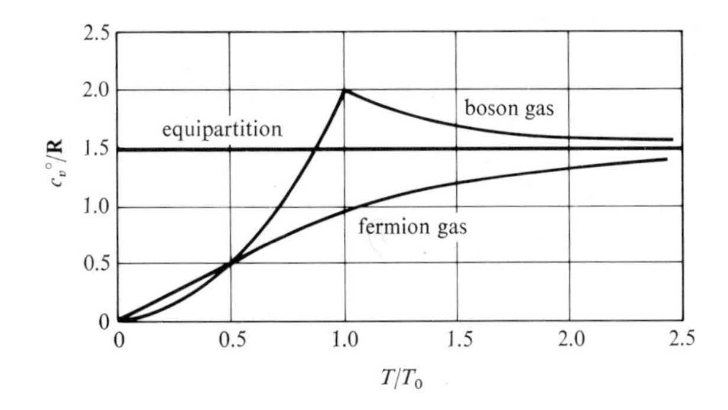

FIGURE 6.5. The specific heat $c_v{}^\circ$ of a monatomic gas near absolute zero. (T_0 denotes a reference temperature which has not been introduced formally in the text.) [From T. Hill, *Introduction to Statistical Thermodynamics*, p. 444, Addison Wesley, 1960.]

gas ($c_v{}^\circ/\mathbf{R} = 1.5$), a degenerate boson gas, and a degenerate fermion gas. The specific heat of the latter two tends to zero and assures the convergence of the corresponding integral. The asymptotic integration implied in the Sackur–Tetrode equation makes no distinction between these two cases and supplies a single constant for both. The characteristic temperature $T_0 \neq T'$ used in the diagram arises naturally in the complete theory, but its exact form is of no particular concern at the moment. It should be noted that the passage to the limits $P \to 0$ or $v_m \to \infty$ at $T = \text{const}$ cannot be performed because the entropy in those limits becomes infinite.

To preserve the same form for entropy as that given in Equations (6.29a, b),it is useful to rewrite Equations (6.27a, b) to yield

$$s = \tfrac{5}{2}\mathbf{R} + \tfrac{3}{2}\mathbf{R} \ln\left[\frac{2\pi m \mathbf{k} T}{\mathbf{h}^2}\left(\frac{v_\mathrm{m}^*}{\mathbf{N}}\right)^{2/3}\right] + \mathbf{R} \ln \frac{v_\mathrm{m}}{v_\mathrm{m}^*}, \tag{6.31a}$$

or

$$s = \tfrac{5}{2}\mathbf{R} + \tfrac{5}{2}\mathbf{R} \ln\left[\left(\frac{2\pi m}{\mathbf{h}^2}\right)^{3/5} \frac{\mathbf{k}T}{(P^*)^{2/5}}\right] - \mathbf{R} \ln \frac{P}{P^*}. \tag{6.31b}$$

Since the dependence of the entropy of a perfect gas on volume or pressure is explicit, tables normally contain only the *entropy function at constant volume*

$$s_v^\circ(T) = \tfrac{5}{2}\mathbf{R} + \tfrac{3}{2}\mathbf{R} \ln\left[\frac{2\pi m \mathbf{k} T}{\mathbf{h}^2}\left(\frac{v_\mathrm{m}^*}{\mathbf{N}}\right)^{2/3}\right] = \tfrac{5}{2}\mathbf{R} + \mathbf{R} \ln\left[\frac{(2\pi m \mathbf{k} T)^{3/2}}{\mathbf{h}^3} \frac{v_m^*}{\mathbf{N}}\right], \tag{6.32a}$$

or, more frequently, the *entropy function at constant pressure*

$$s_p^\circ(T) = \tfrac{5}{2}\mathbf{R} + \tfrac{5}{2}\mathbf{R} \ln\left[\left(\frac{2\pi m}{\mathbf{h}^2}\right)^{3/5} \frac{\mathbf{k}T}{(P^*)^{2/5}}\right] = \tfrac{5}{2}\mathbf{R} + \mathbf{R} \ln\left[\frac{(2\pi m)^{3/2}(\mathbf{k}T)^{5/2}}{\mathbf{h}^3 P^*}\right]. \tag{6.32b}$$

When $P^* = 1$ atm is stipulated, the value $s_p^\circ(T)$ is called *standard entropy*. The standard entropy function for monatomic gases is, normally, evaluated in this way.† The tabulation for polyatomic gases requires the inclusion of the contributions from the internal degrees of freedom.

The reader will have noticed that the preceding equations have been written in a manner to preserve dimensional homogeneity and to ensure that the arguments of logarithmic functions remain dimensionless. Some authors write Equations (6.27a, b) in the following manner:

$$s = \tfrac{5}{2}\mathbf{R} + \mathbf{R} \ln\left[\frac{(2\pi m \mathbf{k})^{3/2}}{\mathbf{h}^3 \mathbf{N}}\right] + \tfrac{3}{2}\mathbf{R} \ln T + \mathbf{R} \ln v_\mathrm{m},$$

or

$$s = \tfrac{5}{2}\mathbf{R} + \mathbf{R} \ln\left[\frac{(2\pi m)^{3/2}\mathbf{k}^{5/2}}{\mathbf{h}^3}\right] + \tfrac{5}{2}\mathbf{R} \ln T - \mathbf{R} \ln P.$$

These forms are inadvisable because they are conducive to numerical errors when units are changed.

6.9. Internal Degrees of Freedom

The molecular partition function Z_int can be separated into a product of partition functions, each of which corresponds to a statistically independent degree of freedom. We first subdivide the internal partition function for a

† See Table XVIb in J. Kestin, *A Course in Thermodynamics*, Vol. II, Blaisdell, Waltham, Massachusetts, 1969.

molecule into a rotational term Z_r, a vibrational term Z_v, and an electronic term Z_{el}, so that

$$Z_{int} = Z_r Z_v Z_{el}, \tag{6.33}$$

and take each term in turn.

6.9.1. Rotational Degrees of Freedom

The evaluation of the rotational partition function for a diatomic molecule is based on Equation (3.37). It is useful to introduce the rotational characteristic temperature by the definition

$$\Theta_r = \frac{\mathbf{h}^2}{8\pi^2 I \mathbf{k}}, \tag{6.34}$$

and to introduce the ratio

$$\xi = \frac{\Theta_r}{T} \quad \left(= \frac{\mathbf{h}^2}{8\pi^2 I \mathbf{k} T} \right). \tag{6.34a}$$

Since the energy levels are $\varepsilon_j = j(j+1)\mathbf{k}T\xi$, with degeneracy $g_j = 2j+1$, we find that

$$Z_r = \sum_{j=0}^{\infty} (2j+1)\, e^{-\xi j(j+1)} \qquad \text{(diatomic)}. \tag{6.35}$$

The rotational partition function for a diatomic gas cannot be evaluated in closed form. However, an examination of Table II shows that for all gases listed, except for hydrogen and its isotopes, the characteristic temperatures are very low. Thus in normal circumstances ξ is very small, and the sum in Equation (6.35) can be replaced by an integral. This integral need not be evaluated if only the internal energy and the specific heats are required, because in the *classical limit* the equipartition values of $\frac{1}{2}\mathbf{R}T$ for energy per degree of freedom in rotation must be found as a result. Nevertheless, in order to calculate the other thermodynamic functions, we may record the result explicitly. As we pointed out in Section 3.10, the symmetry requirements for homonuclear molecules impel us to divide the result by 2. It is customary to introduce here the *symmetry number* σ (=1 for heteronuclear and =2 for homonuclear diatomic molecules) and to write

$$Z_r = \frac{1}{\sigma}\int_0^{\infty} (2j+1)\, e^{-\xi j(j+1)}\, dj = \frac{1}{\sigma\xi}; \tag{6.36}$$

or equivalently

$$Z_r = \frac{T}{\sigma\Theta_r} = \frac{8\pi^2 I \mathbf{k} T}{\sigma \mathbf{h}^2} \qquad \text{(classical limit)}. \tag{6.36a}$$

Application of Equations (5.54) through (5.59) gives

$$f_r = -\mathbf{R}T \ln\left(\frac{T}{\sigma\Theta_r}\right), \tag{6.37}$$

$$u_r^\circ = \mathbf{R}T, \tag{6.37a}$$

$$s_r = \mathbf{R} \ln\left(\frac{eT}{\sigma\Theta_r}\right), \tag{6.37b}$$

$$\mu_r = -\mathbf{R}T \ln\left(\frac{T}{\sigma\Theta_r}\right), \tag{6.37c}$$

$$c_{v,r}^\circ = c_{p,r}^\circ = \mathbf{R}, \tag{6.38}$$

$$\gamma^\circ = \frac{\frac{5}{2}\mathbf{R} + \mathbf{R}}{\frac{3}{2}\mathbf{R} + \mathbf{R}} = \frac{7}{5} = 1.4 \qquad \text{(classical limit; diatomic)}. \tag{6.38a}$$

(It is noted that the symmetry number σ enters into the expressions for entropy, Helmholtz function, and chemical potential, but is absent in those for energy and specific heat.) As expected, all the preceding terms are functions of temperature alone.

The energy diagram and that of its derivative, the specific heat c_v°, have been plotted in Figures 6.6 and 6.7 for a heteronuclear diatomic gas. These show that the classical values are attained at temperatures T which only slightly exceed the characteristic temperature Θ_r and that the specific heat vanishes at $T \to 0$ as required.

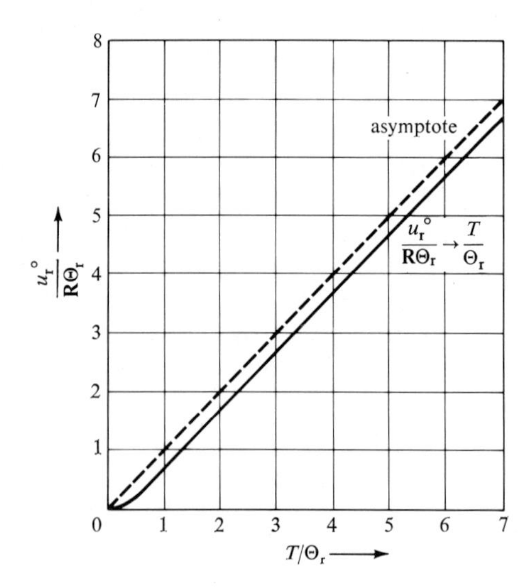

FIGURE 6.6. Rotational internal energy u_r° of a diatomic gas.

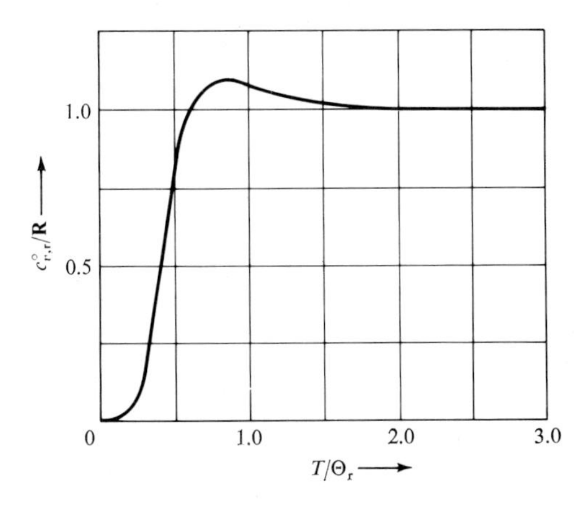

FIGURE 6.7. Rotational specific heat $c^\circ_{v,r}$ of a diatomic gas.

The rotational partition function for more complex molecules has not been evaluated owing to the difficulties described in Section 3.10. In the classical limit, we would obtain

$$Z_r = \frac{1}{\sigma}\frac{8\pi^2}{\mathbf{h}^3}(I_1 I_2 I_3)^{1/2}(2\pi\mathbf{k}T)^{3/2}. \tag{6.39}$$

Here I_1, I_2, I_3 denote the three principal moments of inertia of the molecule. The symmetry number of a polyatomic molecule is defined as the number of rotations, reflections, and inversions which transform the configurations of the molecule into itself. For example, $\sigma = 2$ for H_2O (isosceles triangle), $\sigma = 3$ for NH_3 (equilateral triangular pyramid), and $\sigma = 12$ for CH_4 (regular tetrahedron with carbon atom in the center of mass).

In practice, except for hydrogen and its isotopes, at relatively low temperatures the rotational degrees of freedom of gases are accounted for classically. We then say that all the rotational degrees of freedom have been "fully excited." Nevertheless, Figures 6.6 and 6.7 demonstrate that adding the classical rotational terms to the translational terms in the expression for entropy still produces its absolute value with respect to absolute zero, except for a small correction which results from the integral of specific heat taken from $T = 0$ to $T = \Theta_r$.

6.9.2. Hydrogen and its Isotopes

Because of their small moments of inertia, the rotational characteristic temperatures of the isotopes of hydrogen are relatively high. Therefore, at low temperatures we may expect these gases to have their rotational degrees "frozen" and to behave like classical monatomic gases with $c_v^\circ \approx 3\mathbf{R}/2$. The situation is further complicated by the fact that hydrogen and deuterium occur in two different quantum states depending on the relative orientation of the spins of the two protons. The proton has spin $\hbar/2$ and the molecule as a whole may have spin 0 or $\hbar$. In hydrogen there exists 1 molecule of spin 0 (parahydrogen) to every 3 molecules of spin $\hbar$ (orthohydrogen). Considerations of symmetry† lead to the conclusion that parahydrogen quantum states have j even, and those of orthohydrogen have j odd. Hence we find that

$$Z_r = \sum_{\text{even } j} (2j+1)e^{-\xi j(j+1)} + 3 \sum_{\text{odd } j} (2j+1)e^{-\xi j(j+1)}.$$

Similar considerations apply to deuterium. The resulting specific heats have been plotted in Figure 6.8, from which it is seen that the contribution of the rotational degrees of freedom in hydrogen or deuterium is practically nonexistent at about 50 and 30 K, respectively.

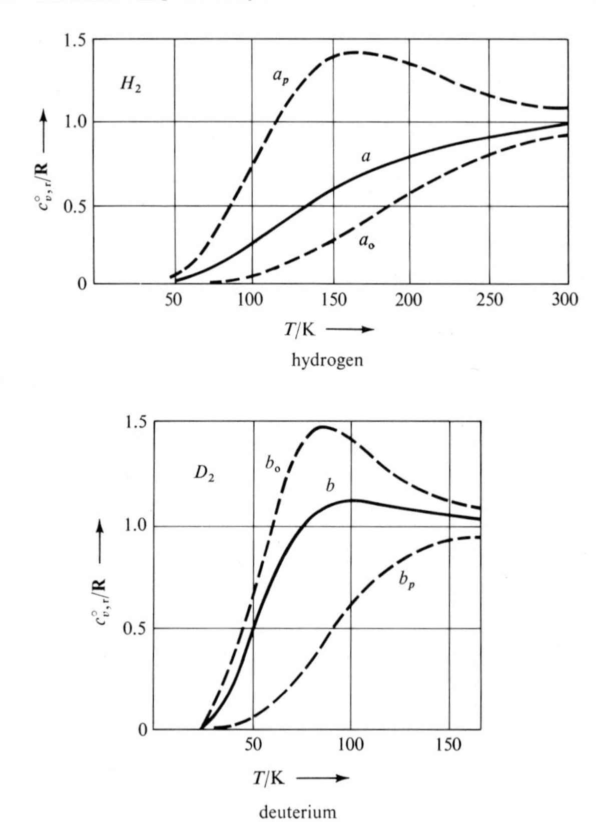

FIGURE 6.8. Rotational specific heat of hydrogen and deuterium: a, hydrogen $= \frac{1}{4}$ para $+ \frac{3}{4}$ ortho; a_p, parahydrogen (spin 0); a_o, orthohydrogen (spin $\hbar$); b, deuterium $= \frac{1}{3}$ para $+ \frac{2}{3}$ ortho; b_p, paradeuterium (spin 0); b_o, orthodeuterium (spin $\hbar$).

† See A. H. Wilson, *Thermodynamics and Statistical Mechanics*, p. 139, Cambridge Univ. Press, London and New York, 1957.

6.9.3. Vibrational Degrees of Freedom

The calculation of the vibrational partition function, Z_v, for a polyatomic molecule is analogous to that for the rotational partition function, except for the circumstance that now the energy levels must be taken from Equation (3.43b) instead of Equation (3.37) and that account must be taken of the number of normal modes: $f_v = 3n - 5$ in a linear molecule and $f_v = 3n - 6$ in a nonlinear molecule (Section 2.3). As long as the vibration amplitudes are small enough and unaffected by rotation, the energy of each mode can be taken to be additive to that of any other, which means that every normal mode k contributes a *multiplicative* term Z_{vk} to the vibrational partition function, Z_v, for the molecule. Hence, we may write

$$Z_v = \prod_{k=1}^{f_v} Z_{vk}. \tag{6.40}$$

In turn, the energy levels of a single normal mode are given by the equation

$$\varepsilon_{vk} = (\tfrac{1}{2} + v)\mathbf{h}\nu_k \qquad (v = 0, 1, \ldots). \tag{6.41}$$

To each frequency we may assign a characteristic vibrational temperature

$$\Theta_{vk} = \frac{\mathbf{h}\nu_k}{\mathbf{k}}. \tag{6.41a}$$

Characteristic vibrational temperatures for a number of diatomic and polyatomic gases can be found listed in Table II. We note that vibrational characteristic temperatures, unlike rotational ones, are relatively large. Consequently, the classical limit of $\mathbf{R}T$ in energy for every vibrational mode (that is, $\frac{1}{2}\mathbf{R}T$ per square term in the Hamiltonian) is reached only at temperatures which are well above normal. Thus, for practical applications, it is necessary to evaluate the complete partition function for each vibrational mode, noting that in some cases several distinct normal modes may possess identical characteristic frequencies. This leads to groups of identical characteristic vibrational temperatures (for example, $\Theta_v = 960$ K occurs twice in CO_2) whose contributions must be counted with the appropriate multiplicity. The characteristic temperatures of vibration are determined spectroscopically, as reported in Section 3.15.

The vibrational partition function for a single normal mode can be evaluated in closed form if it is noted that the sum

$$\begin{aligned} Z_{vk} &= \sum_{v=0}^{\infty} \exp\left[-(v + \tfrac{1}{2})\frac{\mathbf{h}\nu_k}{\mathbf{k}T}\right] \\ &= \exp(-\tfrac{1}{2}\mathbf{h}\nu_k/\mathbf{k}T) \sum_{v=0}^{\infty} \exp\left(-\frac{v\mathbf{h}\nu_k}{\mathbf{k}T}\right) \end{aligned} \tag{6.42}$$

represents an infinite geometric series whose terms tend to zero as $v \to \infty$. The upper limit $v = \infty$ has been inserted for the same reason that induced us to admit an analogous upper limit in Equations (6.12) for the tranlational partition function. In doing so, we have implied that the potential function from Figure 3.16 is assumed to be parabolic even for extremely high values of the quantum number v and that there is no upper limit to it. In fact, for a sufficiently high value of the vibrational quantum number, the interatomic potential will cease to be parabolic thus making it necessary to modify Equation (6.41). Eventually, as explained in Section 3.11, the molecule will dissociate, and this possibility sets an upper limit to v. Nevertheless, the resulting error is negligible, since the spurious terms are extremely small at temperatures which are even quite close to the onset of appreciable dissociation. It must, however, be remembered that our equation ceases to be applicable when dissociation becomes important.

With the preceding approximation, the infinite sum in Equation (6.42) can be evaluated in closed form when we obtain

$$Z_{vk} = \frac{e^{-\xi_k/2}}{1 - e^{-\xi_k}} = \frac{1}{2 \sinh(\xi_k/2)}, \tag{6.42a}$$

where

$$\xi_k = \frac{\Theta_{vk}}{T}. \tag{6.42b}$$

In some applications the zero-point energy is disregarded, and the assumption is made that $\varepsilon_{vk} = 0$ in the ground state. In such cases

$$\varepsilon_{vk} = v\mathbf{h}\nu_k \qquad (v = 0, 1, 2, \ldots)$$

so that

$$Z'_{vk} = \frac{1}{1 - e^{-\xi_k}}. \tag{6.42c}$$

Thus, the term contributed by each vibrational mode per molecule to the system's partition function is

$$\begin{aligned} -\mathbf{k}T \ln Z_{vk} &= \mathbf{k}T[\tfrac{1}{2}\xi_k + \ln(1 - e^{-\xi_k})] \\ &= \mathbf{k}T(\ln \sinh(\xi_k/2) + \ln 2). \end{aligned} \tag{6.43}$$

The corresponding contribution to the specific heat, $c_v{}^\circ$, of one mol of gas follows by differentiation from Equation (5.58) and is

$$c_{v,\,vk} = \mathbf{R}\,\frac{\xi_k{}^2\, e^{\xi_k}}{(e^{\xi_k} - 1)^2} = \mathbf{R}\mathscr{E}(\xi_k). \tag{6.43a}$$

The expression

$$\mathscr{E}(x) = \frac{x^2\,\mathrm{e}^x}{(\mathrm{e}^x - 1)^2} = \frac{(\tfrac{1}{2}x)^2}{\sinh^2(\tfrac{1}{2}x)} \tag{6.43b}$$

is known as the Einstein function. This and related functions have been tabulated in great detail;† a short extract is given in Table III.

The application of Equations (5.54) through (5.59) to the form in Equation (6.43) leads to the following terms per single normal mode per mol:

$$f^\circ_{vk} = \tfrac{1}{2}\mathbf{R}\Theta_{vk} + \mathbf{R}T\ln(1 - \mathrm{e}^{-\Theta_{vk}/T}), \tag{6.44}$$

$$u^\circ_{vk} = \tfrac{1}{2}\mathbf{R}\Theta_{vk} + \mathbf{R}\frac{\Theta_{vk}}{\mathrm{e}^{\Theta_{vk}/T} - 1}, \tag{6.44a}$$

$$s^\circ_{vk} = -\mathbf{R}\ln(1 - \mathrm{e}^{-\Theta_{vk}/T}) + \mathbf{R}\frac{\Theta_{vk}/T}{\mathrm{e}^{\Theta_{vk}/T} - 1}, \tag{6.44b}$$

$$\mu^\circ_{vk} = \tfrac{1}{2}\mathbf{R}\Theta_{vk} + \mathbf{R}T\ln(1 - \mathrm{e}^{-\Theta_{vk}/T}), \tag{6.44c}$$

$$c^\circ_{v,\,vk} = c^\circ_{p,\,vk} = \mathbf{R}\frac{(\Theta_{vk}/T)^2\,\mathrm{e}^{\Theta_{vk}/T}}{(\mathrm{e}^{\Theta_{vk}/T} - 1)^2} = \mathbf{R}\mathscr{E}\left(\frac{\Theta_{vk}}{T}\right). \tag{6.45}$$

The diagrams in Figures 6.9 and 6.10 represent the variation of the vibrational internal energy u°_{vk} and of the vibrational specific heat $c^\circ_{v,\,vk}$ of a single normal mode with temperature. As expected, the classical limits are reached

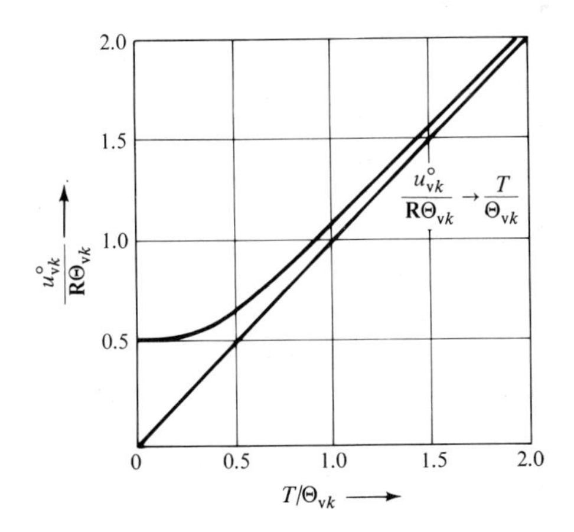

FIGURE 6.9. Vibrational internal energy of a single normal mode.

† J. Hilsenrath and G. G. Ziegler, *Tables of Einstein Functions*, U.S. National Bureau of Standards Monograph 49, U.S. Govt. Printing Office, Washington, D.C., 1962.

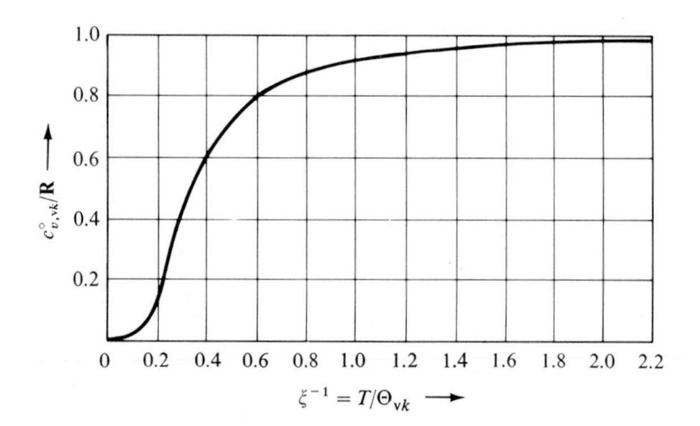

FIGURE 6.10. Vibrational specific heat of a single normal mode.

for $T \gg \Theta_{vk}$. The diagram for energy has been drawn so as to include the zero-point energy $\frac{1}{2}\mathbf{R}\Theta_{vk}$.

At temperatures $T \ll \Theta_{vk}$ the respective mode remains completely "unexcited" or "frozen" and contributes nothing to the thermodynamic properties. At temperatures $T \gg \Theta_{vk}$ the full contribution comes into play, the mode is "fully excited" and demands its full share in equipartition. At intermediate temperatures, the mode is "weighted" and forces the specific heats to become temperature-dependent.

6.9.4. The Electronic Partition Function

We now revert to the introductory remarks made in Section 6.7 and consider the last term contributed by the internal degrees of freedom—the electronic quantum states—assuming, as before, that they do not interact with those accounted for previously.

In cases where the energy difference between the first excited electronic state and the ground state is not large, say, of the order of $\mathbf{k}T$, the contribution to the properties resulting from this source may not be negligible. Generally, it is sufficient to account for the first two states, the ground state of energy $\varepsilon_{\text{el},0}$ and degeneracy g_0, and the first excited state of energy $\varepsilon_{\text{el},1}$ and degeneracy g_1. In such cases, the particle electronic partition function can be written

$$Z'_{\text{el}} = g_0\, e^{-\varepsilon_{\text{el},0}/\mathbf{k}T} + g_1\, e^{-\varepsilon_{\text{el},1}/\mathbf{k}T}. \tag{6.46}$$

It is convenient to write this partition function in the form of the product

$$Z'_{\text{el}} = Z_{\text{el}}\, g_0\, e^{-\varepsilon_{\text{el},0}/\mathbf{k}T},$$

where

$$Z_{\text{el}} = 1 + \frac{g_1}{g_0} e^{-\Delta\varepsilon_{\text{el}}/\mathbf{k}T} \quad \text{with} \quad \Delta\varepsilon_{\text{el}} = \varepsilon_{\text{el},1} - \varepsilon_{\text{el},0}. \tag{6.47}$$

The term $g_0 \exp(-\varepsilon_{el,0}/\mathbf{k}T)$ is usually added to the translational partition function. Since the energy $\varepsilon_{el,0}$ can be adjoined to the arbitrary constant in the full expression for energy, we may put $\varepsilon_{el,0} = 0$ and $Z'_{el}/Z_{el} = g_0$. This will add the constant factor g_0 which was introduced in Section 6.7 in anticipation. This factor affects only the expression for entropy and the related potentials. In particular, it has no influence on the thermal equation of state.

Thus the electronic contribution is confined to the effect of the partition function Z_{el} from Equation (6.47). We shall restrict ourselves to writing the equation for entropy and specific heat only:

$$s_{el} = \mathbf{R} \ln\left(1 + \frac{g_1}{g_0} e^{-\Delta\varepsilon_{el}/\mathbf{k}T}\right) + \mathbf{R} \frac{\Delta\varepsilon_{el}/\mathbf{k}T}{1 + (g_0/g_1) e^{\Delta\varepsilon_{el}/\mathbf{k}T}}, \tag{6.48a}$$

and

$$c^\circ_{v,el} = \mathbf{R} \frac{(\Delta\varepsilon_{el}/\mathbf{k}T)^2}{[1 + (g_1/g_0) e^{-\Delta\varepsilon_{el}/\mathbf{k}T}][1 + (g_0/g_1) e^{\Delta\varepsilon_{el}/\mathbf{k}T}]}. \tag{6.48b}$$

The diagram of $c^\circ_{v,el}$ has been sketched in Figure 6.11a for nitric oxide, NO, for which $g_0 = g_1 = 2$ and $\Delta\varepsilon_{el} = (178\text{ K})\mathbf{k}$; hence we have

$$c^\circ_{v,el} = \mathbf{R} \frac{(178\text{ K}/T)^2}{2 + e^{-178\text{ K}/T} + e^{178\text{ K}/T}}.$$

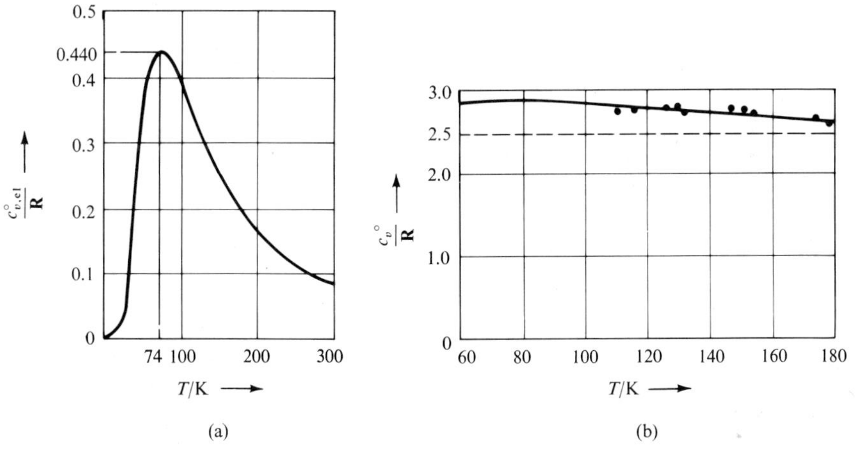

FIGURE 6.11. The electronic specific heat of nitric oxide (NO) with $g_0 = g_1 = 2$; $\Delta\varepsilon = (178\text{ K})\mathbf{k}$: a, electronic contribution; b, total specific heat c_v°. (A. Eucken and L. d'Or.)

The curve exhibits a sharp maximum at $T = 74$ K and $c^\circ_{v,el}/\mathbf{R} = 0.440$, and the electronic contribution is seen to be significant. Figure 6.11b represents the total specific heat

$$c_v^\circ = c^\circ_{v,tr} + c^\circ_{v,tr} + c^\circ_{v,el} = \tfrac{5}{2}\mathbf{R} + \tfrac{2}{2}\mathbf{R} + c^\circ_{v,el},$$

the vibrational degree of freedom being frozen, since $\Theta_v = 2690$ K (Table II). The theoretical result has been compared with the measurements performed by A. Eucken and L. d'Or.†

As far as normal gases are concerned, the only molecule of interest which possesses a degenerate ground state is that of oxygen (O_2) with $g_0 = 3$. Owing to the very large value of $\Delta\varepsilon_{el}$ for it, the electronic degrees of freedom need not be taken into account below about 1500 K.

6.10. Summarizing Remarks

The equations given in the preceding three sections allow us to calculate the properties of any perfect gas as functions of pressure and temperature, provided that the molecular structure and the characteristic temperatures are known from spectroscopic observations. Direct calorimetric measurements have fully confirmed the spectroscopically derived numerical data, and the view is taken that the spectroscopic data are more precise than those obtained calorimetrically. Thus the properties of real gases in the asymptotic limit of very low pressures are today tabulated with the aid of the equations of statistical mechanics. The numerical data in most standard tables have been obtained in this way.

The thermodynamic properties for each gas occur in the form of a sum of three contributions: the translational contribution, the rotational contribution, and the vibrational contribution. In exceptional cases, a fourth, electronic term is required. The vibrational term consists of as many terms as there are normal vibrational modes. The translational portion always corresponds to the classical limit and yields constant specific heats. Except for hydrogen and its isotopes at temperatures below about 80 K, the rotational degrees of freedom are also fully excited, and the classical limit is taken. Thus the rotational degrees of freedom also contribute constant values to the specific heats. The vibrational degrees of freedom possess high characteristic temperatures, and the full quantum-mechanical expressions must be used for them. Thus in most gases the temperature dependence of the specific heats is fully accounted for by the internal, molecular vibrations. In complex molecules, this temperature dependence may lead to the appearance of several steps, the vertical distance from step to step increasing by **R** each time. The steps become very pronounced when two neighboring, characteristic vibrational temperatures differ appreciably.

The temperature as well as the pressure dependence of all the other thermodynamic properties is given explicitly by the equations, the variation with temperature being a consequence of the temperature dependence of the specific heats. The Sackur–Tetrode formula, that is, in effect, the factorial

† *Nachr. Akad. Wiss. Göttingen, Math. Physik Kl.* (1932), 107.

correction term, assures us that the entropy of a perfect gas at any pressure and temperature contains an absolute constant which causes the entropy to vanish in the limit when $T \to 0$ at $P = \text{const}$. Since all specific heats vanish when $T \to 0$, the entropy integral converges.

As example, Figure 6.12 shows the various terms which combine to form

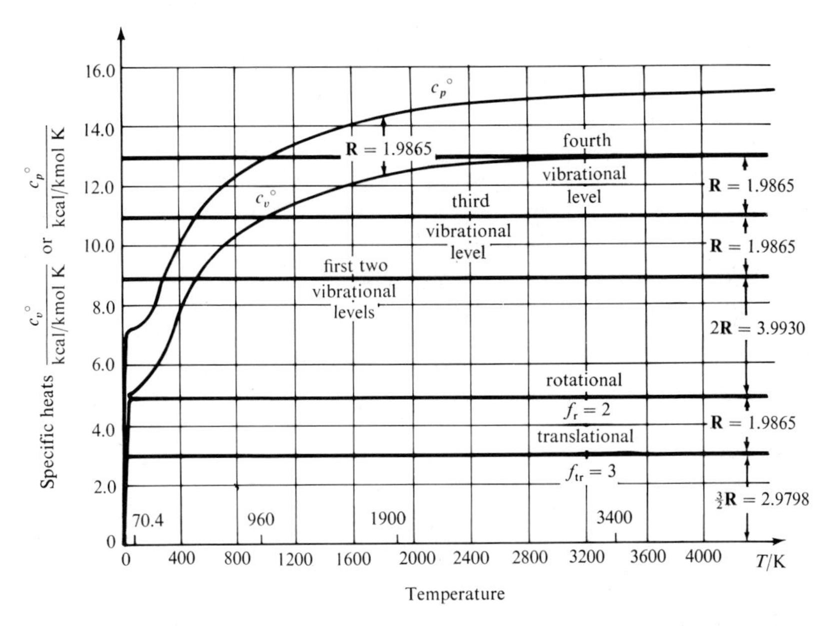

FIGURE 6.12. Specific heats of carbon dioxide: $\Theta_{v1} = \Theta_{v2} = 960$ K; $\Theta_{v3} = 1900$ K; $\Theta_{v4} = 3400$ K; $\Theta_r = 70.4$ K. (Dissociation disregarded.)

the specific heats of carbon dioxide (CO_2), whose molecule is linear with $f_v = 3 \times 3 - 5 = 4$ normal modes of vibration. The lowest vibrational characteristic temperature $\Theta_{v1} = \Theta_{v2} = 960$ K corresponds to two normal modes, and the remaining two are $\Theta_{v3} = 1900$ K and $\Theta_{v4} = 3400$ K. The specific heats c_p° and c_v° tend to zero as $T \to 0$ in agreement with the Third law. The highest value of the specific heat is $c_v^\circ = 6.5\ \mathbf{R} = 12.908$ kcal/kmol K. For the reader's convenience, the most important working equations have been collected in Table 6.4 for easy reference.

6.11. Mixtures of Chemically Inert Perfect Gases

For ease of expression we may assume that an isolated vessel of volume V contains N_a molecules of gas a and N_b molecules of gas b, the extension to an arbitrary number of species presenting no difficulties. We assume that the

TABLE 6.4

Summary of Formulas for Perfect, Nondegenerate Gases[a]

$(Pv_{\mathrm{m}} = \mathbf{R}T)$

Property (per mol)	Translational (classical limit)	Rotational[b] (classical limit)	Vibrational (quantized; per normal mode)	Electronic (lowest level only)
1. Energy $u^\circ(T)$	$\frac{3}{2}\mathbf{R}T$	$\mathbf{R}T$	$\frac{1}{2}\mathbf{R}\Theta_{\mathrm{v}k} + \dfrac{\mathbf{R}\Theta_{\mathrm{v}k}}{e^{\Theta_{\mathrm{v}k}/T} - 1}$	0
2. Helmholtz function $f^\circ(T, v_{\mathrm{m}})$	$-\mathbf{R}T \ln\left[\left(\dfrac{2\pi m\mathbf{k}T}{\mathbf{h}^2}\right)^{3/2} \dfrac{e v_{\mathrm{m}}}{\mathbf{N}}\right]$	$-\mathbf{R}T \ln\left(\dfrac{T}{\sigma\Theta_{\mathrm{r}}}\right)$	$\frac{1}{2}\mathbf{R}\Theta_{\mathrm{v}k} + \mathbf{R}T \ln(1 - e^{-\Theta_{\mathrm{v}k}/T})$	$-\mathbf{R}T \ln g_0$
3. Chemical potential $\mu^\circ(T, P)$	$-\mathbf{R}T \ln\left[\dfrac{(2\pi m)^{3/2}(\mathbf{k}T)^{5/2}}{\mathbf{h}^3 P}\right]$	$-\mathbf{R}T \ln\left(\dfrac{T}{\sigma\Theta_{\mathrm{r}}}\right)$	$\frac{1}{2}\mathbf{R}\Theta_{\mathrm{v}k} + \mathbf{R}T \ln(1 - e^{-\Theta_{\mathrm{v}k}/T})$	$-\mathbf{R}T \ln g_0$
4. Entropy $s^\circ(T, P)$	$\mathbf{R} \ln\left[\left(\dfrac{2\pi m}{\mathbf{h}^2}\right)^{3/2} \dfrac{(e\mathbf{k}T)^{5/2}}{P}\right]$	$\mathbf{R} \ln\left(\dfrac{eT}{\sigma\Theta_{\mathrm{r}}}\right)$	$-\mathbf{R} \ln(1 - e^{-\Theta_{\mathrm{v}k}/T}) + \dfrac{\mathbf{R}(\Theta_{\mathrm{v}k}/T)}{e^{\Theta_{\mathrm{v}k}/T} - 1}$	$\mathbf{R} \ln g_0$
5. Specific heat $c_v^\circ(T)$	$\frac{3}{2}\mathbf{R}$	$\mathbf{R}$	$\mathbf{R}\mathscr{E}(\Theta_{\mathrm{v}k}\, T) = \dfrac{\mathbf{R}(\Theta_{\mathrm{v}k}/T)^2 e^{\Theta_{\mathrm{v}k}/T}}{(e^{\Theta_{\mathrm{v}k}/T} - 1)^2}$	0

[a] $c_p^\circ = c_v^\circ + \mathbf{R}$; at $T = 0$ the entropy $s(0) = 0$ for all values of pressure or volume; at $T = 0$ all specific heats tend to zero ($c_v^\circ \to 0$ and $c_p^\circ \to 0$).

[b] Diatomic molecule.

density of the mixture is small enough for the molecules to interact only through collisions, so that the components may be treated as perfect gases. We further assume that the molecules are inert, which means that the numbers N_a and N_b remain constant because each molecule preserves its identity upon collision. In the absence of interactions, the translational quantum states ε_i of molecules a are independent of those of molecules b and are determined by the familiar equation

$$\varepsilon_i = \frac{\mathbf{h}^2}{8m_a V^{2/3}} (s_x{}^2 + s_y{}^2 + s_z{}^2) \qquad (s_x, s_y, s_z = 1, 2, 3, \ldots),$$

with an analogous equation

$$\varepsilon_j = \frac{\mathbf{h}^2}{8m_b V^{2/3}} (s_x{}^2 + s_y{}^2 + s_z{}^2) \qquad (s_x, s_y, s_z = 1, 2, 3, \ldots)$$

valid for the second species. Similar pairs of equations are valid for the internal degrees of freedom of each group of molecules, but these additional equations are independent of the volume V of the vessel. A particular energy level of the system, E_ν, can therefore, be represented as the sum

$$E_\nu = E_i + E_j \tag{6.49}$$

of energy levels produced by each species separately.

The physical assumption contained in Equation (6.49) implies that the system partition function

$$\mathscr{Z} = \sum_{\nu=1}^{k} \mathrm{e}^{-E_\nu/\mathbf{k}T} \tag{6.50}$$

can be represented in the form of the product

$$\mathscr{Z} = \mathscr{Z}_a \mathscr{Z}_b, \tag{6.51}$$

where

$$\mathscr{Z}_a = \sum_{i=1}^{k'} e^{-E_i/\mathbf{k}T} \qquad \text{and} \qquad \mathscr{Z}_b = \sum_{j=1}^{k''} \mathrm{e}^{-E_j/\mathbf{k}T}. \tag{6.51a, b}$$

In passing from system partition functions to particle partition functions, it is necessary to recognize that particles a or particles b are not distinguishable among themselves, but that particles a *are* distinguishable from particles b. Hence we must divide each particle partition function by the appropriate factorial separately:

$$\mathscr{Z}_a = \frac{Z_a^{N_a}}{N_a!} \qquad \text{and} \qquad \mathscr{Z}_b = \frac{Z_b^{N_b}}{N_b!}. \tag{6.52}$$

It follows that

$$\ln \mathscr{Z} = (N_a \ln Z_a - \ln N_a!) + (N_b \ln Z_b - \ln N_b!) . \qquad (6.53a, b)$$

It is hardly necessary explicitly to put down all the steps to perceive that the additivity of the logarithms of the partition functions $\mathscr{Z}_a$ and $\mathscr{Z}_b$ implies that

> *the thermodynamic properties of the mixture are computed in the form of sums of the respective thermodynamic properties of the inert species present, on the condition that it is supposed that each species occupies the* whole *volume V alone.*

The only essential step omitted is to show that the Lagrangian multipliers β_a and β_b must be the same, as was implied in Equations (6.51a, b) where we have put

$$\beta_a = \beta_b = 1/\mathbf{k}T,$$

with T denoting the common temperature of the mixture. This condition is intuitively obvious, but if it should not appear so to the reader he may revert to Section 5.11.4 and model a proof on the argument developed there. The number and nature of the undetermined Lagrangian multipliers depends on the nature of the constraints. The additivity of energy produces a single equation of constraint and hence a single, and therefore common, multiplier β. Conservation of mass now produces two separate equations and so two multipliers α_a and α_b which, however, can be eliminated. The existence of a single multiplier β implies that the velocities of the molecules of each species separately arrange themselves in accordance with the Maxwellian distribution characteristic for the common temperature T.

The preceding displayed conclusion contains in it Dalton's law of partial pressures

$$P = P_a + P_b,$$

where

$$P_a = \frac{n_a \mathbf{R}T}{V} \quad \text{and} \quad P_b = \frac{n_b \mathbf{R}T}{V},$$

with

$$n_a = \frac{N_a}{\mathbf{N}} \quad \text{and} \quad n_b = \frac{N_b}{\mathbf{N}}.$$

Similarly, the displayed assertion expresses the Gibbs–Dalton rule for the computation of the entropy of mixtures

$$S(T, V) = n_a s_a(T, V) + n_b s_b(T, V) . \qquad (6.54)$$

The explanation of Gibbs's paradox† is contained in Equation (6.52). As long as the otherwise identical gases being mixed are distinguishable, we must use *separate* factors $N_a!$ and $N_b!$, even if the difference is only that of isotopic composition or spin orientation. However, when the particles are of one kind, we would have to write

$$\mathscr{Z} = \frac{Z_a^{N_a} Z_b^{N_b}}{(N_a + N_b)!}. \tag{6.55}$$

Employing Equation (5.48e), which reads

$$S = \mathbf{k} \ln \mathscr{Z} + \frac{U}{T},$$

to the two cases, we obtain two different formulas. In the case of a mixture of indistinguishable molecules ($a \equiv b$), we have

$$S_{\text{ind}} = \mathbf{k} N_a \ln Z_a + \mathbf{k} N_b \ln Z_b + \frac{U}{T} - \mathbf{k} \ln[(N_a + N_b)!],$$

whereas in the case of distinguishable molecules ($a \neq b$), we would write

$$S_{\text{d}} = \mathbf{k} N_a \ln Z_a + \mathbf{k} N_b \ln Z_b + \frac{U}{T} - \mathbf{k} \ln(N_a!\, N_b!).$$

We must now compare

$$-\ln(N_a + N_b)! \qquad \text{with} \qquad -\ln(N_a!\, N_b!).$$

Now it is known (or easily proved) that‡

$$N_a! N_b! < (N_a + N_b)!,$$

which shows that

$$S_{\text{ind}} < S_{\text{d}}.$$

The higher entropy of a mixture of operationally distinguishable elements—the entropy of mixing—is thus seen to be a direct consequence of their distinguishability.

† For a discussion of the Gibbs paradox, see J. Kestin, *A Course in Thermodynamics*, Vol. I, p. 578, Blaisdell, Waltham, Massachusetts, 1966.

‡ For example, for $N_a = 3$ and $N_b = 4$, we have

$$N_a! N_b! = (1 \cdot 2 \cdot 3)(1 \cdot 2 \cdot 3 \cdot 4) \qquad \text{but} \qquad (N_a + N_b)! = (1 \cdot 2 \cdot 3 \cdot 4) \cdot 5 \cdot 6 \cdot 7,$$

the integers outside the last pair of parentheses being each larger than those in the first pair of parentheses after the first sign of equality.

We can compute the entropy of mixing by noting that $N_a = n_a \mathbf{N}$ and $N_b = n_b \mathbf{N}$, where $\mathbf{N}$ is the Avogadro number. Hence we have

$$\Delta S = S_{\mathrm{d}} - S_{\mathrm{ind}} = \mathbf{k}(n_a + n_b)\mathbf{N} \ln(n_a + n_b) - \mathbf{k}n_a \mathbf{N} \ln n_a - \mathbf{k}n_b \mathbf{N} \ln n_b .$$

Since $\mathbf{kN} = \mathbf{R}$, we obtain

$$\Delta S = n_a \mathbf{R} \ln(n_a + n_b) + n_b \mathbf{R} \ln(n_a + n_b) - n_a \mathbf{R} \ln n_a - n_b \mathbf{R} \ln n_b ,$$

and

$$\Delta S = -n_a \mathbf{R} \ln[n_a/(n_a + n_b)] - n_b \mathbf{R} \ln[n_b/(n_a + n_b)].$$

Since

$$(n_a + n_b)\mathbf{R} = PV/T \qquad \text{and} \qquad x_a = n_a/(n_a + n_b) \qquad \text{with} \qquad x_b = n_b/(n_a + n_b) ,$$

we have

$$\Delta S = -\frac{PV}{T}(x_a \ln x_a + x_b \ln x_b) > 0 \qquad \text{for} \quad x_a < 1, x_b < 1 .$$

6.12. Reacting Perfect Gases. Law of Mass Action

When the components in a mixture react chemically, the numbers of molecules N_a and N_b are not constant because different species may appear. However, at equilibrium these numbers remain constant on the average, even though the structure of the molecules changes during *individual* collisions. Thus when chemical equilibrium prevails, the expressions for the properties of the mixture may be formed in the same way as in the case of inert mixtures, the thermodynamic properties for the mixture being equal to the sums of the respective quantities for the components evaluated as if each of the latter alone occupied the common volume V. The difficulty resides in the determination of the unknown average numbers of molecules of the components from a given, initial, nonequilibrium composition. This is the principal problem discussed in classical thermodynamics when a study of chemical equilibria is undertaken.† In the case of perfect gases, the same results can be obtained with the aid of a statistical argument.

To be specific and to simplify our language, we stipulate provisionally that a vessel of given volume V and temperature T contains two chemically reacting gaseous elements A and B which can combine to produce the compound $\mathrm{A}_a\mathrm{B}_b$ according to the chemical formula

$$a\mathrm{A} + b\mathrm{B} = \mathrm{A}_a\mathrm{B}_b . \tag{6.56}$$

† For a complete macroscopic treatment of chemical equilibria see J. Kestin, *A Course in Thermodynamics*, Vol. II, Chapter 21, Blaisdell, Waltham, Massachusetts, 1969.

The restricted results obtained for this case can be generalized to the most complex case by inspection. We assume that the species A, B, and $A_a B_b$ are statistically independent corresponding to the fact that the atoms and molecules of a perfect gas interact on collision only, and apply the reasoning of Section 5.11 to the system of molecules directly. A molecule may disintegrate into its constituent atoms or a chance encounter of a atoms A and b atoms B may produce a molecule $A_a B_b$, depending on circumstances, with energy being conserved in each case. In a state of equilibrium the statistically most probable distributions n_a^*, n_b^*, n_{ab}^* prevail at all times, and encounters which result in combination are as frequent as those which result in disintegration. In general, we shall denote the respective occupation numbers in a distribution by n_a, n_b, n_{ab}, using the subscripts in the capacity of running indices. We shall denote the total numbers of atoms by N_A and N_B, with N_a, N_b, N_{ab} denoting the numbers of particles of each species in a distribution.

It is clear that the detailed effects of gas degeneracy may be left out of account and that it is sufficient to employ corrected Maxwell–Boltzmann statistics by the insertion of the factorials $N_a!$, $N_b!$, and $N_{ab}!$. In contrast with inert gases, the numbers N_a, N_b, N_{ab} may change from distribution to distribution, though the most probable numbers N_a^*, N_b^*, N_{ab}^* are fixed when equilibrium has been attained.

The energy for a given distribution is

$$E = \sum_{a=1}^{k_a} n_a \varepsilon_a + \sum_{b=1}^{k_b} n_b \varepsilon_b + \sum_{ab=1}^{k_{ab}} n_{ab} \varepsilon_{ab} . \tag{6.57}$$

Here the levels ε_a, ε_b, ε_{ab} are determined by the external constraints, and the existence of a single equation for energy shows that the Lagrangian multiplier $\beta = 1/\mathbf{k}T$ must be the same for all species in analogy with an inert mixture.

The energy levels may not be chosen arbitrarily and must be *normalized*. It is convenient to assume that the energy of the lowest energy quantum state (ground state) of every *atom* is zero, so that

$$\varepsilon_{a,0} = \varepsilon_{b,0} = 0 \qquad \text{(ground state)}, \tag{6.57a}$$

and to adopt the convention that the *binding energy* of the molecule $A_a B_b$, denoted here by ε_f (for *energy of formation*), is assigned to *its* ground state. This energy is usually negative because the dissociation of a molecule requires the *addition* of energy ε_f for the disintegration of every molecule. By contrast, the process of ionization results in ions whose energy of formation is positive, again because the ionization of an atom requires the addition of energy ε_f to produce an ion and an electron. The difference is merely one of convenience in setting the reference level: separated atoms in the first case and the combined ion–electron in the second case.

In accordance with the preceding convention, we agree to denote the energy levels of each species by the symbols ε_a, ε_b, and ε_{ab} when they are measured from the three respective ground states, and this necessitates the separate inclusion of the energy of formation. Hence we have

$$E = \sum_{a=1}^{k_a} n_a \varepsilon_a + \sum_{b=1}^{k_b} n_b \varepsilon_b + \sum_{ab=1}^{k_{ab}} n_{ab}(\varepsilon_{ab} + \varepsilon_f)$$

or

$$E = \sum_{a=1}^{k_a} n_a \varepsilon_a + \sum_{b=1}^{k_b} n_b \varepsilon_b + \sum_{ab=1}^{k_{ab}} n_{ab} \varepsilon_{ab} + N_{ab} \varepsilon_f . \tag{6.57b}$$

Conservation of mass does not impose separate conditions on each species as was the case with inert mixtures. Statistically, this is the single essential difference between inert and reacting gaseous mixtures; it produces the law of mass action and allows us to recognize that, according to our microscopic model, reacting gases must obey the Gibbs–Dalton law for entropy as well as Dalton's law of partial pressures. Conservation of mass now leads to the equations

$$\sum_{a=1}^{k_a} n_a + \sum_{ab=1}^{k_{ab}} a n_{ab} = N_A , \qquad \sum_{b=1}^{k_b} n_b + \sum_{ab=1}^{k_{ab}} b n_{ab} = N_B ,$$

or

$$N_a + aN_{ab} = N_A , \qquad N_b + bN_{ab} = N_B . \tag{6.57c}$$

There is no need to enter into a full statistical argument if it is recognized that the system partition function, like that of an inert mixture, separates into the product

$$\mathscr{Z} = \mathscr{Z}_a \mathscr{Z}_b \mathscr{Z}_{ab} = \frac{Z_a^{N_a}}{N_a!} \frac{Z_b^{N_b}}{N_b!} \frac{Z_{ab}^{N_{ab}}}{N_{ab}!} \exp(-N_{ab} \varepsilon_f / \mathbf{k}T) , \tag{6.58}$$

with the last exponential term arising from the normalization introduced in Equation (6.57b). The particle partition functions Z_a, Z_b, Z_{ab} are computed with respect to their own ground states. In each case, the partition function appears in terms of the common volume V and temperature T, the volume itself appearing as a factor in the translational component as explained in Section 6.4. Therefore, we may put

$$Z_a = V\xi_a , \qquad Z_b = V\xi_b , \qquad \text{and} \qquad Z_{ab} = V\xi_{ab} , \tag{6.59}$$

where ξ_a, ξ_b, ξ_{ab} are functions of temperature alone.

To shorten the derivation, we suppose that the system is maintained at a constant temperature, T, and a constant pressure, P, and exploit the fact that in such circumstances the Gibbs function

$$G = F + PV = -\mathbf{k}T \ln \mathscr{Z} + PV \tag{6.60}$$

must be a minimum with respect to a variation in the number of molecules but subject to the constraints imposed by the stoichiometric equation. Since the pressure, P, is determined from Dalton's law, that is, from

$$PV = (N_a + N_b + N_{ab})\mathbf{k}T, \tag{6.61}$$

we compute the *maximum* of

$$\begin{aligned} -\frac{G}{\mathbf{k}T} &= \ln \mathscr{Z} - (N_a + N_b + N_{ab}) \\ &= N_a \ln Z_a + N_b \ln Z_b - N_{ab} \ln Z_{ab} - N_{ab}\, \varepsilon_f/\mathbf{k}T \\ &\quad - (\ln N_a! + \ln N_b! + \ln N_{ab}!) - (N_a + N_b + N_{ab}). \end{aligned} \tag{6.62}$$

Instead of introducing undetermined multipliers, it is simpler to utilize Equations (6.57c) in order to eliminate the variables N_a and N_b in favor of the single variable N_{ab}, and to note that

$$V = \frac{\mathbf{k}T}{P}\,[N_{\mathrm{A}} + N_{\mathrm{B}} + (1 - a - b)N_{ab}]. \tag{6.63}$$

Evaluating

$$\left[\frac{\partial}{\partial N_{ab}}\left(-\frac{G}{\mathbf{k}T}\right)\right]_{T,\,P,\,N_{\mathrm{A}},\,N_{\mathrm{B}}} = 0, \tag{6.64}$$

we are led to an equation for the most probable value N_{ab}^* for N_{ab}. In the course of this derivation we make use of Stirling's approximation in the form

$$\frac{\mathrm{d} \ln x!}{\mathrm{d}x} = \frac{\mathrm{d}}{\mathrm{d}x}(x \ln x - x) = \ln x,$$

and utilize the relations (6.57c) to introduce the most probable values N_a^*, N_b^*. In this manner, we would write

$$\begin{aligned} &-a \ln \xi_a - b \ln \xi_b + \ln \xi_{ab} + a \ln N_a^* + b \ln N_b^* - \ln N_{ab}^* \\ &\quad + (1 - a - b)\ln \mathbf{k}T/P + (1 - a - b)\ln(N_a^* + N_b^* + N_{ab}^*) - \varepsilon_f/\mathbf{k}T = 0, \end{aligned} \tag{6.65}$$

and transform to

$$\frac{N_{ab}^*}{(N_a^*)^a (N_b^*)^b} = \frac{\xi_{ab}}{(\xi_a)^a (\xi_b)^b}\, V^{(1-a-b)} \exp(-\varepsilon_f/\mathbf{k}T). \tag{6.66}$$

The right-hand side of this form of the *law of mass action* contains the function of temperature

$$\tilde{K}_p(T) = \frac{\xi_{ab}}{(\xi_a)^a (\xi_b)^b} \exp(-\varepsilon_f/\mathbf{k}T) \tag{6.67}$$

which can be computed from spectroscopic data and tabulated. In actual applications, we prefer to introduce the mol fractions

$$x_a = \frac{N_a^*}{N^*}, \qquad x_b = \frac{N_b^*}{N^*}, \qquad x_{ab} = \frac{N_{ab}^*}{N^*},$$

where

$$N^* = N_a^* + N_b^* + N_{ab}^*.$$

With these substitutions, Equation (6.66) reduces to the normally encountered form of the *law of mass action*:

$$\frac{x_{ab}}{x_a{}^a x_b{}^b} = K_p(T)\left(\frac{P^*}{P}\right)^{(1-a-b)}, \tag{6.68}$$

where P^* is an arbitrary reference pressure. In the course of this derivation we would show that

$$\left(\frac{V}{N^*}\right)^{(1-a-b)} \tilde{K}_p(T) = \left(\frac{\mathbf{k}T}{P^*}\right)^{(1-a-b)} \left(\frac{P^*}{P}\right)^{(1-a-b)} \tilde{K}_p(T)$$

and introduce an alternative *standard* equilibrium constant

$$K_p(T) = \left(\frac{\mathbf{k}T}{P^*}\right)^{(1-a-b)} \tilde{K}_p(T) = \left(\frac{\mathbf{k}T}{P^*}\right)^{(1-a-b)} \frac{\xi_{ab}}{(\xi_a)^a(\xi_b)^b} \exp(-\varepsilon_f/\mathbf{k}T) \tag{6.68a}$$

which is the one habitually listed in tables of thermochemical data.

In some applications it is preferable to express the law of mass action in terms of the partial pressures

$$P_a = N_a^* \mathbf{k}T/V,$$

and so on for P_b and P_{ab}. Returning to Equation (6.66), we can easily show that

$$\frac{P_{ab}}{P_a{}^a P_b{}^b} (P^*)^{-(1-a-b)} = K_p(T). \tag{6.69}$$

The preceding equations were obtained for the case when the chemical formula has the simplified form given to it in Equation (6.56). However, there is no difficulty in rewriting the law of mass action for the most general form of a chemical reaction, namely

$$\sum_{i=1}^{\alpha} \nu_i \mathrm{A}_i = 0. \tag{6.70}$$

In doing so, it is sufficient to notice that the factors 1, a, and b represent the stoichiometric coefficients ν_i for Equation (6.56), and that $1 - a - b$ must be replaced by $\sum \nu_i = \Delta\nu$. Hence, by inspection,

$$\Delta\varepsilon_{\mathrm{f}} = \sum_{i=1}^{\alpha} \nu_i \varepsilon_{\mathrm{f}i}, \tag{6.71}$$

and

$$\prod_{i=1}^{\alpha} x_i^{\nu_i} = K_p(T)\left(\frac{P^*}{P}\right)^{\Delta\nu}, \tag{6.72}$$

with

$$K_p(T) = \left(\frac{\mathbf{k}T}{P^*}\right)^{\Delta\nu} \prod_{\nu_i=1}^{\alpha} \xi_i^{\nu_i} \exp(-\Delta\varepsilon_{\mathrm{f}}/\mathbf{k}T), \tag{6.73}$$

or

$$(P^*)^{-\Delta\nu} \prod_{i=1}^{\alpha} P_i^{\nu_i} = K_p(T). \tag{6.74}$$

6.12.1. Dissociation

The reaction of dissociation of a homonuclear diatomic molecule

$$A_2 = 2A$$

is particularly simple, and the equilibrium equation can be written explicitly in microscopic terms. For the diatomic molecule, we have

$$\xi_{\mathrm{m}} = \frac{g_{\mathrm{m}}(2\pi m_{\mathrm{m}}\mathbf{k}T)^{3/2}}{\mathbf{h}^3}\left(\frac{T}{2\Theta_{\mathrm{r}}}\right)\frac{1}{(1-\mathrm{e}^{-\Theta_{\mathrm{v}}/T})}, \tag{6.75}$$

as seen from Equations (6.24a), (6.36a) with $\sigma = 2$, and (6.43b). Similarly we have for the atom

$$\xi_{\mathrm{a}} = \frac{g_{\mathrm{a}}(2\pi m_{\mathrm{a}}\mathbf{k}T)^{3/2}}{\mathbf{h}^3}. \tag{6.75a}$$

The normalization constant is equal to the difference between the energy of the atoms ($=0$) and the negative dissociation energy ε_{d}. Hence we obtain

$$\Delta\varepsilon_{\mathrm{f}} = \varepsilon_{\mathrm{d}}.$$

Introducing the numbers of mols of the atoms n_a and of the molecules n_{m}, together with the *mol fraction* x of the atoms

$$x = \frac{n_{\mathrm{a}}}{n_{\mathrm{a}} + n_{\mathrm{m}}},$$

with $1 - x$ denoting the *mol fraction* of the molecules, we have

$$\frac{(N_{\mathrm{a}}^*)^2}{N_{\mathrm{m}}^*} = \frac{x^2}{1-x}N^* = \frac{x^2}{1-x}\frac{PV}{\mathbf{k}T},$$

where

$$N^* = N_m^* + N_a^* .$$

Substitution into Equation (6.72) gives the final result

$$\frac{x^2}{1-x} = \frac{g_a^2}{g_m}\left(\frac{\pi M_a}{N}\right)^{3/2} \frac{(\mathbf{k}T)^{5/2}}{\mathbf{h}^3 P} \frac{1-e^{-\Theta_v/T}}{T/2\Theta_r} e^{-\varepsilon_d/\mathbf{k}T} . \qquad (6.76)$$

In deriving Equation (6.76) we have introduced the molecular masses according to the following relations:

$$M_a = m_a \mathbf{N}, \qquad M_m = m_m \mathbf{N}, \qquad \text{and} \qquad M_m = 2M_a,$$

where **N** is the Avogadro number.

Figure 6.13 represents the variation of the mol fraction x of atomic hydrogen for the dissociation reaction

$$H_2 = 2H,$$

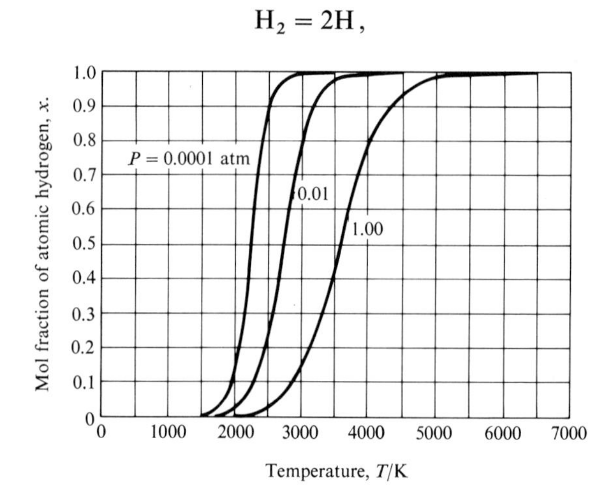

FIGURE 6.13. The dissociation of molecular hydrogen according to the equation $H_2 = 2H$. Plot of Equation (6.76) with $M_a = 1.008$gr/gmol, $g_a = 2$, $g_m = 1$, $\Theta_v = 6100$ K, $\Theta_r = 85.4$ K, $\varepsilon_d = 1.714 \times 10^{-22}$ kcal.

with $M_a = 1.008$ gr/gmol, $g_a = 2$, $g_m = 1$, $\Theta_v = 6100$ K, $\Theta_r = 85.4$ K, and $\varepsilon_d = 1.714 \times 10^{-22}$ kcal. It is useful to verify that Le Chatelier's principle of spite† (principle of Le Chatelier and Braun) is satisfied and to note that dissociation is completed at lower temperatures as the pressure is lowered. For example, at $P = 0.0001$ atm dissociation is 99% complete at $T \approx 2800$ K. At

† For a discussion of Le Chatelier's principle of spite, see J. Kestin, *A Course in Thermodynamics*, Vol. II, Section 21.10, Blaisdell, Waltham, Massachusetts, 1969.

the other extreme, there is less than 1% dissociation below $T \approx 1500$ K at $P = 0.0001$ atm.

6.12.2. Ionization: The Saha Equation

The ionization of a monatomic gas into an ion and a free electron provides another, and perhaps even simpler, example where recourse to statistical formulas results in an explicit working equation. The process is described by the "chemical" formula

$$\mathrm{A} = \mathrm{A}^+ + \mathrm{e}, \tag{6.77}$$

in which e denotes a free electron. Representing the ionization energy by

$$\varepsilon_\mathrm{i} = \mathbf{k}\Theta_\mathrm{i}, \tag{6.77a}$$

where Θ_i is a *characteristic temperature of ionization*, we put

$$\Delta\varepsilon/\mathbf{k}T = \Theta_\mathrm{i}/T. \tag{6.77b}$$

The electronic partition functions of all three species may be approximated by their ground-state degeneracies: g_a for A, g_i for A^+, and $g_\mathrm{e} = 2$ for the electrons corresponding to the two spin orientations which are possible in them.

We assume that the ionized gas resulted from $N^\circ = n^\circ\mathbf{N}$ atoms of a neutral gas, and denote the *degree* or *extent of ionization* by ϕ. Hence we have

$$N_\mathrm{a}^* = (1 - \phi)N^\circ, \qquad N_\mathrm{i}^* = \phi N^\circ, \qquad \text{and} \qquad N_\mathrm{e}^* = \phi N^\circ.$$

At a given equilibrium pressure P and temperature T, we must have

$$PV = (N_\mathrm{a}^* + N_\mathrm{i}^* + N_\mathrm{e}^*)\mathbf{k}T = (1 + \phi)N^\circ\mathbf{k}T. \tag{6.78}$$

The species present may be assumed to perform translatory motions only, and the difference between the partition functions for the ions and neutral atoms may be disregarded.

Substitutions similar to those performed in the preceding section allow us to write

$$\frac{\phi^2}{1-\phi}\cdot\frac{PV}{(1+\phi)\mathbf{k}T} = \frac{(2\pi\mathbf{m}_\mathrm{e}\mathbf{k}T)^{3/2}V}{\mathbf{h}^3}\left(\frac{2g_\mathrm{i}}{g_\mathrm{a}}\right)\mathrm{e}^{-\Theta_\mathrm{i}/T},$$

or finally

$$\frac{\phi^2}{1-\phi^2} = C\,\frac{T^{5/2}}{P}\,\mathrm{e}^{-\Theta_\mathrm{i}/T}, \tag{6.79a}$$

where

$$C = \frac{(2\pi\mathbf{m}_\mathrm{e})^{3/2}\mathbf{k}^{5/2}}{\mathbf{h}^3}\left(\frac{2g_\mathrm{i}}{g_a}\right). \tag{6.79b}$$

Equation (6.79a) together with (6.79b) is known as the *Saha equation*† in honor of the scientist who derived it originally with the aid of a macroscopic argument.

If necessary, the partition functions for the internal motions may be introduced, and the equations can be extended to include double ionization.‡ Figure 6.14 represents the ionization equilibrium of helium; the appearance of the diagram is very similar to that for dissociation.

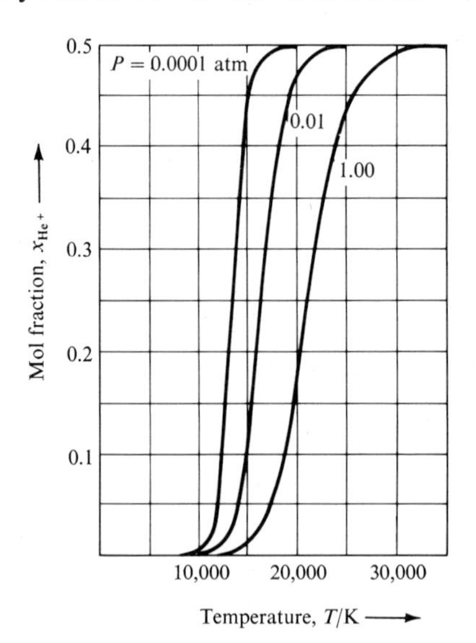

FIGURE 6.14. Ionization equilibrium of helium. Plot of the Saha equation (6.79a,b) with $\mathbf{m}_e = 9.1055 \times 10^{-28}$ gr, $g_i = 2$, $g_a = 1$, $\Theta_i = 28.53 \times 10^4$ K. [Note: $x_{He+} = \phi/(1+\phi)$.]

6.12.3 Plasmas

The examples of Figures 6.13 and 6.14 demonstrate that increasing the temperature of a gas, particularly at lower pressures, causes it progressively to decompose into simpler constituents. At lower temperatures the gas dissociates and subsequently undergoes ionization. Each such process occurs quite rapidly over a relatively narrow temperature interval, and the simpler

† M. N. Saha, Ionization in the solar chromosphere, *Phil. Mag.* **40** (1920), 472. See also M. J. Lighthill, Dynamics of a Dissociating Gas, *J. Fluid Mech.* **2** (1957), 1.

‡ A. B. Cambel, D. P. Duclos, and T. P. Anderson, *Real Gases*, p. 48, Academic Press, New York, 1963. A detailed calculation of the properties of ionized helium according to the reactions $He = He^+ + e$ and $He^+ = He^{2+} + e$ is contained in W. J. Lick and H. W. Emmons, *Thermodynamic Properties of Helium to 50,000* °K, Harvard Univ. Press, Cambridge, Massachusetts, 1962.

structure persists at higher temperatures, becoming simpler still when a new process sets in. At very high temperatures the gas radiates, and this is equivalent to the emission of photons throughout its volume.

A mixture of neutral atoms, positive and negative ions, free electrons, and photons is called a *plasma*. A plasma is sometimes described as the fourth state of matter—fourth, that is, in addition to the solid, liquid, and gaseous states of aggregation.

Moderately high temperatures can be achieved by combustion with pure oxygen, with temperatures of about 3000 K possible in acetylene flames. Above about 10,000 K all gases exist in the form of atoms, the molecules having dissociated at lower temperatures. Temperatures of about 4000 K can be attained in carbon arcs in which the sublimation of carbon sets an upper limit. Somewhat higher temperatures can be attained with metallic electrodes. The seeding of plasmas (cesium fluoride) allows us to attain temperatures of the order of 10,000 K. Similar temperatures can be created for short periods of time in shock tubes.

When the plasma is protected by a vapor film, it is possible to reach temperatures as high as 50,000 K in suitably arranged electric arcs. Even higher temperatures of very short duration (order 1 μsec) have been attained with the aid of high voltage discharges and exploding metallic wires. The shock waves produced by nuclear fission explosions lead to temperatures of the order of 300,000 K, the temperature inside the bomb itself reaching 10^8 K. The substance of which the fixed stars are made also exists at temperatures of the order of 10^6 K.

At the very highest temperatures the plasma consists of nuclei, free electrons, and photons, and the very high particle velocities create favorable conditions for the occurrence of the nuclear reactions of fission and, eventually, fusion. Since plasmas contain free charged particles, they are good electric conductors and their flow can be affected by electric and magnetic fields.

6.13. Spectroscopic and Calorimetric Entropy of a Gas

The evaluation of the standard entropy of a gas with the aid of the equations given on line 4 in Table 6.4 requires only the knowledge of the characteristic vibrational temperatures, Θ_v, and, possibly, of the rotational characteristic temperature, Θ_r, together with the symmetry number, σ. All these quantities are, essentially, calculated on the basis of spectroscopic data, and the result is normalized with respect to absolute zero, as required by the Third law of thermodynamics. For this reason, the standard entropy of a gas is also known as its *spectroscopic entropy*. The same quantity can be obtained by the direct integration of the contributions to the entropy from the specific heats, c, as well as the latent heats, l, of a substance heated from the solid state at absolute zero through all phase transformations, including allotropic transformations, melting and evaporation. Such data are usually obtained calorimetrically, and the entropy computed in this manner is known as the *calorimetric entropy* of the gas. Evidently, the calorimetric entropy must agree with the spectroscopic entropy.

Referring to Figure 6.15 we can write the calorimetric entropy as

$$s_a' = \int_0^{T_s} \frac{c_s}{T}\,dT + \frac{l_s(T_s)}{T_s} + \int_{T_s}^{T} \frac{c_p(P^*, T)}{T}\,dT \tag{6.80a}$$

for a standard pressure, P^*, which is lower than that at the triple point, P_3. Alternatively, at $P^{**} > P_3$, we would write

$$s_a'' = \int_0^{T_m} \frac{c_s}{T}\,dT + \frac{l_m(T_m)}{T_m} + \int_{T_m}^{T_e} \frac{c_l}{T}\,dT + \frac{l_e(T_e)}{T_e} + \int_{T_e}^{T} \frac{c_p(P^{**}, T)}{T}\,dT; \tag{6.80b}$$

in these equations we have neglected the contributions from allotropic transformations for the sake of simplicity.

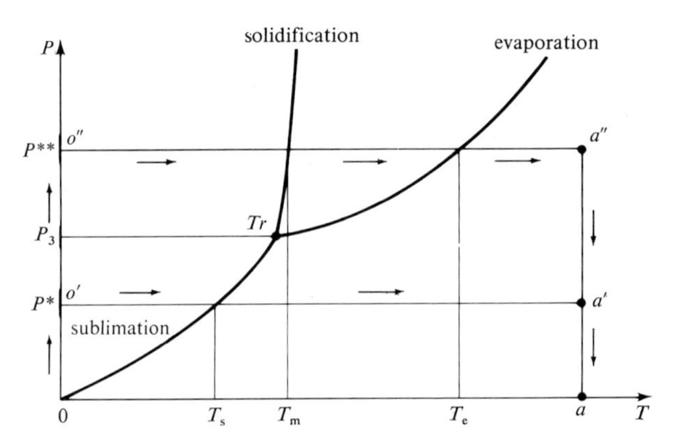

FIGURE 6.15. Path of integration for calorimetric entropy.

The spectroscopic entropy can be written as

$$s(P, T) = \tfrac{5}{2}\mathbf{R} + \mathbf{R}\ln\left[\frac{g_0(2\pi m)^{3/2}(\mathbf{k}T)^{5/2}}{\mathbf{h}^3 P}\right] + s_{\text{int}}(T) + \int_0^P \left[\frac{\mathbf{R}}{P} - \left(\frac{\partial v}{\partial T}\right)_P\right] dP, \tag{6.80c}$$

where the last term corrects for real-gas effects. As already stated, this equation must yield the same result as Equation (6.80a) or (6.80b). For this reason, a comparison between calorimetric and spectroscopic entropy has been used as a test for the validity of the Third law of thermodynamics, since independent sets of experimental data are used. Alternatively, we may regard such a comparison as an experimental test of the validity of the Sackur–Tetrode equation. Many specific cases have been studied in this way and complete agreement can be reported, except in cases where the calorimetric entropy has included a transition from a constrained to an unconstrained state of equilibrium.† We have listed several such comparative values in Table 6.5; the comparison fully substantiates the theory.

† Such a comparison can be found in K. K. Kelly, *U.S. Bur. Mines Bull.* **477** (1950).

TABLE 6.5

Comparison between Calorimetric and Spectroscopic Entropy at $T^* = 25°C$ and $P^* = 1$ atm

Substance	N_2	O_2	H_2	CO	CO_2	NH_3
s_{spec}	45.78	49.03	31.23	47.31	51.07	45.94
s_{calor}	45.9	49.1	29.74	46.2	51.11	45.91

6.14. Absolute Vapor-Pressure Curve

The fact that the Sackur–Tetrode equation represents the *absolute* entropy of a gas at the standard pressure, P^*, and given temperature, T, except for a small correction for nonideal behavior, can be used to obtain absolute forms of the vapor-pressure lines for sublimation and evaporation. This form greatly reduces the amount of experimental calorimetric information normally required to establish such phase boundaries. This possibility can be used to great advantage for the sublimation curves of most solids and for the evaporation curves of helium-3 and helium-4, since, as the reader may know, the latter do not sublimate and can exist in the liquid phase right down to absolute zero.

We shall denote the properties of the vapor by subscript v and those of the condensed phase by c; we utilize the general form of the Clausius–Clapeyron equation

$$g_v = g_c, \tag{6.81}$$

which is valid along the phase boundary. The above two quantities can be calculated easily with the aid of the general scheme employed in Section 6.13 for spectroscopic and calorimetric entropy, together with corrections for nonideal gas behavior. Referring to Figure 6.16, we compute the chemical potential of the condensed phase at a_c as well as that for the vapor phase at a_v, making sure that the component quantities are properly normalized.

We assume that the two phases can exist in equilibrium at $P = 0$ and $T = 0$, and denote the latent heat of evaporation or sublimation in the limit $T \to 0$ by l_0. Supposing that the

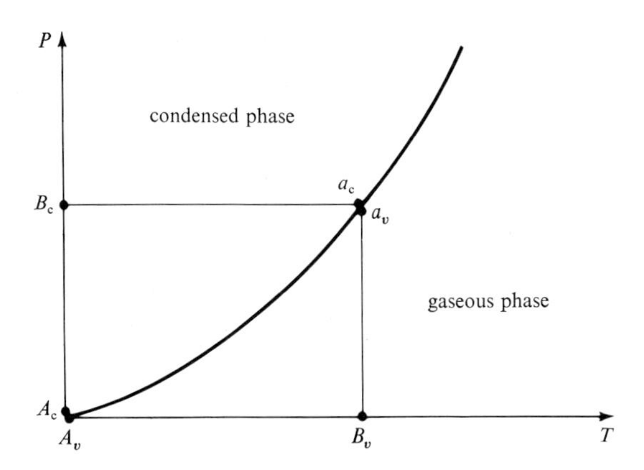

FIGURE 6.16. Absolute form of the vapor-pressure curve.

enthalpy at A_v is $h_v(0, 0) = 0$, we must put $h_c(0, 0) = -l_0$. As far as the condensed phase is concerned, we proceed along $A_c B_c a_c$ and note that along $T = 0$ we have $g = h$ because $Ts = 0$. Hence at B_c we obtain

$$g_c(P, 0) = -l_0 + \int_0^P \left(\frac{\partial g}{\partial P}\right)_T \mathrm{d}P = -l_0 + \int_0^P v_c(P, 0)\,\mathrm{d}P.$$

Integration along $B_c a_c$ gives us

$$g_c(P, T) = g_c(P, 0) + \int_0^T \left(\frac{\partial g}{\partial T}\right)_P \mathrm{d}T = -l_0 + \int_0^P v_c(P, 0)\,\mathrm{d}P - \int_0^T s_c(P, T)\,\mathrm{d}T,$$

where the substitutions $v = (\partial g/\partial P)_T$ and $-s = (\partial g/\partial T)_P$ have been made.

For the vapor, we employ Equation (6.25c),

$$g^{\mathrm{id}}(P, T) = -\mathbf{R}T \ln\left[g_0 \frac{(2\pi m)^{3/2}(\mathbf{k}T)^{5/2}}{\mathbf{h}^3 P}\right] + g_{\mathrm{int}},$$

in which the contribution from the internal degeres of freedom, g_{int}, has been added for completeness. This is equivalent to an integration from B_v to a_v in the perfect-gas plane. If the solid yields a monatomic gas or if the vapor-pressure curve for helium is computed, the term g_{int} vanishes. Thus,

$$g_v(P, T) = \mathbf{R}T \ln\left(\frac{P}{P^*}\right) + g_{\mathrm{int}} + \int_0^P \left[v_v(P, T) - \frac{\mathbf{R}T}{P}\right] \mathrm{d}P, \tag{6.83}$$

where

$$P^* = \frac{g_0 (2\pi m)^{3/2}(\mathbf{k}T)^{5/2}}{\mathbf{h}^3}. \tag{6.83a}$$

Substitution into Equation (6.81) yields the following absolute form of the integrated Clausius–Clapeyron equation. :

$$\ln\left(\frac{P}{P^*}\right) = -\frac{l_0}{\mathbf{R}T} + \frac{1}{\mathbf{R}T}\int_0^P v_c(P, 0)\,\mathrm{d}P - \frac{1}{\mathbf{R}T}\int_0^T s_c(P, T)\,\mathrm{d}T$$
$$- \frac{g_{\mathrm{int}}}{\mathbf{R}T} - \frac{1}{\mathbf{R}T}\int_0^P \left[v_v(P, T) - \frac{\mathbf{R}T}{P}\right] \mathrm{d}P. \tag{6.84}$$

Sometimes it is possible to employ a somewhat different, approximate equation in which the condensed phase is treated as incompressible and in which the specific heats appear explicitly. Now it is supposed that at B_c we may put

$$g_c(P, 0) = h_c(P, 0) = -l_0,$$

thus neglecting the effect of pressure on the incompressible condensed phase. Along $B_c a_c$, we may set

$$h_c(P, T) = -l_0 + \int_0^T c_c(\theta)\,\mathrm{d}\theta,$$

$$s_c(P, T) = \int_0^T c_c(\theta) \frac{\mathrm{d}\theta}{\theta}$$

(θ is a dummy variable of integration). The vapor pressure curve is then†

$$\ln\left(\frac{P}{P^*}\right) = -\frac{l_0}{\mathbf{R}T} - \frac{g_{int}}{\mathbf{R}T} + \frac{1}{\mathbf{R}T}\int_0^T c_c(\theta)\,d\theta - \frac{1}{\mathbf{R}}\int_0^T c_c(\theta)\frac{d\theta}{\theta}. \tag{6.84a}$$

The specific heat of the condensed phase may be taken from experiment or calculated statistically. If warranted, the corrections employed in the derivation of Equation (6.83) may be introduced.

An experimental verification of the absolute vapor-pressure equation with respect to temperatures measured on the thermodynamic scale would be equivalent to a comparison between the calorimetric and spectroscopic entropy of the substance.

PROBLEMS FOR CHAPTER 6‡

6.1. Calculate the average number of molecules contained in a volume $V = 1\ \text{m}^3$ of a perfect gas at $P = 1000, 1, 10^{-3}$ mm Hg and ice-point temperature.

6.2. A cylinder of volume V contains N gaseous molecules in thermodynamic equilibrium. Calculate the probability $\pi(N_1)$ that a subvolume $V_1 = V/k$ will contain $N_1 = N/k$ molecules.

6.3. Show that the Hamiltonian form in Equation (6.15b) leads to the general form for the classical partition function

$$\mathscr{Z} = \frac{1}{N!\mathbf{h}^{fN}}\prod_{i=1}^{fN}(2\pi m_i \mathbf{k}T)^{1/2}\int_V e^{-\beta\phi(\mathbf{q})}\,d\mathbf{q}.$$

Specialize this equation to the case of a perfect monatomic gas, and show that

$$\mathscr{Z} = \frac{1}{N!}\left[\frac{(2\pi m\mathbf{k}T)^{3/2}V}{\mathbf{h}^3}\right]^N = \frac{1}{N!}Z_{tr}^{3N}.$$

6.4. Calculate the following moments: $\int_{-\infty}^{\infty} qm\rho(\mathbf{v})\,d\mathbf{v}$, with $q = 1$, $\mathbf{v}$, $\frac{1}{2}\mathbf{v}^2$, $\frac{1}{2}(\mathbf{v}-\mathbf{u})^2$, for the distribution function

$$\rho(\mathbf{v}) = N\left(\frac{m}{2\pi\mathbf{k}T}\right)^{3/2}\exp\left[-\frac{m(\mathbf{v}-\mathbf{u})^2}{2\mathbf{k}T}\right],$$

and explain the physical significance of each.

In carrying out the above vector integrations, it is useful to note that if $\mathbf{c}$ is a

† Frequently, the last two terms are combined into one:

$$-T\int_0^T\int_0^\theta c_c(\theta')\,d\theta'\frac{d\theta}{\theta^2} = \int_0^T c_c(\theta)\,d\theta - T\int_0^T c_c(\theta)\frac{d\theta}{\theta}.$$

We put $\xi\eta = \int \xi\,d\eta + \int \eta\,d\xi$ with $\xi = 1/T$ and $\eta = \int c_c\,dT$, so that $d\xi = -dT/T^2$ and $d\eta = c_c\,dT$. Finally, we divide by $\mathbf{R}T$.

‡ Problems numbered in boldface type are considered relatively easy; they are recommended for assigning in an elementary first course.

vector with components u, v, and w, and if the function $\phi(\mathbf{c})$ in the integral $\int_{\mathbf{c}} \phi(\mathbf{c})\, d\mathbf{c}$ is odd in u, v, or w, then the integral taken over the whole of **c**-space vanishes.

6.5. Determine the fraction of particles of helium and xenon, in an equimolar mixture of helium and xenon, respectively, which move with absolute velocities that are less than 10, 100, 1000, 1500 m/sec, if the gas is at $P=1$ atm and $T=80$; 500 K.

6.6. Compute the number Ω of distributions of N molecules when: (a) all molecules have the same velocity $+u$; (b) an equal number ($\frac{1}{2}N$) of molecules have velocities $+u$ and $-u$, respectively; (c) an equal number ($\frac{1}{6}N$) of molecules have velocities $\pm u$, $\pm v$, $\pm w$, respectively.

Apply Stirling's approximation to each case, and prove that the last distribution is the most probable.

6.7. Calculate the mean and rms absolute velocity of a molecule of hydrogen and xenon at $T=2$; 300; 3000 K.

6.8. Compute the partition function, and hence establish expressions for the thermal equation of state, the molar specific heat, c_v, and the molar entropy, s, for: (a) a perfect gas of classical point-particles; (b) a perfect gas of molecules whose ellipsoids of inertia are rotationally symmetrical ("symmetric tops"). Note that the kinetic energy for such a molecule is

$$\varepsilon_k = \frac{1}{2m}(p_x{}^2 + p_y{}^2 + p_z{}^2) + \frac{1}{2}\left[\frac{p_\theta{}^2}{A} + \frac{(p_\phi - p_\psi \cos\theta)^2}{A \sin^2\theta} + \frac{p_\psi{}^2}{C}\right];$$

here m is the mass of the molecule, A and C are the principal moments of inertia about the center of mass, whose coordinates are at x, y, z. θ, ϕ, ψ are the Euler angles, each of which may be considered to vary from 0 to 2π; (c) a perfect gas of classical particles for which the energy ε is proportional to the magnitude of its momentum, $\varepsilon = c|\mathbf{p}|$. (*Note:* This problem occurs in the theory of relativistic gases, phonons, and photons.)

6.9. Calculate the specific heat, c_v, and the entropy, s, of a gas consisting of classical point-particles in a one-dimensional potential given by $\phi(x) = ax^2 + bx^3 + cx^4$. Express the above quantities as power series in $\xi = b^2/\beta a^3$ and $\eta = c/\beta a^2$, and compute the first two terms.

6.10. When a gas is whirled in a centrifuge, its molecules are acted on by a centripetal force $F = m\omega^2 r$. Find the density, ρ, of this gas as a function of r. Are the particles concentrated at the center or at the walls of the centrifuge? (Assume that the pressure field is a function of r only.)

6.11. The vibrational frequency and rotational constant of HCl are $\nu/\mathbf{c} = 2990$ cm^{-1} and $B = \mathbf{h}/8\pi^2 I\mathbf{c} = 10.6\ \text{cm}^{-1}$. Prepare a graph of $c_v = c_v(T)$ in the range $T = 5$ to 4000 K.

6.12. The velocity of sound in a perfect gas is $a = (\gamma RT)^{1/2}$ where $R = \mathbf{R}/M$. Calculate this velocity in water vapor (H_2O) and methane (CH_4) at $T = 300$ and 1000 K. Do not use ready-made tables, but calculate $\gamma(T)$ from first principles.

Identify the gas (not necessarily H_2O or CH_4) for which the velocity of sound is highest at $T = 300$ and 1000 K.

6.13. The rotation of a diatomic molecule may be viewed classically as the motion of a mass point on a spherical surface. Find the classical partition function for such a mass-point, and relate the mass of the point and the radius of the sphere to the properties of the diatomic molecules so represented.

SOLUTION: The kinetic energy of a particle of mass m expressed in spherical coordinates is

$$\mathscr{H} = \frac{p_x^2 + p_y^2 + p_z^2}{2m} = \frac{1}{2m} p_r^2 + \frac{1}{2mr^2}\left(p_\theta^2 + \frac{2}{\sin^2\theta} p_\phi^2\right).$$

If the particle is constrained to move on a sphere of radius a we may set $p_r = 0$, $r = a$ to obtain

$$\mathscr{H}(\theta, \phi) = \frac{1}{2ma^2}\left(p_\theta^2 + \frac{1}{\sin^2\theta} p_\phi^2\right).$$

The single-particle partition function is

$$Z = \frac{1}{\mathbf{h}^2} \int_0^{2\pi} \mathrm{d}\phi \int_0^{\pi} \mathrm{d}\theta \int_{-\infty}^{\infty} \mathrm{d}p_\phi \int_{-\infty}^{\infty} \mathrm{d}p_\theta \exp\left[-\frac{\beta}{2ma^2}\left(\frac{p_\phi^2}{\sin^2\theta} + p_\theta^2\right)\right]$$

$$= \frac{8\pi^2 ma^2\mathbf{k}T}{\mathbf{h}^2} = \frac{8\pi^2 I\mathbf{k}T}{\mathbf{h}^2}.$$

The quantity $I = ma^2$ represents the moment of inertia of the corresponding diatomic molecule.

6.14. A system consists of N identifiable particles, each of which can be in either of two states with respective energies

$$\varepsilon_1 = \varepsilon_0 \qquad \text{and} \qquad \varepsilon_2 = 2\varepsilon_0 .$$

(a) If $\mathbf{N}$ is Avogadro's number, show that the energy E, the entropy S, and the specific heat c_x of the system are

$$E = \mathbf{N}\varepsilon_0 \frac{(1 + 2\,\mathrm{e}^{-\beta\varepsilon_0})}{(1 + \mathrm{e}^{-\beta\varepsilon_0})},$$

$$S = \mathbf{R}\ln(1 + \mathrm{e}^{-\beta\varepsilon_0}) + \mathbf{R}\beta\varepsilon_0 \frac{\mathrm{e}^{-\beta\varepsilon_0}}{1 + \mathrm{e}^{-\beta\varepsilon_0}},$$

$$c_x = \mathbf{R}\frac{(\beta\varepsilon_0)^2\,\mathrm{e}^{-\beta\varepsilon_0}}{(1 + \mathrm{e}^{-\beta\varepsilon_0})^2},$$

where $\beta = 1/\mathbf{k}T$ and the subscript x denotes that all variables other than T are to be held constant.

(b) Obtain an expression for entropy as a function of energy, and show that it is equivalent to

$$S/\mathbf{R} = \left[\frac{E}{\mathbf{N}\varepsilon_0} - 1\right] \ln\left[\frac{2 - (E/\mathbf{N}\varepsilon_0)}{(E/\mathbf{N}\varepsilon_0) - 1}\right] - \ln[2 - (E/\mathbf{N}\varepsilon_0)].$$

(c) Plot the function $(S/\mathbf{R})$ *versus* $(E/\mathbf{N}\varepsilon_0)$ for the interval

$$(E_{\min}/\mathbf{N}\varepsilon_0) = 1 \leq (E/\mathbf{N}\varepsilon_0) \leq 2 = (E_{\max}/\mathbf{N}\varepsilon_0).$$

Note that the entropy possesses a maximum and that, in accord with the Third law, it is zero at the lowest energy of the system.

(d) Recalling the relation

$$\left(\frac{\partial S}{\partial E}\right)_x = \frac{1}{T},$$

demonstrate that the system under consideration can exist in a state of *negative absolute* temperature. In particular, show that

$$\lim_{E \to E_{\min}} \left(\frac{\partial S}{\partial E}\right)_V \to +\infty \qquad (T \to 0^+ \text{ K}),$$

$$\lim_{E \to (\bar{E} - \delta)} \left(\frac{\partial S}{\partial E}\right)_V \to 0^+ \qquad (T \to +\infty \text{ K}),$$

$$\lim_{E \to (\bar{E} + \delta)} \left(\frac{\partial S}{\partial E}\right)_V \to 0^- \qquad (T \to -\infty \text{ K}),$$

$$\lim_{E \to E_{\max}} \left(\frac{\partial S}{\partial E}\right)_V \to -\infty \qquad (T \to 0^- \text{ K}),$$

where $\bar{E} = \frac{3}{2}\mathbf{N}\varepsilon_0$ and δ is a small positive number. Mark these temperatures on the entropy diagram drawn for part 3. It is noteworthy that if the First and Second laws of thermodynamics are to be applied to negative temperature systems intact, that is, without changing the conventions established for positive absolute temperatures, then negative absolute temperatures must be regarded as "hotter" than infinite positive absolute temperature. As an example, cooling would be required to reduce a temperature of -1 to -300 K.

(e) Plot $E/\mathbf{N}\varepsilon_0$, $S/\mathbf{R}$, and $c_x/\mathbf{R}$ as functions of $(-\varepsilon_0/\mathbf{k}T)$ for the interval

$$-6 \leq (-\varepsilon_0/\mathbf{k}T) \leq +6.$$

Indicate the directions of "hotter" and "colder" on this plot.

Note: For a complete discussion of thermodynamics and statistical mechanics at negative absolute temperatures and a description of physically realizable systems which can possess negative absolute temperatures, the student may wish to study: N. F. Ramsey, Thermodynamics and statistical mechanics at negative absolute temperatures, *Phys. Rev.* **103** (1956), 20; or L. P. Bazarov, *Thermodynamics*, (F. Immirzi, transl., and A. E. J. Hayes, ed.), Chapter X, Pergamon Press, New York, 1964.

6.15. Suppose that the system of particles described in Problem 6.14 had available to it an infinite number of equally spaced energy levels $\varepsilon_m = m\varepsilon_0$, with $m = 1, 2, 3, \ldots, \infty$; that is, there is no upper bound on the energy of the system. Show that the entropy S of this system as a function of energy E is given by the expression

$$\frac{S}{\mathbf{R}} = \frac{E}{N\varepsilon_0} \ln \frac{E}{N\varepsilon_0} - \left(\frac{E}{N\varepsilon_0} - 1\right) \ln\left(\frac{E}{N\varepsilon_0} - 1\right).$$

Demonstrate that the entropy increases monotonically with energy or that the system under consideration cannot have states of negative absolute temperature.

Note: Consult Problems 4.25 and 4.26 for the evaluation of the two infinite series which arise during the solution of this problem.

6.16. The energy levels of the hydrogen atom are $E_n = -\mathbf{m_e e^4}/8\varepsilon_0{}^2 n^2 \mathbf{h}^2$ ($n = 1, 2, \ldots$). The degeneracy of the state n is $g = 2n^2$. Evaluate the partition function of the hydrogen atoms, considering only these levels, that is, disregard any ionized states of the atom. The sum or integral you obtain will turn out to be divergent. Using the fact that the average separation of the proton and electron in the nth state is $r' = n^2\mathbf{h}^2\varepsilon_0/\pi\mathbf{m_e e}^2$, discuss the untenable assumptions that were made in arriving at the preceding result.

In a more rigorous treatment, the series is truncated at a finite value of n. Choose an appropriate value for it and numerically estimate the partition function at a specified temperature.

Note: This problem has been treated in: R. H. Fowler, *Statistical Mechanics*, 2nd ed., Chapter XIV, Cambridge Univ. Press, London and New York, 1936.

6.17. Consider an ideal monatomic gas at temperature T which is confined in a container of infinite vertical extent and cross-sectional area A. A uniform gravitational field (intensity g) acts vertically in the negative z direction, and it is assumed that the density of the molecules is small enough so that interactions between the molecules can be disregarded.

(a) Show that the Hamiltonian for a single molecule of mass m is $\mathscr{H} = \mathbf{p}^2/2m + mgz$; hence, show that the probability density is

$$\rho(\mathbf{p}, \mathbf{r}) = \frac{1}{(\mathbf{k}T)^{5/2}} \frac{g}{A} \frac{1}{m^{1/2}(2\pi)^{3/2}} \mathrm{e}^{-\mathbf{p}^2/2m\mathbf{k}T} \mathrm{e}^{-mgz/\mathbf{k}T}$$

(b) Compute the average potential energy of a molecule.

(c) Compute the average kinetic energy of a molecule per degree of freedom. Compare this result with that obtained in part (b) and comment on the equipartition of energy for this situation.

(d) Relate the probability $\pi(z)$ of finding a molecule between z and $z + \mathrm{d}z$, irrespective of its x, y, or momentum coordinates, to the probability $\pi(0)$ of finding a molecule between 0 and $\mathrm{d}z$. Compare this result with the classical barometric formula: $\rho = \rho_0 \exp(-Mg\,z/RT)$.

6.18. Revert to Problem 6.17. How does the density of the gas vary with height, z? At the surface of the earth, $z = 0$, the mol fractions of N_2 and O_2 are approximately $x_{N_2} = 0.78$ and $x_{O_2} = 0.21$. What is the relative concentration of N_2 with respect to that of O_2 at a height $z = 300$ km where the temperature is $T = 260$ K.

6.19. J. Perrin† measured the distribution of suspended Brownian particles in a gravitational field as a function of their elevation, z. He assumed that these particles were distributed according to the barometric formula

$$n(z) = A \exp(-mgz/\mathbf{k}T),$$

where g is the gravitational acceleration. After correcting for buoyancy, the particles had an effective mass of $m = 8.2 \times 10^{-15}$ gr when the system was at $T = 293$ K. Perrin's observations can be summarized as

z (cm)	Average no. of particles, n
0.0000	100
0.0030	47
0.0060	23
0.0090	12

Determine Boltzmann's constant $\mathbf{k}$.

6.20. Show that the isentropic equation of state $PV^{5/3} = \text{const}$ holds for a perfect gas of molecules without internal structure, independently of the statistics that the molecules obey.

6.21. The Hamiltonian function for a relativistic ideal gas is

$$\mathscr{H} = \sum_{i=1}^{N} (\mathbf{p}_i{}^2\mathbf{c}^2 + m^2\mathbf{c}^4)^{1/2} .$$

Which of its thermodynamic properties, U, c_v, P are the same as that of a non-relativistic gas?

6.22. Derive the perfect-gas equation of state using the grand canonical ensemble with the assumption that the particles have no internal structure.

SOLUTION: The grand canonical partition function, Equation (5.71), may be written in the form (see Problem 5.15) $\Xi = \sum_{N=0}^{\infty} \mathscr{Z}_N \exp(\beta\mu N)$. Using Equation (6.22), we may write $\mathscr{Z}_N = (\lambda V)^N/N!$ with $\lambda = (2\pi m\mathbf{k}T/\mathbf{h}^2)^{3/2}$. This shows that the grand-canonical partition function Ξ is a simple geometric sum

$$\Xi = \sum_{N=0}^{\infty} \frac{(\lambda V e^{\beta\mu})^N}{N!} = \exp(\lambda V e^{\beta\mu}) .$$

† J. Perrin, *Ann. Chemie et Physique* **18** (1909), 1.

Therefore, $PV/\mathbf{k}T = \ln \Xi = \lambda V \exp(\beta\mu)$. Now $N = \mathbf{k}T\, \partial \ln \Xi/\partial\mu = \lambda V \exp(\beta\mu)$, and finally, $PV = N\mathbf{k}T$.

6.23. At what temperature would quantum effects become important in the translational partition function for a quantity of oxygen or hydrogen contained in a volume $V = 1$ cm^3?

6.24. When an emitter of radiation moves with respect to the observer, the observed frequency ν is given by $\nu = \nu_0(1 + \mathbf{v}\cdot\mathbf{r}/\mathbf{c})$, where ν_0 is the frequency observed when the emitter is at rest relative to the source, $\mathbf{v}$ is the velocity of the source, and $\mathbf{r}$ a unit vector in the direction from the emitter to the observer. Calculate the distribution in frequency of a particular line in the emission spectrum of a gaseous system if the gas is maintained at temperature T? What is the mean square difference, $\langle(\nu - \nu_0)^2\rangle$, of the observed frequency with respect to the value ν_0 for this line?

6.25. (a) The mechanical model of a thermodynamic system consists of N non-interacting particles. We imagine that the system is in contact with an infinitely large reservoir of temperature T. Show that the probability density for the system to have an energy confined between E and $E + \mathrm{d}E$ is

$$\rho(E) = CE^r \exp(-E/\mathbf{k}T) \qquad \text{with} \quad r = \tfrac{3}{2}N - 1\,.$$

Here C denotes the normalization constant which should be determined explicitly.

(b) Show that the most probable value of energy, $\langle E\rangle$, is approximately equal to the average energy, $\bar{E}$, that is, that

$$\bar{E} = \tfrac{3}{2}N\mathbf{k}T \approx \langle E\rangle.$$

(c) Sketch the function $\rho(E)$ for $N = 5$; 100; 1000 and interpret the diagrams in physical terms.

(d) How does this result relate to the principle of equipartition?

6.26. The rotational partition function for a diatomic gas,

$$Z_r = \frac{1}{\sigma}\sum_{j=0}^{\infty} (2j + 1)\, \mathrm{e}^{-\xi J(J+1)},$$

cannot always be evaluated accurately by replacing the sum by an integral because, for some diatomic gases, ξ $(=\Theta_r/T)$ is not small enough to satisfy the criterion established in Section 6.9.1.

Apart from direct summation, which often must be resorted to when one of the atoms is hydrogen, a better approximation than

$$Z_r = 1/\sigma\xi$$

can be obtained for the sum by considering more terms in the Euler–Maclaurin expansion.

Starting with the full expansion

$$\sum_m^n f_i = \int_m^n f(x)\,dx + \tfrac{1}{2}(f_m + f_n) + \sum_{k=1}^{\infty} \frac{B_{2k}}{(2k)!}\,[f_n^{(2k-1)} - f_m^{(2k-1)}]$$

$$= \int_m^n f(x)\,dx + \tfrac{1}{2}(f_m + f_n) + \tfrac{1}{12}(f_n' - f_m') - \tfrac{1}{720}(f_n''' - f_m''')$$

$$+ \tfrac{1}{30240}(f_n^{\mathrm{v}} - f_m^{\mathrm{v}}) \pm \cdots,$$

where B_k are the Bernoulli numbers 7†; and

$$f_n^{(2k-1)} = \frac{d^{2k-1}}{dx^{2k-1}} f(x)\bigg|_{x=n},$$

show that a better approximation to the sum is

$$Z_r = \frac{1}{\sigma\xi}\,(1 + \tfrac{1}{3}\xi + \tfrac{1}{15}\xi^2 + \cdots).$$

Evaluate the correction term at $T = 10$, 100, and 300 K for H_2 ($\Theta_r/K = 85.4$, $\sigma = 2$).

6.27. What is the fraction of N_2 molecules that exist in excited vibrational states at $T = 300$ K? The characteristic vibrational temperature of N_2 is $\Theta_v = 3340$ K.

6.28. Give an account of the classical theory of the specific heats of a polyatomic gas whose molecules are treated as an assembly of harmonic oscillators and rigid rotors. Would you expect significant changes in this theory if a more realistic description of the molecules were adopted (anharmonic oscillators coupled to the rotations)? In particular, can this elaboration possibly account for the fact that the specific heats are functions of temperature? At what temperatures does the simple classical theory lead to values which are in accord with experiment?

6.29. Prove that the ratio of specific heats, γ, for the most complex polyatomic molecule cannot be less than unity, approaching it as the number of degrees of freedom grows large. Show also that the largest possible value for any perfect gas is $\gamma = \frac{5}{3}$.

6.30. (a) Show that the vibrational characteristic temperature of a diatomic heteronuclear molecule of reduced mass $\mu = m_1 m_2/(m_1 + m_2)$, whose atoms obey the Morse potential (Problems 2.12 and 2.14) is

$$\Theta_v = \frac{\mathbf{h}}{2\pi\mathbf{k}} \sqrt{\frac{1}{\mu}\left(\frac{d^2\phi}{dr^2}\right)_{r=r_m}},$$

which gives

$$\Theta_v = \frac{\mathbf{h}}{\pi\mathbf{k}} \left(\frac{c}{r_0}\right) \sqrt{\frac{\varepsilon}{m_a}}$$

for a homonuclear molecule of atomic mass m_a.

† *Handbook of Mathematical Functions*, (M. Abramowitz and I. A. Stegun, eds.), U.S. National Bureau of Standards Applied Mathematics Series 55, U.S. Govt. Printing Office, Washington, D.C., 1964.

(b) Evaluate the vibrational contribution to the specific heat at $T = 300$, 1000, and 2500 K of the gases for which data are given in the accompanying table, and compare the results with those in a standard table, adding appropriate translational and rotational contributions.

	r_m (Å)	c/r_0 (1/Å)	ε_0 (eV)	$10^{24} m_a$ (gr)
H_2	0.7417	2.0054	4.477	1.67
N_2	1.0976	8.1923	9.757	23.25
O_2	1.2074	2.6812	5.115	26.56

Note: See Table XVI in J. Kestin, *A Course in Thermodynamics*, Vol. II, Blaisdell, Waltham, Massachusetts, 1969.

6.31. Solve Problem 6.30 for the Lennard–Jones potential for the case of the gases listed in the table:

	r_0 (Å)	$\varepsilon_0/\mathbf{k}$ (K)
N_2	3.70	95.05
O_2	3.58	117.50

Note: $r_m = 2^{1/6} r_0 = 1.1225\, r_0$.

6.32. Calculate the molar thermodynamic properties f, u, μ, c_v, and s for argon (a perfect gas) at standard conditions ($P = 1$ atm; $T = 298.15$ K), and compare the results with tabulated values. For argon, $m = 6.64 \times 10^{-23}$ gr.

6.33. Given that the characteristic vibrational temperature for oxygen is $\Theta_v = 2230$ K, compute the specific heat $c_p{}^\circ$ for it at $T = 500$; 1000; 1500; 2000; 2500; and 3000 K; compare with the values in the following table:

T (K)	$c_p{}^\circ$ (kcal/kmol K)
500	7.429
1000	8.335
1500	8.739
2000	9.024
2500	9.280
3000	9.518

Explain why the agreement between your calculation and the standard table of values is not quite perfect.

6.34. The energy difference between the lowest electronic state and the first excited state of the sodium atom is $\Delta E/\mathbf{hc} = 16{,}956\ \text{cm}^{-1}$. Estimate the fraction of excited Na atoms at $T = 5000$ K.

6.35. To a very high degree of approximation, the energy levels of the Morse potential are given by the equation

$$\varepsilon_n = \mathbf{h}\nu(n + \tfrac{1}{2})[1 - \gamma(n + \tfrac{1}{2})] - D$$

where γ is a small quantity and D is a constant. Calculate the effects of the anharmonic term $\gamma(n + \frac{1}{2})^2$ on the vibrational specific heat by computing $c_{v,\text{v}}$ to first order in γ.

6.36. Prove that the Lagrangian multipliers β_a and β_b must be equal for a mixture of nonreacting gases at equilibrium, as discussed in Section 6.11.

6.37. For gaseous reactions it is often assumed that a plot of ln $K_p(T)$ against T^{-1} is a straight line. Is this true? Is this true for a hydrogen plasma?

6.38. Consider a reacting gas mixture with five species and two simultaneous chemical reactions

$$\nu_\text{A}\,\text{A} + \nu_\text{B}\,\text{B} + \nu_\text{C}\,\text{C} = 0$$

$$\nu_\text{A}\text{A} + \nu_\text{D}\,\text{D} + \nu_\text{E}\,\text{E} = 0$$

Derive the equations which are necessary to determine the equilibrium composition of the mixture.

6.39. Determine the circumstances under which the equilibrium composition of a reacting mixture is independent of the system's pressure.

6.40. Show that the law of mass action may be written in the form

$$\prod_{i=1}^{\alpha} (N_i^*)^{\nu_i} = K(V, T) \equiv \prod_{i=1}^{\alpha} (Z_i)^{\nu_i} \exp(-\Delta\varepsilon_\text{f}/\mathbf{k}T),$$

where N_i^* is the equilibrium number of molecules of species i. Show also that the same form of the law of mass action results if the Helmholtz free energy is rendered a minimum subject to the constraints that the temperature and the volume are constant.

6.41. Prove the statement following Equation (6.55), that is, that

$$N_a!N_b! < (N_a + N_b)!\,.$$

6.42. Compute the mol fraction of oxygen atoms in O_2 gas at $T = 6000$ K and $P = 0.01$ atm. You may set $g_\text{a} = 9$, $g_\text{m} = 3$.

6.43. Compute the extent of ionization of atomic hydrogen at $P = 0.01$ atm and $T = 10^4$ K if the ionization potential is $\Delta\varepsilon = 13.60$ eV per atom.

Note: 1 eV $= 1.60206 \times 10^{-12}$ erg.

6.44. Using the data given in Table II, estimate the equilibrium constant, K_p, for the reaction

$$N_2 + O_2 = 2NO.$$

Here $g_{O_2} = 3, g_{H_2} = 1, g_{NO} = 4.$

6.45. Using the data given in Table II, estimate the equilibrium constant, K_p, for the reaction

$$H_2 + D_2 = 2HD.$$

6.46. Compute the mol fraction, x, of doubly ionized helium in helium gas at $T = 10^4$ K and $P = 0.1$ atm. The first and second ionization potentials of helium are $\Delta\varepsilon_1 = 24.6$ eV and $\Delta\varepsilon_2 = 54.4$ eV. The degeneracy factors are $g_{He} = 1$; $g_{He^+} = 2$; $g_{He^{2+}} = 1$. (*Hint:* This is an example of a system which sustains the two simultaneous chemical reactions:

$$He = He^+ + e^-$$
$$He^+ = He^{2+} + e^-.$$

The condition that the gas, as a whole, remains electrically neutral is

$$N^*_{He^+} + 2N^*_{He^{2+}} = N_e^*.)$$

LIST OF SYMBOLS FOR CHAPTER 6

Latin letters

A	Constant; normalization factor
A	Chemical species
a	Constant; stoichiometric coefficient
B	Chemical species
b	Stoichiometric coefficient
C	Constant in Saha equation (6.79a, b)
c_p	Specific heat at constant pressure
c_v	Specific heat at constant volume
E	Energy
$\mathscr{E}$	Einstein function
F	Helmholtz function
f	Number of degrees of freedom; molar Helmholtz function
G	Gibbs function
g	Degeneracy of energy level; molar Gibbs function
g_0	Degeneracy factor
$\mathscr{H}$	Hamiltonian
h	Planck's constant
ħ	**h**/2π
I	Moment of inertia
I_1, I_2, I_3	Principal moments of inertia

j	Rotational quantum number
K	Number of independent elements; equilibrium constant
K_p	Standard equilibrium constant, Equation (6.68)
$\tilde{K}_p$	Spectroscopic equilibrium constant, Equation (6.67)
$\mathbf{k}$	Boltzmann's constant
l	Latent heat of a phase change
l_0	Latent heat at absolute zero
M	Molar mass
m	Mass of molecule; constant
$\mathbf{m}_e$	Mass of an electron
N	Number of molecules
$\mathbf{N}$	Avogadro's number
n	Occupation number; number of mols
P	Pressure
p	Generalized momentum
$\mathbf{p}$	Represents momenta p_x, p_y, p_z
$\mathbf{R}$	Universal gas constant
$\mathbf{r}$	Represents coordinates x, y, z
S, S'	Entropy
s	Specific entropy; quantum number
$s_p{}^\circ$	Standard entropy
T	Temperature
$\tilde{T}$	Arbitrary temperature scale
T'	Characteristic temperature
u	Specific internal energy
V	Volume
v	Specific volume; quantum number; velocity
v_m	Molar volume
v^*	Velocity for maximum in Maxwellian velocity distribution function
$v_i{}^*$	Half-width in velocity distribution function
$\mathbf{v}$	Represents velocities v_x, v_y, v_z
x	Mol fraction
x, y, z	Coordinates
Z	Partition function for molecule, per degree of freedom, etc.
$\mathscr{Z}$	Partition function for system

Greek letters

α	Coefficient of thermal expansion; Lagrangian multiplier
β	$(=1/\mathbf{k}T)$; Lagrangian multiplier in method of most probable distribution
γ	Ratio of specific heats
ε_f	Energy of formation
ε_{vk}	Vibrational energy level of kth oscillator
Θ	Characteristic temperature
Θ_i	Characteristic temperature of ionization
Θ_r	Rotational characteristic temperature
Θ_v	Vibrational characteristic temperature
κ	Compressibility

λ Lagrangian multiplier
μ Chemical potential
ν Frequency; stoichiometric coefficient
ξ Dimensionless ratio of characteristic temperature to absolute temperature; also Z/V
ρ Probability density in terms of energy and position
ρ' Marginal probability density in terms of energy (energy probability distribution function)
ρ_1 Marginal probability density for momentum (momentum distribution function)
ρ_2 Marginal probability density for velocity (velocity distribution function)
ρ_v Marginal probability density for absolute velocity (Maxwellian velocity distribution function)
ρ_v^* Maximum value of ρ_v
σ Symmetry number
ϕ Potential energy; extent of reaction
Ω Number of microstates in Γ-space

Subscripts
a Refers to gas a of binary mixture
av Average value
cMB Corrected Maxwell–Boltzmann statistics
c Critical point
cl Classical theory
d Distinguishable particles
el Electronic degrees of freedom
ind Indistinguishable particles
int Internal degrees of freedom
k Refers to kth energy level
MB Maxwell–Boltzmann statistics
m Maximum value; molar quantity
p At constant pressure
rms Root-mean-square value
r Rotational degrees of freedom
tr Translational degrees of freedom
v At constant volume
v Vibrational degrees of freedom

Superscripts
BE Bose–Einstein statistics
FD Fermi–Dirac statistics
MB Maxwell–Boltzmann statistics
$\circ$ Perfect gas or standard state
$*$ Reference value

Special symbol
$\langle\ \rangle$ Expectation value

PART 2

Applications

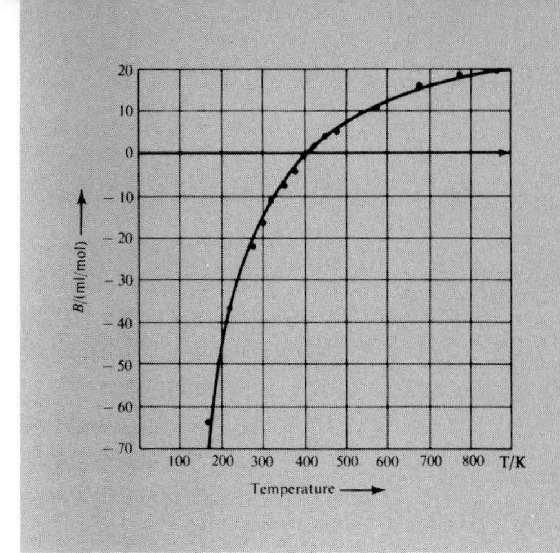

CHAPTER 7

PROPERTIES OF REAL GASES

7.1. Introductory Remarks

In the preceding chapter we have been concerned with systems consisting of statistically independent elements; the exact quantum-mechanical energy levels were known, and we were able to compute the thermodynamic properties to any desired degree of approximation. As a rule, however, real systems are not adequately described by such simple mechanical models, and we must now study systems whose elements interact with one another. In a real gas which consists of nonpolar molecules, the interaction can be described by one of the simple intermolecular force potentials discussed in Section 2.3.2; the interaction is more complex if the molecules are polar.

In this chapter we propose to calculate the thermodynamic properties of nonpolar monatomic gases thus neglecting possible contributions from internal motions. Even this goal cannot be carried out completely, and we shall be forced to rely on heuristic approximations. This will lead to a theory whose validity is restricted to gases of moderate density, that is, to densities which are fractions of that at the critical point.

The scheme of approximation which we shall introduce presently will allow us to expand the thermodynamic properties of gases in powers of nr_0^3, that is, in powers of the ratio of the volume of a molecule ($\sim r_0^3$) to that available per molecule in the space occupied by the gas, $v \sim 1/n$. Finally, we shall discuss the van der Waals equation of state which provides a qualitative description of the properties of gases near liquefaction and up to the critical point, devoting additional space to a discussion of the mathematical nature of the critical point itself.

7.2. Quantum or Classical Partition Function

In order to evaluate the partition function for any system we need to know the energy levels of the system. Since the elements of the system are not distinguishable, we must establish whether they obey Bose–Einstein or Fermi–Dirac statistics. However, if the ratio of the de Broglie wavelength, λ, to the average distance between the molecules, d, becomes very small, molecules can be treated as distinguishable and corrected Maxwell–Boltzmann statistics prove to be adequate. We can formulate this condition by recalling that the de Broglie wavelength

$$\lambda = \frac{\mathbf{h}}{mu}. \tag{7.1}$$

Although all possible velocities, u, are represented in an assembly of molecules at some equilibrium temperature, T, it is sufficient to base the criterion on the average velocity

$$u = \left(\frac{2\pi \mathbf{k} T}{m}\right)^{1/2},$$

that is, on the average wavelength

$$\lambda_{\text{th}} = \mathbf{h}/(2\pi m \mathbf{k} T)^{1/2}. \tag{7.2}$$

This average wavelength is known as the *thermal* wavelength of the particles, and the estimate is valid if T is not too close to absolute zero.

The average volume, v, available to one of N molecules of gas filling a volume V is $v = V/N$, and the average intermolecular distance becomes

$$d \approx v^{1/3}. \tag{7.3}$$

Thus, if

$$d \gg \lambda_{\text{th}} \tag{7.4}$$

or if

$$\frac{\mathbf{h}}{v^{1/3}(2\pi m \mathbf{k} T)^{1/2}} \ll 1, \tag{7.4a}$$

it is possible to treat the particles as distinguishable. If, however,

$$d \approx \lambda_{\text{th}} \qquad \text{or} \qquad d < \lambda_{\text{th}}, \tag{7.4b, c}$$

Bose–Einstein or Fermi–Dirac statistics must be invoked.

The data in Table 6.2 serve to show that the inequality (7.4) is satisfied under a wide range of conditions; it breaks down at very low temperatures, at extremely high densities, and for very light molecules like He and H_2 even at moderate temperatures and densities. It follows that the theory which is about to be developed, although widely applicable, is nevertheless restricted.

Before the actual partition function is written, it is necessary to compare the separation between successive energy levels and the basic energy measure—$\mathbf{k}T$. It has been shown in Section 6.3 that the former is much smaller than the latter for the translational levels of a perfect gas. It now seems reasonable to *assume* that the introduction of intermolecular forces into our considerations does not change this relationship, at least in the range of moderate densities and temperatures of interest to us here. Thus, we conclude that we may use the *classical* partition function, first introduced as Equation (6.15a),

$$\mathscr{Z} = \frac{1}{\mathbf{h}^{3N} N!} \int_{\Gamma\text{-space}} \mathrm{e}^{-\beta \mathscr{H}(\mathbf{q}, \mathbf{p})} \, \mathrm{d}\mathbf{q} \, \mathrm{d}\mathbf{p}, \tag{7.5a}$$

where

$$\mathscr{H}(\mathbf{q}, \mathbf{p}) = \sum_{i=1}^{3N} \frac{p_i^2}{2m} + \phi(\mathbf{q}), \tag{7.5b}$$

with

$$\phi(\mathbf{q}) = \sum_{i=1}^{N} \sum_{j>i}^{N} \varepsilon_2(|\mathbf{r}_i - \mathbf{r}_j|). \tag{7.5c}$$

Here, as usual, $\mathbf{q}$, $\mathbf{p}$ stand for all the generalized coordinates and momenta, and ε_2 is the so-called pair-potential (such as the Lennard-Jones discussed in Section 2.3.2) acting between particles i and j in the gas; its value depends only on the distance, $|\mathbf{r}_i - \mathbf{r}_j|$, between them. Equation (7.5c) *assumes* that the total potential energy can be represented as a simple sum of the potential energies of *pairs* of molecules only. We shall revert to this problem in the next section.

7.3. The Configurational Partition Function

The partition function, Equation (7.5a), can be written

$$\mathscr{Z} = \frac{1}{\mathbf{h}^{3N} N!} \int_{-\infty}^{+\infty} \mathrm{d}p_1 \cdots \int_{-\infty}^{+\infty} \mathrm{d}p_{3N} \int_V \mathrm{d}\mathbf{r}_1 \cdots \int_V \mathrm{d}\mathbf{r}_N \exp\left[-\beta\left(\sum_{i=1}^{3N} \frac{p_i^2}{2m} + \phi\right)\right]. \tag{7.6}$$

Since in our classical approximation the Hamiltonian separates into the sum of a kinetic and a potential energy term, the momentum integrations can be performed immediately, yielding

$$\mathscr{Z} = \frac{1}{N!}\left(\frac{2\pi m\mathbf{k}T}{\mathbf{h}^2}\right)^{3N/2} Q_N(V, \beta), \tag{7.7}$$

where

$$Q_N(V, \beta) = \int_V d\mathbf{r}_1 \cdots \int_V d\mathbf{r}_N \exp(-\beta\phi). \tag{7.8}$$

The quantity $Q_N(V, \beta)$ is called the *configurational partition function.* Insofar as Q_N differs from V^N, it represents the deviation from the ideal gas behavior of the system.

A full evaluation of the configurational partition function for a given potential would give the classical approximation to the thermodynamic functions for all temperatures and densities. In particular, it would be possible to base the classical theory of the gaseous, liquid, and solid states on the configurational partition function, since the contribution of the translational energy to the partition function is so simple. In practice, however, it is very difficult to evaluate the configurational partition function for a reasonable potential. As a result, it is necessary to devise approximations suited to the physical situation under consideration. We shall restrict the evaluation of the configurational partition function, Q_N, to the case of a dilute or moderately dense gas. Thus, if n is the number density of the gas, and r_0 is the effective diameter of the molecules, we stipulate that

$$n r_0^3 \ll 1. \tag{7.9}$$

In these circumstances, we may think of each gas molecule as moving freely most of the time, and interacting with another molecule only occasionally. Any three molecules will interact simultaneously with each other even less frequently, and so on for higher-order collisions. In attempting to devise a method for the evaluation of Q_N we shall be guided by our intuitive feeling that the frequency, or probability, of an interaction decreases as the number of particles interacting *simultaneously* increases.

The method that we now adopt is due to N. G. van Kampen.† This method has the unique feature that it allows us to use only the canonical ensemble, and avoids introducing the grand canonical ensemble; it is also very simple, and we hope, intuitively clear.‡

† *Physica* **27** (1961), 783.

‡ For a more standard treatment based on the grand canonical ensemble see G. E. Uhlenbeck and G. W. Ford, *Lectures in Statistical Mechanics*, p. 43, Am. Math. Soc., Providence, Rhode Island, 1963.

We begin by writing the total potential energy as the sum of pair potentials,† as we already did in Equation (7.5c)

$$Q_N = \int_V d\mathbf{r}_1 \cdots \int_V d\mathbf{r}_N \exp\left[-\beta \sum_{i=1}^{N} \sum_{j=i+1}^{N} \varepsilon_2(|\mathbf{r}_i - \mathbf{r}_j|)\right], \tag{7.10}$$

and introduce the abbreviation

$$\Phi_{ij} = \exp[-\beta\varepsilon_2(|\mathbf{r}_i - \mathbf{r}_j|)]. \tag{7.11}$$

This allows us to express the configurational partition function as an integral of a product of the Φ-functions:

$$Q_N = \int_V d\mathbf{r}_1 \cdots \int_V d\mathbf{r}_N \, \Phi_{12} \Phi_{13} \Phi_{14} \cdots \Phi_{23} \Phi_{24} \cdots \Phi_{N-1,N}. \tag{7.12}$$

7.4. First Approximation to Configurational Partition Function

Our intention now is to provide a method by which the potential energy, Q_N, can be expanded into a power series in nr_0^3. Equation (7.12) expresses Q_N in terms of the mutual interactions of all molecules in the gas. If the gas is dilute, we might imagine that the interaction of a pair of molecules is more probable than an interaction of three or more molecules. We shall therefore construct our first approximation to Q_N by considering each of the

$$\binom{N}{2} = \tfrac{1}{2}N(N-1)$$

interacting pairs in Equation (7.12) separately, as if any pair were unaffected by the other molecules. Consequently, the first approximation may be written as

$$Q_N^{(1)} = V^N \left[\frac{1}{V^2} \int d\mathbf{r}_1 \int d\mathbf{r}_2 \, \Phi_{12}\right] \left[\frac{1}{V^2} \int d\mathbf{r}_1 \int d\mathbf{r}_3 \, \Phi_{13}\right] \left[\frac{1}{V^2} \int d\mathbf{r}_1 \int d\mathbf{r}_4 \, \Phi_{14}\right] \cdots$$
$$\times \left[\frac{1}{V^2} \int d\mathbf{r}_2 \int d\mathbf{r}_3 \, \Phi_{23}\right] \cdots \left[\frac{1}{V^2} \int d\mathbf{r}_{N-1} \int d\mathbf{r}_N \, \Phi_{N-1,N}\right]. \tag{7.13}$$

† Strictly speaking, the potential of a molecule surrounded by a large number of additional molecules is not equal to the sum of the pair potentials as *assumed* here. To see this, we first imagine that a third molecule has been placed near a pair of them. The resulting interaction would not only add a supplementary term but would also affect the field of forces of the first pair and so on. At the present time, all statistical theories are based on the hypothesis of the *additivity of pair potentials* which is justified by the fact, emphasized in Section 2.3.2, that the potential of two molecules decreases to a small value at a relatively short range. For a fuller discussion of this point see H. Margenau and N. R. Kestner, *Theory of Intermolecular Forces*, in *International Series of Monographs in Natural Philosophy*, Volume XVIII, Chap. 5, Pergamon Press, New York, 1969.

The factor V^{2N-N^2} is inserted to ensure that $Q_N^{(1)}$ has the dimension V^N. The preceding substitution introduces an error because the interaction of three particles cannot be described in terms of bimolecular interactions in all regions of configuration space, and thus ignores the presence of the third molecule. In order to appreciate the difference, we write Q_3 explicitly as

$$Q_3 = \int d\mathbf{r}_1 \int d\mathbf{r}_2 \int d\mathbf{r}_3 \, \Phi_{12} \Phi_{13} \Phi_{23} \tag{7.14a}$$

and approximate it by

$$Q_3^{(1)} = V^{-3}\left[\int d\mathbf{r}_1 \int d\mathbf{r}_2 \, \Phi_{12}\right]\left[\int d\mathbf{r}_1 \int d\mathbf{r}_3 \, \Phi_{13}\right]\left[\int d\mathbf{r}_2 \int d\mathbf{r}_3 \, \Phi_{23}\right]. \tag{7.14b}$$

We write Q_3 explicitly as

$$Q_3 = \int d\mathbf{r}_1 \int d\mathbf{r}_2 \int d\mathbf{r}_3 \exp\{-\beta[\varepsilon_2(|\mathbf{r}_1 - \mathbf{r}_2|) + \varepsilon_2(|\mathbf{r}_1 - \mathbf{r}_3|) + \varepsilon_2(|\mathbf{r}_2 - \mathbf{r}_3|)]\}. \tag{7.15}$$

It is now convenient to introduce a new coordinate system for the configuration of three particles. We locate the three particles in space by specifying: (a) the position of the center of mass of the system of three particles, $\mathbf{R}$, and (b) the relative separation of any two pairs of particles, say, $\boldsymbol{\rho}_{12}$ and $\boldsymbol{\rho}_{13}$, noting that

$$\mathbf{R} = \tfrac{1}{3}(\mathbf{r}_1 + \mathbf{r}_2 + \mathbf{r}_3) \tag{7.16a}$$

and that

$$\boldsymbol{\rho}_{12} = \mathbf{r}_1 - \mathbf{r}_2 \tag{7.16b}$$

$$\boldsymbol{\rho}_{13} = \mathbf{r}_1 - \mathbf{r}_3. \tag{7.16c}$$

Substituting these new variables into Equation (7.15), we obtain

$$Q_3 = \int d\mathbf{R} \int d\boldsymbol{\rho}_{12} \int d\boldsymbol{\rho}_{13} \exp\{-\beta[\varepsilon_2(|\boldsymbol{\rho}_{12}|) + \varepsilon_2(|\boldsymbol{\rho}_{12}|) + \varepsilon_2(|\boldsymbol{\rho}_{12} - \boldsymbol{\rho}_{13}|)]\}. \tag{7.17}$$

The new integrand is independent of the location of the center of mass, $\mathbf{R}$, and integration with respect to it produces a factor V, leading to

$$Q_3 = V \int d\boldsymbol{\rho}_{12} \int d\boldsymbol{\rho}_{13} \exp\{-\beta[\varepsilon_2(|\boldsymbol{\rho}_{12}|) + \varepsilon_2(|\boldsymbol{\rho}_{13}|) + \varepsilon_2(|\boldsymbol{\rho}_{12} - \boldsymbol{\rho}_{13}|)]\}. \tag{7.17a}$$

Let us now compare the expression for Q_3, Equation (7.17a), with the explicit representation of $Q_3^{(1)}$, as

$$Q_3^{(1)} = V^{-3}\left\{\int d\mathbf{r}_1 \int d\mathbf{r}_2 \exp[-\beta\varepsilon_2(|\mathbf{r}_1 - \mathbf{r}_2|)]\right\}$$
$$\times \left\{\int d\mathbf{r}_1 \int d\mathbf{r}_3 \exp[-\beta\varepsilon_2(|\mathbf{r}_1 - \mathbf{r}_3|)]\right\}$$
$$\times \left\{\int d\mathbf{r}_2 \int d\mathbf{r}_3 \exp[-\beta\varepsilon_2(|\mathbf{r}_2 - \mathbf{r}_3|)]\right\}. \quad (7.18)$$

The difference between Equation (7.15) for Q_3 and (7.18) for $Q_3^{(1)}$ is that we have considered in (7.18) each possible pair of molecules in isolation and eliminated the effect due to the mutual interactions of three particles. Since each of the integrals in Equation (7.18) extends only over the coordinates of two particles, we can simplify them. For example, the first one can be put in the form

$$\int d\mathbf{r}_1 \int d\mathbf{r}_2 \exp[-\beta\varepsilon_2(|\mathbf{r}_1 - \mathbf{r}_2|)] = \int d\mathbf{R}_{12} \int d\boldsymbol{\rho}_{12} \exp[-\beta\varepsilon_2(|\boldsymbol{\rho}_{12}|)]$$
$$= V \int d\boldsymbol{\rho}_{12} \exp[-\beta\varepsilon_2(|\boldsymbol{\rho}_{12}|)], \quad (7.19)$$

where

$$\mathbf{R}_{12} = \tfrac{1}{2}(\mathbf{r}_1 + \mathbf{r}_2) \quad (7.20a)$$
$$\boldsymbol{\rho}_{12} = \mathbf{r}_1 - \mathbf{r}_2. \quad (7.20b)$$

Consequently,

$$Q_3^{(1)} = \int d\boldsymbol{\rho}_{12} \exp[-\beta\varepsilon_2(|\boldsymbol{\rho}_{12}|)] \int d\boldsymbol{\rho}_{13} \exp[-\beta\varepsilon_2(|\boldsymbol{\rho}_{13}|)]$$
$$\times \int d\boldsymbol{\rho}_{23} \exp[-\beta\varepsilon_2(|\boldsymbol{\rho}_{23}|)]. \quad (7.21)$$

The integrals (7.17a) and (7.21) become equal to each other only in the case when ε_2 is identically zero.

The first approximation, $Q_N^{(1)}$, can be evaluated explicitly by integration which yields

$$Q_N^{(1)} = V^N\left\{\frac{1}{V}\int_V d\boldsymbol{\rho} \exp[-\beta\varepsilon_2(|\boldsymbol{\rho}|)]\right\}^{N(N-1)/2}$$
$$= V^N\left\{1 + \frac{1}{V}\int_V d\boldsymbol{\rho}(\exp[-\beta\varepsilon_2(|\boldsymbol{\rho}|)] - 1)\right\}^{N(N-1)/2}. \quad (7.22)$$

Here, we have utilized the fact that

$$\frac{1}{V}\int_V d\boldsymbol{\rho} = 1.$$

7.5. The Second Virial Coefficient

The physical significance of our approximation, $Q_N^{(1)}$, is not readily apparent from its form in Equation (7.22). This emerges from the fact that our interest is confined to systems which contain a very large number, N, of molecules, so that an asymptotic form valid for a constant number density $n = N/V$ with $N \to \infty$ and $V \to \infty$ must be obtained. For this purpose, we note that Equation (7.22) can be simplified to

$$Q_N^{(1)} = V^N\left(1 + \frac{\frac{1}{2}N(N/V)}{\frac{1}{2}N^2}\int_V d\boldsymbol{\rho}\{\exp[-\beta\varepsilon_2(|\boldsymbol{\rho}|)] - 1\}\right)^{N^2/2},$$

where 1 has been neglected with respect to N, and where a factor of unity has been inserted in front of the integral. Recalling that

$$\lim_{N\to\infty}(1 + a/N^2)^{N^2} = e^a,$$

provided that

$$\lim_{N\to\infty} a/N^2 = 0,$$

we see that

$$Q_N^{(1)} = V^N \exp\left(\tfrac{1}{2}nN\int_V d\boldsymbol{\rho}\{\exp[-\beta\varepsilon_2(|\boldsymbol{\rho}|)] - 1\}\right), \tag{7.23}$$

on condition that the following limit can be established:

$$\lim_{N\to\infty}\frac{\frac{1}{2}N(N/V)}{\frac{1}{2}N^2}\int_V d\boldsymbol{\rho}\{\exp[-\beta\varepsilon_2(|\boldsymbol{\rho}|)] - 1\} = 0. \tag{7.23a}$$

The diagram in Figure 7.1 illustrates the variation of the integrand

$$f(|\boldsymbol{\rho}|) = \exp[-\beta\varepsilon_2(|\boldsymbol{\rho}|)] - 1, \tag{7.24}$$

also known as the *Mayer function*,† with $|\boldsymbol{\rho}|$ for a pair potential $\varepsilon_2(|\boldsymbol{\rho}|)$ which has the general form given in Figure 2.2; the functions $\varepsilon_2(|\boldsymbol{\rho}|)$ and $\exp[-\beta\varepsilon_2(|\boldsymbol{\rho}|)]$ have also been plotted. It will be recalled from Section 2.3 that $\varepsilon_2(|\boldsymbol{\rho}|)$ vanishes at $\boldsymbol{\rho} = r_0$ and becomes very small again for $\boldsymbol{\rho} = \alpha r_0$, where α is of the order of 2 or 3; the same is true about $f(|\boldsymbol{\rho}|)$. This proves that the integral over the Mayer function remains bounded as $V \to \infty$ and establishes the validity of condition (7.23a) and of the approximation in Equation (7.23).

† J. E. Mayer and M. G. Mayer, *Statistical Mechanics*, Wiley, New York, 1940.

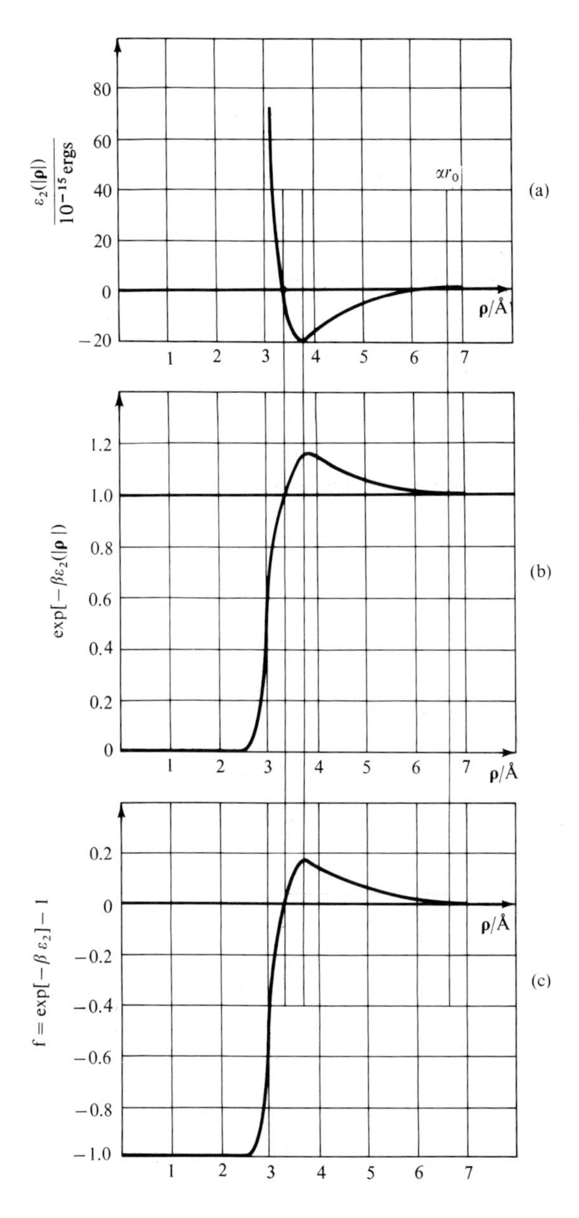

FIGURE 7.1. Variation of a, $\varepsilon_2(|\boldsymbol{\rho}|)$, b, $\exp[-\beta\varepsilon_2(|\boldsymbol{\rho}|)]$, and of c, the Mayer function $f(|\boldsymbol{\rho}|)=\exp[-\beta\varepsilon_2(|\boldsymbol{\rho}|)]-1$ with $\boldsymbol{\rho}$ for a typical pair potential (argon at $T=1000$ K, $\varepsilon/\mathbf{k}=146$ K, $r_0=3.32$ Å, $\alpha=2$).

The preceding argument leads to the following first approximation to the partition function

$$\mathscr{Z} = \frac{V^N}{N!}\left(\frac{2\pi m\mathbf{k}T}{\mathbf{h}^2}\right)^{3N/2} \mathrm{e}^{-nNb(T)}, \tag{7.25a}$$

where

$$b(T) = -\frac{1}{2}\int_V \mathrm{d}\boldsymbol{\rho}\, \mathrm{f}(|\boldsymbol{\rho}|). \tag{7.25b}$$

In turn, reference to Equation (5.57) determines the first approximation to the thermal equation of state,

$$P = n\mathbf{k}T + n^2\mathbf{k}Tb(T). \tag{7.26}$$

For one mol, with $n = \mathbf{N}/v_{\mathrm{m}}$, we see that

$$Pv_{\mathrm{m}} = \mathbf{R}T\left[1 + \frac{\mathbf{N}b(T)}{v_{\mathrm{m}}}\right] = \mathbf{R}T\left[1 + \frac{B(T)}{v_{\mathrm{m}}}\right]. \tag{7.27}$$

Here, $\mathbf{N}$ is the Avogadro number, and $\mathbf{Nk} = \mathbf{R}$ is the universal gas constant. This form convinces us that $b(T)$, defined in Equation (7.25b), is proportional to the second virial coefficient

$$B(T) = \mathbf{N}b(T) \tag{7.27a}$$

of the gas in an expansion in terms of the inverse molar density, v_{m}^{-1}.

The important theoretical result embodied in Equation (7.25b) can be used in two alternative ways. First, assuming that the analytic form of the pair potential $\varepsilon_2(|\boldsymbol{\rho}|)$ is known for any particular gas†, we find it possible to evaluate the second virial coefficient, B, over a range of temperatures for that gas. Secondly, it might appear that the pair potential itself could be computed if the function $B(T)$ were known from precise measurements. This would entail the solution of an integral equation, and the question of the uniqueness of the solution $\mathrm{f}(|\boldsymbol{\rho}|)$ given the function $B(T)$ naturally poses itself. This last problem was investigated by J. B. Keller and B. Zumino‡ and, independently, by E. J. Le Fevre.§ It is found that the required inversion of the

† Strictly speaking this holds only for a monatomic gas whose pair potential is spherically symmetric. Nevertheless, the preceding result is often used with success for diatomic and polyatomic gases whose pair potentials are known to be symmetric. It is interesting to realize that with slight modifications Equation (7.25b) can be made to apply to polar gases. The only essential requirement in its derivation is that the pair potential must vanish for large separations. The angular dependence would be integrated out and would not affect the factor of v_m^{-1}.

‡ *J. Chem. Phys.* **30** (1959), 1351.

§ A transformation of the expression for the classical second virial coefficient. Paper No. 129, Heat Division, Mech. Eng. Research Laboratory, East Kilbride, Glasgow, 1957.

integral is unique only on condition that the integrand is analytic and *monotonic*. We know, however, that the pair-potential function is *not* monotonic, and this demonstrates that its determination from measurements of the second virial coefficient is impossible even in principle. Moreover, a simple assessment of errors would prove to us that the present-day accuracy in the determination of $B(T)$ is insufficient for an accurate representation of $\varepsilon_2(|\boldsymbol{\rho}|)$, even if the latter were monotonic.

Since, generally speaking, the proper forms of the pair potentials have not been discovered even for the simplest of gases, neither of the two preceding approaches is used in actual practice, and an indirect method must be resorted to. In any given case an assumption is made regarding the pair potential and the function $B(T)$ is fitted to its measured results in order to determine the best values of the parameters left free in the assumed form of the potential.

The diagram in Figure 7.2 shows the results of the experimental determination of the second virial coefficient of several monatomic gases from which it is seen that it is negative at low temperatures, becomes equal to zero at a certain temperature, T_B, known as the Boyle temperature, and continues to be positive thereafter having passed through a maximum at a temperature which is of no particular importance. The virial coefficient, B, for polyatomic gases has the same general dependence on temperature. Second virial coefficients

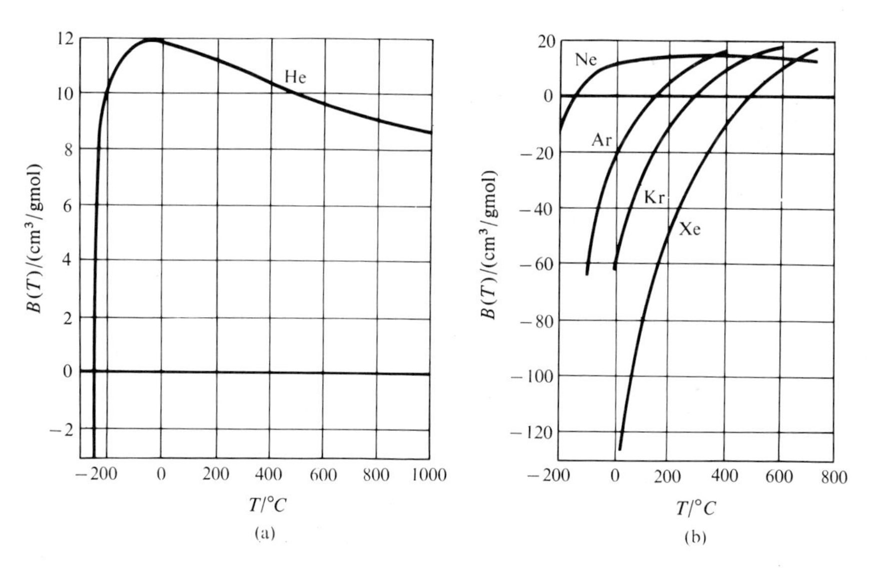

FIGURE 7.2. Second virial coefficient of gases: a, helium; b, other gases. [From Table XXIV of J. A. Beattie, in *Argon, Helium and the Rare Gases*, (G. A. Cook, ed.), p. 273, Wiley (Interscience), New York, 1961.]

are determined from simultaneous measurements of pressure, density, and temperature and must be extracted by properly organized empirical fits.†

Bearing the preceding points in mind, we can now introduce, omitting the algebraic details, the results which are obtained when simple forms of the potential are postulated. For the *rigid-sphere potential*, Figure 2.6, we obtain

$$b = \tfrac{2}{3}\pi\sigma^3 . \tag{7.28}$$

This is a positive constant which is independent of temperature. It is clear that this potential is incapable of representing the properties of gases over a wide range of temperatures. Nevertheless, upon comparing this result with Figure 7.2, it may be concluded that real molecules do approximate rigid spheres at high temperatures.

For the *square-well potential*, Figure 2.7, we find that

$$b(T) = \tfrac{2}{3}\pi\sigma^2 - \tfrac{2}{3}\pi({r_0}^3 - \sigma^3)(e^{\beta\phi_0} - 1), \tag{7.28a}$$

and that the Boyle temperature is

$$T_\mathrm{B} = 1/\mathbf{k}\beta_\mathrm{B}, \tag{7.28b}$$

where β_B is the solution of the transcendental equation

$$(e^{\beta_\mathrm{B}\phi_0} - 1)\left[\left(\frac{r_0}{\sigma}\right)^3 - 1\right] = 1 . \tag{7.28c}$$

With three adjustable constants, this model reproduces the experimental function $B(T)$ quite well, except for the highest temperatures.

In the case of the *Lennard-Jones potential*, the result cannot be put in closed form and must be tabulated.‡ This model (which introduces only two adjustable constants) has been used successfully in many applications. It is superior to the square-well potential, even though it does not reproduce the virial coefficients with an accuracy comparable to that of the most precise measurements.§

† The problems posed by this procedure have been discussed by A. Michels, J. C. Abels, C. A. ten Seldam and W. de Graaf, *Physica* **26** (1960), 381. See also E. A. Mason and T. H. Spurling, *The virial equation of state*, in *The International Encyclopedia of Physical Chemistry and Chemical Physics*, Vol. 2, Topic 10, Pergamon Press, New York, 1969.

‡ Tables of the function $B^*(T^*)$ from Equation (7.29) can be found in: L. F. Epstein and G. M. Roe, *J. Chem. Phys.* **19** (1951), 1320; J. O. Hirschfelder, C. F. Curtiss and R. B. Bird, *Molecular Theory of Gases and Liquids*, Wiley, New York, 1964; H. M. N. Sze and H. W. Hsu, *J. Chem. Eng. Data* **11** (1966), 77 [for six-twelve]; W. E. Rice and J. O. Hirschfelder, *J. Chem. Phys.* **22** (1954), 187 [exp-six].

§ A. Michels, W. de Graaf, and C. A. ten Seldam, *Physica* **26** (1960), 393.

The diagram in Figure 7.3 illustrates how a Lennard-Jones potential is fitted to experimental data The points correspond to the very precise measurements performed on argon by E. Whalley and W. G. Schneider,† and the full curve is a statistically optimal fit to them computed electronically with the aid of tables of the function

$$B^*(T^*) = \frac{B(T)}{(2\pi/3)\mathbf{N}r_0^{\,3}} = 3\int_0^\infty \left\{1 - \exp\left[-\frac{4}{T^*}\left(\frac{1}{r^{12}} - \frac{1}{r^6}\right)\right]\right\} r^2 \,\mathrm{d}r \quad (7.29)$$

where

$$T^* = \mathbf{k}T/\varepsilon. \quad (7.29a)$$

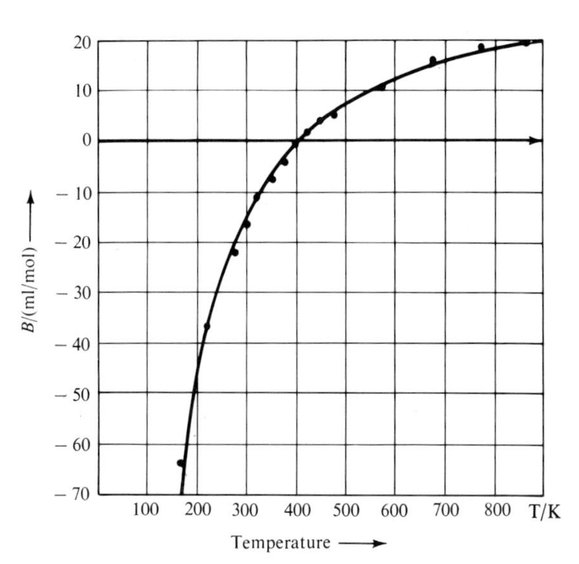

FIGURE 7.3. Second virial coefficients of argon measured by E. Whalley and W. S. Schneider (*J. Chem. Phys.* **23** (1955), 1644) and fitted to a Lennard-Jones potential $\varepsilon_2 = 4\varepsilon[(r_0/r)^{12} - (r_0/r)^6]$ with $\varepsilon/\mathbf{k} = 119.5$ K and $r_0 = 3.76$ Å

● measured; ——— computed.

The calculation established that the sum of the squares of all deviations of measured from calculated points is a minimum for $\varepsilon/\mathbf{k} = 119.5$ K and $r_0 = 3.76$ Å. These two physical constants equip us with a very effective interpolation scheme for the determination of the equilibrium properties of argon.

The reader is reminded that the second virial coefficient at the lower temperatures and for the light molecules (He, H_2) cannot be computed from

† *J. Chem. Phys.* **23** (1955), 1644.

the present, classical approximation, and that quantum corrections must be introduced† for that range.

7.6. Third Virial Coefficient

The next approximation to the configuration integral Q_N from Equation (7.10) cannot be obtained by merely repeating the steps which proved successful in the derivation of the second virial coefficient, B, from the first approximation, $Q_N^{(1)}$. If we were to do this, we would suppose that each of the

$$\binom{N}{3} = \tfrac{1}{6}N(N-1)(N-2)$$

groups of molecules taken three at a time interact in isolation, suggesting that we should put

$$\tilde{Q}_N^{(2)} = \left[\frac{1}{V^3}\int_V d\mathbf{r}_1 \int_V d\mathbf{r}_2 \int_V d\mathbf{r}_3\, \Phi_{12}\Phi_{23}\Phi_{31}\right]\cdots.$$

This expression would contain $\frac{1}{6}N(N-1)(N-2)$ identical multiplicative terms, and would be equivalent to

$$\tilde{Q}_N^{(2)} = \left[\frac{1}{V^3}\int_V d\mathbf{r}_1 \int_V d\mathbf{r}_2 \int_V d\mathbf{r}_3\, \Phi_{12}\Phi_{23}\Phi_{31}\right]^{N(N-1)(N-2)/6}. \tag{7.30}$$

If this were done, the contributions from the interaction of the three pairs in each triplet would also be included. Therefore, the proper second approximation must arise from $\tilde{Q}_N^{(2)}$ by dividing each multiplicative term in Equation (7.30) by

$$\tilde{Q}_3^{(1)} = \left[\frac{1}{V^2}\int_V d\mathbf{r}_1 \int d\mathbf{r}_2\, \Phi_{12}\right]^3, \tag{7.30a}$$

that is, by the contributions from the three pairs in each triplet computed with the aid of Equation (7.22). Hence we put

$$Q_N^{(2)} = \frac{\tilde{Q}_N^{(2)}}{[\tilde{Q}_3^{(1)}]^{N(N-1)(N-2)/6}},$$

or

$$Q_N^{(2)} = \left[\frac{V^3 \int_V d\mathbf{r}_1 \int_V d\mathbf{r}_2 \int_V d\mathbf{r}_3\, \Phi_{12}\Phi_{23}\Phi_{31}}{[\int_V d\mathbf{r}_1 \int_V d\mathbf{r}_2\, \Phi_{12}]^3}\right]^{N(N-1)(N-2)/6}. \tag{7.31}$$

† For further details see J. O. Hirschfelder, C. F. Curtiss, and R. B. Bird, *Molecular Theory of Gases and Liquids*, Wiley, New York, 1964.

If we were to proceed in this manner to successively larger groups of molecules, we would compute the configuration integral Q_N as the *product*

$$Q_N = Q_N^{(1)} Q_N^{(2)} Q_N^{(3)} \cdots, \tag{7.32}$$

in which the factors correspond to pairs, triplets, quadruplets, etc. In order to see more clearly that a product is involved, let us examine the case when $N = 3$, and write, *in extenso*

$$Q_3 = \int_V d\mathbf{r}_1 \int_V d\mathbf{r}_2 \int_V d\mathbf{r}_3 \, \Phi_{12} \Phi_{23} \Phi_{31} .$$

On the other hand,

$$Q_3^{(1)} = \frac{1}{V^3} \int_V d\mathbf{r}_1 \int_V d\mathbf{r}_2 \, \Phi_{12} \int_V d\mathbf{r}_1 \int_V d\mathbf{r}_3 \, \Phi_{13} \int_V d\mathbf{r}_2 \int_V d\mathbf{r}_3 \, \Phi_{23}$$

and

$$Q_3^{(2)} = \frac{V^3 \int_V d\mathbf{r}_1 \int_V d\mathbf{r}_2 \int_V d\mathbf{r}_3 \, \Phi_{12} \Phi_{23} \Phi_{31}}{[\int_V d\mathbf{r}_1 \int_V d\mathbf{r}_2 \, \Phi_{12}]^3} .$$

Clearly, for $N = 3$ Equation (7.32) is satisfied identically. Proceeding to $N = 4$, etc., by induction, we can show that the form of Equation (7.32) is justified for any N.

In order to evaluate $Q_N^{(2)}$ we introduce the center-of-mass coordinates as well as the Mayer function, f, from Equation (7.24), and observe that each of the terms in $\tilde{Q}_N^{(2)}$ may be put in the form

$$\frac{1}{V^2} \int_V d\boldsymbol{\rho}_{12} \int_V d\boldsymbol{\rho}_{13} [1 + \mathrm{f}(|\boldsymbol{\rho}_{12}|) + \mathrm{f}(|\boldsymbol{\rho}_{13}|) + \mathrm{f}(|\boldsymbol{\rho}_{23}|)$$
$$+ \mathrm{f}(|\boldsymbol{\rho}_{12}|)\,\mathrm{f}(|\boldsymbol{\rho}_{13}|) + \mathrm{f}(|\boldsymbol{\rho}_{12}|)\,\mathrm{f}(|\boldsymbol{\rho}_{23}|) + \mathrm{f}(|\boldsymbol{\rho}_{13}|)\,\mathrm{f}(|\boldsymbol{\rho}_{23}|)$$
$$+ \mathrm{f}(|\boldsymbol{\rho}_{12}|)\,\mathrm{f}(|\boldsymbol{\rho}_{13}|)\,\mathrm{f}(|\boldsymbol{\rho}_{23}|)] .$$

Recalling that

$$|\boldsymbol{\rho}_{23}| = |\boldsymbol{\rho}_{12} - \boldsymbol{\rho}_{13}| ,$$

and that

$$\int_V d\boldsymbol{\rho}_{12} \int_V d\boldsymbol{\rho}_{13} \, \mathrm{f}(|\boldsymbol{\rho}_{12}|)\,\mathrm{f}(|\boldsymbol{\rho}_{13}|) = \left[\int d\boldsymbol{\rho} \, \mathrm{f}(|\boldsymbol{\rho}|) \right]^2 ,$$

we calculate

$$Q_N^{(2)} = \left\{ 1 + \frac{\tilde{Q}}{[1 + (1/V) \int_V d\boldsymbol{\rho} \, \mathrm{f}(|\boldsymbol{\rho}|)]^3} \right\}^{N(N-1)(N-2)/6} . \tag{7.33}$$

where

$$\tilde{Q} = (1/V^2) \int_V d\boldsymbol{\rho}_{12} \int_V d\boldsymbol{\rho}_{13} f(|\boldsymbol{\rho}_{12}|) f(|\boldsymbol{\rho}_{13}|) f(|\boldsymbol{\rho}_{23}|) - (1/V^3)[\int_V d\boldsymbol{\rho} f(|\boldsymbol{\rho}|)]^3.$$

In the limit $N \to \infty$, $V \to \infty$ with $n = N/V = \text{const}$, this simplifies to

$$Q_N^{(2)} = \exp[-\tfrac{1}{2} N n^2 c(T)], \tag{7.34}$$

with

$$c(T) = -\frac{1}{3} \int_V d\boldsymbol{\rho}_{12} \int_V d\boldsymbol{\rho}_{13} f(|\boldsymbol{\rho}_{12}|) f(|\boldsymbol{\rho}_{13}|) f(|\boldsymbol{\rho}_{23}|), \tag{7.35}$$

so that

$$Q_N \approx Q_N^{(1)} Q_N^{(2)} = V^N \exp N[-nb(T) - \tfrac{1}{2} n^2 c(T)],$$

from which it follows that

$$P = n\mathbf{k}T[1 + nb(T) + n^2 c(T)], \tag{7.36}$$

or that

$$Pv_{\text{m}} = \mathbf{R}T\left[1 + \frac{B(T)}{v_{\text{m}}} + \frac{C(T)}{v_{\text{m}}^2}\right], \tag{7.36a}$$

where the third virial coefficient

$$C(T) = \mathbf{N}^2 c(T). \tag{7.36b}$$

In the case of *rigid spheres*, an evaluation of the integral in Equation (7.35) gives a constant of the form

$$c = \tfrac{5}{18} \pi^2 \sigma^6. \tag{7.37}$$

As far as the other possible potential functions are concerned, the calculation of $c(T)$ presents considerable difficulties, and the results must be tabulated.†

The third virial coefficient, like the second, is also deduced from the simultaneous measurement of pressure, volume, and temperature by a systematic procedure of fitting progressively higher order polynomials to isotherms.‡ Consequently, the accuracy with which it is known is inferior to that of the second and its use as a means of discovering the form of the pair potential of different gases is limited.

† Tables for $c(T)$ for the Lennard-Jones potential can be found in J. O. Hirschfelder, C. F. Curtiss, and R. B. Bird, Molecular Theory of Gases and Liquids, Wiley, New York, 1964; those for the Buckingham potential are given in A. E. Sherwood and J. M. Prausnitz, *J. Chem. Phys.* **41** (1964), 413.

‡ See also J. Kestin, *A Course in Thermodynamics*, Vol. II, chapter 20, Blaisdell, Waltham, Massachusetts, 1969.

7.7. Higher Approximations and Other Thermodynamic Properties

Although seldom used in practice, the preceding method can be extended to the calculation of higher approximations. Evidently, however, the algebraic complexity increases considerably with each step. A comprehensive statistical and experimental study of the virial equation of state was given by E. A. Mason and T. H. Spurling.†

In order to evaluate the remaining equilibrium properties, it is now possible to write the following, general expression for the partition function. In this equation, we denote the second virial coefficient (per molecule) by b_1, the third by b_2, and so on. Thus

$$\mathscr{Z} = \exp N\left(\frac{3}{2}\ln\frac{2\pi m\mathbf{k}Te^{2/3}}{\mathbf{h}^2 n^{2/3}} - \sum_{k=1}^{\infty}\frac{n^k}{k}b_k\right). \tag{7.38}$$

There is now no further difficulty in obtaining the fundamental equation in the form of the Helmholtz function and in terms of its canonical variables, T and v_{m}. This is

$$f(T, v_{\mathrm{m}}) = f^{\circ}_{\mathrm{tr}}(T) + \mathbf{R}T\sum_{k=1}^{\infty}\frac{1}{k}\left(\frac{\mathbf{N}}{v_{\mathrm{m}}}\right)^k b_k, \tag{7.39}$$

where f°_{tr} is the standard molar Helmholtz function of the translational degrees of freedom of the perfect gas known to us from Equation (6.25). An application of the familiar equations (5.54)–(5.59) produces the following explicit correction terms:

$$u - u^{\circ}_{\mathrm{tr}} = -\mathbf{R}T^2\sum_{k=1}^{\infty}\frac{1}{k}\left(\frac{\mathbf{N}}{v_{\mathrm{m}}}\right)^k\frac{\mathrm{d}b_k}{\mathrm{d}T}; \tag{7.40}$$

$$s - s^{\circ}_{\mathrm{tr}} = -\mathbf{R}\sum_{k=1}^{\infty}\frac{1}{k}\left(\frac{\mathbf{N}}{v_{\mathrm{m}}}\right)^k\left(T\frac{\mathrm{d}b_k}{\mathrm{d}T} + b_k\right); \tag{7.41}$$

$$\mu - \mu^{\circ}_{\mathrm{tr}} = \mathbf{R}T\sum_{k=1}^{\infty}\frac{k+1}{k}\left(\frac{\mathbf{N}}{v_{\mathrm{m}}}\right)^k b_k; \tag{7.42}$$

$$c_v - c^{\circ}_{v,\mathrm{tr}} = -2\mathbf{R}T\sum_{k=1}^{\infty}\frac{1}{k}\left(\frac{\mathbf{N}}{v_{\mathrm{m}}}\right)^k\frac{\mathrm{d}b_k}{\mathrm{d}T} - \mathbf{R}T^2\sum_{k=1}^{\infty}\frac{1}{k}\left(\frac{\mathbf{N}}{v_{\mathrm{m}}}\right)^k\frac{\mathrm{d}^2 b_k}{\mathrm{d}T^2}; \tag{7.43}$$

$$Pv_{\mathrm{m}} - \mathbf{R}T = \mathbf{R}T\sum\left(\frac{\mathbf{N}}{v_{\mathrm{m}}}\right)^k b_k. \tag{7.44}$$

The internal degrees of freedom have not been included because their effect

† The virial equation of state, in *The International Encyclopedia of Physical Chemistry and Chemical Physics*, Topic 10, Vol. 2, Pergamon Press, New York, 1969.

on collisions has not been incorporated into our theory. To do so, it would be necessary to formulate a much more complex theory.† It is worth knowing, however, that the contributions from the internal degrees of freedom can, in practice, continue to be taken to be additive.

The virial expansion of the Helmholtz function conceived as a fundamental equation of state and derived from statistical considerations provides a very useful representation of the corrections to the thermodynamic properties of a perfect gas as its pressure is increased, say along an isotherm. Since our theory appears to be complete, we may expect that it should also account for the fact that at a specified pressure for each temperature the gas condenses into a liquid. It is clear, however, that this process cannot be described by an analytic function, since, at the saturation line, the isotherm acquires a horizontal straight-line portion in the P, v diagram, and becomes nonanalytic. In addition, the mathematical representation becomes complicated by the fact that a metastable state (undercooling) may also occur. This was a problem which eluded understanding for a long time. It was first resolved by L. Onsager‡ who proved that a particular partition function, namely that associated with a two-dimensional Ising model,§ does describe a phase transition. The failure of the present theory to do so is attributed to the *assumption* that an analytic density expansion constitutes an adequate approximation to the fundamental equation. It turns out that this is the case only in the gaseous phase.¶ A completely general treatment of this problem is still outstanding.

7.8. The van der Waals Equation of State

As far as our ability to represent phase transitions is concerned, it has been known for a considerable time that the van der Waals thermal equation of state, supplemented by Maxwell's lever rule, succeeds qualitatively, even though it cannot be quantitively fitted to any pure substance. It is, therefore, interesting to see how this equation can be justified statistically. A rigorous statistical derivation of this equation for a hypothetical one-dimensional gas (a collection of particles constrained to move on a line) has been provided by M. Kac, G. E. Uhlenbeck, and P. C. Hemmer.†† N. G. van Kampen,‡‡ J. L.

† See Mason and Spurling, *ibid.*

‡ *Phys. Rev.* **65** (1944), 117.

§ A two-dimensional array of magnetic spins with nearest-neighbor nteractions accounted for.

¶ C. N. Yang and T. D. Lee, Statistical theory of state and phase transitions. I. Theory of condensation. *Phys. Rev.* **87** (1952), 404.

†† On the van der Waals theory of the vapor liquid equilibrium, *J. Math. Phys.* **4** (1963), 216, 229; **5** (1963), 60, 75.

‡‡ Condensation of a classical gas with long-range attraction, *Phys. Rev.* **135** (1964), A362.

Lebowitz and O. Penrose† and E. Lieb‡ generalized the derivation to include three-dimensional (that is, realistic) gases as well as a wide class of intermolecular force potentials. These authors have demonstrated that the van der Waals equation as well as the lever rule result from a partition function for molecules whose pair potential is similiar to that given in Figure 7.4; it is characterized by strong repulsion at small intermolecular separations and an attractive "tail" which extends to a distance that is much larger than the molecular diameter ("long-range attraction"). Figure 7.4 gives an idea of the type of potential function involved.

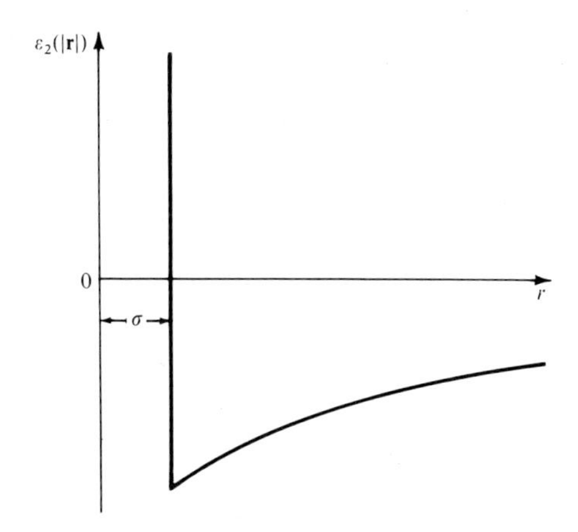

FIGURE 7.4. Pair potential for a van der Waals gas.

The rigorous theory is too complex to be discussed here, and we shall be satisfied with an approximate evaluation of the partition function which leads to the same result.

We assume a pair potential of the form

$$\varepsilon_2(|\mathbf{r}|) = \varepsilon_2^\circ(|\mathbf{r}|) - \frac{C}{V}, \tag{7.45}$$

where ε_2° is the pair potential of a rigid sphere, Figure 2.6, C is a positive constant, and V denotes the volume occupied by the gas. The attractive "tail" has been characterized here by the constant C which is independent of the molecular separation; we shall see later that the inverse dependence on volume assures that the Helmholtz free energy, F, of the gas is an extensive quantity.

† *J. Math. Phys.* **7** (1966), 98.
‡ *J. Math. Phys.* **7** (1966), 1016.

We return to Equation (7.7) which now assumes the form

$$\mathscr{Z} = \frac{1}{N!}\left(\frac{2\pi m\mathbf{k}T}{\mathbf{h}^2}\right)^{3N/2} \int_V d\mathbf{r}_1 \cdots \int_V d\mathbf{r}_N \exp -\beta \sum_{i=1}^{N} \sum_{j=i+1}^{N} \left[\varepsilon_2{}^\circ(|\mathbf{r}_i - \mathbf{r}_j|) - \frac{C}{V}\right]$$
$$= \frac{1}{N!}\left(\frac{2\pi m\mathbf{k}T}{\mathbf{h}^2}\right)^{3N/2} \exp\left[\frac{\beta C}{V}\frac{N(N-1)}{2}\right] Q_N{}^\circ, \tag{7.46}$$

where $Q_N{}^\circ$ is the configurational integral for a gas of rigid-sphere molecules of diameter σ. We can estimate the configurational integral, $Q_N{}^\circ$ by suppos– ing that volume V is filled with spheres each of which displaces a volume $\frac{4}{3}\pi\sigma^3$, since the distance between the centres of two such spheres cannot be less than σ, and by accepting that the volume available to a particle is†

$$V - \tfrac{2}{3}N\pi\sigma^3. \tag{7.47}$$

In other words, the corresponding configurational integral can be approximated by that of a perfect gas whose volume V has been reduced by the *covolume* according to Equation (7.47). Thus we obtain

$$Q_N{}^\circ \approx (V - \tfrac{2}{3}N\pi\sigma^3)^N. \tag{7.47a}$$

Combining Equation (7.47a) with (7.46), we obtain

$$\mathscr{Z} \approx \frac{1}{N!}\left(\frac{2\pi m\mathbf{k}T}{\mathbf{h}^2}\right)^{3N/2} (V - \tfrac{2}{3}N\pi\sigma^3)^N \exp\left(\frac{\beta}{2} CNn\right), \tag{7.48}$$

because for large values of N we may assume that

$$\frac{(N-1)}{V} \approx n.$$

It is now easy to show by means of Equation (7.48) that the Helmholtz function is

$$F = -\mathbf{k}T \ln \mathscr{Z}$$
$$= -\mathbf{k}TN\left\{\ln\left[\left(\frac{2\pi m\mathbf{k}T}{\mathbf{h}^2}\right)^{3/2}\left(\frac{V}{N} - \tfrac{2}{3}\pi\sigma^3\right)\right] + 1 + \frac{\beta}{2}Cn\right\}. \tag{7.49}$$

As announced earlier, this is an extensive quantity owing to the appropriate form adopted for the attractive part of the approximate pair potential.

Employing Equation (7.49), it is possible to deduce the van der Waals form of the thermal equation of state:

$$\left(P + \frac{a}{v_\mathrm{m}{}^2}\right)(v_\mathrm{m} - b) = \mathbf{R}T, \tag{7.50}$$

where

$$a = \tfrac{1}{2}C\mathrm{N}^2, \qquad b = \tfrac{2}{3}\mathrm{N}\pi\sigma^3. \tag{7.51a, b}$$

† See Problem 7.9.

We refrain from discussing the van der Waals equation (7.50), asssuming that its theory is known to the reader.†

7.9. The Law of Corresponding States

7.9.1. The Statistical Form

There is a class of intermolecular force potentials, of which the Lennard-Jones potential defined in Equation (2.5b) is an example, that contain only two parameters which vary from gas to gas. Such potentials are of the form

$$\varepsilon_2(r) = \varepsilon\, \mathrm{f}(r/r_\mathrm{m}) \tag{7.52}$$

which is imposed on them by dimensional considerations. It follows that the thermodynamic properties of substances whose molecules obey such a potential contain three material constants: the depth of the potential well, ε, a measure of the molecular diameter, r_m, and the Boltzmann constant, $\mathbf{k}$, or its equivalent, the universal gas constant $\mathbf{R} = \mathbf{Nk}$.

The classical partition function, $\mathscr{Z}$, of each class of gases whose molecular properties are described by the same function f from Equation (7.52) can be made dimensionless by expressing all lengths in units of r_m and all energies in units of ε. Thus we put

$$\mathbf{r}_i = r_\mathrm{m}\, \mathbf{r}_i^* \qquad \text{and} \qquad \mathbf{k}T/\varepsilon = (\beta\varepsilon)^{-1} = T^*, \tag{7.53}$$

where the starred quantities are dimensionless. In this manner we obtain for one mol:

$$\mathscr{Z} = \frac{1}{\mathbf{N}!}\left(\frac{2\pi m\mathbf{k}T r_\mathrm{m}^2}{\mathbf{h}^2}\right)^{3\mathrm{N}/2} \int_{v_\mathrm{m}/r_\mathrm{m}^3} d\mathbf{r}_1^* \cdots \int_{v_\mathrm{m}/r_\mathrm{m}^3} d\mathbf{r}_\mathrm{N}^* \times \exp\left[-\sum_{i=1}^{N}\sum_{j=i+1}^{N} \mathrm{f}(|\mathbf{r}_i^* - \mathbf{r}_j^*|)/T^*\right] \tag{7.54}$$

and

$$\begin{aligned} F &= -\mathbf{k}T \ln \mathscr{Z} \\ &= -\mathbf{Nk}T \ln\left(\frac{2\pi m\mathbf{k}T r_\mathrm{m}^2}{\mathbf{h}^2}\right)^{3/2} \\ &\quad -\mathbf{k}T \ln \frac{1}{\mathbf{N}!}\int_{v_\mathrm{m}/r_\mathrm{m}^3} d\mathbf{r}_1^* \cdots \int_{v_\mathrm{m}/r_\mathrm{m}^3} d\mathbf{r}_N^* \\ &\quad \times \exp\left[-\sum_{i=1}^{N}\sum_{j=i+1}^{N} \mathrm{f}(|\mathbf{r}_i^* - \mathbf{r}_j^*|)/T^*\right]. \end{aligned} \tag{7.55}$$

† See, for example, J. Kestin, *A Course in Thermodynamics*, Vol. II, p. 236 ff, Blaisdell, Waltham, Massachusetts, 1969.

In particular, the thermal equation of state assumes the form

$$P = \frac{\varepsilon}{\mathbf{N}r_{\mathrm{m}}^{3}}\left(\frac{\mathbf{k}T}{\varepsilon}\right)\frac{\partial}{\partial(v_{\mathrm{m}}/\mathbf{N}r_{\mathrm{m}}^{3})}\ln\frac{1}{N!}\int_{v_{\mathrm{m}}/r_{\mathrm{m}}^{3}} d\mathbf{r}_1^* \cdots \int_{v_{\mathrm{m}}/r_{\mathrm{m}}^{3}} d\mathbf{r}_{\mathbf{N}}^* \times \exp\left[-\sum_{i=1}^{N}\sum_{j=i+1}^{N} \mathrm{f}(|\mathbf{r}_i^* - \mathbf{r}_j^*|)/T^*\right]. \tag{7.55a}$$

It is easy to see now that the integral

$$\int_{v_{\mathrm{m}}/r_{\mathrm{m}}^{3}} d\mathbf{r}_1^* \cdots \int_{v_{\mathrm{m}}/r_{\mathrm{m}}^{3}} d\mathbf{r}_{\mathbf{N}}^* \exp\left[-\sum_{i=1}^{N}\sum_{j=i+1}^{N} \mathrm{f}(|\mathbf{r}_i^* - \mathbf{r}_j^*|)/T^*\right]$$

depends only on the dimensionless quantities T^* and $v_{\mathrm{m}}/r_{\mathrm{m}}^{3}$, and, in terms of these quantities, its value is uniquely determined by the *form* of the function f from Equation (7.52)—it is the same for all gases which obey an identical two-parameter pair potential, ε_2.

If we introduce the reduced partition function

$$\mathscr{Z}^* = \mathscr{Z}\left(\frac{2\pi m\varepsilon r_{\mathrm{m}}^{2}}{\mathbf{h}^2}\right)^{-3\mathbf{N}/2} \tag{7.56}$$

and the reduced Helmholtz function

$$F^* = F/\varepsilon + \frac{3\mathbf{N}T^*}{2}\ln\frac{2\pi m\varepsilon r_{\mathrm{m}}^{2}}{\mathbf{h}^2}, \tag{7.56a}$$

we can easily see that they are both expressed as universal functions of the independent variables

$$T^* = \frac{\mathbf{k}T}{\varepsilon} \quad \text{and} \quad v_{\mathrm{m}}^* = \frac{v_{\mathrm{m}}}{\mathbf{N}r_{\mathrm{m}}^{3}}. \tag{7.56b, c}$$

Similarly, the independent, reduced pressure

$$P^* = \frac{Pr_{\mathrm{m}}^{3}}{\varepsilon} \tag{7.56d}$$

continues to be a universal function of the reduced variables (7.56b, c). It follows that all equilibrium properties of classes of gases can be expressed in terms of universal functions if properly reduced, dimensionless variables are employed. When this is the case, we say that a *law of corresponding states* obtains for this class of systems. Two gases of a class which are in states described by the same reduced parameters are said to be in corresponding states because the relations between them are the same for both gases even though the relations between their pressures, P, temperatures, T, and so on are different. If it is known that two gases belong to the same class, exactly or approximately, it is possible to calculate all properties of gas 2 from those of gas 1 from a knowledge of ε and r_{m} for the second gas.

To illustrate the use of the law of corresponding states, we consider the virial expansion

$$P^* v_m{}^* = T^* \left[1 + \frac{B^*(T^*)}{v_m{}^*} + \frac{C^*(T^*)}{v_m^{*2}} + \cdots \right], \tag{7.57}$$

and verify that

$$B^*(T^*) = (\mathrm{N}/2) \int \{1 - \exp[-\mathrm{f}(|\mathbf{r}^*|)/T^*]\} \, d\mathbf{r}^*$$

is a universal function which depends only on the algebraic form of the function f. Similar expressions would be found for the higher virials, $C^*(T^*)$, $D^*(T^*)$, and so on. Thus, for any member of this class we may write, for example

$$B(T) = r_m{}^3 B^*(T^*). \tag{7.57a}$$

The diagram in Figure 7.5 compares the curve $B(T)/\frac{2}{3}\mathrm{N}\pi r_m{}^3$ calculated on the basis of a Lennard-Jones potential with the experimental results for six gases, all expressed in terms of the reduced temperature, $\mathbf{k}T/\varepsilon$. It is seen that all these gases obey the same law of corresponding states and we may

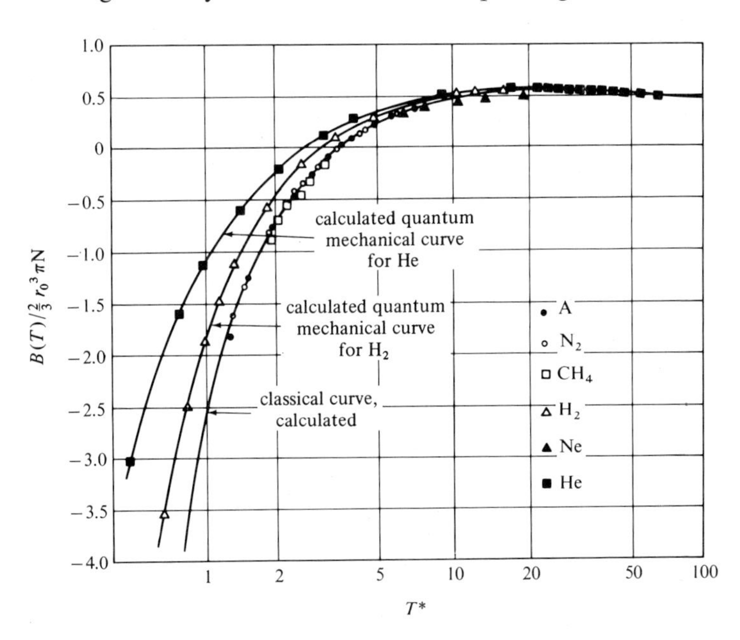

FIGURE 7.5. The reduced second virial coefficient for the Lennard-Jones potential. The classical curve of $B^*(T^*)$ is shown here along with the experimental points for several gases. Also shown are the curves for hydrogen and helium, which have been calculated by quantum mechanics. We can see the extent to which quantum deviations are important for these gases. (From R. J. Lunbeck, Doctoral Dissertation, Amsterdam, 1950.)

conclude that they also obey the same two-parameter pair potential, at least with a good degree of approximation.

In cases when quantum corrections to the partition function become important, it is possible to derive an extended law of corresponding states.† It is then found necessary to introduce an additional variable

$$\Lambda = \frac{\mathbf{h}}{r_{\mathrm{m}}(m\varepsilon)^{1/2}}. \tag{7.58}$$

The diagram in Figure 7.5 illustrates the departure from classical behavior shown by helium and hydrogen at the lower temperatures and demonstrates that quantum corrections for them are important below, say, $T^* = 10$. The theoretical quantum-corrected curves in the diagram have been derived by J. de Boer and R. J. Lunbeck.‡

7.9.2. Use of Critical Parameters

Remembering that phase transitions, and hence the existence of a critical point, are implied in the general form of the partition function, $\mathscr{Z}$, even though we have no reasonable methods of computing them, we can conclude that the critical parameters should be also expressible in the reduced forms

$$P_{\mathrm{cr}}^* = \frac{r_{\mathrm{m}}^3}{\varepsilon} P_{\mathrm{cr}}, \qquad T_{\mathrm{cr}}^* = \frac{\mathbf{k}T_{\mathrm{cr}}}{\varepsilon}, \qquad \text{and} \qquad v_{\mathrm{cr}}^* = \frac{v_{\mathrm{cr}}}{\mathbf{N}r_{\mathrm{m}}^3}. \tag{7.59}$$

If the law of corresponding states were satisfied exactly, the parameter P_{cr}^*, T_{cr}^*, and v_{cr}^* would be the same for all gases. Table 7.1 shows that this is not really the case, proving that even for the simplest gases, the law of corresponding states is only approximate; it operates more and more precisely as the density is reduced, and cannot be relied upon in the neigborhood of the critical state.

In view of Equations (7.59), we can replace the starred dimensionless quantities P^*, v_{m}, T^*, and so on, by the ratios

$$\begin{aligned} \pi &= \frac{P}{P_{\mathrm{cr}}} = \frac{\varepsilon P^*/r_{\mathrm{m}}^3}{\varepsilon P_{\mathrm{cr}}^*/r_{\mathrm{m}}^3} \\ \theta &= \frac{T}{T_{\mathrm{cr}}} = \frac{\varepsilon T^*/\mathbf{k}}{\varepsilon T_{\mathrm{cr}}^*/\mathbf{k}} \\ \phi &= \frac{v_{\mathrm{m}}}{v_{\mathrm{cr}}} = \frac{v_{\mathrm{m}}^*\mathbf{N}r_{\mathrm{m}}^3}{v_{\mathrm{cr}}^*\mathbf{N}r_{\mathrm{m}}^3}, \end{aligned} \tag{7.60}$$

† See for example, J. O. Hirschfelder, C. F. Curtiss, and R. B. Bird, *Molecular Theory of Gases and Liquids*, p. 424 ff, Wiley, New York, 1954.

‡ *Physica* **14** (1948), 520.

TABLE 7.1 [a]

Critical Constants and Reduced Critical Constants for Several Gases

Gas	T_{cr} (K)	v_{cr} (cm^3)	P_{cr} (atm)	T^*_{cr}	v^*_{cr}	P^*_{cr}
Ne	44.5	41.7	25.9	1.25	3.33	0.111
Ar	151	75.2	48	1.26	3.16	0.116
Xe	289.81	120.2	57.89	1.31	2.90	0.132
N_2	126.1	90.1	33.5	1.33	2.96	0.131
CH_4	190.7	99.0	45.8	1.29	2.96	0.126
He	5.3	57.8	2.26	0.52	5.75	0.027
H_2	33.3	65.0	12.8	0.90	4.30	0.064

[a] From J. O. Hirschfelder, C. F. Curtiss, and R. B. Bird, *Molecular Theory of Gases and Liquids*, p. 245, Wiley, New York, 1954.

realizing that the resulting law of corresponding states must be expected to be satisfied less well by actual gases, as already stated.

7.9.3. Comparison with Experiment

The law of corresponding states, expressed in terms of the ratios (7.60), is obeyed within experimental accuracy by the three noble monatomic gases: argon (Ar), krypton (Kr), and xenon (Xe). The same universal relation is satisfied, albeit with a reduced accuracy, if the following gases are added: neon (Ne), nitrogen (N_2), oxygen (O_2), carbon monoxide (CO), and methane (CH_4).†

The degree of validity of the law of corresponding states for this restricted class of gases can be judged with reference to Figures 7.6 and 7.7. These show that all points follow substantially the same vapor-pressure curve as demanded by the law of corresponding states. It is useful to note that the common vapor-pressure line fits the empirical equations‡

$$\frac{\rho' + \rho''}{2\rho_{cr}} = 1 + \frac{3}{4}\left(1 - \frac{T}{T_{cr}}\right), \tag{7.61}$$

$$\frac{\rho' - \rho''}{\rho_{cr}} = \frac{7}{2}\left(1 - \frac{T}{T_{cr}}\right)^{1/3}. \tag{7.62}$$

† See E. A. Guggenheim, *Thermodynamics*, pp. 165 ff, North-Holland Publ., Amsterdam, 1957.

‡ Equation (7.61), referred to as the law of rectilinear diameters, seems to be satisfied quite well by all substances, even though the numerical factors may differ from substance to substance.

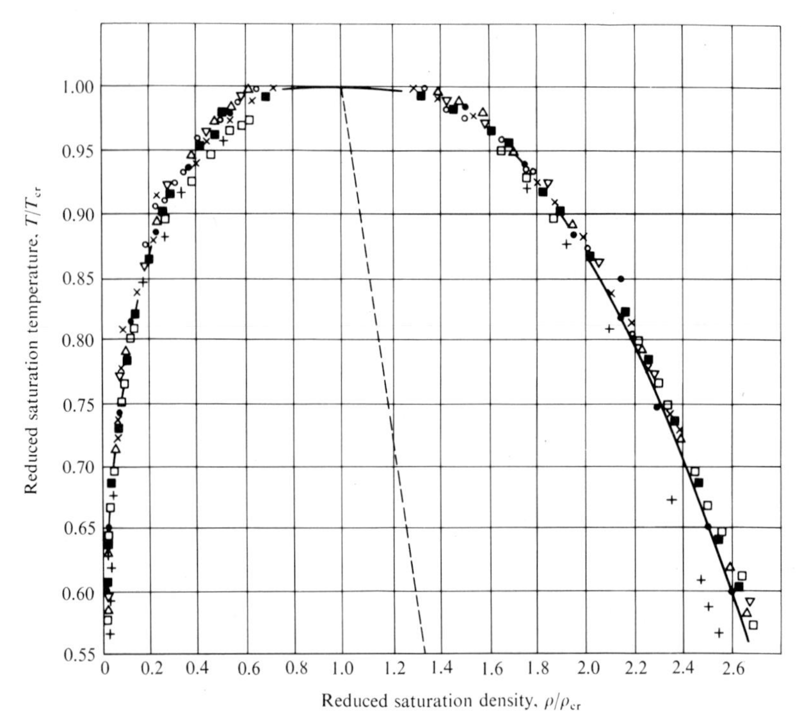

FIGURE 7.6. Vapor-pressure curve in reduced coordinates $\theta = T/T_{cr}$ and $1/\phi = v_{cr}/v$ (after E. A. Guggenheim).

+Ne; ● Ar; □ Kr; × Xe; △ N_2; ▽ O_2; □ CO; ○ CH_4

The mean reduced density $\frac{1}{2}(\rho' + \rho'')/\rho_{cr}$ is seen to trace the straight line shown dotted in the diagram.

The vapor-pressure curve in the $\ln \pi$, $1/\theta$ diagram of Figure 7.7 satisfies the empirical linear relation

$$\ln \pi = \tilde{A} - \frac{\tilde{B}}{\theta}, \tag{7.63}$$

with $A = 5.29$ and $B = 5.31$ which does not quite pass through the critical point ($\pi = \theta = 1$ implying that $\tilde{A} = \tilde{B}$). This equation may be used between the triple and critical points with satisfactory accuracy.

The latent heat of evaporation, l, is nearly constant and is given by another empirical fit:

$$l/\mathbf{R} = \tilde{B}T_{cr}. \tag{7.64}$$

Between the triple and normal boiling points it is preferable to change the constants to $\tilde{A} = 5.13$ and $\tilde{B} = 5.21$.

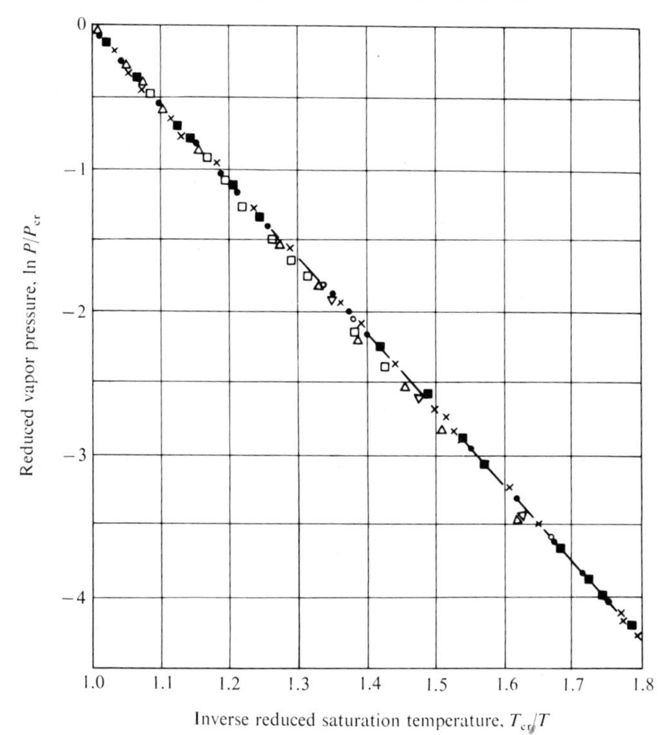

FIGURE 7.7. Vapor-pressure curve in reduced coordinates $\ln \pi$ versus $1/\theta$ (after E. A. Guggenheim).

● Ar; □ Kr; × Xe; △ N_2; ▽ O_2; □ CO; ○ CH_4

Figure 7.8 shows the variation of the second virial B in reduced form as a plot of $\beta = B/v_{cr}$ against $\theta = T/T_{cr}$ for the gases discussed earlier. We see that the law of corresponding states extends its validity to the second virial coefficient for this particular group of gases and note that the empirical correlation equation

$$\frac{B}{v_{cr}} = 0.438 - 0.881\,\frac{T_{cr}}{T} - 0.757\left(\frac{T_{cr}}{T}\right)^2 \tag{7.65}$$

represented by the full line in the graph gives a very good correlation for the relevant experimental material.

Reasonably successful attempts have been made to improve the validity of the law of corresponding states by means of an empirical correction. One such correction replaces the reduced specific volume ϕ by the pseudo-reduced specific volume

$$\bar{\phi} = \frac{P_{cr}\,v}{\mathbf{R}T_{cr}}, \tag{7.66}$$

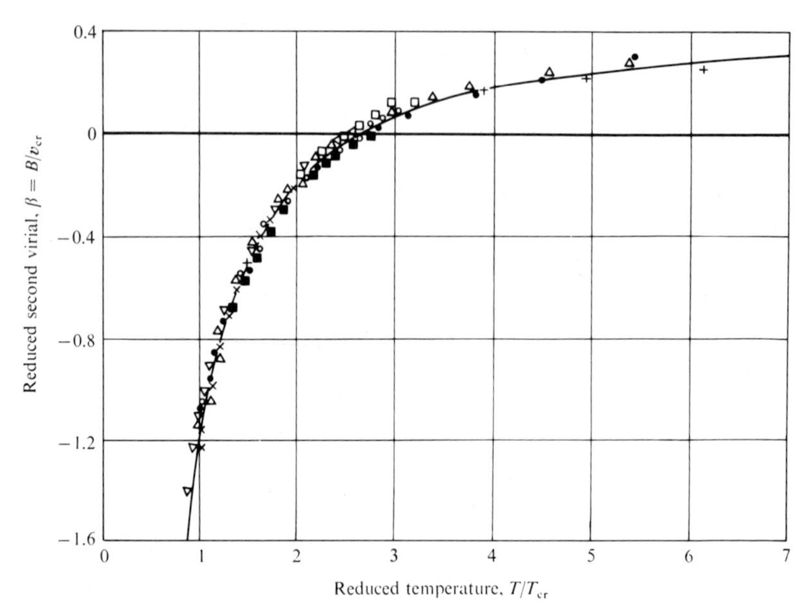

FIGURE 7.8. Second virial coefficient in reduced coordinates $\beta = B/v_{cr}$ versus $\theta = T/T_{cr}$ (after E. A. Guggenheim).
+ Ne; ● Ar; □ Kr; × Xe; △ N_2; ▽ O_2; □ CO; ○ CH_4

in which the group of terms $\mathbf{R}T_{cr}/P_{cr}$ replaces the critical volume, and $\bar{\phi}$ replaces ϕ in the list of variables (7.60). Extensive calculations suggest that the relation between the quantities π, θ, and $\bar{\phi}$, as expressed by the equation of state,

$$f(\pi, \theta, \bar{\phi}) = 0 \tag{7.66a}$$

is more nearly universal than

$$f(\pi, \theta, \phi) = 0 \tag{7.66b}$$

When use is made of this correction, it is customary to represent the thermal equation of state in terms of the compressibility factor

$$z = \frac{Pv_m}{\mathbf{R}T}. \tag{7.67}$$

The use of the pseudo-reduced specific volume as a parameter has the additional advantage that the critical volume, v_{cr}, does not appear in it at all, and experience shows that it is much more difficult to measure the critical volume of a substance than its critical pressure and temperature.

Another attempt at improvement, first proposed by L. Riedel and later developed by K. Pitzer and others,† introduces an additional parameter, the *acentric factor*

$$\omega = -\ln \frac{P_s}{P_{cr}} - 1.000, \tag{7.68}$$

where P_s denotes the saturation pressure at the reduced temperature $\theta = 0.700$. The acentric factor enters the universal relation as an additional parameter, and we write

$$f(\pi, \theta, \phi, \omega) = 0 \tag{7.68a}$$

instead of Equation (7.66a). A detailed representation of this function in the form of tables is given in the reference quoted, in which the reduced pressure and temperature appear as independent variables, and corrections for the acentric factor ω are introduced by calculation.

7.10. Properties of a Pure Substance near the Critical Point

It has been mentioned in Section 7.7 that an analytic form of the Helmholtz free energy, such as, for example, the series expansion given in Equation (7.39), fails to reproduce the nonanalytic isotherms which characterize a two-phase region. As the temperature is increased, the nonanalytic, straight-line segments of isotherms converge to the point of inflection which the critical isotherm exhibits *at* the critical point itself, and the question arises as to how well such a function is able to represent the properties of a pure substance in the *neighborhood* of its critical point. We shall see, presently, that experimental results indicate convincingly that the critical point itself is also nonanalytic.‡

The van der Waals equation is analytic at the critical point and all empirical thermal equations of state are characterized by the same mathematical property. In order to make comparisons with experiment, it is useful, first, to determine the thermal properties of a pure substance around the critical point in a general way, making use, as far as possible, only of the tentative assumption of analyticity and of the well-known properties of the critical isotherm expressed by the conditions

† G. N. Lewis and M. Randall, *Thermodynamics*, 2nd ed. (revised by K. S. Pitzer and L. Brewer), Appendix 1, pp. 605 ff, McGraw-Hill, New York, 1961.

‡ We remind the reader that a function $f(x)$ is said to be analytic at a point $x = x_0$ if it can be expanded into a power series in a neighborhood of x. If this is not the case, the function is nonanalytic. Thus, for example, the function $y = \exp x$ is analytic at $x_0 = 0$, because there $y = 1 + x + \frac{1}{2}x^2 + \cdots$. By contrast, $y = (x - x_0)^{1/2}$ is nonanalytic at $x = x_0$, as is easy to verify by attempting to write out the coefficients of the appropriate Taylor series.

$$\left(\frac{\partial P}{\partial v}\right)_T = \left(\frac{\partial^2 P}{\partial v^2}\right)_T = 0 \qquad (\text{at } T = T_{cr}, v = v_{cr}) \tag{7.69}$$

together with the stability condition

$$\left(\frac{\partial^3 P}{\partial v^3}\right)_T < 0 \qquad (\text{at } T = T_{cr}). \tag{7.69a}$$

In particular, it is useful to obtain the first term of the series expansion of the following thermodynamic properties: (a) the critical isotherm in the neighborhood of the critical point; (b) the isothermal compressibility $\kappa_T = -(1/v)(\partial v/\partial P)_T$ along the critical isochore $v_{cr} = \text{const}$; (c) the shape of the coexistence dome, expressed as a relation between specific volume and temperature; and (d) the specific heat, c_v, along the critical isochore, $v_{cr} = \text{const}$, across the two regions.

It is left to the reader to show that in the neighborhood of the critical point, the isotherms of an analytic equation of state can be expressed as a series starting with

$$P - P_{cr} = -A(T - T_{cr})(v - v_{cr}) - \tfrac{1}{3}B(v - v_{cr})^3 + \cdots, \tag{7.70}$$

(where A and B are adjustable, positive constants) showing that, for $T = T_{cr}$, the difference $P - P_{cr}$ is proportional to the third power of the difference $v - v_{cr}$ in specific volumes.

The diagram in Figure 7.9 represents the variation of $|P - P_{cr}|$ with $|v - v_{cr}|$ for CO_2 in a logarithmic diagram which suggests that experiments point to a relation of the type

$$|P - P_{cr}| \sim |v - v_{cr}|^{\delta} \tag{7.70a}$$

with $\delta \approx 4.4$, instead of the value 3 for a van der Waals gas or, generally, for an analytic equation of state.

Similarly, it can be shown that for an analytic equation of state

$$\kappa_T = \frac{1}{Av_{cr}(T - T_{cr})} \qquad (\text{at } v = v_{cr}). \tag{7.71}$$

On the other hand, the diagram in Figure 7.10 in which the reciprocal of the isothermal compressibility has been plotted, again logarithmically, for CO_2 suggests that

$$\kappa_T \sim |T - T_{cr}|^{-\gamma} \tag{7.71a}$$

with $\gamma = 1.26$ instead of $\gamma = 1$ from Equation (7.71). For a range of gases, this exponent varies from $\gamma = 1.1$ to $\gamma = 1.4$.

Employing general thermodynamic principles together with the lever

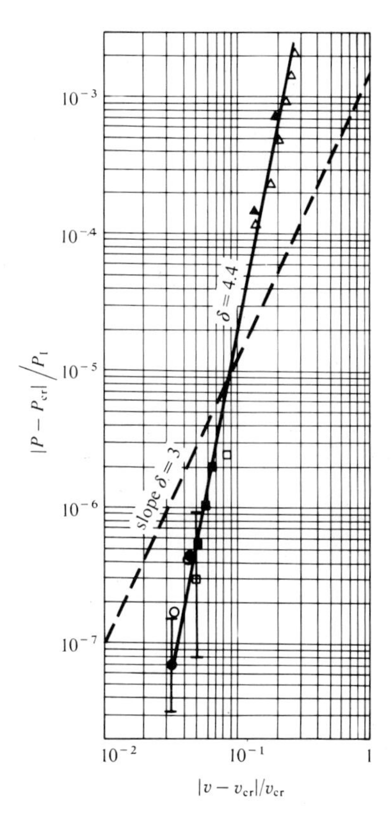

FIGURE 7.9. The critical isotherm of carbon dioxide. The reference pressure P_I is the pressure of a perfect gas at the critical density and temperature. [After P. Heller, *Rept. Progr. Phys.* **30** II (1967), 731.]

rule, we can show that the top portion of the coexistence curve for a gas with an analytic critical point is given as

$$v'' - v' = 2(3A/B)^{1/2}(T_{cr} - T)^{1/2}, \tag{7.72}$$

where v'' refers to the vapor and v' to the liquid, thus showing that the temperature–volume relation is quadratic. A comparison with the experimental data for carbon dioxide is given in Figure 7.11 in which $(v''-v')/2v_{cr}$ has been plotted in terms of $(T_{cr} - T)/T_{cr}$. Once again it is seen that

$$v'' - v' \sim |T_{cr} - T|^{\beta} \tag{7.72a}$$

with $\beta \approx 0.35$ instead of $\beta = 0.5$ for an analytic relation.

As far as the specific heat, c_v, is concerned, the conditions expressed in Equation (7.69) becomes insufficient for its determination, since the P, v, T

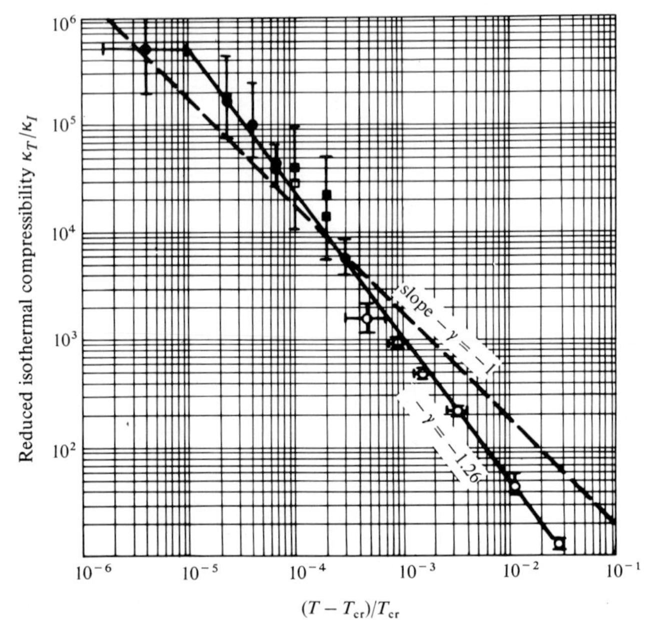

FIGURE 7.10. The isothermal compressibility of carbon dioxide. The reference quantity κ_I is the isothermal compressibility of a perfect gas at the critical density and temperature. [After P. Heller, *Rept. Progr. Phys.* **30** II (1967), 731.]

relation is not a fundamental one. It is, therefore, more convenient to restrict our attention to a van der Waals gas, and to utilize the expression for the Helmholtz free energy from Equation (7.49). In this manner we can calculate

$$c_v = -\left\{2T\left[\frac{\partial}{\partial T}\left(\frac{f}{T}\right)\right]_v + T^2\left[\frac{\partial^2}{\partial T^2}\left(\frac{f}{T}\right)\right]_v\right\} = \tfrac{3}{2}\mathbf{R}, \tag{7.73}$$

in the gaseous phase, finding, that is, essentially, the perfect-gas value.

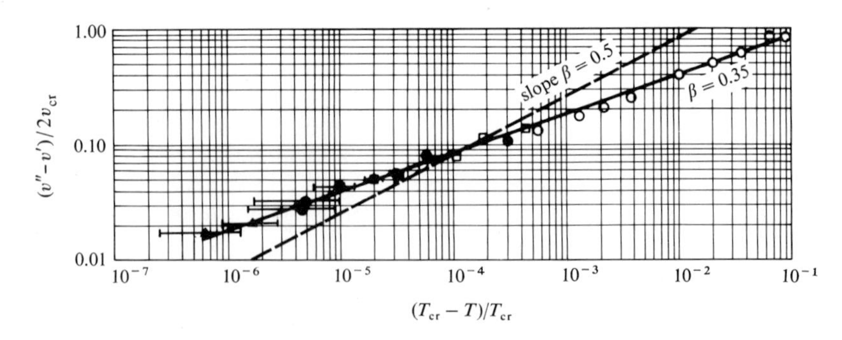

FIGURE 7.11. The coexistence curve for carbon dioxide; v' is the specific volume of the saturated liquid and v'' is the specific volume of the saturated vapor. [After P. Heller, *Rept. Progr. Phys.* **30** II (1967), 731.]

In order to determine c_v on the critical isochore in the two-phase region, we compute the Helmholtz free energy with the aid of Maxwell's lever rule and the Helmholtz free energy in the gaseous region. This procedure is outlined in the book by J. S. Rowlinson and in a report by J. M. H. Levelt Sengers.† As a result, it has been shown that the specific heat, c_v, along the critical isochore in the two-phase region increases with temperature attaining the value

$$c_v = 6\mathbf{R} \qquad (v = v_{cr}, \quad T < T_{cr})$$

at the critical point. We conclude that the specific heat, c_v, of a van der Waals gas suffers a jump discontinuity of $\frac{9}{2}\mathbf{R}$ upon crossing the critical point along the critical isochore. Experimental data show an entirely different kind of behavior. The two alternatives have been sketched in Figure 7.12 which

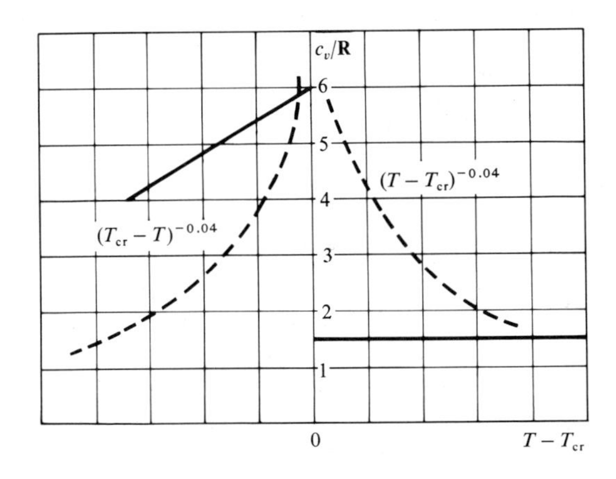

FIGURE 7.12. The specific heat ratio along the critical isochore (not to scale); full lines, van der Waals gas; broken lines, real gas.

shows that $c_v \to \infty$ as $T \to T_{cr}$, and that the two branches can be represented by the common relation

$$c_v \sim |T - T_{cr}|^{-\alpha} \tag{7.73b}$$

with $\alpha = 0.04$.

Considerations of thermodynamic consistency‡ lead to the conclusion

† J. S. Rowlinson, *Liquids and Liquid Mixtures*, pp. 99–100, Butterworths, London, 1959. J. M. H. Levelt Sengers, *Ind. Eng. Chem. Fundam.* **9** (1970), 470.

‡ G. S. Rushbrooke, *J. Chem. Phys.* **39** (1963), 842; and *J. Chem. Phys.* **43** (1965), 3439. See also R. B. Griffiths, *J. Chem. Phys.* **43** (1965), 1958.

that the exponents α, β, γ, and δ are not entirely independent, but must satisfy the relations

$$\begin{aligned} \beta(\delta + 1) &\geq 2 - \alpha \\ \gamma + 2\beta &\geq 2 - \alpha . \end{aligned} \tag{7.74}$$

The experimental values of the exponents satisfy these relations as equalities almost exactly.

The preceding discrepancies between experimental results and the conclusions which follow from analytic forms for the equations of state† are interpreted as providing conclusive proof that the critical point is nonanalytic. Understandably, no general theory leading to a partition function which possesses the above mathematical properties has yet been formulated. However, it is interesting to note that relations of the given type in Equations (7.71a), (7.72a), and (7.73b) have been anticipated in the work of L. Onsager‡ on the so-called Ising model when he studied magnetic transitions from para- to ferro-magnetism (Chapter 11). Analogous mathematical properties are also observed in the study of superfluidity in helium and superconductivity in metals at low temperatures. It is now accepted that the nonanalytic character of all phase transitions is related to the range of the interparticle forces, so that the essential unity of all these processes can be made understandable only in statistical terms. In the absence of a rigorous theory, much of this explanation is conjectural.

As we have seen, the van der Waals equation is only appropriate for systems having an attractive intermolecular potential which extends over a range which is large compared to the average distance between molecules. However, with the exception of those for solids, the experimentally determined potentials extend over much shorter distances. We may then suppose that the difference between the observed properties near the critical point and those predicted by the van der Waals equation (or, probably, any analytic equation) is due mainly to the different ranges of the intermolecular potential in the two cases.

However, when a gas condenses, the particles are correlated over distances which are large compared with the range of the intermolecular potential. That is to say, particles which are separated by such distances interact in larger than binary groups. This may be seen from the fact that drops of liquid start to form as the critical point is approached, and that macroscopic regions

† More complete summaries of experimental results can be found in P. Heller, Experimental Investigations of Critical Phenomena. *Rep. Prog. Phys.* **30** II (1967) 731, and in L. P. Kadanoff *et al*, Static phenomena in critical points: theory and experiment, *Rev. Mod. Phys.* **39** (1967), 395.

‡ *Phys. Rev.* **65** (1944), 117.

of the gas condense at the same time. It is generally felt that these long-range correlations contribute the crucial factors in determining the properties of the gas near the critical point. For a van der Waals gas, the range of the intermolecular potential is always larger than the range of the correlations and one set of properties results. For real gases, the range of the potential is much shorter than the range of the correlations and this leads to a qualitatively different set of properties.

It is assumed that the radius over which the motion of a large number of particles is correlated provides the most essential parameter in the problem; it is then possible to show,† largely on dimensional grounds, that the thermodynamic properties of a pure substance near the critical point can be represented in the form of a single, so-called *scaling law*. This asserts that the composite dimensionless variable

$$\Gamma = \frac{\rho_{cr}[g(\rho, T) - g(\rho_{cr}, T)]}{P_{cr}\, \Delta\rho\, |\Delta\rho|^{\delta-1}} \tag{7.75}$$

should be a unique function of the dimensionless group

$$x = \frac{T - T_{cr}}{T_{cr}\, |\Delta\rho|^{1/\beta}}, \tag{7.75a}$$

where

$$\Delta\rho = \frac{\rho - \rho_{cr}}{\rho_{cr}}, \tag{7.75b}$$

and where g denotes the specific Gibbs function. This scaling law is satisfied with a high degree of precision by the experimental results for a large number of pure substances. It must, however, be noted that the relation $\Gamma = \Gamma(x)$ is not a fundamental one. Consequently, the scaling law by itself does not determine all thermodynamic properties of a pure substance in the neighborhood of the critical point.

PROBLEMS FOR CHAPTER 7

7.1. Compute the second virial coefficient for a gas obeying the Sutherland potential

$$\varepsilon_2(r) = \begin{cases} \infty & r < \sigma \\ -c(\sigma/r)^6 & r > \sigma \end{cases}$$

in the range of temperatures where $\mathbf{k}T \gg c$. Determine the thermal equation of state, and derive expressions for the internal energy, u, and specific heat, c_v, to the same degree of approximation.

† M. Vicentini-Missoni, J. M. H. Levelt Sengers, and M. S. Green, *J. Res. Nat. Bur. Std.* (*U.S.*) **73A** (1969), 563; and *Phys. Rev. Lett.* **22** (1969), 389.

7.2. Compute the second virial coefficient, B, and the Joule–Thomson coefficient, μ, for a gas obeying the square-well potential illustrated in Figure 2.7. Discuss the Joule–Thomson effect for this gas.

7.3. Compute the virial coefficients for a gas that obeys the van der Waals equation of state.

7.4. The internal energy of a classical system may be regarded as the average of the Hamiltonian $\mathscr{H}_N(\mathbf{r}_1, \mathbf{p}_1, \mathbf{r}_2, \mathbf{p}_2, \ldots, \mathbf{r}_N, \mathbf{p}_N)$ over the canonical ensemble. Show that the entropy may be regarded as the average of $-\mathbf{k} \ln(p_N N! \mathbf{h}^{3N})$, that is,

$$S = -\frac{\mathbf{k}}{\mathscr{Z}} \int \exp(-\beta \mathscr{H}_N) \ln \frac{\exp(-\beta \mathscr{H}_N)}{\mathscr{Z}} \, \mathrm{d}\mathbf{q} \, \mathrm{d}\mathbf{p} - \mathbf{k} \ln(N! \, \mathbf{h}^{3N}) .$$

What is the physical meaning of p_N?

7.5. Show that the Joule–Thomson inversion temperature satisfies the relation

$$\frac{\mathrm{d}}{\mathrm{d}T}\left(\frac{B}{T}\right) = 0$$

in cases when the pressure is low. Would an inversion temperature be expected in a gas of molecules which interact with purely repulsive forces?

7.6. A classical system is in contact with a heat bath at temperature T. Show that this system satisfies a generalized principle of equipartition. In other words, prove that

$$\frac{1}{\mathscr{Z}} \int r_{i,\alpha} \frac{\partial \mathscr{H}}{\partial r_{i,\alpha}} \exp(-\beta \mathscr{H}) \, \mathrm{d}\mathbf{q} \, \mathrm{d}\mathbf{p} = \mathbf{k}T ,$$

where $r_{i,\alpha}$ is the component of the position coordinate of the ith particle measured in the direction denoted by α.

7.7. Examine whether the properties of a class of gases which obey the exp-six potential can be described by a law of corresponding states.

7.8. Prove that the second virial coefficient of a gas which obeys a spherically symmetrical intermolecular force potential is proportional to the integral

$$\mathscr{I} = \int_0^\infty r \frac{\mathrm{d}\varepsilon}{\mathrm{d}r} \mathrm{e}^{-\varepsilon(r)/\mathbf{k}T} 4\pi r^2 \, \mathrm{d}r .$$

7.9. The estimate of the configurational partition function for a gas of hard-sphere molecules (of diameter σ) given as Equation (7.47a) may be obtained in the following way. If the gas consists of one molecule, the configurational partition function is simply V. If there are two molecules, one can be anywhere in the vessel and the other is restricted to a volume $V - \frac{4}{3}\pi\sigma^3$. If there are s molecules in the gas, we may suppose that each molecule restricts the volume available to the others, and write

$$Q_s \approx V(V - v_0)(V - 2v_0)(V - 3v_0) \cdots [V - (s-1)v_0]$$

where $v_0 = \frac{4}{3}\pi\sigma^3$. Show that $Q_N \approx (V - \frac{1}{2}Nv_0)^N$ if $V > N\frac{4}{3}\pi\sigma^3$.

SOLUTION: According to the above estimate, we may write

$$Q_N \cong V^N \prod_{s=1}^{N-1} (1 - sv_0/V);$$

or

$$\ln Q_N \approx N \ln V + \sum_{s=1}^{N-1} \ln(1 - sv_0/V).$$

If $V \gg Nv_0$, we may approximate $\ln(1 - sv_0/V)$ by $-sv_0/V$, and write

$$\ln Q_N \approx N \ln V - \sum_{s=1}^{N-1} sv_0/V \approx N \ln V - N^2 v_0/2V$$

or $\ln Q_N \approx N \ln(V - Nv_0/2)$.

7.10. Suppose that the thermal equation of state of a gas is analytic at the critical point. Show that the appropriate series expansion for the pressure has the form

$$P = P_{cr} - A(T - T_{cr})(v - v_{cr}) - \tfrac{1}{3}B(v - v_{cr})^3 + \cdots$$

for T and v near the critical point; here A and B are constants. (Why must these be positive?)

SOLUTION: In the neighborhood of the critical point, we may expand the pressure as a Taylor series in powers of $(T - T_{cr})$ and $(v - v_{cr})$:

$$\begin{aligned} P - P_{cr} = & \left(\frac{\partial P}{\partial v}\right)_{cr}(v - v_{cr}) + \left(\frac{\partial P}{\partial T}\right)_{cr}(T - T_{cr}) \\ & + \frac{1}{2}\left(\frac{\partial^2 P}{\partial v^2}\right)_{cr}(v - v_{cr})^2 + \left(\frac{\partial^2 P}{\partial T \partial v}\right)_{cr}(v - v_{cr})(T - T_{cr}) \\ & + \frac{1}{2}\left(\frac{\partial^2 P}{\partial T^2}\right)_{cr}(T - T_{cr})^2 + \frac{1}{6}\left(\frac{\partial^3 P}{\partial v^3}\right)_{cr}(v - v_{cr})^3 + \cdots . \end{aligned}$$

At the critical point

$$\left(\frac{\partial P}{\partial v}\right)_{cr} = \left(\frac{\partial^2 P}{\partial v^2}\right)_{cr} = 0 \text{ and } \left(\frac{\partial^3 P}{\partial v^3}\right)_{cr} < 0.$$

Therefore,

$$(P - P_{cr}) = -\tfrac{1}{3}B(v - v_{cr})^3 - A(v - v_{cr})(T - T_{cr}) + \cdots$$

where $A > 0$ and $B > 0$, since thermodynamic stability requires that $-(\partial P/\partial v)_T > 0$ for $T > T_{cr}$. We have disregarded terms which depend on $(T - T_{cr})$, since such terms are not important for Problems 7.11 and 7.12, or for the discussion in the text.

7.11. Show that the isothermal compressibility $\kappa_T = -(1/v)(\partial v/\partial P)_T$ is proportional to $(T - T_{cr})^{-1}$ near the critical point for a gas with an analytic equation of state.

7.12. Show that the equation of state

$$P = P_{cr} - A(T - T_{cr})(v - v_{cr}) - \tfrac{1}{3}B(v - v_{cr})^3$$

and the Maxwell lever rule lead to the result that the dome of the coexistence curve for a gas with an analytic critical point is described by

$$v'' - v' = 2(3A/B)^{1/2}(T_{cr} - T)^{1/2}$$

where v'' and v' are the specific volumes for the vapor and liquid, respectively.

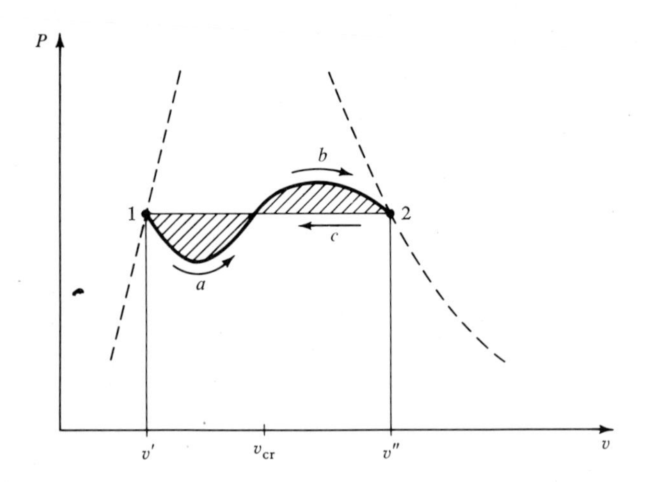

SOLUTION: Above the critical temperature, the isotherms corresponding to the equation $P = -A(T - T_{cr})(v - v_{cr}) - \tfrac{1}{3}B(v - v_{cr})^3$ are monotonically decreasing functions of v. However, below the critical temperature each of the isotherms displays a maximum and a minimum and possesses a segment along which $(\partial P/\partial v)_T > 0$. The Maxwell lever rule replaces the curved isotherm by a straight, condensation isotherm determined by the requirement that the chemical potential must be the same in the liquid and in the gas phase. Referring to the figure, this requires that

$$\oint d\mu = \oint v\, dP = 0$$

along path $1ab2c1$ (at $T = \text{const}$). If we write this equation in the form

$$\int_{v'}^{v''} v\left(\frac{\partial P}{\partial v}\right)_T dv = \int_{v'}^{v''} (v - v_{cr})\left(\frac{\partial P}{\partial v}\right)_T dv = 0,$$

since $P_1 = P_2$, and substitute the equation for P, we obtain $(v'' - v_{cr}) = (v_{cr} - v')$. Now we use the fact that the pressures are equal in each phase to prove the result

$$|v - v_{cr}| = (3A/B)^{1/2}\,|T_{cr} - T|^{1/2}$$

7.13. It is possible to make use of the first few virial coefficients of a gas approximately to predict the critical temperature and critical specific volume of a gas. Show that these are determined by the equations

$$v_{cr}^2 + 2B(T_{cr})\,v_{cr} + 3C(T_{cr}) = 0 \qquad 2v_{cr}^2 + 6B(T_{cr})\,v_{cr} + 12C(T_{cr}) = 0$$

in terms of the second and third virial coefficients. What assumptions must be made to derive these equations, and what can be said about the signs of the virial coefficients if these equations are to have solutions?

Discuss the validity of this estimate in the light of the fact that the critical point is nonanalytic.

7.14. The virial equation of state for a gaseous mixture may be written as

$$\frac{PV}{\mathrm{R}T} = \sum_i n_i + \sum_{i,\,j} n_i n_j B_{ij}(T) + \sum_{i,\,j,\,k} n_i n_j n_k C_{ijk}(T) + \cdots$$

where B_{ii}, C_{iii}, are the virial coefficients of a pure system composed of species i, and the other coefficients are characteristic for the mixture, since they depend on the forces acting between the molecules of different species. Calculate the change in the entropy of mixing n_1 moles of species 1, n_2 moles of species 2, ..., at constant temperature; take into account only terms up to and including the second generalized virial coefficient.

7.15. Show that the pressure, P, of a gas composed of hard-sphere molecules varies linearly with the temperature, T.

7.16. Show that the product of the Mayer f-functions which appear in Equation (7.35) for the third virial coefficient, $\mathrm{f}(|\boldsymbol{\rho}_{12}|)\,\mathrm{f}(|\boldsymbol{\rho}_{13}|)\,\mathrm{f}(|\boldsymbol{\rho}_{23}|)$, vanishes whenever one particle does not interact with either of the other two.

7.17. Show explicitly that the configurational partition function for a system of $N = 4$ interacting particles may be put in the form $Q_4 = Q_4^{(1)} Q_4^{(2)} Q_4^{(3)}$ where $Q_4^{(1)}$ and $Q_4^{(2)}$ are appropriate specializations of $Q_N^{(1)}$ and $Q_N^{(2)}$ defined by Equations (7.13) and (7.31), respectively.

7.18. Show that it is possible to generalize van Kampen's method for the calculation of the virial coefficients of a gas whose particles interact pairwise to include the case of multiple interactions. For example, consider the case where the potential energy of the system may be written in the form

$$\phi = \sum_{1 \le i < j \le N} \varepsilon_2(\mathbf{r}_i, \mathbf{r}_j) + \sum_{1 \le i < j < k \le N} \varepsilon_3(\mathbf{r}_i, \mathbf{r}_j, \mathbf{r}_k).$$

Show that the resulting second virial coefficient remains unaffected by the presence of ε_3 in the potential energy, but that the third virial coefficient $C(T)$ is modified to the form

$$C(T) = -\frac{1}{3}\int d\boldsymbol{\rho}_{12} \int d\boldsymbol{\rho}_{13}\,\mathrm{f}(|\boldsymbol{\rho}_{12}|)\,\mathrm{f}(|\boldsymbol{\rho}_{13}|)\,\mathrm{f}(|\boldsymbol{\rho}_{23}|)$$

$$-\frac{1}{3V}\int d\mathbf{r}_1 \int d\mathbf{r}_2 \int d\mathbf{r}_3\, \mathrm{g}(\mathbf{r}_1, \mathbf{r}_2, \mathbf{r}_3)\,\Phi_{12}\Phi_{13}\Phi_{23}$$

where

$$g(\mathbf{r}_1, \mathbf{r}_2, \mathbf{r}_3) = \exp[-\beta\varepsilon_3(\mathbf{r}_1, \mathbf{r}_2, \mathbf{r}_3)] - 1,$$

and

$$\Phi_{ij} = \exp[-\beta\varepsilon_2(\mathbf{r}_i, \mathbf{r}_j)].$$

7.19. Does the law of corresponding states apply to gases composed of hard-sphere molecules?

7.20. Derive Equations (7.40)–(7.44) in the text.

7.21. Derive the van der Waals equation (7.50) from Equation (7.49) and verify the relation in Equation (7.51).

7.22. Show that the second virial coefficient may have a maximum at some temperature. Show that a maximum occurs if the potential energy $\varepsilon_2(|\mathbf{r}_{12}|)$ satisfies the conditions that (a) $\varepsilon_2(|\mathbf{r}_{12}|) < 0$ for a range of $|\mathbf{r}_{12}|$ and (b) $\int \varepsilon_2(|\mathbf{r}_{12}|)\, d\mathbf{r}_{12} > 0$.

SOLUTION: Consider the derivative of the second virial coefficient

$$\frac{dB(T)}{dT} = -\frac{\mathbf{N}}{2\mathbf{k}T^2}\int d\boldsymbol{\rho}\, \varepsilon_2(|\boldsymbol{\rho}|) \exp[-\beta\varepsilon_2(|\boldsymbol{\rho}|)]$$

where $\boldsymbol{\rho} = \mathbf{r}_{12}$. If $\varepsilon_2(|\boldsymbol{\rho}|) < 0$ for some values of $\boldsymbol{\rho}$, then at very low temperatures $dB(T)/dT$ is very large in magnitude and positive in sign. However, at very high temperature dB/dT becomes negative if

$$\frac{1}{\mathbf{k}T^2}\int d\boldsymbol{\rho}\, \varepsilon_2(|\boldsymbol{\rho}|) > 0.$$

It follows that under conditions (a) and (b) dB/dT passes through zero from positive to negative values, and hence $B(T)$ has a maximum.

7.23. The scaling law theory of critical phenomena asserts that the inequalities expressed in Equation (7.74) should be satisfied as equalities. Check these assertions using the values of the critical exponents α, β, γ, and δ (a) as provided by experiment and (b) as provided by the analytic equation of state.

Note: The analytic equation of state predicts that c_v remains finite near the critical point, although it becomes discontinuous at the critical point itself. In such circumstances one may set the critical exponent, α, equal to zero. [*Reference:* M. E. Fisher, The theory of equilibrium critical phenomena, *Rept. Progr. Phys.* **30**, II (1967), 615.]

LIST OF SYMBOLS FOR CHAPTER 7

Latin letters

A	Positive coefficient in Taylor expansion of pressure about the critical value
$\tilde{A}$	Constant in empirical vapor-pressure relation

a	Constant in van der Waals equation of state
B	Second virial coefficient; positive coefficient in Taylor expansion of pressure about the critical value
$\tilde{B}$	Constant in empirical vapor-pressure relation
b	Specific second virial coefficient; constant in van der Waals equation of state
b_k	General specific virial coefficient
C	Third virial coefficient; constant in van der Waals potential
c	Specific third virial coefficient
c_v	Specific heat at constant volume
c_p	Specific heat at constant pressure
d	Molecular separation
f	Mayer f-function; dimensionless form of equation of state; dimensionless form of pair potential
f	Molar Helmholtz function
g	Molar Gibbs function
$\mathcal{H}$	Hamiltonian
$\mathbf{h}$	Planck's constant
k	Denotes general term in virial series, Equation (7.38)
$\mathbf{k}$	Boltzmann's constant
l	Latent heat of evaporation
m	Mass of a molecule
N	Number of particles
$\mathbf{N}$	Avogadro's number
n	Number of molecules per unit volume
P	Pressure
P_I	Pressure of ideal gas at critical density and temperature
P_s	Saturated vapor pressure
$\mathbf{p}$	Represents momentum vectors $\mathbf{p}_1, \ldots, \mathbf{p}_N$
Q_N	Configurational partition function of a system comprised of N particles
$Q_N^{(1)}$	First approximation to configurational partition function
$Q_N^{(2)}$	Second approximation to configurational partition function
Q_N°	Configurational partition function for hard sphere system
$\mathbf{q}$	Represents coordinate vectors $\mathbf{r}_1, \ldots \mathbf{r}_N$
$\mathbf{R}$	Perfect-gas constant; center of mass coordinate
r	Molecular separation
r_m	Molecular diameter
r_0	Separation at zero value of the potential
$\mathbf{r}$	Represents coordinates x, y, z
s	Molar entropy
T	Temperature
T_B	Boyle temperature
u	Molar internal energy; average speed of a molecule
V	Volume
v	Specific volume
v_m	Molar volume

x Dimensionless variable in scaling law equation of state
$\mathscr{Z}$ Partition function for a collection of particles
z Compressibility factor

Greek letters

α Exponent of c_v along critical isochore near critical point
β $(\mathbf{k}T)^{-1}$; exponent of coexistence curve near critical point
β_B $(\mathbf{k}T_B)^{-1}$
Γ Dimensionless form of scaling law equation of state
γ Exponent of isothermal compressibility along critical isochore near critical point
δ Exponent of critical isotherm near critical point
ε Energy parameter in Lennard-Jones potential
ε_2 Intermolecular potential energy (pair potential)
θ Reduced temperature
κ_I Isothermal compressibility of ideal gas at critical density and temperature
κ_T Isothermal compressibility
Λ Dimensionless variable in corresponding state treatment of quantum gases
λ Wavelength
λ_{th} Thermal wavelength
μ Chemical potential
π Reduced pressure
ρ Mass density
$\boldsymbol{\rho}$ Separation of two particles
σ Distance of closest approach of two molecules
Φ Φ-function $[=\exp(-\beta\varepsilon_2)]$
ϕ Potential energy of a collection of particles; reduced volume
ϕ_0 Depth of square-well potential
$\bar{\phi}$ Pseudo reduced specific volume, Equation (7.66)
ω Acentric factor

Superscripts

$*$ Denotes reduced or dimensionless quantity
$\circ$ Denotes standard reference value
$'$ Denotes liquid properties
$''$ Denotes vapor properties

Subscripts

cr Denotes values at critical point
tr Denotes translational degrees of freedom

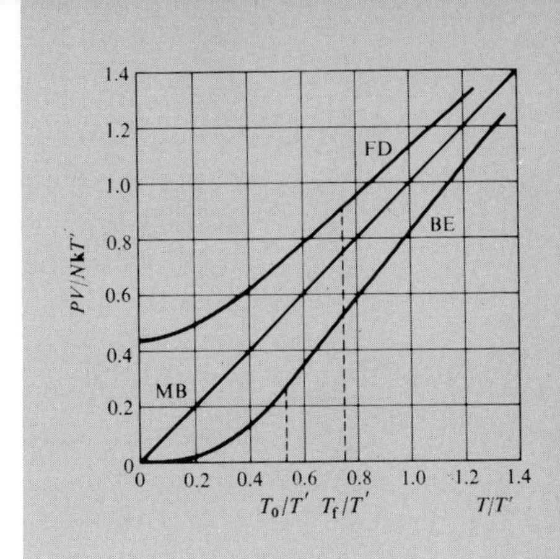

CHAPTER 8

DEGENERATE PERFECT GASES

8.1. Prefatory Remarks

The statistical theory for the thermodynamic properties of gases, developed in Chapters 6 and 7, is restricted to distinguishable molecules when the classical approximation to the partition function can be used. There are, however, many important physical situations in which this approximation becomes inadequate. We have seen, for example, that quantum effects are important at low temperatures in the case of hydrogen and helium. Consequently, it must be assumed that these gases are composed of *indistinguishable* particles, at least at low temperatures. Another example of a system in which the particles are indistinguishable, even at ordinary temperatures, is provided by the collection of electrons in a metal. The electrons that participate in the conduction of electricity move like molecules in a gas, as was first suggested by P. Drude,† and many of the thermodynamic properties of simple metals such as silver, gold, and copper were first computed theoretically by A. Sommerfeld and H. Bethe who assumed that the conduction electrons in such metals constitute a gas of indistinguishable fermions. In order to demonstrate the importance of the indistinguishability for electrons in a metal, we compute the "thermal" de Broglie wavelength, λ_{th}, for them,

$$\lambda_{\text{th}} = \frac{\mathbf{h}}{(2\pi m \mathbf{k} T)^{1/2}}, \tag{8.1}$$

† *Ann. Physik* **1** (1900), 566.

and the average distance between the electrons, $(V/N)^{1/3}$. We remind the reader that if

$$\lambda_{\text{th}} \ll (V/N)^{1/3}\,, \tag{8.2}$$

we may assume, to a good approximation, that the particles are distinguishable. In silver, at room temperature, each atom contributes one electron to the electron gas, and the number density becomes

$$N/V \approx 10^{24}\ \text{electrons/cm}^3\,,$$

or

$$(V/N)^{1/3} \approx 10^{-8}\ \text{cm}\,. \tag{8.3}$$

Given that the mass of an electron is $\mathbf{m}_e \approx 10^{-27}$gr, we calculate that

$$\lambda_{\text{th}} \approx 4 \times 10^{-7}\text{cm} \tag{8.4a}$$

at room temperature, $T \approx 300$ K, and that

$$\frac{\lambda_{\text{th}}}{(V/N)^{1/3}} \approx 40\,. \tag{8.4b}$$

This signifies that the conduction electrons in silver must be treated as a gas composed of indistinguishable particles. Similarly, at temperatures below 5 K, the atoms in gaseous helium are also indistinguishable. However, in contrast to electrons which obey Fermi–Dirac statistics, the atoms of helium obey Bose–Einstein statistics. The reader will recall that the rules which must be applied in order to determine whether a particle should be treated as a boson (symmetric wave function) or fermion (antisymmetric wave function) were discussed in Section 3.9.

This chapter will extend our discussion of the thermodynamic properties of gases to include systems composed of indistinguishable particles obeying either Fermi–Dirac or Bose–Einstein statistics. We confine our study to systems of noninteracting particles, since the discussion of systems of interacting particles is too complex to be included in this book.†

8.2. The Quantum-Mechanical Partition Function

The discussion in Chapter 5 led us to the conclusion that the thermodynamic properties of any system can be determined by evaluating the partition function, $\mathscr{Z}$. This was given in Equation (5.32) in terms of the energy levels E_j for the system in microstate j, that is, by

† See, for example, D. Pines, *The Many-Body Problem*, Benjamin, New York, 1961.

$$\mathscr{Z} = \sum_{j=1}^{k'} \exp(-\beta E_j),$$

where the summation is extended over all k' possible microstates of the system. For a system that consists of N independent, indistinguishable or distinguishable particles, the energy levels E_j are determined by the energy states available to a single particle, ε_s, and by the number of particles that have energy ε_s, when the system is in microstate j. The latter is the occupation number n_s^j for particles in state j. Given a set of occupation numbers n_s^j, the energies E_j are determined from the sum

$$E_j = \sum_{s=1}^{k} n_s^j \varepsilon_s, \tag{8.5}$$

where k now denotes the number of energy levels of a particle. The occupation numbers are, in turn, restricted by two conditions: (a) the total number of particles distributed among the various states, s, must be equal to the total number of particles in the system, N, or

$$\sum_{s=1}^{k} n_s^j = N \qquad \text{for all} \quad j; \tag{8.6}$$

and (b) the occupation numbers must be consistent with the quantum statistics that the particles obey. For particles that obey Fermi–Dirac statistics, *at most* one particle may occupy a single particle level, so that n_s^j can only assume the values 0 or 1. For particles that obey Bose–Einstein statistics, there is no restriction on the number of particles in a state s, except that stated in (a) above. The latter is also true for Maxwell–Boltzmann statistics.

The summation over all possible microstates in the partition function is equivalent to the summation over all possible distributions of the occupation numbers. This is true for both fermions and bosons and is due to the fact that the specification of the occupation numbers n_s^j for any quantum state produces a single microstate when particles are indistinguishable. Accordingly, the index j may be used to denote a particular distribution of occupation numbers, and we write the partition function as

$$\mathscr{Z} = \sum_{j} \exp\left(-\beta \sum_{s=1}^{k} n_s^j \varepsilon_s\right). \tag{8.7}$$

The average quantities of the system, such as the internal energy, the average occupation number, and so on, can now be related to the partition function, Equation (8.7). The average number of particles, $\bar{n}_s$, with single-particle energy ε_s, that is, the occupation number of state s, is given by

$$\bar{n}_s = \mathscr{Z}^{-1} \sum_{j} n_s^j \exp\left(-\beta \sum_{s=1}^{k} n_s^j \varepsilon_s\right). \tag{8.8}$$

The average energy, E, is

$$E = \mathscr{Z}^{-1} \sum_j \left(\sum_{s=1}^{k} n_s{}^j \varepsilon_s \right) \exp\left(-\beta \sum_{s=1}^{k} n_s{}^j \varepsilon_s \right)$$

$$= \sum_{s=1}^{k} \bar{n}_s \varepsilon_s, \tag{8.9}$$

and so on. It is clear from Equation (8.9) that the internal energy of the system can be expressed in terms of the single-particle energy levels, ε_s, and the average number of particles in each of those levels, $\bar{n}_s$.

The pressure can also be expressed in terms of the average occupation number as

$$P = \mathscr{Z}^{-1} \sum_j \left(\sum_{s=1}^{k} - \frac{\partial \varepsilon_s}{\partial V} n_s{}^j \right) \exp\left(-\beta \sum_{s=1}^{k} n_s{}^j \varepsilon_s \right)$$

$$= \sum_{s=1}^{k} \left(- \frac{\partial \varepsilon_s}{\partial V} \right) \bar{n}_s. \tag{8.10}$$

8.3. The Kronecker Delta Function

A set of occupation numbers for the single-particle energy levels, consistent with the requirements imposed by particle statistics and with the requirement that the total number of particles is N, define a single microstate of the system of N particles. Therefore, the summation over all microstates of the system is the sum over all consistent sets of occupation numbers. In order to express the summation, we first imagine that no attention is being paid to the total number of particles. Thus, in Fermi–Dirac statistics we may have $n_s{}^j = 0$ or 1, and all microstates would be enumerated by the sum

$$\sum_{n_1=0}^{1} \sum_{n_2=0}^{1} \cdots \sum_{n_k=0}^{1} d_{n_i} = 2^k \qquad (d_{n_i} = 1).$$

However, this sum includes spurious terms, such as, for example, the cases when $N = 0$ $(n_1 = n_2 = \cdots = n_k = 0)$ or $N = k$ $(n_1 = n_2 = \cdots n_k = 1)$, and so on. In order to remove these from the count, it is convenient to introduce a factor, δ, which vanishes for all distributions which do not correspond to exactly N particles. Thus, we define a Kronecker delta function, denoted by $\delta(N, \sum_{s=1}^{k} n_s)$, and stipulate that

$$\delta\left(N, \sum_{s=1}^{k} n_s\right) = \begin{cases} 1 & \text{if } N = \sum_{s=1}^{k} n_s \\ 0 & \text{otherwise.} \end{cases} \tag{8.11}$$

With this device we can correctly enumerate the microstates, subject to both the requirements of Fermi–Dirac statistics as well as the limitation on the number of particles, and write for the total number of microstates,

$$\Omega_{\text{FD}} = \sum_{n_1=0}^{1} \sum_{n_2=0}^{1} \cdots \sum_{n_k=0}^{1} \delta\left(N, \sum_{s=1}^{k} n_s\right). \tag{8.12a}$$

Similarly, the partition function can be expressed by the sum

$$\mathscr{Z}_{\text{FD}} = \sum_{n_1=0}^{1} \sum_{n_2=0}^{1} \cdots \sum_{n_k=0}^{1} \delta\left(N, \sum_{s=1}^{k} n_s\right) \exp\left(-\beta \sum_{s=1}^{k} \varepsilon_s n_s\right). \tag{8.12b}$$

We no longer require the superscript j that appears in Equation (8.7), since the occupation numbers now adequately label the microstates. Given a set of occupation numbers, we can determine the corresponding value of the Kronecker function, at least in principle, even though a tedious enumeration would be called for. However, as we shall see presently, the need to perform such a detailed calculation can be circumvented.

In order to enumerate the microstates for a system of bosons, we utilize the fact that there are no restrictions on the number of particles in a single-particle energy level. If we were to remove the restriction that the number of particles in the system of bosons is N, then the number of microstates would be

$$\sum_{n_1=0}^{\infty} \sum_{n_2=0}^{\infty} \cdots \sum_{n_k=0}^{\infty} d_{n_i} \qquad (d_{n_i} = 1).$$

With the restriction on N imposed, this becomes

$$\Omega_{\text{BE}} = \sum_{n_1=0}^{\infty} \sum_{n_2=0}^{\infty} \cdots \sum_{n_k=0}^{\infty} \delta\left(N, \sum_{s=1}^{k} n_s\right).$$

Since the delta function vanishes if there are more than N particles, the sum is also equal to

$$\Omega_{\text{BE}} = \sum_{n_1=0}^{N} \sum_{n_2=0}^{N} \cdots \sum_{n_k=0}^{N} \delta\left(N, \sum_{s=1}^{k} n_s\right). \tag{8.13a}$$

The partition function for a system of N bosons can now be expressed as

$$\mathscr{Z}_{\text{BE}} = \sum_{n_1=0}^{N} \cdots \sum_{n_k=0}^{N} \delta\left(N, \sum_{s=1}^{k} n_s\right) \exp\left(-\beta \sum_{s=1}^{k} n_s \varepsilon_s\right). \tag{8.13b}$$

8.4. The Average Occupation Numbers

A systematic evaluation of the expressions (8.12b) or (8.13b) for the partition functions cannot be accomplished, as already mentioned, because the restriction on the occupation numbers prevents the summations from being easily performed in a general way. Several methods have been invented to evaluate the restricted sums that appear in the expressions for the Fermi–Dirac and Bose–Einstein partition functions.† We shall circumvent these difficulties by exploiting the fact that the number of particles, N, is very large.

8.4.1. The Fermi–Dirac Distribution

We begin our calculation of $\bar{n}_s$ for fermions by giving the explicit representation of the average value as

$$\bar{n}_b = \frac{\sum_{n_1=0}^{1} \sum_{n_2=0}^{1} \cdots \sum_{n_k=0}^{1} n_b \, \delta(N, \sum_{s=1}^{k} n_s) \exp(-\beta \sum_{s=1}^{k} n_s \varepsilon_s)}{\sum_{n_1=0}^{1} \cdots \sum_{n_k=0}^{1} \delta(N, \sum_{s=1}^{k} n_s) \exp(-\beta \sum_{s=1}^{k} n_s \varepsilon_s)}. \tag{8.14}$$

Let us now concentrate our attention for the moment on the denominator in Equation (8.14) and sum over the allowed values of the occupation numbers for an arbitrary state b. If $n_b = 0$, we must have

$$\delta\left(N, \sum_{s=1}^{k} n_s\right) = \delta\left(N, \sum_{s \neq b} n_s\right),$$

whereas for $n_b = 1$, we see that

$$\delta\left(N, \sum_{s=1}^{k} n_s\right) = \delta\left(N, 1 + \sum_{s \neq b} n_s\right).$$

Consequently, for $n_b = 0$, we find that

$$\delta\left(N, \sum_{s=1}^{k} n_s\right) \exp\left(-\beta \sum_{s=1}^{k} n_s \varepsilon_s\right) = \delta\left(N, \sum_{s \neq b} n_s\right) \exp\left(-\beta \sum_{s \neq b} n_s \varepsilon_s\right),$$

whereas for $n_b = 1$, we may write

$$\delta\left(N, \sum_{s=1}^{k} n_s\right) \exp\left(-\beta \sum_{s=1}^{k} n_s \varepsilon_s\right) = \delta\left(N, 1 + \sum_{n \neq b} n_s\right) \exp(-\beta \varepsilon_b) \exp\left(-\beta \sum_{s \neq b} n_s \varepsilon_s\right).$$

In this manner we can put the denominator of Equation (8.14) in the form

$$\begin{aligned} &\sum_{n_1=0}^{1} \cdots \sum_{n_k=0}^{1} \delta\left(N, \sum_{s=1}^{k} n_s\right) \exp\left(-\beta \sum_{s=1}^{k} n_s \varepsilon_s\right) \\ &= \sum_{n_1=0}^{1} \cdots \sum_{n_{b-1}=0}^{1} \sum_{n_{b+1}=0}^{1} \cdots \sum_{n_k=0}^{1} \left[\delta\left(N, \sum_{s \neq b} n_s\right)\right. \\ &\quad \left. + \delta\left(N, 1 + \sum_{s \neq b} n_s\right) \exp(-\beta \varepsilon_b)\right] \exp\left(-\beta \sum_{n \neq b} n_s \varepsilon_s\right). \end{aligned} \tag{8.15}$$

† See, for example, A. Sommerfeld, "Lectures on Theoretical Physics," Vol. V, *Thermodynamics and Statistical Mechanics* (J. Kestin, transl.), Academic Press, New York, 1956.

In this form, we have simply rearranged the summation in order to indicate more clearly the sum over the possible occupation numbers for state b. We can further simplify the expression in Equation (8.15) if we introduce a new symbol, $\mathscr{Z}_N^{(b)}$, defined by

$$\mathscr{Z}_N^{(b)} = \sum_{n_1=0}^{1} \cdots \sum_{n_{b-1}=0}^{1} \sum_{n_{b+1}=0}^{1} \cdots \sum_{n_k=0}^{1} \delta\left(N, \sum_{s\neq b} n_s\right) \exp\left(-\beta \sum_{s\neq b} \varepsilon_s n_s\right). \tag{8.16a}$$

Hence, we may write

$$\sum_{n_1=0}^{1} \cdots \sum_{n_k=0}^{1} \delta\left(N, \sum_{s=1}^{k} n_s\right) \exp\left(-\beta \sum_{s=1}^{k} n_s \varepsilon_s\right) = \mathscr{Z}_N^{(b)} + \mathscr{Z}_{N-1}^{(b)} \exp(-\beta\varepsilon_b). \tag{8.16b}$$

The quantity $\mathscr{Z}_N^{(b)}$ is, in fact, the partition function for a system of N fermions, none of which has the energy ε_b. Similarly, the quantity $\mathscr{Z}_{N-1}^{(b)}$ is the partition function for $N-1$ fermions, none of which can have energy ε_b.

The numerator in Equation (8.14) can be expressed in terms of the special partition function $\mathscr{Z}_{N-1}^{(b)}$ as

$$\begin{aligned}
&\sum_{n_1=0}^{1} \cdots \sum_{n_k=0}^{1} n_b\, \delta\left(N, \sum_{s=1}^{k} n_s\right) \exp\left(-\beta \sum_{s=1}^{k} n_s \varepsilon_s\right) \\
&\quad = \sum_{n_1=0}^{1} \cdots \sum_{n_{b-1}=0}^{1} \sum_{n_{b+1}=0}^{1} \cdots \sum_{n_k=0}^{1} \delta\left(N, 1 + \sum_{s\neq b} n_s\right) \\
&\qquad \times \exp(-\beta\varepsilon_b) \exp\left(-\beta \sum_{s\neq b} n_s \varepsilon_s\right) \\
&\quad = \mathscr{Z}_{N-1}^{(b)} \exp(-\beta\varepsilon_b).
\end{aligned} \tag{8.17}$$

We now combine expressions (8.16b) and (8.17) to obtain $\bar{n}_b$ as

$$\bar{n}_b = \frac{\mathscr{Z}_{N-1}^{(b)} \exp(-\beta\varepsilon_b)}{\mathscr{Z}_N^{(b)} + \mathscr{Z}_{N-1}^{(b)} \exp(-\beta\varepsilon_b)} = \left[1 + \frac{\mathscr{Z}_N^{(b)}}{\mathscr{Z}_{N-1}^{(b)}} \exp(\beta\varepsilon_b)\right]^{-1}. \tag{8.18}$$

If the single-particle energy levels are closely spaced, and if the number of particles, N, in the system is very large, we may approximate the ratio $\mathscr{Z}_N^{(b)}/\mathscr{Z}_{N-1}^{(b)}$ by the ratio of the unrestricted partition functions, $\mathscr{Z}_N/\mathscr{Z}_{N-1}$. In other words, we assume that the ratio is not sensitive to whether any particular energy level b is occupied or not. If we now define a parameter α by

$$e^{\alpha} = \mathscr{Z}_N/\mathscr{Z}_{N-1}, \tag{8.19}$$

then, for fermions, the occupation numbers $\bar{n}_b$ are given by

$$\bar{n}_b = [1 + \exp(\alpha + \beta\varepsilon_b)]^{-1}. \tag{8.20a}$$

We note that $\bar{n}_b \leq 1$, in accordance with Pauli's exclusion principle for fermions. The parameter α can be determined by the requirement that the average occupation numbers must also satisfy the restriction on the total number, N, of particles, or

$$\sum_b \bar{n}_b = \sum_b [1 + \exp(\alpha + \beta\varepsilon_b)]^{-1} = N. \tag{8.20b}$$

The preceding derivation, though rigorous, may appear difficult to grasp on first reading, and it is worth knowing that the same result can be obtained by the method of the most probable distribution. This alternative derivation is discussed in Problem 8.2.

By virtue of its definition, the parameter α can be given a simple physical interpretation. If we consider the partition functions $\mathscr{Z}_{N-n}$ and $\mathscr{Z}_N$, where $n/N \ll 1$, we may employ the Taylor-series expansion

$$\ln \mathscr{Z}_{N-n} = \ln \mathscr{Z}_N - n \frac{\partial \ln \mathscr{Z}_N}{\partial N} + \frac{n^2}{2} \frac{\partial^2 \ln \mathscr{Z}_N}{\partial N^2} + \cdots, \tag{8.21a}$$

or

$$\ln \frac{\mathscr{Z}_{N-n}}{\mathscr{Z}_N} = -n \frac{\partial \ln \mathscr{Z}_N}{\partial N} + \frac{n^2}{2} \frac{\partial^2 \ln \mathscr{Z}_N}{\partial N^2} + \cdots. \tag{8.21b}$$

Using the relation between the partition function, $\mathscr{Z}_N$, and the Helmholtz free energy for a system of N particles,

$$F_N = -\mathbf{k}T \ln \mathscr{Z}_N$$

we find that

$$\ln \frac{\mathscr{Z}_{N-n}}{\mathscr{Z}_N} = \left(\frac{n}{\mathbf{k}T} \frac{\partial F_N}{\partial N}\right) - \frac{n^2}{2\mathbf{k}T} \left(\frac{\partial^2 F_N}{\partial N^2}\right) + \cdots. \tag{8.21c}$$

The chemical potential, μ, is defined as $\partial F_N / \partial N$ and we can relate α to the chemical potential, since

$$-\alpha = \frac{\mu}{\mathbf{k}T} - \frac{1}{2\mathbf{k}T} \frac{\partial \mu}{\partial N} + \cdots. \tag{8.22}$$

If, in a particular case, the chemical potential does not depend strongly on the number of particles in the system, or, if

$$\frac{1}{\mu} \frac{\partial \mu}{\partial N} = \frac{\partial \ln \mu}{\partial N} \ll 1, \tag{8.23}$$

then the higher terms in (8.22) can be neglected, and we discover that

$$\alpha = -\mu/\mathbf{k}T. \tag{8.24}$$

While Equation (8.24) is restricted to systems in which μ is insensitive to the number of particles, the validity of Equations (8.20a) and (8.20b) is circumscribed merely by the assumption that the ratio $\mathscr{Z}_{N-1}^{(b)}/\mathscr{Z}_N^{(b)}$ can be replaced by the ratio $\mathscr{Z}_{N-1}/\mathscr{Z}_N$ for any quantum state b. Therefore, for those situations where it is possible to determine α by using Equations (8.20a) and (8.20b), we can verify whether or not the inequality (8.23) is satisfied. It is, in fact, correct in all known cases to equate

$$\alpha = -\mu/\mathbf{k}T,$$

and the distribution of fermions may be written

$$\bar{n}_b = \{\exp[\beta(\varepsilon_b - \mu)] + 1\}^{-1}. \tag{8.25}$$

8.4.2. The Bose–Einstein Distribution

The calculation of the occupation numbers for ideal boson systems generally proceeds along the lines of the previous section, except for certain difficulties that were not present in the fermion case. These are connected with the fact that any number of bosons can occupy any quantum state. If we focus our attention on the bth quantum state, we can use Equations (8.10) and (8.11) to write $\bar{n}_b$ for bosons as

$$\bar{n}_b = \frac{\sum_{n_1=0}^{N}\cdots\sum_{n_k=0}^{N} n_b\,\delta(N,\sum_{s=1}^{k} n_s)\exp(-\beta\sum_{s=1}^{k} n_s\varepsilon_s)}{\sum_{n_1=0}^{N}\cdots\sum_{n_k=0}^{N}\delta(N,\sum_{s=1}^{k} n_s)\exp(-\beta\sum_{s=1}^{k} n_s\varepsilon_s)}. \tag{8.26}$$

We define a "restricted" partition function $\mathscr{Z}_{N-n_b}^{(b)}$ by the relation

$$\mathscr{Z}_{N-n_b}^{(b)} = \sum_{n_1=0}^{N}\cdots\sum_{n_{b-1}=0}^{N}\sum_{n_{b+1}=0}^{N}\cdots\sum_{n_k=0}^{N}\delta\left(N, n_b + \sum_{s\neq b} n_s\right)\exp\left(-\beta\sum_{s\neq b} n_s\varepsilon_s\right). \tag{8.27}$$

Another form for $\mathscr{Z}_{N-n_b}^{(b)}$ can be given if we utilize the fact that the Kronecker delta satisfies the relation

$$\delta\left(N, n_b + \sum_{s\neq b} n_s\right) = \delta\left(N - n_b, \sum_{s\neq b} n_s\right).$$

It is clear that this delta function restricts the maximum number of particles in any quantum state to $N - n_b$, and $\mathscr{Z}_{N-n_b}^{(b)}$ can be written

$$\mathscr{Z}_{N-n_b}^{(b)} = \sum_{n_1=0}^{N-n_b}\cdots\sum_{n_{b-1}=0}^{N-n_b}\sum_{n_{b+1}=0}^{N-n_b}\cdots\sum_{n_k=0}^{N-n_b}\delta\left(N-n_b, \sum_{s\neq b} n_s\right)\exp\left(-\beta\sum_{s\neq b} n_s\,\varepsilon_s\right). \tag{8.27a}$$

The restricted partition function $\mathscr{Z}_{N-n_b}^{(b)}$ is now the partition function for a system of $N - n_b$ bosons none of which can be in quantum state b. In a manner identical with that for fermions, we obtain the average occupation number, $\bar{n}_b$, as

$$\bar{n}_b = \frac{\mathrm{K}_{\mathrm{num}}}{\mathrm{K}_{\mathrm{denom}}} \tag{8.28}$$

with

$$\mathrm{K}_{\mathrm{num}} = \mathscr{Z}_{N-1}^{(b)}[\exp(-\beta\varepsilon_b)] + 2\mathscr{Z}_{N-2}^{(b)}[\exp(-2\beta\varepsilon_b)] + 3\mathscr{Z}_{N-3}^{(b)}[\exp(-3\beta\varepsilon_b)] + \cdots + N\mathscr{Z}_0^{(b)}[\exp(-N\beta\varepsilon_b)] \tag{8.28a}$$

and

$$\mathrm{K}_{\mathrm{denom}} = \mathscr{Z}_N^{(b)} + \mathscr{Z}_{N-1}^{(b)}[\exp(-\beta\varepsilon_b)] + \mathscr{Z}_{N-2}^{(b)}[\exp(-2\beta\varepsilon_b)] + \cdots + \mathscr{Z}_0^{(b)}[\exp(-N\beta\varepsilon_b)] \tag{8.28b}$$

and note that

$$\mathscr{Z}_0^{(b)} = 1 .$$

In order to simplify Equation (8.28) it becomes necessary to make a somewhat more extensive set of assumptions than previously, when only one term appeared in the numerator and two in the denonimator, Equation (8.18). Before doing this, we divide the numerator and denominator by $\mathscr{Z}_N^{(b)}$ to obtain

$$\bar{n}_b = \frac{(\mathscr{Z}_{N-1}^{(b)}/\mathscr{Z}_N^{(b)})[\exp(-\beta\varepsilon_b)] + 2(\mathscr{Z}_{N-2}^{(b)}/\mathscr{Z}_N^{(b)})[\exp(-2\beta\varepsilon_b)] + \cdots}{1 + (\mathscr{Z}_{N-1}^{(b)}/\mathscr{Z}_N^{(b)})[\exp(-\beta\varepsilon_b)] + (\mathscr{Z}_{N-2}^{(b)}/\mathscr{Z}_N^{(b)})[\exp(-2\beta\varepsilon_b)] + \cdots} . \tag{8.29}$$

We now *assume* that the series which appear in the numerator and denominator are so rapidly convergent that only n terms, with $n \ll N$,† make significant contributions to the sum. We assume, further, that $\mathscr{Z}_{N-n}^{(b)}/\mathscr{Z}_N^{(b)}$ can be replaced by $\mathscr{Z}_{N-n}/\mathscr{Z}_N$, arguing again that no noticeable effect will result from excluding quantum state b from the count. If all of these conditions are satisfied, then

$$\bar{n}_b = \frac{\exp[-(\alpha + \beta\varepsilon_b)] + 2\exp[-2(\alpha + \beta\varepsilon_b)] + 3\exp[-3(\alpha + \beta\varepsilon_b)] + \cdots}{1 + \exp[-(\alpha + \beta\varepsilon_b)] + \exp[-2(\alpha + \beta\varepsilon_b)] + \cdots} . \tag{8.30}$$

If $\alpha + \beta\varepsilon_b > 0$, these series converge rapidly and may be replaced by

$$\bar{n}_b = \frac{\sum_{n_b=0}^{\infty} n_b \exp[-n_b(\alpha + \beta\varepsilon_b)]}{\sum_{n_b=0}^{\infty} \exp[-n_b(\alpha + \beta\varepsilon_b)]} = [\exp(\alpha + \beta\varepsilon_b) - 1]^{-1} . \tag{8.31}$$

† We remind the reader that N is of the order of 10^{24}, so n may be quite large and still satisfy the condition that $n/N \ll 1$.

Here the restriction that $\alpha + \beta\varepsilon_b > 0$ is obviously quite important, since the numerator as well as the denominator in Equation (8.30) diverges as $\alpha + \beta\varepsilon_b \to 0$. In fact, for small values of $\alpha + \beta\varepsilon_b$, $\bar{n}_b$ can become quite large. We shall see in Section 8.8 that such difficulties do arise in systems of bosons, and that they are related to the phenomena of condensation that occur even in an ideal boson gas.

We have replaced $\mathscr{Z}_{N-n}/\mathscr{Z}_N$ by $\exp(-n\alpha) = \exp(n\mu\beta)$ assuming that $d \ln \mu/dN \ll 1$, as before. Assuming, further, the validity of Equation (8.31), we can determine the exponent α or the chemical potential, μ from the condition that

$$N = \sum_b \{\exp[\beta(\varepsilon_b - \mu)] - 1\}^{-1}. \tag{8.32}$$

8.4.3. Some Further Properties of the Occupation Numbers

The expressions for the average occupation numbers for fermions and bosons may be written in the compact notation

$$\bar{n} = [\exp(\alpha + \beta\varepsilon_s) \pm 1]^{-1}, \qquad \frac{\text{FD}}{\text{BE}} \tag{8.33}$$

where the upper sign refers to fermions and the lower sign to bosons, as indicated. We suppose that all the possible values of ε_s are greater than or equal to zero, the ground state having been normalized to $\varepsilon_0 = 0$. If we consider the ground state and examine the dependence of $\bar{n} = \bar{n}_0$ on α, we conclude that in the case of fermions, α can range from $-\infty$ to $+\infty$, without leading to an infinite value for $\bar{n}_0$, whereas in the case of bosons, α can only range from 0 to $+\infty$.† For $\alpha < 0$, $\bar{n}_0$ becomes negative for bosons and Equation (8.31) no longer follows from Equation (8.30). This relation has been plotted in Figure 8.1. The curves labeled FD and BE correspond to the upper and lower signs, in Equation (8.33), respectively, whereas the curve labeled MB represents the corrected Maxwell–Boltzmann distribution from Equation (5.30). For large positive values of α, the Fermi–Dirac as well as the Bose–Einstein distribution approaches the Maxwell–Boltzmann distribution equation,

$$\bar{n} = \exp(-\alpha)\exp(-\beta\varepsilon_s), \qquad \text{MB}.$$

In this limit, the distinction between the three different statistics is eliminated, and the results are identical with those for distinguishable particles. This result can be easily understood in physical terms. When α is large and positive, we must have $\bar{n} \ll 1$, that is, on the average, at most one particle occupies any quantum state. Since indistinguishability affects the count of the number of

† The value $\alpha = 0$ itself is excluded for bosons.

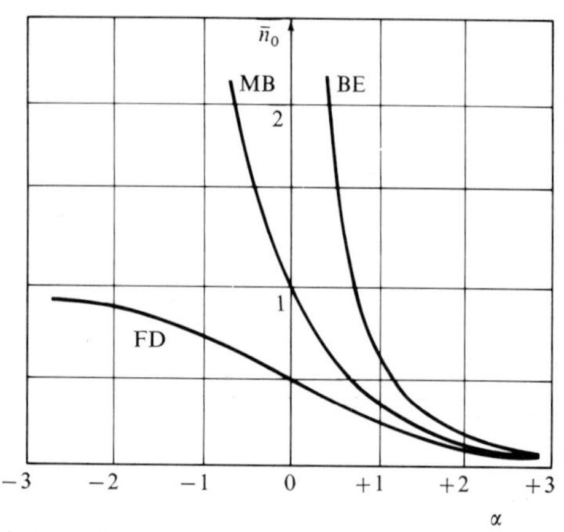

FIGURE 8.1 The dependence of the average occupation number of the ground state, n_0, on α for fermions and bosons. The ground state has been normalized to $\varepsilon_0 = 0$. [From T. Hill, *Introduction to Statistical Thermodynamics*, p. 438, Addison-Wesley, Reading, Massachusetts, 1960.]

quantum states of the system only if there are several particles per energy level, its effect vanishes when, on the average, most states remain unoccupied. We utilized this observation in a heuristic manner in Section 6.2. As α approaches $-\beta\varepsilon_s$, the boson distribution diverges, but the fermion distribution exists for $\alpha \to -\infty$, with $\bar{n} \to 1$ in this limit.

We can use the form of $\bar{n}$ given by Equation (8.33) to obtain expressions for the various thermodynamic properties. The internal energy E is

$$E = \sum_s \varepsilon_s[\exp(\alpha + \beta\varepsilon_s) \pm 1]^{-1}, \qquad \frac{\text{FD}}{\text{BE}}; \tag{8.34a}$$

the pressure, P, is

$$P = \sum_s -\frac{\partial \varepsilon_s}{\partial V}[\exp(\alpha + \beta\varepsilon_s) \pm 1]^{-1}, \qquad \frac{\text{FD}}{\text{BE}}; \tag{8.34b}$$

finally, the partition function $\mathscr{Z}$ must be such that the relations

$$E = -\left[\frac{\partial \ln \mathscr{Z}}{\partial \beta}\right]_V, \qquad \bar{n}_b = -\left[\frac{\partial \ln \mathscr{Z}}{\partial(\beta\varepsilon_b)}\right]_{V,\beta},$$

and

$$\beta P = \left[\frac{\partial \ln \mathscr{Z}}{\partial V}\right]_\beta$$

are all satisfied.

An expression for $\mathscr{Z}$ which conforms to these requirements is

$$\ln \mathscr{Z} = \pm \sum_b \ln[1 \pm \exp(-\alpha - \beta\varepsilon_b)] + N\alpha + \text{const}, \qquad \frac{\text{FD}}{\text{BE}}, \tag{8.34c}$$

This form can also be obtained by a more direct evaluation.† The latter also yields the value of the constant, and the complete expression is

$$\ln \mathscr{Z} = \pm \sum_{b} \ln[1 \pm \exp(-\alpha - \beta\varepsilon_s)] + N\alpha, \qquad \frac{\text{FD}}{\text{BE}}. \tag{8.35}$$

8.5. The Perfect Quantum Gas

It is recalled that the energy levels in the preceding equations relate to a single particle. Consequently, the magnitudes of these levels and the spacing between them are determined by the Schrödinger equation for a single particle, and are independent of the statistics. In the remainder of this section we shall confine attention to a *perfect quantum gas* without internal degrees of freedom. In this case, the energy levels, ε_s, are given by Equation (6.9a) as

$$\varepsilon_s = \frac{\mathbf{h}^2({s_x}^2 + {s_y}^2 + {s_z}^2)}{8mV^{2/3}} \qquad (s_x = 0, 1, 2, \ldots;\ s_y = 0, 1, 2, \ldots;\ s_z = 0, 1, 2, \ldots). \tag{8.36}$$

We can now easily show that the energy becomes

$$E = \sum_{s_x=0}^{\infty} \sum_{s_y=0}^{\infty} \sum_{s_z=0}^{\infty} \frac{\mathbf{h}^2({s_x}^2 + {s_y}^2 + {s_z}^2)}{8mV^{2/3}} \times \left\{\exp\left[\alpha + \frac{\beta\mathbf{h}^2({s_x}^2 + {s_y}^2 + {s_z}^2)}{8mV^{2/3}}\right] \pm 1\right\}^{-1}, \qquad \frac{\text{FD}}{\text{BE}} \tag{8.37a}$$

and that the number of particles in the gas is

$$N = \sum_{s_x=0}^{\infty} \sum_{s_y=0}^{\infty} \sum_{s_z=0}^{\infty} \left\{\exp\left[\alpha + \frac{\beta\mathbf{h}^2}{8mV^{2/3}}({s_x}^2 + {s_y}^2 + {s_z}^2)\right] \pm 1\right\}^{-1}, \qquad \frac{\text{FD}}{\text{BE}}. \tag{8.37b}$$

If the spacing between successive levels is small compared to $\mathbf{k}T = 1/\beta$, the sums may be replaced by integrals, and the approximation is valid provided that

$$\frac{\varepsilon_{s+1} - \varepsilon_s}{\mathbf{k}T} \approx \frac{\mathbf{h}^2}{8mV^{2/3}\mathbf{k}T} \ll 1.$$

† See, for example, the method of C. G. Darwin and R. H. Fowler in A. Sommerfeld, "Lectures on Theoretical Physics," Vol. V, *Thermodynamics and Statistical Mechanics* (J. Kestin, transl.), p. 259, Academic Press, New York, 1956.

Thus

$$E = \int_0^\infty ds_x \int_0^\infty ds_y \int_0^\infty ds_z \frac{\mathbf{h}^2 s^2}{8mV^{2/3}} \left[\exp\left(\alpha + \frac{\beta \mathbf{h}^2 s^2}{8mV^{2/3}}\right) \pm 1\right]^{-1}, \qquad \frac{\text{FD}}{\text{BE}}, \tag{8.38a}$$

and

$$N = \int_0^\infty ds_x \int_0^\infty ds_y \int_0^\infty ds_z \left[\exp\left(\alpha + \frac{\beta \mathbf{h}^2 s^2}{8mV^{2/3}}\right) \pm 1\right]^{-1}, \qquad \frac{\text{FD}}{\text{BE}}, \tag{8.38b}$$

where

$$s^2 = s_x^{\,2} + s_y^{\,2} + s_z^{\,2}.$$

A change of variables from s_x, s_y, s_z to the variables of spherical coordinates s, θ, ϕ given by

$$\begin{aligned} s_z &= s \cos\theta \\ s_y &= s \sin\theta \sin\phi \\ s_x &= s \sin\theta \cos\phi \end{aligned} \tag{8.39}$$

allows us to write

$$E = \int_0^{\pi/2} d\phi \int_0^{\pi/2} \sin\theta \, d\theta \int_0^\infty s^2 \frac{\mathbf{h}^2 s^2}{8mV^{2/3}} \left[\exp\left(\alpha + \frac{\beta \mathbf{h}^2 s^2}{8mV^{2/3}}\right) \pm 1\right]^{-1} ds, \qquad \frac{\text{FD}}{\text{BE}} \tag{8.40a}$$

and

$$N = \int_0^{\pi/2} d\phi \int_0^{\pi/2} \sin\theta \, d\theta \int_0^\infty s^2 \left[\exp\left(\alpha + \frac{\beta \mathbf{h}^2 s^2}{8mV^{2/3}}\right) \pm 1\right]^{-1} ds, \qquad \frac{\text{FD}}{\text{BE}}. \tag{8.40b}$$

The limits on the new variables are dictated by the requirement that the original integration extends over the octant $s_x \geq 0$, $s_y \geq 0$, $s_z \geq 0$. The final form in which we shall put these integrals is obtained by introducing the variable

$$\varepsilon = \frac{\mathbf{h}^2 s^2}{8mV^{2/3}}.$$

In this manner, we obtain

$$E = \int_0^\infty n(\varepsilon) D(\varepsilon) \varepsilon \, d\varepsilon \tag{8.41a}$$

and

$$N = \int_0^\infty n(\varepsilon) D(\varepsilon) \, d\varepsilon, \tag{8.41b}$$

where

$$n(\varepsilon) = [\exp(\alpha + \beta\varepsilon) \pm 1]^{-1}, \qquad \frac{\text{FD}}{\text{BE}} \tag{8.41c}$$

and

$$D(\varepsilon) = \frac{4\pi V}{\mathbf{h}^3}(2m^3\varepsilon)^{1/2}. \tag{8.41d}$$

The quantity $D(\varepsilon)\,\mathrm{d}\varepsilon$ may be interpreted as the number of quantum states in the energy interval $\mathrm{d}\varepsilon$. The quantity $D(\varepsilon)$ defined by Equation (8.41d) is called the *density of states for a perfect quantum gas.* The expression for $D(\varepsilon)$ can also be obtained by considering the number $\mathrm{d}\Omega$ of quantum states in the volume $\mathbf{dr}\,\mathbf{dp}$ of μ-space. We know that

$$\mathrm{d}\Omega = \frac{\mathbf{drdp}}{\mathbf{h}^3} = \frac{\mathbf{dr}\sin\theta_p\,\mathrm{d}\theta_p\,\mathrm{d}\phi_p\,p^2\,\mathrm{d}p}{\mathbf{h}^3}. \tag{8.42}$$

For a perfect gas,

$$\varepsilon = \frac{\mathbf{p}^2}{2m}, \tag{8.43}$$

and

$$\mathrm{d}\Omega = \frac{\mathbf{dr}\sin\theta_p\,\mathrm{d}\theta_p\,\mathrm{d}\phi_p(2m)^{3/2}\varepsilon^{1/2}\,\mathrm{d}\varepsilon}{2\mathbf{h}^3}. \tag{8.44}$$

The subscript p in θ_p and ϕ_p serves to remind us that the coordinates are drawn in momentum space. To obtain the total number of quantum states in the interval $\mathrm{d}\varepsilon$, we integrate expression (8.44) over the volume of the container and over all possible directions of the momentum. Thus,

$$D(\varepsilon)\,\mathrm{d}\varepsilon = \int_V \mathbf{dr}\int_0^\pi \sin\theta_p\,\mathrm{d}\theta_p\int_0^{2\pi}\mathrm{d}\phi_p\frac{(2m)^{3/2}\varepsilon^{1/2}\,\mathrm{d}\varepsilon}{2h^3} = \frac{4\pi V}{\mathbf{h}^3}(2m^3\varepsilon)^{1/2}\,\mathrm{d}\varepsilon, \tag{8.45}$$

which is identical with (8.41d).

8.6. The Weakly Degenerate Gas

We are now prepared to discuss the properties of perfect quantum gases for Bose–Einstein as well as Fermi–Dirac statistics. As our first example, we derive an expression for the pressure of a quantum gas without internal degrees of freedom, under circumstances where the parameter

$$y = \frac{N}{V}\left[\frac{\mathbf{h}^2}{2\pi m\mathbf{k}T}\right]^{3/2} = \frac{\lambda_{\text{th}}^3}{V/N} \tag{8.46}$$

is small. This makes it possible to express the deviations of the behavior of the gas under consideration from that of a classical gas which obeys corrected Maxwell–Boltzmann statistics and to put it in the form of a power series in y. To simplify expression, we use the phrase that perfect gases for which the corrections introduced by Fermi–Dirac or Bose–Einstein statistics are important exhibit *degeneracy*. The term " degeneracy " used in this context has no connection to the same word which describes the multiplicity of states per energy level. The expansion parameter, y, is called the *degeneracy parameter*, and when its value is small, as assumed now, the gas is called *weakly degenerate*. The opposite case of a *strongly degenerate gas*, which is said to exist for large values of y, will be studied in Sections 8.7 and 8.8.

Reverting to the equations of the preceding section, we can write explicitly

$$E = \frac{4\pi V}{\mathbf{h}^3}(2m^3)^{1/2}\int_0^\infty \frac{\varepsilon^{3/2}\,\mathrm{d}\varepsilon}{\exp(\alpha+\beta\varepsilon)\pm 1}, \qquad \frac{\text{FD}}{\text{BE}} \tag{8.47a}$$

$$N = \frac{4\pi V}{\mathbf{h}^3}(2m^3)^{1/2}\int_0^\infty \frac{\varepsilon^{1/2}\,\mathrm{d}\varepsilon}{\exp(\alpha+\beta\varepsilon)\pm 1}, \qquad \frac{\text{FD}}{\text{BE}} \tag{8.47b}$$

and

$$PV = \tfrac{2}{3}E, \qquad \text{FD and BE} \tag{8.47c}$$

as we already know, quite generally, from Section 6.3. The integrals that appear in Equations (8.47) cannot be expressed in terms of simple functions, and we confine further attention to four limiting cases. The simplest corresponds to large values of α when the behavior of the degenerate gas closely approximates that of a Maxwell–Boltzmann gas, Figure 8.1. Accordingly, we must expect that the gas will now be weakly degenerate.

It is convenient to introduce the abbreviation

$$z = \mathrm{e}^\alpha$$

and to express Equations (8.47a)–(8.47c) in terms of the expansion

$$\frac{1}{\exp(\alpha+\beta\varepsilon)\pm 1} = \frac{\exp(-\beta\varepsilon)}{z}\left[1 \pm \frac{\exp(-\beta\varepsilon)}{z}\right]^{-1}$$

$$= \frac{\exp(-\beta\varepsilon)}{z}\sum_{n=0}^{\infty}(\mp 1)^n z^{-n}\exp(-n\beta\varepsilon), \qquad \frac{\text{FD}}{\text{BE}}. \tag{8.48}$$

In this manner

$$E = \frac{4\pi V}{\mathbf{h}^3}(2m^3)^{1/2}\frac{1}{z}\int_0^\infty \varepsilon^{3/2}\,\mathrm{d}\varepsilon \sum_{n=0}^{\infty}(\mp 1)^n z^{-n}\exp[-(n+1)\beta\varepsilon], \qquad \frac{\text{FD}}{\text{BE}} \tag{8.48a}$$

and

$$N = \frac{4\pi V}{\mathbf{h}^3}(2m^3)^{1/2}\frac{1}{z}\int_0^\infty \varepsilon^{1/2}\,d\varepsilon \sum_{n=0}^{\infty}(\mp 1)^n z^{-n}\exp[-(n+1)\beta\varepsilon], \qquad \frac{\text{FD}}{\text{BE}}. \tag{8.48b}$$

In order to evaluate the terms in the z^{-1} expansion, we must consider integrals of the form

$$I_t(s) = \int_0^\infty \varepsilon^t\, e^{-s\varepsilon}\, d\varepsilon\,.$$

In particular, we need

$$I_{1/2}(s) = \int_0^\infty \varepsilon^{1/2}\, e^{-s\varepsilon}\, d\varepsilon, \tag{8.49}$$

and note that the integral which appears in Equation (8.48a), $I_{3/2}(s)$, can be obtained from it by differentiation, since

$$I_{3/2}(s) = -\frac{d}{ds} I_{1/2}(s). \tag{8.50}$$

The substitution $x^2 = \varepsilon$ permits us to write

$$I_{1/2}(s) = 2\int_0^\infty x^2\, e^{-sx^2}\, dx = \frac{1}{2}\left(\frac{\pi}{s^3}\right)^{1/2}, \tag{8.51a}$$

whence we obtain that

$$I_{3/2}(s) = \frac{3}{4}\left(\frac{\pi}{s^5}\right)^{1/2}. \tag{8.51b}$$

Finally, we compute

$$N = \frac{V}{\mathbf{h}^3}(2\pi m\mathbf{k}T)^{3/2}\left[\frac{1}{z} \mp \frac{1}{2^{3/2}z^2} + \frac{1}{3^{3/2}z^3} \mp \cdots\right], \qquad \frac{\text{FD}}{\text{BE}} \tag{8.52a}$$

and

$$E = \frac{3\pi^{3/2}V}{\mathbf{h}^3}(2m^3)^{1/2}(\mathbf{k}T)^{5/2}\left[\frac{1}{z} \mp \frac{1}{2^{5/2}z^2} + \frac{1}{3^{5/2}z^3} \mp \cdots\right], \qquad \frac{\text{FD}}{\text{BE}}. \tag{8.52b}$$

In order to obtain the thermal equation of state, it is necessary to eliminate z from these two expressions and to substitute E as a function of N into Equation (8.47c). To this end, we solve Equation (8.52a) for z in terms of N and substitute into Equation (8.52b).

In detail, we notice that Equation (8.52a) expresses $n = N/V$ as a power series in z^{-1}, and seek to invert it in terms of n. Thus, we postulate that

$$z^{-1} = a_1 n + a_2 n^2 + a_3 n^3 + \cdots \tag{8.53}$$

and substitute into Equation (8.52a) in order to determine the unknown coefficients $a_1, a_2, a_3, \ldots$ by comparing like powers of n. Consequently, we find that

$$n = \left(\frac{2\pi m\mathbf{k}T}{\mathbf{h}^2}\right)^{3/2} \left[(a_1 n + a_2 n^2 + \cdots) \mp \frac{1}{2^{3/2}}(a_1 n + a_2 n^2 + \cdots)^2 + \cdots\right],$$

and evaluate

$$a_1 = \left[\frac{2\pi m\mathbf{k}T}{\mathbf{h}^2}\right]^{-3/2}, \tag{8.54a}$$

$$a_2 \mp \frac{1}{2^{3/2}} a_1^2 = 0, \tag{8.54b}$$

or

$$a_2 = \pm \frac{1}{2^{3/2}} \left(\frac{2\pi m\mathbf{k}T}{\mathbf{h}^2}\right)^{-3}, \quad \text{and so on.} \tag{8.54c}$$

This leads to

$$\begin{aligned} E &= \frac{3}{2} N\mathbf{k}T\left[1 \pm \frac{n}{2^{5/2}(2\pi m\mathbf{k}T/\mathbf{h}^2)^{3/2}} + \cdots\right] \\ &= \frac{3}{2} N\mathbf{k}T[1 \pm 2^{-5/2} y + \cdots], \quad \frac{\text{FD}}{\text{BE}}, \end{aligned} \tag{8.55}$$

and to

$$PV = N\mathbf{k}T[1 \pm 2^{-5/2} y + \cdots], \quad \frac{\text{FD}}{\text{BE}}, \tag{8.56}$$

or, per mol,

$$Pv_m = \mathbf{R}T[1 \pm 2^{-5/2} y + \cdots], \quad \frac{\text{FD}}{\text{BE}}. \tag{8.56a}$$

The expressions for the thermodynamic properties of a weakly degenerate quantum gas have been summarized in Table 8.1 (at the end of this chapter) to first order in the degeneracy parameter y. In Figure 8.2 we have plotted the pressure as a function of temperature along an isochore and compared it with the corresponding isochore for a nondegenerate perfect gas. At a given density and temperature, the pressure of a Fermi gas is somewhat higher than the classical value. This effect can be understood in physical terms if it is recalled that the exclusion principle forces a larger number of molecules into states of higher momentum. Consequently, the pressure increases owing to an increased average transfer of momentum to the walls. In a boson gas, on the

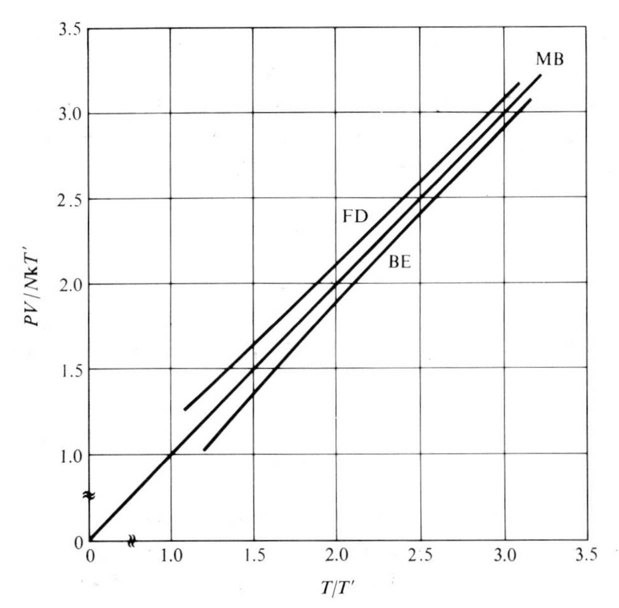

FIGURE 8.2. Thermal equation of state for weakly degenerate and classical gases compared (same mass and density). Note: $y = (1/v)(\mathbf{h}^2/2\pi m\mathbf{k}T)^{3/2} = (T'/T)^{3/2}$ with $T' = (\mathbf{h}^2/2\pi m\mathbf{k})(N/V)^{2/3}$.

other hand, a larger number of molecules is present in lower-energy states, thus decreasing the transfer of momentum to the walls. We can express both phenomena graphically by saying that fermions experience effective repulsion, whereas bosons are subject to an effective attraction between the molecules.

The diagram in Figure 8.3 shows a plot of the specific heat ratio $c_v/\mathbf{R}$ and compares it with the constant value of $c_v/\mathbf{R} = 1.5$ for a nondegenerate gas. A more complete diagram of this kind was already introduced as Figure 6.5.

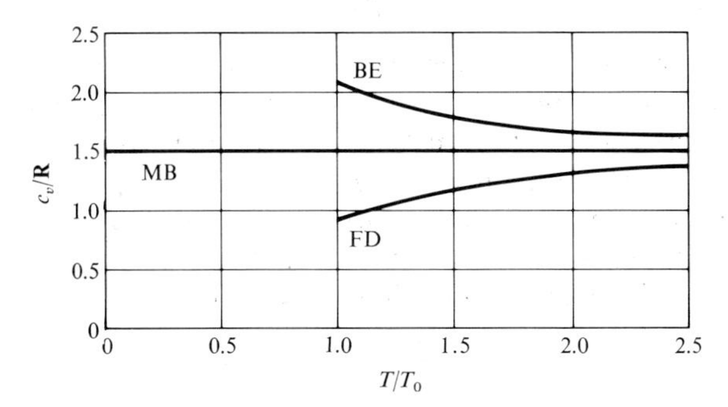

FIGURE 8.3. Specific heat ratio, $c_v/\mathbf{R}$, for weakly degenerate gases in terms of the degeneracy parameter $y^{-2/3} = (2\pi m\mathbf{k}T/\mathbf{h}^2)(V/N)^{2/3}$; here T_0 is the Bose–Einstein condensation temperature, $T_0 = 0.527\ T'$, with $T' = (\mathbf{h}^2/2\pi m\mathbf{k})(N/V)^{2/3}$.

The departures due to degeneracy are too small to be observed in gases under normal conditions. In a dilute gas, the degeneracy parameter y is of the order of 10^{-5}, increasing in the case of dense gases to values of the order of 10^{-2}. However, even in such cases the effects of degeneracy cannot be observed because they are masked by the much stronger influence of the intermolecular forces.

8.7. The Strongly Degenerate Fermi Gas

We have seen earlier in Section 8.1 that the electrons in silver are characterized by the value

$$y = \left[\frac{\lambda_{\text{th}}}{(V/N)^{1/3}}\right]^3 \approx 10^5$$

of the degeneracy parameter even at room temperature. We conclude, therefore, quite generally, that the electrons in a metal constitute a strongly degenerate gas. In order to treat this case, we now assume that

$$e^{\alpha} \ll 1 ,$$

indicating that α is negative and large in absolute magnitude. Recalling that $\alpha = -\mu/\mathbf{k}T$, we note that $\mu/\mathbf{k}T$ must now be large, which restricts us to the study of fermions. We shall consider the case of bosons in the next section.

Although a temperature of 300 K does not seem to be very low, the characteristic temperature, $\mu/\mathbf{k}$, for the electrons in a metal is of the order of 420,000 K, and room temperature can be said to be "quite close" to absolute zero. Accordingly, let us now examine the Fermi distribution in the limit of low temperatures, and substitute $-\mu/\mathbf{k}T$ for α. Thus, the average number of particles in a state of energy ε_s is given by

$$\bar{n}(\varepsilon_s) = [\exp \beta(\varepsilon_s - \mu) + 1]^{-1} .$$

The diagram in Figure 8.4 represents this distribution for several positive values of temperature and illustrates the fact that a wide range of energy levels $\varepsilon_s < \mu$ is occupied by a particle since $\bar{n}(\varepsilon_s) \approx 1$ for them. Similarly, for energies much larger than μ, a wide range of energies remains unoccupied with $\bar{n}(\varepsilon_s) \approx 0$; most of the change in $\bar{n}(\varepsilon_s)$ occurs rapidly and quite close to μ. In the limit of $T = 0$ all of the change occurs abruptly at $\mu(T = 0) = \mu_0$ and the distribution becomes a step function when the limit $\beta = \infty$ is reached. In this limit

$$\bar{n}(\varepsilon_s) = \begin{cases} 1 & \text{if} \quad \varepsilon_s < \mu_0 \\ 0 & \text{if} \quad \varepsilon_s > \mu_0 . \end{cases} \tag{8.57}$$

The limit when $\beta = \infty$ and $T = 0$ represents the case of complete degeneracy.

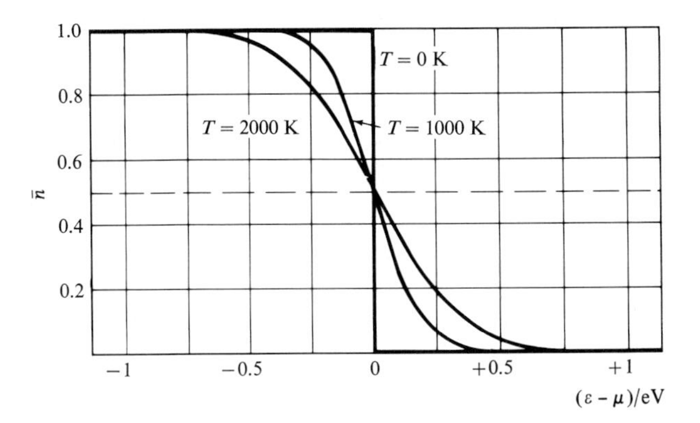

FIGURE 8.4. The Fermi distribution function in terms of $\varepsilon - \mu$ for several positive temperatures T. In the limit of $T = 0$, the distribution is represented by a step function. [After T. Hill *Introduction to Statistical Thermodynamics*, p. 438, Addison-Wesley, Reading, Massachusetts, 1960.]

All energy levels below $\varepsilon_s = \mu_0$ are occupied by a single particle each, in accordance with Pauli's exclusion principle, whereas all energy levels above it are empty.

In the case of electrons, it is necessary to recognize that one level can be occupied by two of them, one for each of the two possible spin orientations. Such double occupancy is allowed by Pauli's principle because an electron quantum state is specified by its spin as well as its energy, leading to distinct wave functions. This means that the formula for the density of states in terms of energy must now acquire a factor of 2. Therefore, the energy and number of particles for an ideal electron gas are

$$N = 2\left(\frac{4\pi V}{\mathbf{h}^3}\right)(2m^3)^{1/2} \int_0^\infty \frac{\varepsilon_s^{1/2}\, d\varepsilon_s}{\exp[\beta(\varepsilon_s - \mu)] + 1}, \tag{8.58a}$$

and

$$E = 2\left(\frac{4\pi V}{\mathbf{h}^3}\right)(2m^3)^{1/2} \int_0^\infty \frac{\varepsilon_s^{3/2}\, d\varepsilon_s}{\exp[\beta(\varepsilon_s - \mu)] + 1}. \tag{8.58b}$$

These, with the exception of the factors of 2, are identical with Equations (8.47a) and (8.47b). Equations (8.55) and (8.56), however, are corrected by replacing y by $y' = y/2$ (Problem 8.8).

At absolute zero, the equations for E and N assume the forms

$$N = 2\left(\frac{4\pi V}{\mathbf{h}^3}\right)(2m^3)^{1/2} \int_0^{\mu_0} \varepsilon_s^{1/2}\, d\varepsilon_s = \frac{4}{3}\left(\frac{4\pi V}{\mathbf{h}^3}\right)(2m^3)^{1/2}\mu_0^{3/2} \tag{8.59a}$$

and

$$E = 2\left(\frac{4\pi V}{\mathbf{h}^3}\right)(2m^3)^{1/2} \int_0^{\mu_0} \varepsilon_s^{3/2}\, d\varepsilon_s = \frac{4}{5}\left(\frac{4\pi V}{\mathbf{h}^3}\right)(2m^3)^{1/2}\mu_0^{5/2}. \tag{8.59b}$$

The chemical potential at absolute zero, μ_0, is then

$$\mu_0 = \frac{\mathbf{h}^2}{8m}\left(\frac{3N}{\pi V}\right)^{2/3}, \tag{8.60}$$

and the energy E at $T = 0$ is

$$E = \tfrac{3}{5} N \mu_0 . \tag{8.61}$$

The pressure still satisfies the perfect-gas law $PV = \frac{2}{3}E$, or

$$P = \frac{\mathbf{h}^2}{5m}\left(\frac{3}{8\pi}\right)^{2/3}\left(\frac{N}{V}\right)^{5/3}. \tag{8.62}$$

Thus at absolute zero the pressure of a Fermi gas is proportional to the $\frac{5}{3}$ power of the density.

In order to extend these results, we assume that

$$\exp(-\beta\mu) \ll 1 .$$

Suppose for the moment that μ does not sensitively depend upon the temperature so that we can approximate $\mu/\mathbf{k}T$ by $\mu_0/\mathbf{k}T$. This means that $\mu/\mathbf{k}T$ is large at low temperatures if

$$T \ll \frac{\mu_0}{\mathbf{k}}.$$

For temperatures $T \ll \mu_0/\mathbf{k}$ the distribution $\bar{n}(\varepsilon_s)$ still resembles a step function but is smoothed out somewhat, because the electrons just start to fill the higher energy states as the temperature rises. For example, in the case of silver, with $\mathbf{m}_e \approx 10^{-27}$ gr, Equation (8.60) would show that

$$\frac{\mu_0}{\mathbf{k}} \approx 420{,}000\ \mathrm{K},$$

as already stated, in a preliminary way. We conclude, once again, that the electron gas is effectively at absolute zero when the metal is kept at room temperature.

In order to calculate the thermodynamic properties, it is natural to seek expansions for N and E in terms of powers of the small parameter $\mathbf{k}T/\mu$. In the process, it becomes necessary to evaluate definite integrals of the general form

$$I = \int_0^\infty \frac{f(\varepsilon_s)\, d\varepsilon_s}{\exp[\beta(\varepsilon_s - \mu)] + 1} \tag{8.63}$$

where

$$f(\varepsilon_s) = \varepsilon_s^{1/2} \qquad \text{or} \qquad f(\varepsilon_s) = \varepsilon_s^{3/2}.$$

Employing the substitution

$$\varepsilon_s - \mu = z/\beta$$

we are led to the integrals

$$I = \frac{1}{\beta}\int_{-\beta\mu}^{\infty} \frac{f(z/\beta + \mu)}{e^z + 1} dz = \frac{1}{\beta}\left[\int_0^{\infty} \frac{f(z/\beta + \mu)}{e^z + 1} dz + \int_0^{\beta\mu} \frac{f(\mu - z/\beta)}{e^{-z} + 1} dz\right]. \tag{8.64}$$

The second integral on the right-hand side must be handled with some care since the integrand does not approach zero for large values of z, whereas the upper limit of integration, $\beta\mu$, is very large. To make this integral easier to treat we put

$$\frac{1}{e^{-z} + 1} = 1 - \frac{1}{e^z + 1}$$

and obtain

$$I = \frac{1}{\beta}\left[\int_0^{\beta\mu} f(\mu - z/\beta)\, dz + \int_0^{\infty} \frac{f(\mu + z/\beta)}{e^z + 1} dz - \int_0^{\beta\mu} \frac{f(\mu - z/\beta)}{e^z + 1} dz\right]. \tag{8.65}$$

In the last integral on the right-hand side we can replace the upper limit $\beta\mu$ by infinity, since the integrand tends to zero rapidly for large values of z. Thus

$$I = \int_0^{\mu} f(\varepsilon)\, d\varepsilon + \frac{1}{\beta}\left[\int_0^{\infty} \frac{f(\mu + z/\beta) - f(\mu - z/\beta)}{e^z + 1} dz\right]. \tag{8.66}$$

We now expand the integrand in the second integral in powers of β^{-1},† obtaining

$$f(\mu + z/\beta) - f(\mu - z/\beta) = 2z \frac{df(\mu)}{d\mu} \beta^{-1} + \frac{1}{3} z^3 \frac{d^3 f(\mu)}{d\mu^3} \beta^{-3} + \cdots, \tag{8.67}$$

and proceed to derive a temperature expansion for I, not forgetting that μ is also a function of T. Thus,

$$I = \int_0^{\mu} f(\varepsilon_s)\, d\varepsilon_s + 2\beta^{-2} f'(\mu) \int_0^{\infty} \frac{z\, dz}{e^z + 1} + \frac{1}{3}\beta^{-4} f'''(\mu) \int_0^{\infty} \frac{z^3\, dz}{e^z + 1} + \cdots. \tag{8.68}$$

The definite integrals which have now appeared are not elementary, but can be evaluated.‡ For our purposes, only the following results need to be recorded:

† The resulting expansions are asymptotic, not convergent. See R. Weinstock, *Am. J. Phys.* **37** (1969), 1273.

‡ The integral $A_n = \int_0^{\infty} x^{n-1}(e^x + 1)^{-1} dx$ is related to the Riemann zeta function, $\zeta(n)$, by the identity $A_n = (n-1)!(1 - 2^{1-n})\zeta(n)$. The Riemann zeta function is discussed and tabulated in E. Jahnke, F. Emde, and F. Lösch, *Tables of Higher Transcendental Functions*, p. 37, McGraw-Hill, New York, 1960.

$$\int_0^\infty \frac{z\,dz}{e^z+1} = \frac{\pi^2}{12} \tag{8.69a}$$

$$\int_0^\infty \frac{z^3\,dz}{e^z+1} = \frac{7\pi^4}{120}. \tag{8.69b}$$

Consequently,

$$I = \int_0^\mu f(\varepsilon_s)\,d\varepsilon_s + \frac{\pi^2}{6}\beta^{-2}f'(\mu) + \frac{7\pi^4}{360}\beta^{-4}f'''(\mu) + \cdots. \tag{8.70}$$

Using appropriate forms for $f(\varepsilon_s)$, we are led to the expansions

$$N = \frac{8\pi V}{3\mathbf{h}^3}(2m\mu)^{3/2}\left(1 + \frac{\pi^2}{8}\beta^{-2}\mu^{-2} + \cdots\right), \tag{8.71}$$

and

$$E = \frac{4\pi V}{5m\mathbf{h}^3}(2m\mu)^{5/2}\left(1 + \frac{5\pi^2}{8}\beta^{-2}\mu^{-2} + \cdots\right). \tag{8.72}$$

It should be noticed that the corrections to the results for $T = 0$ K are proportional to $(\beta\mu)^{-2}$, which is assumed to be small. Again, in order to compute the thermodynamic properties, we need to solve Equation (8.71) for μ in terms of N to express E in terms of it.

Equation (8.71) can be written

$$1 = \left(\frac{\mu}{\mu_0}\right)^{3/2}\left[1 + \frac{\pi^2}{8}(\beta\mu)^{-2} + \cdots\right], \tag{8.73}$$

and the expansion for μ in powers of $(\beta\mu_0)^{-1}$ can be assumed to be of the form

$$\mu = \mu_0[1 + a(\beta\mu_0)^{-1} + b(\beta\mu_0)^{-2} + \cdots]. \tag{8.74a}$$

Hence

$$1 = \left[1 + \frac{3a}{2}(\beta\mu_0)^{-1} + \left(\frac{3b}{2} + \frac{3a^2}{8}\right)(\beta\mu_0)^{-2} + \cdots\right]\left[1 + \frac{\pi^2}{8}(\beta\mu_0)^{-2} + \cdots\right] \tag{8.74b}$$

which leads to

$$a = 0, \qquad b = -\frac{\pi^2}{12}$$

or

$$\mu = \mu_0\left[1 - \frac{\pi^2}{12}(\beta\mu_0)^{-2} + \cdots\right]. \tag{8.75}$$

Equation (8.75) verifies the assumption on which our calculations have been based, namely that $\beta\mu$ is very large as long as $\beta\mu_0$ is large. Now, the energy of the electron gas becomes

$$E = \frac{3}{5} N\mu_0 \left[1 + \frac{5\pi^2}{12} (\beta\mu_0)^{-2} + \cdots\right]. \tag{8.76}$$

From Equation (8.76) we can obtain the following expression for the specific heat at constant volume,

$$c_v = \left(\frac{\partial E}{\partial T}\right)_V = \frac{\pi^2}{2} \frac{N}{\mu_0} \mathbf{k}^2 T + \cdots, \tag{8.77}$$

and the pressure is

$$P = \frac{2}{3} \frac{E}{V} = \frac{2}{5} \frac{N}{V} \mu_0 \left[1 + \frac{5\pi^2}{12} (\beta\mu_0)^{-2} + \cdots\right]. \tag{8.78}$$

The molar specific heat, Equation (8.77), is of special interest, because it can be written

$$c_{v,\mathrm{m}} = \left(\frac{\pi^2}{3} \frac{\mathbf{k}T}{\mu_0}\right) \frac{3}{2} \mathbf{R} + \cdots. \tag{8.79}$$

Thus the specific heat of a strongly degenerate perfect electron gas is smaller than the classical specific heat 1.5$\mathbf{R}$ by a factor $\pi^2\mathbf{k}T/3\mu_0$. This means that the contribution to the specific heat of a metal due to the electrons is negligibly small at low temperatures. As $T \to 0$, the specific heat approaches zero, in accord with the Third law of thermodynamics. Similarly, the entropy of the degenerate Fermi gas is given by

$$s = \frac{\pi^2}{2} \frac{N\mathbf{k}^2 T}{\mu_0}, \tag{8.80}$$

and also approaches zero as $T \to 0$ K.

The results contained in Equations (8.76)–(8.80) can be understood in terms of the occupation of the energy levels by the electrons, illustrated in Figure 8.4. At absolute zero, all levels below $\varepsilon = \mu_0$ are fully occupied, with one electron residing in each state. Therefore, the energy at absolute zero differs from zero, and the energy per particle is $3\mu_0/5$. Evidently, these values are measured with respect to the energy of the ground state for one particle which has been normalized at $\varepsilon_0 = 0$. Similarly, the pressure is not zero, since the particles have a residual kinetic energy. As the temperature is increased, only those electrons whose energies are near μ_0 become excited to higher-energy states. Most of the electrons remain unaffected by the rise in temperature, except for the small number whose energy is close to μ_0. Thus, the specific heat remains small in comparison with its classical value.

Table 8.1 at the end of this chapter summarizes the formulas for the thermodynamic properties of a strongly degenerate fermion gas, and the relations are also displayed graphically. Figure 8.5 contains a plot of pressure along the isochore in terms of temperature, which may be compared with Figure 8.2. It is seen that the two approximations deviate in the middle range of temperatures where the gas is neither weakly nor strongly degenerate. The complete isochore cannot be represented analytically and must be

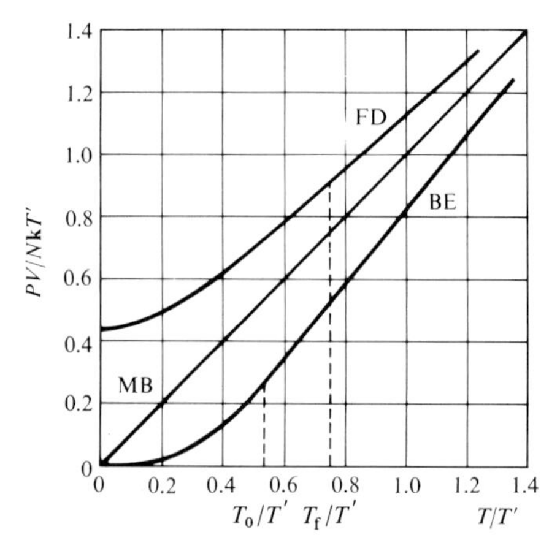

FIGURE 8.5. Isochores of degenerate electron and boson gases compared with the isochore of a classical perfect gas. Note: $T' = (\mathbf{h}^2/2\pi m n\mathbf{k})(N/V)^{2/3}$; T_0 is the Bose condensation temperature (0.527 T'); T_f is the Fermi temperature ($\mu_0/\mathbf{k} = 0.76\ T'$).

evaluated numerically.† The diagram also contains a plot of the same isochore for a strongly degenerate boson gas which will be discussed in the next section. Figure 8.6 shows the variation of the energy together with the corresponding variation for a classical gas. The present curve deviates strongly at the low temperatures and fails to reach zero as $T \to 0$. The same diagram contains an additional plot of the specific heat, c_v, which may be compared with Figure 6.5 earlier in the book. As required by the Third law, the specific heat vanishes as $T \to 0$, thus removing one of the flaws of the analysis of Section 6.8. The variation of the degeneracy parameter

$$y = \frac{\mathbf{N}}{v_\mathrm{m}} \left(\frac{2\pi m\mathbf{k}T}{\mathbf{h}^2}\right)^{-3/2},$$

† See J. McDougall and E. C. Stoner, *Phil. Trans. Roy. Soc. Ser. A*, **237** (1938), 67

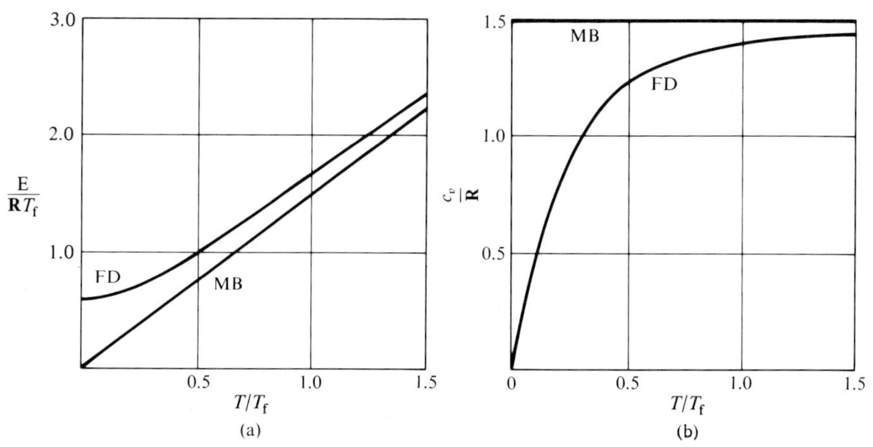

FIGURE 8.6. a, The energy ratio, $E/\mathbf{R}T_f$, and b, the specific heat ratio, $c_v/\mathbf{R}$, for an electron gas (a highly degenerate fermiun gas). Note: $T_f = \mu_0/\mathbf{k}$. [From R. Kubo, *Statistical Mechanics*, p. 243, North-Holland Publ., Amsterdam, 1965.]

with temperature ratio T/T' is seen plotted in Figure 8.7 from which it is seen that $y = 1$ at $T = T'$. Below this temperature, the specific heat tends to zero and deviates markedly from its classical value, whereas above it, the classical value prevails. Generally, the temperature

$$T' = \frac{1}{\mathbf{k}}\left(\frac{\mathbf{N}}{v_m}\right)^{2/3}\frac{\mathbf{h}^2}{2\pi m}$$

depends on v_m.

8.8. The Degenerate Boson Gas. The Bose–Einstein Condensation

As our final illustration of the properties of degenerate quantum gases, we consider a perfect boson gas at low temperatures. It was first pointed out by A. Einstein† that near absolute zero an ideal boson gas exhibits a kind of phase transition which is akin to condensation, since most of the particles are then in the ground state. Liquid ^{4}He is also a collection of bosons (see Section 3.9), even though it cannot be treated as a perfect gas owing to the large effects of interactions. Nevertheless, liquid ^{4}He undergoes a phase transition, the λ transition, at temperatures which are below, but close to, the critical point at 5.25 K. Along the vapor-pressure line, the λ transition is at 2.173 K decreasing to 1.760 K at the solidification line. Above the λ point,

† *Berliner Berichte* **2** (1924), 261; **3** (1925), 3.

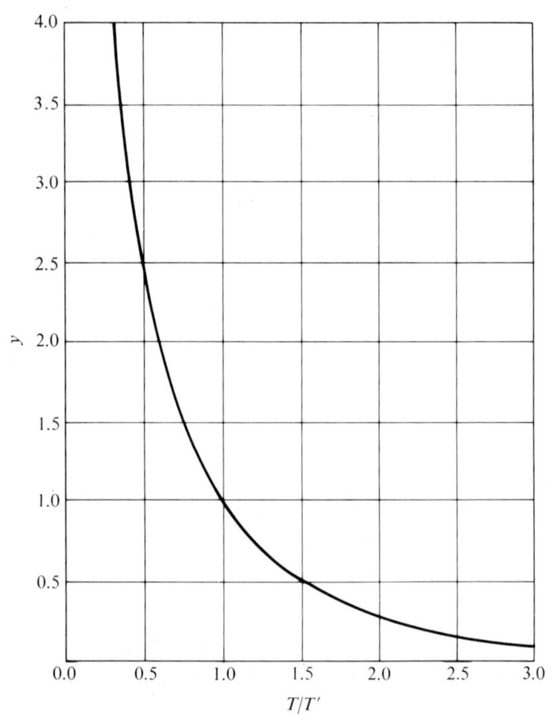

FIGURE 8.7. Variation of degeneracy factor y with temperature ratio T/T'. The curve is given by $y = (T'/T)^{3/2}$. Note : $T' = (\mathbf{h}^2/2\pi m\mathbf{k})(N/V)^{2/3}$.

^{4}He behaves like a normal liquid, but acquires the property of superfluidity below it.† The basic mechanism responsible for the λ transition is the same as that of Bose–Einstein condensation which we are about to discuss.

We begin this discussion by considering the mean occupation number

$$\bar{n}(\varepsilon_s) = \{\exp[\beta(\varepsilon_s - \mu)] - 1\}^{-1}, \tag{8.81}$$

and suppose that the lowest energy available to the particles is $\varepsilon_s = 0$. Consequently, we must have

$$\mu \leq 0. \tag{8.82}$$

In a system containing N particles, the chemical potential, μ, is determined by the relation

$$N = \frac{4\pi V}{\mathbf{h}^3}(2m^3)^{1/2} \int_0^\infty \frac{\varepsilon^{1/2}\, d\varepsilon_s}{\exp[\beta(\varepsilon_s - \mu)] - 1}. \tag{8.83}$$

† For a fuller discussion of the phase transitions in ^{4}He see, for example, J. Wilks, *The Properties of Liquid and Solid Helium*, Oxford Univ. Press, London and New York, 1967; and F. London, *Superfluids*, Wiley, New York, 1954.

By way of introduction we must examine first some of the mathematical properties of the integral in Equation (8.83), conceived of as a function of the chemical potential, μ. For $\mu > 0$, the integral diverges, signifying that positive values of μ cannot occur. When $\mu \leq 0$, the integral converges, and, as μ approaches closer to zero, the integral increases in value. For $\mu = 0$, the integral still exists, and its largest finite value occurs for $\mu = 0$. In this limit

$$N(\mu = 0) = N_{\max} = \frac{4\pi V}{\mathbf{h}^3}(2m^3)^{1/2}\beta^{-3/2}\int_0^\infty \frac{z^{1/2}\,\mathrm{d}z}{\mathrm{e}^z - 1}$$

$$= \frac{4\pi V\sqrt{\pi}}{2\mathbf{h}^3}(2m^3)^{1/2}\beta^{-3/2}\zeta(\tfrac{3}{2}), \tag{8.84}$$

where $\zeta(\frac{3}{2})$ is the Riemann zeta function†

$$\zeta(\tfrac{3}{2}) = \frac{2}{\sqrt{\pi}}\int_0^\infty \frac{x^{1/2}\,\mathrm{d}x}{\mathrm{e}^x - 1} = 2.612. \tag{8.85}$$

Equation (8.84) implies that the *largest* number of bosons that can be accommodated in a volume V is finite and proportional to $T^{3/2}$. According to Equation (8.84), no bosons can be accommodated in a volume V at absolute zero itself. This conclusion is in apparent contradiction to the known fact that any quantum state can be filled with an arbitrary number of bosons. On the other hand, Equation (8.83) asserts that the number of states in the energy interval $\mathrm{d}\varepsilon_s$ about ε_s is proportional to $\varepsilon_s^{1/2}\,\mathrm{d}\varepsilon_s$. As $\varepsilon_s \to 0$, this number vanishes and we must conclude that Equation (8.84) *fails to count any of the particles in the ground state*.

The paradox is resolved by recognizing that Equation (8.83), as well as the subsequent analysis, contain a *mathematical* approximation. Since the energy levels are discrete, the correct representation of the number of particles must occur in the form of the sum

$$N = \sum_{\varepsilon_s}\{\exp[\beta(\varepsilon_s - \mu)] - 1\}^{-1}$$

$$= \frac{1}{\exp(-\beta\mu) - 1} + \sum_{\varepsilon_s \neq 0}\frac{1}{\exp[\beta(\varepsilon_s - \mu)] - 1}. \tag{8.83a}$$

Equation (8.83) approximated this sum by an integral on the assumption that the discrete set of energy levels could be replaced by a continuum. We conclude that the correct representation in Equation (8.83a) may be put in the form

$$N = N_0 + N', \tag{8.86a}$$

† See, for example, E. Jahnke, F. Emde and F. Lösch, *Tables of Higher Transcendental Functions*, p. 37, McGraw-Hill, New York, 1960.

where N_0 is the number of particles in the ground state $\varepsilon_0 = 0$, and N' is the number of particles in states $\varepsilon_s > 0$. On passing to the integral representation, with its implied continuum of levels, we tacitly assume that the number of particles in any particular level, including $\varepsilon_0 = 0$, tends to zero. Thus it becomes clear that Equation (8.83) contains a proper approximation for the second sum in Equation (8.83a), that is for N', and that N_0 is lost in the count. For large negative values of μ, N_0 becomes very small and the particles contained in the ground state contribute only negligible quantities to the thermodynamic properties of the system. However, when $-\mu$ approaches zero, this contribution becomes essential. We can account for these circumstances by writing

$$N = \frac{1}{\exp(-\beta\mu) - 1} + \frac{4\pi V}{\mathbf{h}^3}(2m^3)^{1/2}\int_0^\infty \frac{\varepsilon_s^{1/2}\,d\varepsilon_s}{\exp[\beta(\varepsilon_s - \mu)] - 1}. \quad (8.86b)$$

Equation (8.86b) determines the variation of the chemical potential with temperature, T, and density, N/V, and hence, also, the remaining thermodynamic properties. The pressure variation with temperature along the isochore of Figure 8.5 was obtained in this way. The calculations are tedious and must be performed numerically.† For this reason we can do no more than to summarize the results for the reader. Below a certain temperature, T_0, which depends on density, the chemical potential is very close to zero and a large fraction of the total number, N, of particles resides in the ground state. At this temperature, the rate of change of specific heat with temperature, c_v, becomes discontinuous, as shown in Figure 8.8 (see also Figure 6.5), which indicates a

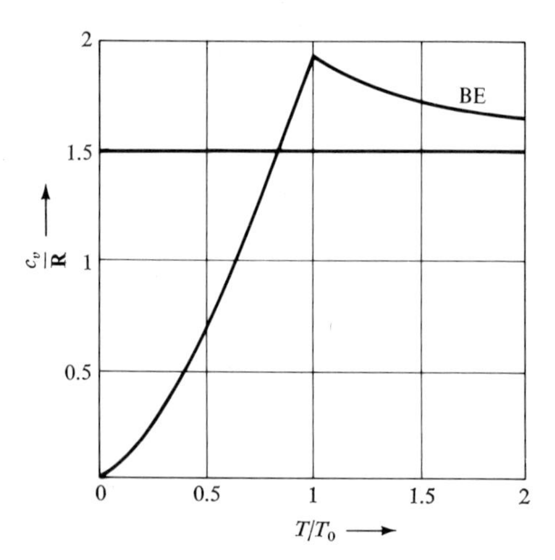

FIGURE 8.8. The specific heat, c_v, of a boson gas ($T_0 = 3.13$ K for $v_m = 27.6$ cm^3/gmol). [From F. London, *Superfluids* Vol. II, p. 41, Wiley, New York, 1954.]

† J. E. Robinson, *Phys. Rev.* **83** (1951), 678. See also E. W. Ng, C. J. Devine, and R. F. Tooper, *Math. Comp.* **23** (1969), 639.

third-order phase transition. In this context, the phase transition is commonly described as the Bose–Einstein condensation.

In order to understand, at least qualitatively, the reasons for this condensation, we refer to Equation (8.84) which represents the maximum number of particles that can exist in states $\varepsilon_s > 0$. At high values of temperature, this maximum becomes considerably larger than the number of particles, N, in the system, and we must infer that almost all of the particles exist in excited states. This situation corresponds to a solution of Equation (8.86) for large absolute values of μ (which is itself negative). As the temperature is reduced at constant density, eventually a temperature

$$T_0 = \frac{\mathbf{h}^2}{2\pi m\mathbf{k}}\left[\frac{N}{V\zeta(\frac{3}{2})}\right]^{2/3} \tag{8.87}$$

will be attained for which $N = N_{\max}$. Since the chemical potential at this temperature still differs from zero, the actual number of particles in the excited states is less than $N_{\max} = N$, and an exact evaluation shows that the number of particles in the ground state is still negligible, being given by

$$N_0 = N\left[1 - \left(\frac{T}{T_0}\right)^{3/2}\right] \qquad (T < T_0), \tag{8.88}$$

as can be seen upon comparing Equations (8.86) and (8.84) with $\mu \approx 0$. As the temperature is decreased below T_0 more and more particles "condense" into the ground state.

We can use a similar approximation ($\mu \approx 0$) to calculate the energy

$$E = \frac{4\pi V}{\mathbf{h}^3}(2m^3)^{1/2}\int_0^\infty \frac{\varepsilon_s^{3/2}\,d\varepsilon}{\exp(\beta\varepsilon_s) - 1} = 0.770\,N\mathbf{k}T\left(\frac{T}{T_0}\right)^{3/2}, \tag{8.89}$$

the specific heat

$$c_v = 1.925\,N\mathbf{k}\left(\frac{T}{T_0}\right)^{3/2}, \tag{8.90}$$

or the thermal equation of state

$$PV = 0.513\,N\mathbf{k}T\left(\frac{T}{T_0}\right)^{3/2}. \tag{8.91}$$

Again, in accordance with the Third law, we observe that $c_v \to 0$ as $T \to 0$. The isochore in Figure 8.5 and the specific heat in Figure 8.8 conform to these equations at the lower end of the temperature scale.

To complete the picture, in Figure 8.9 we have plotted the fraction N_0/N of particles in the ground state, as given by the approximate equation (8.88). The diagram in Figure 8.10 illustrates the variation of the chemical potential

TABLE 8.1

Properties of Ideal Degenerate Gases

Property	Boson gas weakly[a] degenerate	Fermi gas: Weakly degenerate	Fermi gas: Strongly[b] degenerate	Classical gas
Pressure	$\frac{N\mathbf{k}T}{V}[1-2^{-5/2}y+\cdots]$	$\frac{N\mathbf{k}T}{V}[1+2^{-5/2}y+\cdots]$	$\frac{2}{5}\frac{N}{V}\mu_0\left[1+\frac{5}{12}\left(\frac{\mathbf{k}T}{\mu_0}\right)^2+\cdots\right]$	$\frac{N\mathbf{k}T}{V}$
Energy	$\frac{3}{2}N\mathbf{k}T[1-2^{-5/2}y+\cdots]$	$\frac{3}{2}N\mathbf{k}T[1+2^{-5/2}y+\cdots]$	$\frac{3}{5}N\mu_0\left[1+\frac{5\pi}{12}\left(\frac{\mathbf{k}T}{\mu_0}\right)^2+\cdots\right]$	$\frac{3}{2}N\mathbf{k}T$
c_v	$\frac{3}{2}N\mathbf{k}[1+2^{-7/2}y+\cdots]$	$\frac{3}{2}N\mathbf{k}[1-2^{-7/2}y+\cdots]$	$\frac{\pi^2}{2}N\mathbf{k}\left(\frac{\mathbf{k}T}{\mu_0}\right)+\cdots$	$\frac{3}{2}N\mathbf{k}$
s	$N\mathbf{k}[\frac{5}{2}-\ln y-2^{-7/2}y+\cdots]$	$N\mathbf{k}[\frac{5}{2}-\ln y+2^{-7/2}y+\cdots]$	$\frac{\pi^2}{2}N\mathbf{k}\left(\frac{\mathbf{k}T}{\mu_0}\right)$	$N\mathbf{k}(\frac{5}{2}-\ln y)$

$$y=\frac{N}{V}\left(\frac{2\pi m\mathbf{k}T}{\mathbf{h}^2}\right)^{-3/2}, \qquad \mu_0=\frac{\mathbf{h}^2}{8m}\left(\frac{3N}{\pi V}\right)^{2/3}$$

[a] The strongly degenerate boson gas condenses. See Section 8.8.
[b] The formulas listed here pertain to fermions with spin $\frac{1}{2}$.

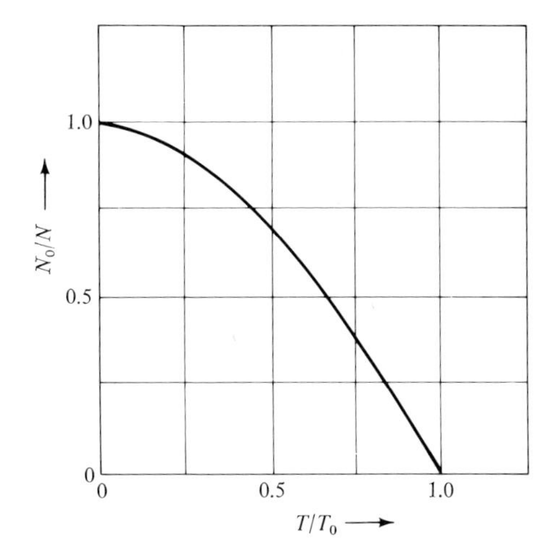

FIGURE 8.9. The number of bosons in the ground state as a function of temperature. [From F. London, *Superfluids*, Vol. II, p. 41, Wiley, New York, 1954.]

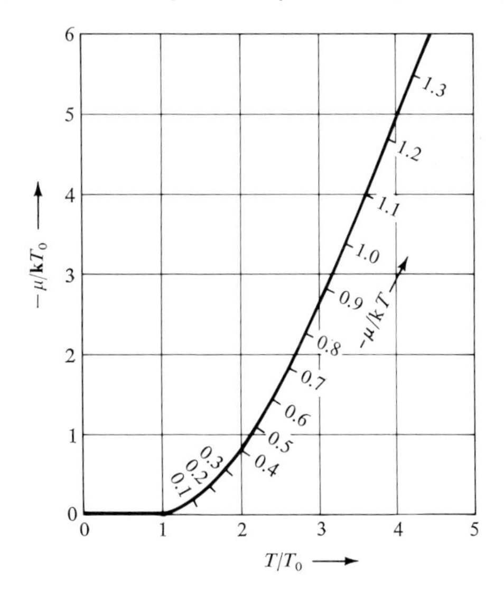

FIGURE 8.10 The chemical potential of a boson gas. [From R. Kubo, *Statistical Mechanics*, p. 249. North-Holland Publ., Amsterdam, 1965.]

as computed numerically from Equation (8.86). Finally, Figure 8.11 shows a network of isotherms drawn for $m = 7 \times 10^{-24}$ gr.

If we applied Equation (8.87) to ^{4}He, substituting $\rho = Nm/V = 0.146$ gr/cm^3 for liquid helium I, we would conclude that the λ transition should occur at $T_0 = 3.13$ K along the vapor-pressure curve. This is of the same

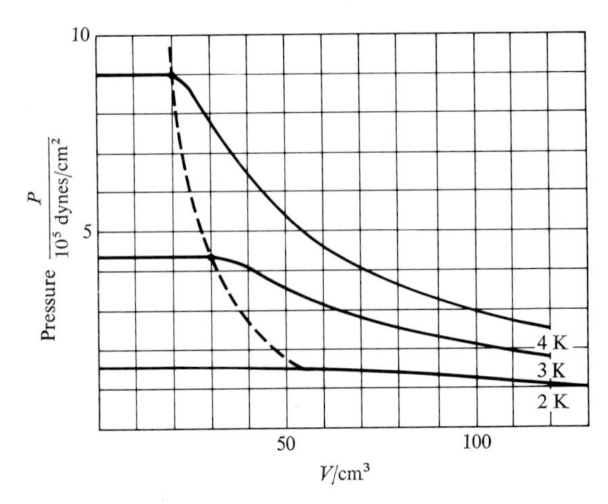

FIGURE 8.11. Isotherms for a perfect boson (^{4}He) at low temperature ($m = 7 \times 10^{-24}$ gr). [From B. Kahn, On the theory of the equation of state in *Studies in Statistical Mechanics*, Vol. III (J. de Boer and G. E. Uhlenbeck, eds.), North-Holland Publ., Amsterdam, 1965.]

order of magnitude as the actual transition temperature of 2.173 K quoted earlier. Good agreement must not be expected because liquid helium cannot be treated as a perfect gas. Nevertheless, the closeness of the two results convinces us that ^{4}He atoms behave like bosons at low temperatures. Moreover, experimental results on liquid ^{4}He lead to the empirical formula†

$$N_0 = N\left[1 - \left(\frac{T}{T_\lambda}\right)^{5.6}\right],$$

giving additional support to this contention. By contrast, the isotope ^{3}He, otherwise almost identical with ^{4}He, is a fermion. Hence no λ transition can be expected to occur in it, in full accord with observations.

PROBLEMS FOR CHAPTER 8

8.1. (Bose–Einstein statistics.) Particles with integral spin (bosons) satisfy the requirement that the wave function for a system of such particles is symmetric. Consequently the number of bosons, which are indistinguishable, present in a given quantum state is not restricted by the exclusion principle. The number of microstates per distribution was computed in Section 4.1.3.

(a) Show that $\Omega_v^{\text{BE}} = 1$ if $g_i = 1$ with $i = 1, 2, \ldots, k$.

† J. Wilks, *The Properties of Liquid and Solid Helium*, p. 53, Oxford Univ. Press, London and New York, 1967; or C. T. Lane, *Superfluid Physics*, p. 50, McGraw-Hill, New York, 1962.

(b) Applying the Stirling approximation together with the simplification that $g_i - 1 \approx g_i$, show that

$$\ln \Omega_\nu^{\mathrm{BE}} = \sum_{i=1}^{k} \left[g_i \ln\left(\frac{g_i + n_i^\nu}{g_i}\right) + n_i^\nu \ln\left(\frac{g_i + n_i^\nu}{n_i^\nu}\right)\right].$$

(c) Using the method of undetermined multipliers (or otherwise), show that for the most probable distribution we must have

$$\left(\frac{n_i^*}{g_i}\right)_{\mathrm{BE}} = \frac{1}{\mathrm{e}^\alpha \, \mathrm{e}^{\beta \varepsilon_i} - 1}.$$

Here α is the multiplier for N and $-\beta$ is the multiplier for E.

8.2. (Fermi–Dirac statistics.) Systems of particles with half-integral spins (fermions) must satisfy the requirement that their wave functions are antisymmetric (exclusion principle). The number of microstates per distribution was computed in Section 4.1.3.

(a) Show that $n_i^\nu \leq g_i$.

(b) Derive the relation

$$\ln \Omega_\nu^{\mathrm{FD}} = \sum_{i=1}^{k} \left[-g_i \ln\left(\frac{g_i - n_i^\nu}{g_i}\right) + n_i^\nu \ln\left(\frac{g_i - n_i^\nu}{n_i^\nu}\right)\right].$$

(c) Show that

$$\left(\frac{n_i^*}{g_i}\right)_{\mathrm{FD}} = \frac{1}{\mathrm{e}^\alpha \, \mathrm{e}^{\beta \varepsilon_i} + 1}.$$

(See Problem 8.1.)

8.3. (Corrected Maxwell–Boltzmann statistics.) Maxwell–Boltzmann statistics deal with a collection of distinguishable elements and are, therefore, quantum-statistically incorrect for gases. It was shown in Section 6.7 that this defect can be corrected by the introduction of the divisor $N!$, so that

$$\Omega_\nu^{\mathrm{CMB}} = \prod_{i=1}^{k} \frac{g_i^{n_i^\nu}}{n_i^\nu!}.$$

(a) Show that this correction has no effect on the most probable distribution

$$\left(\frac{n_i^*}{g_i}\right)_{\mathrm{CMB}} = \mathrm{e}^{-\alpha} \, \mathrm{e}^{-\beta \varepsilon_i}.$$

(b) Show that the corresponding BE and FD distributions tend to the above when $g_i \gg n_i^\nu$.

8.4. Show that the expressions for $\ln \Omega_\nu^{\mathrm{BE}}$ and $\ln \Omega_\nu^{\mathrm{FD}}$ from Problems 8.1 and 8.2 go over into the corresponding expression for $\ln \Omega_\nu^{\mathrm{CMB}}$ for $n_i^\nu \ll g_i$. Obtain this result by a series expansion in terms of the parameter $x = n_i^\nu / g_i$.

Explain why the condition

$$\frac{n_i^*}{g_i} \ll 1$$

is likely to be satisfied in a dilute gas.

8.5. (a) Apply Boltzmann's principle (discussed in Section 5.10) to show that

$$\frac{S}{\mathbf{k}} = \sum_i \mp g_i \ln[1 \mp \exp(-\alpha - \beta\varepsilon_i)] + N\alpha + \beta E, \qquad \frac{\text{BE}}{\text{FD}}$$

with the upper signs referring to BE and the lower sign referring to FD systems. (Sum over energy levels.)

(b) The equation in part (a) can be solved for $E = E(S, N, \alpha, \beta)$. However, as we know from classical thermodynamics, $E = E(S, V, N)$, which leads to the conclusion that both α and β must be regarded as functions of S, V, and N: $\alpha = \alpha(S, V, N)$; $\beta = \beta(S, V, N)$. Recognizing this, and, further, that $T = (\partial E/\partial S)_{V,N}$, differentiate the above equation *implicitly* with respect to S, holding V and N constant; hence, prove that $\beta = 1/\mathbf{k}T$ for both BE and FD statistics.

(c) Implicitly differentiate the expression for entropy with respect to N, holding S and V constant. Recognizing that $\mu = (\partial E/\partial N)_{V,S}$, where μ is the chemical potential, demonstrate that

$$\alpha = -\mu/\mathbf{k}T$$

for both BE and FD statistics.

(d) Show that if α is large enough

$$\mu = -\mathbf{k}T \ln \frac{Z}{N},$$

where Z is the single-particle partition function

$$Z = \sum g_i \, e^{-\beta\varepsilon_i}.$$

8.6. Having evaluated in Problem 8.5 the undetermined multipliers α and β for a system of bosons or fermions, the most probable distributions for these particles (Problems 8.1 and 8.2) become

$$\frac{n_i^*}{g_i} = \frac{e^{\beta(\mu - \varepsilon_i)}}{1 \mp e^{\beta(\mu - \varepsilon_i)}}, \qquad \frac{\text{BE}}{\text{FD}}$$

with $\beta = 1/\mathbf{k}T$.

(a) Show that when the condition

$$e^{\beta(\mu - \varepsilon_i)} \ll 1 \tag{1}$$

is satisfied, the most probable distributions for bosons and fermions transform to the corrected Maxwell–Boltzmann distribution (Problem 8.3). Note that condition (a) also implies that the system is dilute, or $n_i^*/g_i \ll 1$, so that this result is in accord with the result in Problem 8.4.

(b) Utilizing the result of part (d) of Problem 8.5, show that condition (1) is satisfied when

$$N/Z \ll 1. \tag{2}$$

(c) Apply condition (2) to a perfect monatomic gas whose partition function is given by Equation (6.12) in the text ($Z = Z_{\text{tr}}^3$); hence, derive the criterion given by Equation (6.11a) for the applicability of corrected Maxwell–Boltzmann statistics to a perfect monatomic gas.

8.7. Consider a system consisting of two particles each of which can be in any one of three states 0, ε, and 3ε. The system is placed in contact with a heat reservoir at temperature T. Compute the partition function, $\mathscr{Z}$, for (a) distinguishable Maxwell–Boltzmann particles; (b) fermions; (c) bosons.

8.8. Consider a perfect boson or fermion gas composed of particles with spin s. Carry through the arguments of Section 8.6 to obtain expansions for the properties of these systems when they are weakly degenerate. Show that Equations (8.55) and (8.56) retain their form if y is replaced by $y' = y/g$ where $g = 2s + 1$. How should the formulas listed in Table 8.1 be modified for these systems?

8.9. Compute the grand canonical partition function for a collection of noninteracting bosons, and for a collection of noninteracting fermions. Obtain an expression for the average number of particles in the system $\langle N \rangle$, and show that the pressure, P, may be expressed as

$$P = \pm (\mathbf{k}T/V) \sum_{s=1}^{k} \ln\{1 \pm \exp[\beta(\mu - \varepsilon_s)]\}, \qquad \frac{\text{FD}}{\text{BE}}.$$

SOLUTION: We begin by considering a collection of noninteracting bosons. The grand canonical partition function Ξ_{BE} may be written, with the aid of Equations (5.71), (5.74b), and (8.13), and Problem 5.15, as

$$\begin{aligned}\Xi_{\text{BE}} &= \sum_{N=0} \mathscr{Z}_{\text{BE}}^{(N)} \exp(\beta\mu N) \\ &= \sum_{N=0}^{\infty} \sum_{n_1=0}^{\infty} \sum_{n_2=0}^{\infty} \cdots \sum_{n_k=0}^{\infty} \delta\left(N, \sum_{s=1}^{k} n_s\right) \exp\left(-\beta \sum_{s=1}^{k} \varepsilon_s n_s + \beta\mu N\right).\end{aligned}$$

If we interchange the sum over N and the k summations over n_i $i = 1, \ldots, k$, we obtain

$$\Xi_{\text{BE}} = \sum_{n_1=0}^{\infty} \cdots \sum_{n_k=0}^{\infty} \exp\left[-\beta \sum_{s=1}^{k} (\varepsilon_s - \mu) n_s\right].$$

The summations are now independent and may be evaluated as summations over a geometrical series, provided $\varepsilon_s > \mu$ for all s. Then

$$\Xi_{\text{BE}} = \prod_{s=1}^{k} \{1 - \exp[\beta(\mu - \varepsilon_s)]\}^{-1}.$$

In a similar way we may obtain Ξ_{FD} as

$$\Xi_{\text{FD}} = \sum_{N=0}^{\infty} \sum_{n_1=0}^{1} \sum_{n_2=0}^{1} \cdots \sum_{n_k=0}^{1} \delta(N, \sum n_s) \exp(-\beta \sum_s \varepsilon_s n_s + \beta N\mu)$$

$$= \sum_{n_1=0}^{1} \cdots \sum_{n_k=0}^{1} \exp\left[-\beta \sum_{s=1}^{k} (\varepsilon_s - \mu) n_s\right]$$

$$= \prod_{s=1}^{k} \{1 + \exp[\beta(\mu - \varepsilon_s)]\}.$$

Using the relation $PV = \mathbf{k}T \ln \Xi$, we obtain the expression for P quoted above. Moreover,

$$\langle N \rangle = \sum_s \bar{n}_s = \sum_s \{\exp[\beta(\varepsilon_s - \mu)] \pm 1\}^{-1}, \qquad \frac{\text{FD}}{\text{BE}}$$

8.10. Show that

$$P - \langle N \rangle \mathbf{k}T/V = \begin{cases} \leq 0 & \text{for a perfect boson gas} \\ \geq 0 & \text{for a perfect fermion gas.} \end{cases}$$

[H. Falk, *Am. J. Phys.* **36** (1968), 454.]

SOLUTION: We may use the expressions for P and $\langle N \rangle$ given by the previous problem. First we consider the case of noninteracting bosons. Then

$$P = (\mathbf{k}T/V) \sum_{s=1}^{k} \ln(1 + b_s),$$

and $\langle N \rangle = \sum_{s=1}^{k} b_s$, where

$$b_s = [\exp \beta(\varepsilon_s - \mu) - 1]^{-1}.$$

Then $P - \langle N \rangle \mathbf{k}T/V = (\mathbf{k}T/V) \sum_s (\ln(1 + b_s) - b_s)$. Now $b_s \geq \ln(1 + b_s)$ whenever $b_s > -1$ and since, in our case $b_s \geq 0$, we have proved the statement in the case of bosons.

For fermions we write $P = -(\mathbf{k}T/V) \sum_{s=1}^{k} \ln(1 - f_s)$ and $\langle N \rangle = \sum_{s=1}^{k} f_s$, where

$$f_s = [\exp \beta(\varepsilon_s - \mu) + 1]^{-1}.$$

Then

$$P - \langle N \rangle \mathbf{k}T/V = -(\mathbf{k}T/V) \sum_{s=1}^{k} [\ln(1 - f_s) + f_s].$$

For $0 \leq f_s \leq 1$, the sum $\ln(1 - f_s) + f_s$ is always negative. This proves the second part of the assertion.

8.11. The pressure of a noninteracting boson or fermion gas is given in the canonical ensemble by Equation (8.34b) as

$$P = \sum_s -(\partial \varepsilon_s / \partial V)[\exp \beta(\varepsilon_s - \mu) \pm 1]^{-1}, \qquad \frac{\text{FD}}{\text{BE}}$$

and in the grand canonical ensemble by

$$P = \pm (\mathbf{k}T/V) \sum_{s=1}^{k} \ln[1 \pm \exp \beta(\mu - \varepsilon_s)], \qquad \frac{\text{FD}}{\text{BE}}.$$

What is the relation between these two expressions for the pressure? Do they agree if expression (8.36) is used for the energy levels ε_s?

8.12. Calculate the mean and maximum velocities of an electron in an electron gas at $T = 0$ K.

8.13. What is the pressure of the electron gas in silver at $T = 0$ K?

8.14. Give numerical estimates of the Fermi energy, μ_0, and of $T_f = \mu_0/\mathbf{k}$ for (a) electrons in a typical metal such as Ag or Cu, (b) nucleons, for example, neutrons and protons, in a heavy nucleus, (c) ^{3}He atoms in ^{3}He gas, in which the volume available to each atom is 46 Å^3. Treat the particles as free fermions.

8.15. Show that there is no Bose–Einstein condensation for hypothetical one- and two-dimensional perfect boson gases.

8.16. Show that the equation of state of an ideal boson gas may be written in the form of the virial expansion

$$\frac{Pv}{\mathbf{k}T} = 1 - \left(\frac{1}{8}\sqrt{2}\right)\left(\frac{\lambda^3}{v}\right) + \left(\frac{1}{8} - \frac{2\sqrt{3}}{27}\right)\left(\frac{\lambda^3}{v}\right)^2 + \cdots$$

where

$$\lambda = \left(\frac{\mathbf{h}^2}{2\pi m\mathbf{k}T}\right)^{1/2}.$$

8.17. Calculate the third virial coefficient for an ideal Fermi gas. How does it compare with that of an ideal boson gas obtained in Problem 8.16?

8.18. The expression $c_v/T = 6.11 \times 10^{-4}$ J/gmol K^2 describes to a good approximation the measurements of the electronic specific heat of silver. Using this expression, calculate the value of the Fermi energy and the effective number of free electrons per atom for silver. The atomic weight of silver is $M = 107.88$ kg/kg atom, and its density is $\rho = 10.5$ gr/cm^3.

8.19. Compute the ratio of the de Broglie wavelength, λ, to the crystal spacing, d, for a metal at $T = 0$ K, assuming that there exists one free electron per atom.

8.20. Prove that the specific heat, c_v, for an ideal, two-dimensional boson gas is identical with that for an ideal, two-dimensional fermion gas. [*Reference:* R. M. May, *Phys. Rev.* **135** (1964), A1515.]

8.21. Compute the isothermal compressibility of a strongly degenerate perfect fermion gas.

8.22. The electron density in a white dwarf star is $z = 10^{30}/\text{cm}^3$, and the temperature of the star is $T = 10^7$ K. The electrons in the star move with speeds near the velocity of light, **c**, and in these circumstances it is permitted to write the energy of an electron as $\varepsilon = \mathbf{c}p$, where p is the magnitude of the electron's momentum. Assuming that this relation is valid for all values of p, compute the electron pressure, P, in the star.

8.23. Derive Equation (8.88) in the text.

8.24. The density of liquid helium is $\rho = 0.146$ gr/cm³. Calculate the temperature, T, of the Bose–Einstein condensation for an ideal boson gas at this density.

LIST OF SYMBOLS FOR CHAPTER 8

Latin letters

A_n	Definite integral related to Riemann zeta function
a	Constant
b	Constant
c_v	Specific heat at constant volume
D	Density of quantum states
E	Energy
E_j	Energy of system in quantum state j
F	Helmholtz function
f	Arbitrary function
h	Planck's constant
I	Definite integral
k	Number of single particle energy levels
k'	Total number of microstates
k	Boltzmann's constant
m	Mass of a particle
$\mathbf{m}_e$	Mass of electron
N	Number of particles
N_0	Number of particles in ground state
N	Avogadro's number
n	Number of particles in a quantum state
P	Pressure
p	Linear momentum with components p_x, p_y, p_z
R	Universal gas constant
S	Entropy
s_x, s_y, s_z	Quantum numbers for a free particle
T	Temperature
T'	Reference temperature for a quantum gas at which the degeneracy parameter is unity
T_0	Condensation temperature of an ideal boson gas
T_f	Fermi temperatures ($\mu_0/\mathbf{k}$)

T_λ	Temperature of lambda point in liquid helium
V	Volume
v_m	Molar volume
y	Degeneracy parameter
Z	Single-particle partition function
$\mathscr{Z}$	Partition function for a collection of particles
$\mathscr{Z}_N^{(b)}$	Restricted partition function for N particles
z	Fugacity $(=\mathrm{e}^{\alpha}=\mathrm{e}^{-\beta\mu})$

Greek letters

α	Ratio of partition functions, $\mathscr{Z}_N/\mathscr{Z}_{N-1}$
β	$(\mathbf{k}T)^{-1}$
δ	Kronecker delta function
ε	Single-particle energy level
ζ	Riemann zeta function
θ	Angle
λ	Wavelength; normal liquid to superfluid phase transition in liquid helium
λ_{th}	Thermal de Broglie wavelength
μ	Chemical potential; six-dimensional phase space
μ_0	Fermi energy
ρ	Mass density
ϕ	Angle
Ω	Number of microstates

Superscripts

$'$ $''$ $'''$	Derivatives
$\bar{}$	Average values

Subscripts

BE	Denotes Bose–Einstein statistics
FD	Denotes Fermi–Dirac statistics
p	Denotes momentum space

Special symbol

$\frac{\mathrm{FD}}{\mathrm{BE}}$	Upper (lower) signs apply to fermions (bosons)

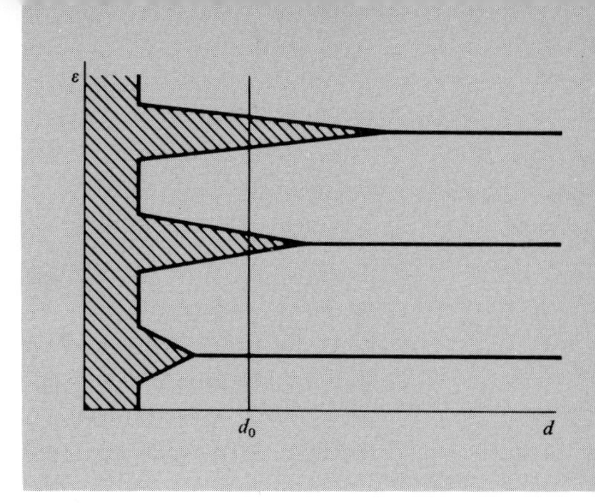

CHAPTER 9

PROPERTIES OF SOLIDS

9.1. Prefatory Remarks

In this chapter we shall consider the applications of quantum statistics to crystalline solids, and devote our attention to two aspects of solid-state physics. These will be concerned with the evaluation of the contribution to the thermodynamic properties of solids made by the motion of the crystal lattice, that is, by the lattice vibrations which result from the displacement of the center of mass of each atom in the crystal from its position in mechanical equilibrium. Secondly, we shall calculate the contribution due to the electrons in the solid.

Each of these types of motion, the nuclear and the electronic, gives rise to its own characteristic effects. The lattice vibrations, for example, cause the specific heat to vary as T^3 at low temperatures. The theory of the low-temperature specific heats of a crystal, due to A. Einstein and P. Debye, marked one of the early triumphs of quantum theory. In fact, the theory formulated by Einstein was the first example of the application of quantum ideas to material systems, as opposed to radiation. The electronic properties of the crystal most easily manifest themselves in its electrical conductivity in an externally applied electric field. The distinction between insulators, semiconductors, and conductors can be understood by studying the energy levels available to the electrons in the solid and the occupation numbers of these levels according to Fermi–Dirac statistics.

We shall begin by analyzing the motion of perfect crystalline solids and proceed to present the simpler theory due to Einstein. From this, we shall develop Debye's more advanced theory of the thermodynamic properties of a crystal. As an application of the Debye theory, we shall calculate the specific heats and the coefficient of thermal expansion of crystals. In the course of this study, we shall introduce the reader to a new entity, the *phonon*, which is the quantum-mechanical particle associated with the vibrational waves that propagate in the crystal.

Next, we shall give an exposition of the band theory of solids, which will provide us with a description of the differences in electrical and thermodynamic properties that exist among solids. Finally, as an example of the electron theory of metals, we shall study thermal emission.

9.2. The Properties of Crystalline Solids

Recapitulating our earlier references, we recall that an unstrained, crystalline solid subjected to unifom pressure is pictured in the form of a *regular* array of atoms, the regularity being provided by the crystal lattice. The existence of impurities or dislocations exerts a negligible effect on the thermodynamic properties of interest at the moment, but their effect on the mechanical† or electrical properties‡ may be considerable. Each atom is assumed to perform oscillations, known as *lattice vibrations*, with respect to an equilibrium position in the lattice; the elastic restoring force is provided by all other atoms. This field is dominated by the "nearest neighbors," but the effect of "long-range" interactions is present, thus causing the whole system to behave like a giant, single molecule. Although the atoms are free to diffuse, the states associated with definite lattice points are localizable, and the elements of the system are regarded as distinguishable.

Even with these simplifying assumptions, the Schrödinger equation cannot be solved in all its generality, and we are forced to introduce additional, heuristic approximations. The introduction of such simplifications may still lead to considerable complexity of detail which is beyond the scope of this course (but not of contemporary science). For this reason, we shall illustrate

† See J. Kestin and J. Rice, Paradoxes in the application of thermodynamics to strained solid materials. *A Critical Review of Thermodynamics*, (E. B. Stuart, B. Gal-Or, and A. J. Brainard, eds.) Mono Book Corp., Baltimore, 1970, p. 275.

Comprehensive treatments: A. H. Cottrell, *Dislocations in Plastic Flow in Crystals*, Oxford Univ. Press, London and New York, 1953; J. C. Fisher, W. G. Johnston, R. Thomson, and T. Vreeland, Jr., eds. *Dislocations and Mechanical Properties of Crystals*, Wiley, New York, 1957; A. R. Verma, *Crystal Growth and Dislocations*, Academic Press, New York, 1953.

‡ A. H. Wilson, *The Theory of Metals*, Cambridge Univ. Press, London and New York, 1953.

the procedure on the example of the simplest theory due to A. Einstein† and then the more complex and successful theory due to P. Debye,‡ as already stated. Both theories assume that the oscillations are of small amplitude and, therefore, harmonic, and confine themselves to isotropic solids made of atoms of one kind, that is, essentially to the solid phases of pure elements. In spite of this limitation, the results are often used to correlate the properties of more complex materials. The simplifying feature of these theories is the assumption that all subsystems of the crystal, that is, its normal modes of vibration, are statistically independent. It follows that the partition function can now be represented in the form of a product of K independent, but not necessarily identical, partition functions—one for each statistically independent normal mode. Thus,

$$\mathscr{Z} = \prod_{k=1}^{\mathbf{K}} Z_k \tag{9.1a}$$

where

$$Z_k = \sum_{v=1}^{k'} \exp[-\varepsilon_{vk}/\mathbf{k}T] \tag{9.1b}$$

as shown in detail in Section 5.12. Here k' denotes the number of energy levels (denoted by k in Section 5.12) over which the sum must extend. However, since the exponential terms decrease rapidly, we may put $k' \to \infty$. Every independent element is assumed to consist of a single harmonic oscillator. The class of available energy levels of such an oscillator is given by the equation

$$\varepsilon_{vk} = (\tfrac{1}{2} + v)\mathbf{h}\nu_k \qquad (v = 0, 1, 2, \ldots) \tag{9.2}$$

as is known from Section 3.11.

The problem now reduces to the formulation of heuristic, physically "reasonable" assumptions which determine the number of systems K and the range of frequencies ν_k. For each frequency ν_k, there exists an infinity of energy levels ε_{vk} which arises when the quantum number v ranges over the integers 0, 1, ...

The theory due to Einstein assumes that $K = 3\mathbf{N}$, that is, that there are as many identical systems as there are degrees of freedom in one mol of the system—hence the Avogadro number, **N**. The additional assumption is then made that ν_k is the same for all degrees of freedom. This is the simplest

† *Ann. Phys.* **22** (1907), 180; **34** (1911), 170.

‡ *Ann. Phys.* **39** (1912), 789. See also M. Born and Th. von Kármán, *Z. Phys.* **13** (1912), 297. A comprehensive presentation of the whole subject is given in C. Kittel, *Introduction to Solid-State Physics*, Wiley, New York, 1956. See also R. A. Smith, *Wave Mechanics of Crystalline Solids*, Wiley, New York, 1961.

possible set of assumptions. The theory due to Debye assumes that K is equal to the number of normal modes of the system of $\mathbf{N}$ atoms treated as an elastic continuum and heuristically prescribes a *range* of frequencies ν_k.

The Einstein Solid

Corresponding to the preceding set of assumptions, an *Einstein solid* is imagined to consist of N atoms which are free to perform small harmonic oscillations completely independently of each other. It is postulated that the motion of each atom can be decomposed into three harmonic oscillations of identical frequency ν_E, each performed in one of three mutually perpendicular directions of a Cartesian system of coordinates. This reduces the class of available quantum states, Equation (9.2), to a single infinity:

$$\varepsilon_v = (\tfrac{1}{2} + v)\mathbf{h}\nu_E \qquad (v = 0, 1, 2, \ldots). \tag{9.3}$$

Hence Equation (9.1a) assumes the form

$$\mathscr{Z} = Z^{\mathbf{N}}, \tag{9.4a}$$

and the *partition function of one atom* is

$$Z = \left\{\sum_{v=0}^{\infty} \exp[-(\tfrac{1}{2} + v)\mathbf{h}\nu_E \beta]\right\}^3 \qquad (\beta = 1/\mathbf{k}T),$$

or

$$Z = \exp(-\tfrac{3}{2}\beta\mathbf{h}\nu_E)\left[\sum_{v=0}^{\infty} \exp(-v\mathbf{h}\nu_E \beta)\right]^3. \tag{9.4b}$$

The infinite sum can be evaluated explicitly if it is noted that it constitutes a geometric progression whose terms tend to zero as $v \to \infty$. To do this expeditiously, it is convenient to introduce two abbreviations. First, we notice that the quantity

$$\Theta_E = \frac{\mathbf{h}\nu_E}{\mathbf{k}} \tag{9.4c}$$

has the dimension of a temperature and represents a constant for each particular substance; it is called the *characteristic* (Einstein) *temperature* of the solid. Secondly, we notice that the dimensionless exponent can be written

$$\xi = \beta\mathbf{h}\nu_E = \frac{\mathbf{h}\nu_E}{\mathbf{k}T} = \frac{\Theta_E}{T}. \tag{9.4d}$$

Thus, the infinite sum in Equation (9.4b) can be represented as

$$\sum_{v=0}^{\infty} \exp(-v\xi) = \frac{1}{1 - \exp(-\xi)}$$

and

$$\ln \mathscr{Z} = \mathbf{N}\{-\tfrac{3}{2}\,\xi - 3\ln[1-\exp(-\xi)]\}\,. \tag{9.5}$$

Application of Equation (5.52) leads us immediately to the molar internal energy

$$u = \tfrac{3}{2}\mathbf{R}\Theta_{\mathrm{E}} + 3\mathbf{R}T\,\frac{\Theta_{\mathrm{E}}/T}{\exp[\Theta_{\mathrm{E}}/T] - 1}\,. \tag{9.6}$$

Here we substituted the expression from Equation (9.4d), and recognized that $\mathbf{kN} = \mathbf{R}$. The first term in this equation, the zero-point energy, represents a constant which is of no particular importance because an arbitrary constant may always be added to internal energy. The second term represents the variation of the internal energy of a crystal with temperature.

By differentiation with respect to temperature, we obtain the molar specific heat

$$c_v = 3\mathbf{R}\mathscr{E}(\xi)\,. \tag{9.7}$$

where

$$\mathscr{E}(\xi) = \frac{\xi^2\,\mathrm{e}^{\xi}}{(\mathrm{e}^{\xi} - 1)^2} \tag{9.7a}$$

is known as the *Einstein function*, and where ξ, defined in Equation (9.4d), is the reciprocal dimensionless temperature.

The Einstein function is of marginal importance in the theory of crystalline solids because Einstein's theory has been superseded by the more successful theory due to Debye. However, it plays a part in the theory of gases as we know from Equation (6.45), and for this reason we have included a short table, Table III, at the end of the book.

Equation (9.7) shows that the specific heat, c_v, of an isotropic solid crystal depends on the ratio of its temperature T to the characteristic temperature Θ_{E} only, and that it is independent of volume or pressure. Moreover, if the ratio T/Θ_{E}, or its reciprocal ξ, is employed as the independent variable, the relation of c_v to ξ is universal, that is, the same for all solids. Two solids 1 and 2 which are at temperatures T_1 and T_2 yielding the same values of ξ, so that

$$\frac{\Theta_{\mathrm{E}1}}{T_1} = \frac{\Theta_{\mathrm{E}2}}{T_2},$$

are said to be in corresponding states, and Equation (9.7) expresses the *principle of corresponding states* for Einstein solids.

To form an idea of the temperature dependence represented by the Einstein function, it is useful to investigate its behavior in the two limiting cases when $\xi \to 0$ and $\xi \to \infty$. The case when $\xi \to 0$ corresponds to very high temperatures, $T \gg \Theta_E$, as seen from Equation (9.4d). The opposite case, $\xi \to \infty$, corresponds to very low temperatures, $T \ll \Theta_E$, including absolute zero in the limit. As $\xi \to 0$, we have

$$\mathscr{E}(\xi) \to 1 \qquad \text{as} \quad \xi \to 0 \qquad (T \gg \Theta_E),$$

whereas as $\xi \to \infty$, we obtain

$$\mathscr{E}(\xi) \approx \xi^2 e^{-\xi} \to 0 \qquad \text{as} \quad \xi \to \infty \qquad (T \ll \Theta_E).$$

The latter expression can be obtained by multiplying the numerator and denominator of Equation (9.7a) by $\exp(-2\xi)$ and neglecting $\exp(-\xi)$ with respect to unity in the latter.

The universal equation for the specific heat of an Einstein solid is seen plotted in Figure 9.1 in terms of the ratio $\xi^{-1} = T/\Theta_E$, that is, effectively, in

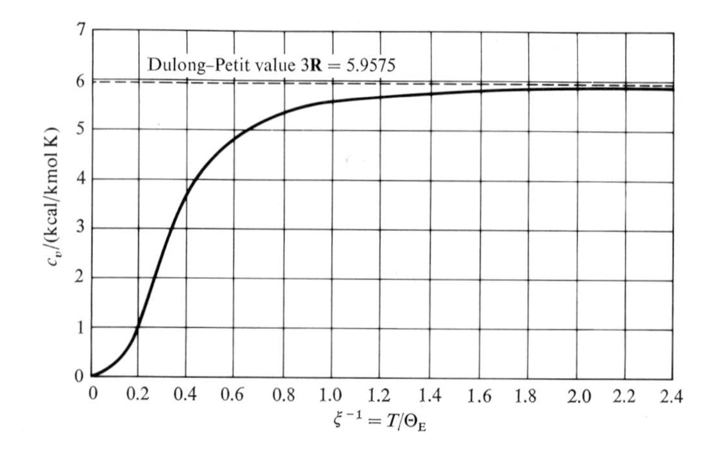

FIGURE 9.1. The universal law of corresponding states for an Einstein solid, Equation (9.7).

terms of temperature. At temperatures which are very high compared with the characteristic temperature, Θ_E, the specific heat of all solids tends to a constant

$$c_v = 3\mathbf{R} = (24.943 \pm 0.001)\frac{\text{J}}{\text{kmol K}} = 5.9575\frac{\text{kcal}}{\text{kmol K}}. \qquad (9.7\text{b})$$

This is, in fact, observed, and the corresponding discovery was first made empirically by P. L. Dulong and A. T. Petit as early as 1819, long before the advent of statistical thermodynamics. For this reason, the preceding statement is known as the *Dulong–Petit rule*. Except for very hard crystals, notably diamond, the characteristic temperature Θ_E is rather low, and the Dulong–Petit value of $3\mathbf{R}$ is measured at normal temperatures, departures occurring only at cryogenic temperatures.

The Dulong–Petit rule was confirmed by classical mechanics which predicted just this constant value for c_v; however, it failed to account for the existing dependence of the specific heat on temperature at low enough temperatures. Historically, this failure was fundamental for statistical thermodynamics, and Einstein's theory, completed in 1907, played an important part in the absorption of quantum mechanics into its fabric. The Dulong–Petit rule provides an example of the *Principle of equipartition* of energy among degrees of freedom. We have discussed this principle in Section 6.5.

At very low temperatures, the theory leads to the conclusion that the specific heat, c_v, of all solid crystals tends to zero as $(\Theta_E/T)^2 \exp(-\Theta_E/T)$. The vanishing of the specific heat as $T \to 0$ is expected from the Third law, but experiments at very low temperatures suggest that the rate of decrease in c_v is different, pointing to an inadequacy in the theory.

The constant frequency, ν_E, in Einstein's theory cannot be determined from any principles and must be deduced for each solid by fitting the results of measurements to Equation (9.7). If the theory were adequate, a single value of frequency ν_E, that is, a single value of the characteristic temperature Θ_E, chosen for every solid separately, should correlate measurements on all solids into one universal expression. In this respect, the theory is only partially successful; experiments point to the existence of a universal relation but of a different form, as will be discussed in the next section.

Since c_v is a function of temperature alone, the theory is consistent with the approximation that solids are incompressible. The thermal equation of state could be derived with the aid of Equation (5.51a), but since $\ln \mathscr{Z}$ does not contain the volume V we obtain the singular result $P = 0$.

9.3. Debye's Theory

An improved approximation to the properties of solids was furnished by Debye in 1912. Although further refinements have been obtained in more recent times† at the cost of increased complication, Debye's theory is adequate for most practical applications to a large class of solids.

We recall from Section 2.3.6 that every low-amplitude vibrational motion of a complex mechanical system, such as a block of solid consisting of $\mathbf{N}$ interacting atoms, can be represented as a sum of independent harmonic oscillations, each corresponding to one *natural* or *normal mode* of vibration.

† G. F. Newell, *J. Chem. Phys.* **21** (1953), 1877.

The number of normal modes, each characterized by a different frequency, is equal to the number of degrees of freedom of the system. Deducting six degrees of freedom, three each for translation and rotation of the block as a whole, we are left with $f = 3\mathbf{N} - 6$. However, owing to the large magnitude of $\mathbf{N}$, we may put $f \approx 3\mathbf{N}$. In this manner, for statistical purposes, it is possible to regard the system as a collection of $K = 3\mathbf{N}$ harmonic oscillators, each of frequency ν_k, leading to the group of energy levels anticipated in Equation (9.2). Now the independent elements are no longer associated with lattice sites but with normal modes. Consequently, the partition function is obtained in the form of a product (9.1a) of partition functions

$$Z_k = \frac{\exp(-\frac{1}{2}\beta \mathbf{h}\nu_k)}{1 - \exp(-\beta \mathbf{h}\nu_k)} \tag{9.8}$$

of the individual normal modes.

9.3.1. The Vibrational Motion of an Elastic Solid

To proceed further and to evaluate the partition function we must know the normal-mode frequencies of the lattice. As an example of a very simple lattice, let us consider a hypothetical one-dimensional crystal consisting of a chain of particles of mass m bound by elastic springs of force constant k to their nearest neighbors. We suppose, further, that the equilibrium distance between neighbors is a and that the equilibrium position of the nth atom is na. If y_n denotes the displacement of the nth atom from its equilibrium position, then the equation of motion for y_n is

$$m \frac{d^2 y_n}{dt^2} = k(y_{n+1} - y_n) - k(y_n - y_{n-1}). \tag{9.9a}$$

To obtain the normal-mode vibrations for this lattice, we seek solutions of the form

$$y_n = A \cos 2\pi(\nu t - na/\lambda), \tag{9.9b}$$

each of which represents a wave propagating in the lattice with a frequency ν and a wavelength λ. Upon substitution, we obtain a *dispersion relation*, that is, a relation between the velocity and the wavelength of a normal mode. In this case the dispersion relation is

$$\nu = \frac{1}{\pi}\left(\frac{k}{m}\right)^{1/2} \sin\frac{\pi a}{\lambda}. \tag{9.10}$$

We see, first of all, that there is a maximum allowed normal-mode frequency, ν_m, equal to $\pi^{-1}(k/m)^{1/2}$. The velocity of propagation is

$$c = \nu\lambda = \frac{1}{\pi}\left(\frac{k}{m}\right)^{1/2} \lambda \sin\frac{\pi a}{\lambda}. \tag{9.11}$$

For waves for which the wavelength is much greater than the lattice spacing ($\lambda \gg a$), the velocity is independent of wavelength and is given by

$$c = a\left(\frac{k}{m}\right)^{1/2} \qquad (\lambda \gg a). \tag{9.11a}$$

In order to find the actual characteristic vibration of the lattice, we need to establish the boundary conditions imposed at the ends of the chain. The simplest conditions, mathematically, are so-called periodic boundary conditions expressed as

$$y_{n+\mathbf{N}} = y_n. \tag{9.12}$$

These lead to the equation

$$\lambda_l = \frac{\mathbf{N}a}{|l|} \tag{9.13}$$

where l is an integer, which may be positive or negative. For even $\mathbf{N}$

$$-\tfrac{1}{2}\mathrm{N} \leq l \leq \tfrac{1}{2}\mathrm{N}. \tag{9.13a}$$

The corresponding frequencies are then

$$\nu_l = \frac{1}{\pi}\left(\frac{k}{m}\right)^{1/2} \sin\frac{\pi|l|}{\mathbf{N}}. \tag{9.14}$$

The physical significance of the periodic boundary conditions can be grasped if we imagine that the linear chain has been bent into a circle of large radius and, hence, negligible curvature, the two ends being joined together, as shown in Figure 9.2. This leads to $y_1 = y_{\mathbf{N}+1}$ and hence to Equation (9.12). It is implied that the normal vibrations of such a closed linear chain are not materially different from those of a free chain.

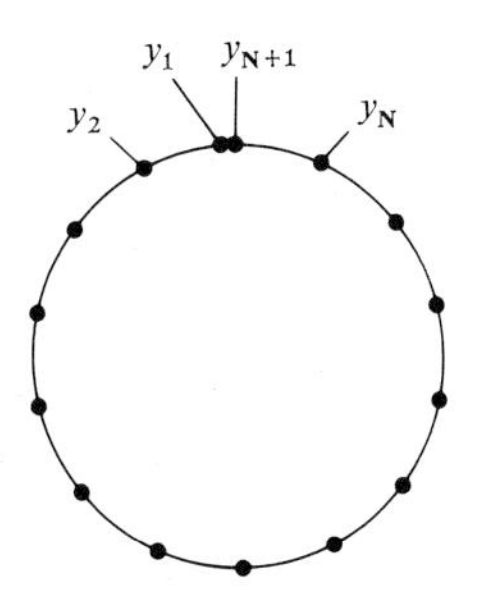

FIGURE 9.2 Significance of periodic boundary conditions.

As far as wavelengths which are small compared to the size of the chain, $\lambda \ll Na$, are concerned, we intuitively expect that the effect of the boundary conditions is unimportant, since only waves for which λ is of the order of the size of the crystal should "feel" the effects of the boundaries.

A finite three-dimensional crystal also possesses a set of frequencies whose values depend on the crystal structure as well as on the boundary conditions. Debye's theory is based on the assumption that the velocity of propagation of the normal-mode waves is independent of frequency and that the frequencies are insensitive to the boundary conditions. Consequently, periodic boundary conditions may be imposed on the crystal without affecting their usefulness when applied to a crystal with fixed boundaries. The one-dimensional example shows that these assumptions are valid for wavelengths larger than the lattice spacing and smaller than the crystal size. Thus Debye's assumptions amount to neglecting the effects of the details of the crystal structure as well as those of the crystal boundaries.

We imagine a monochromatic wave of wavelength λ propagating in a cube of side L where the displacement at any point $\mathbf{r} = (x, y, z)$ in the cube is given by the real part of

$$\mathbf{Y}(x, y, z) = \mathbf{A} \exp[2\pi i(vt - \mathbf{f} \cdot \mathbf{r})]\,; \tag{9.15}$$

here $\mathbf{f}$ is a vector in the direction of propagation of the wave, its magnitude being λ^{-1}. The periodic boundary condition requires that

$$\mathbf{Y}(x + L, y, z) = \mathbf{Y}(x, y + L, z) = \mathbf{Y}(x, y, z + L) = \mathbf{Y}(x, y, z)\,, \tag{9.16}$$

leading to

$$\mathrm{f}_x L = n_1, \qquad \mathrm{f}_y L = n_2\,, \qquad \mathrm{f}_z L = n_3\,, \tag{9.16a, b, c}$$

where n_1, n_2, and n_3 are integers of either sign. The wavelength λ is

$$\lambda = |\mathbf{f}|^{-1} = L(n_1{}^2 + n_2{}^2 + n_2{}^3)^{-1/2}\,, \tag{9.16d}$$

Assuming a constant velocity of propagation, c, the frequencies of the vibrations are

$$v = c\lambda^{-1} = \left(\frac{c}{L}\right)(n_1{}^2 + n_2{}^2 + n_3{}^2)^{1/2}\,. \tag{9.17}$$

For the same reasons as those which exist in the linear chain discussed earlier, a three-dimensional crystal must also possess a maximum frequency, v_m. This puts an upper limit on the normal modes present, and its value is determined by the requirement that the total number of degrees of freedom is 3N.

Owing to the large number of frequencies v_k, we can approximate the 3N values by a continuous spectrum, as suggested by Figure 9.3. In the figure,

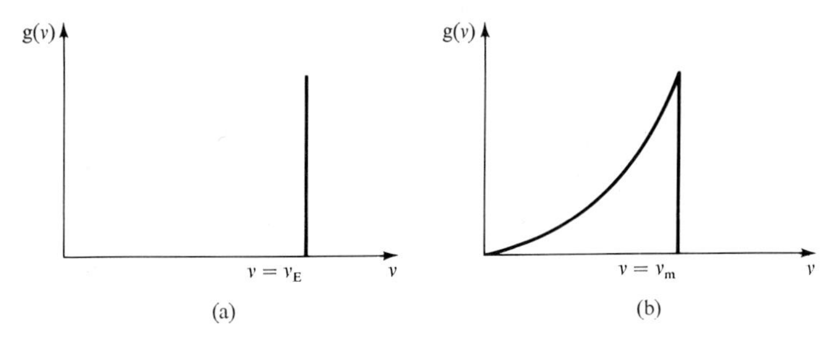

FIGURE 9.3. Frequency spectrum: a, Einstein; b, Debye.

the continuous function g(ν) represents the density of the frequency distribution, chosen to make

$$g(\nu)\,d\nu \tag{9.18}$$

equal to the number of frequencies whose value is contained between ν and $\nu + d\nu$. By contrast, the Einstein model assumed a single frequency ν_E. The maximum frequency ν_m in the Debye model is so adjusted as to render

$$\int_0^{\nu_m} g(\nu)\,d\nu = 3\mathbf{N}. \tag{9.19}$$

In order to find g(ν), we notice that $\nu(L/c)$ represents the distance to a point (n_1, n_2, n_3) in a space where n_1, n_2, n_3 are the coordinates. In this space there is one frequency per cell of unit volume. If $(n_1^2 + n_2^2 + n_3^2)^{1/2} = n$ is large, a spherical shell of volume $(4\pi n^2\,dn)$ will contain $(4\pi n^2\,dn)$ frequencies. This formula is obviously incorrect for small n, or large wavelengths, but we have already agreed to tolerate a comparable approximation for large wavelengths in using the periodic boundary conditions. Therefore, for large frequencies, their number in the range ν to $\nu + d\nu$ is

$$4\pi n^2\,dn = 4\pi\frac{L^3}{c^3}\,\nu^2\,d\nu = \frac{4\pi V}{c^3}\,\nu^2\,d\nu,$$

where $V = L^3$ is the volume of the crystal. In a solid there are transverse and longitudinal waves with different velocities of propagation; their existence is connected with the factor **A** in Equation (9.15). In a longitudinal wave the atoms of the crystal are displaced in the direction of the propagation of the wave, so that **A** is parallel to **f**. In a transverse wave the atomic displacements are perpendicular to the direction of propagation of the wave, and **A** is at a right angle to **f**. There are, in fact, two mutually perpendicular transverse waves, each at right angles to **f**. We can say that the transverse waves are polarized† in two mutually perpendicular directions. In Debye's theory it is

† The idea of polarization is explained in more detail in Section 10.2.

assumed that the velocity of propagation c_l, of the longitudinal wave is independent of frequency. In isotropic solids, the velocity c_t of propagation of transverse waves is independent of polarization as well as of frequency. Bearing in mind that the transverse modes contribute twice as many degrees of freedom as the longitudinal modes, we may write

$$g(\nu)\,d\nu = V\left(\frac{1}{c_l^3} + \frac{2}{c_t^3}\right)4\pi\nu^2\,d\nu. \tag{9.20}$$

Since these are 3**N** modes, the maximum vibrational frequency, ν_m, is obtained from the equation

$$3\mathbf{N} = 4\pi V\left(\frac{1}{c_l^2} + \frac{2}{c_t^3}\right)\int_0^{\nu_m}\nu^2\,d\nu = \frac{4\pi}{3}\left(\frac{1}{c_l^3} + \frac{2}{c_t^3}\right)V{\nu_m}^3. \tag{9.21a}$$

It is preferable to write this as

$$3\mathbf{N} = \frac{4\pi}{c^3}V{\nu_m}^3, \tag{9.21b}$$

where c represents the average velocity of wave propagation, defined by

$$\frac{3}{c^3} = \frac{2}{c_t^3} + \frac{1}{c_l^3}, \tag{9.21c}$$

and so to obtain

$$\nu_m = \left(\frac{3\mathbf{N}c^3}{4\pi V}\right)^{1/3}. \tag{9.21d}$$

9.3.2. The Thermodynamic Properties of a Debye Solid

In the expression for the logarithm of the partition function, the sum over k arising from the product is replaced by an integral from $\nu = 0$ to $\nu = \nu_m$. Since there are $g(\nu)\,d\nu$ frequencies in the range $\langle \nu, \nu + d\nu\rangle$, we must write

$$\begin{aligned}\ln Z &= \int_{\ln Z(0)}^{\ln Z(\nu_m)} d(\ln Z)\\ &= \int_0^{\nu_m} g(\nu)[\ln Z(\nu)]\,d\nu\\ &= \int_0^{\nu_m} g(\nu)\left[\ln\frac{\exp(-\frac{1}{2}\beta\mathbf{h}\nu)}{1-\exp(-\beta\mathbf{h}\nu)}\right]d\nu.\end{aligned} \tag{9.22}$$

If the contribution from the zero-point energy is neglected, the total energy, E, of the solid is

$$E = \frac{12\pi V}{c^3}\int_0^{\nu_m}\frac{\mathbf{h}\nu^3}{\exp(\beta\mathbf{h}\nu) - 1}\,d\nu. \tag{9.23}$$

In general, the molar energy u and specific heat c_v are given by

$$u = \frac{9\mathbf{R}T}{\xi_m{}^3} \int_0^{\xi_m} \frac{\xi^3 \, d\xi}{e^\xi - 1} + \phi(v_m), \tag{9.24a}$$

and

$$c_v = 3\mathbf{R}\mathscr{D}(\xi_m) \qquad \text{with} \quad \xi_m = \frac{\mathbf{h}\nu_m}{\mathbf{k}T}, \tag{9.24b}$$

respectively. Here $\mathscr{D}(\xi_m)$ is the *Debye function*, defined as

$$\mathscr{D}(\xi_m) = \frac{3}{\xi_m{}^3} \int_0^{\xi_m} \frac{\xi^4 \, e^\xi \, d\xi}{(e^\xi - 1)^2}. \tag{9.24c}$$

The function $\phi(v_m)$ of the molar specific volume arises from the zero-point energy and is of no immediate interest to us. The argument ξ in this function is related to the maximum frequency ν_m and to the Debye characteristic temperature

$$\Theta_D = \frac{\mathbf{h}\nu_m}{\mathbf{k}} \tag{9.24d}$$

through the definition

$$\xi_m = \frac{\Theta_D}{T}. \tag{9.24e}$$

The final result, Equation (9.24b), is quite analogous to that of Einstein in Equation (9.7), except that the numerical values are different. Table IV at the end of the book lists some values of the Debye function which can be compared with Table III. In particular, we are still led to the existence of a universal relation for all solids, provided that the Debye characteristic temperature, Θ_D, has been determined for each one of them.

At $T \gg \Theta_D$, we have $\mathscr{D}(\Theta_D) \to 1$, and the Dulong–Petit value of $c_v = 3\mathbf{R} \approx 6$ kcal/kmol K is obtained once more. At very low temperatures we have

$$\mathscr{D}(\xi_m) \to \frac{4\pi^4}{5} \xi_m^{-3},$$

so that

$$c_v \approx 77.93 \times 3\mathbf{R}\left(\frac{T}{\Theta_D}\right)^3 \approx 464.4 \frac{\text{kcal}}{\text{kmol K}} \left(\frac{T}{\Theta_D}\right)^3 \qquad (T \geq \Theta_D). \tag{9.25}$$

This is the famous *cube law* which shows that the specific heat, c_v, of crystals decreases as T^3 when the temperature decreases to absolute zero.

Figure 9.4 shows a comparison between the Debye and Einstein theories

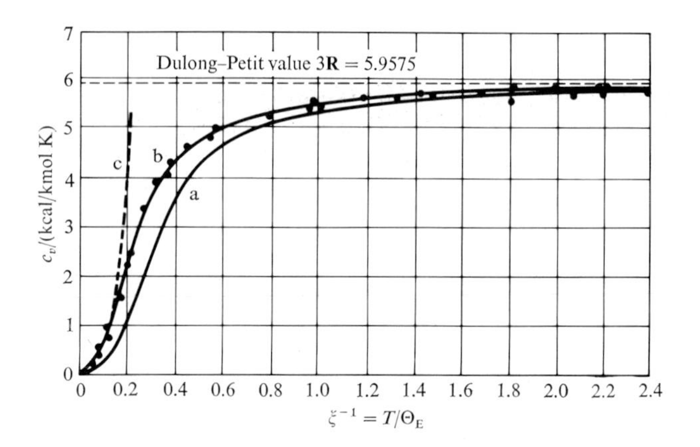

FIGURE 9.4. Comparison between the theories of Einstein and Debye and experiment: a, Einstein's theory; b, Debye's theory; c, Debye's T^3-law.

of specific heats together with a small number of experimental points. The latter have been reduced with the aid of empirically determined Debye temperatures.† These are usually obtained from specific heats directly rather than from the elastic equation for v_m.

9.4. Phonons

In the previous sections we regarded the normal modes of vibration of a crystal lattice as a collection of independent, distinguishable quantum-mechanical oscillators and obtained the appropriate quantum-mechanical partition function. It is now interesting to explore the quantum-mechanical features of the model in more detail. Our method in the previous section was to associate with a propagating wave the energy given by Equation (9.2) as

$$\varepsilon_{vk} = (\tfrac{1}{2} + v)\mathbf{h}\nu_k,$$

where ν_k is the classical frequency of the kth mode. According to the general principles of quantum mechanics, as discussed in Section 3.2, we can also associate with a wave whose wavelength is λ, a momentum $\mathbf{p}$, given by

$$\mathbf{p} = \mathbf{hf}, \tag{9.26a}$$

where

$$|\mathbf{f}| = \lambda^{-1}, \tag{9.26b}$$

† See article by G. T. Furukawa and T. B. Douglas in *American Institute of Physics Handbook*, 2nd ed., pp. 4–61 and 4–62, McGraw-Hill, New York, 1963.

and where the direction of the vector **f** is collinear with that of the propagation of the wave. In this manner, the quantum-mechanical theory presents us simultaneously with a wave picture of the thermal motion of the crystal as well as a particle picture. The "particles" associated with the waves are called *phonons*; they propagate in the lattice with definite energies and momenta and do so in the direction of motion. We can take it that each of these phonons carries an energy

$$\varepsilon = \mathbf{h}\nu \tag{9.27a}$$

and a momentum

$$\mathbf{p} = \mathbf{hf}. \tag{9.27b}$$

If the energy of a mode is, according to (9.2), $(\frac{1}{2} + v)\mathbf{h}\nu_k$, we can neglect the zero-point energy $\frac{1}{2}\mathbf{h}\nu_k$, and say that there are v phonons excited with the energy $\mathbf{h}\nu_k$.

In the previous section we have supposed that the different normal modes of the crystal were independent of one another, or, in other words, that the crystal was perfect and that the motion was harmonic. Physically this meant also that it was possible to excite in the system a single normal mode without at the same time exciting any others. In the phonon picture the independence of the normal modes is equivalent to saying that the phonons do not interact with each other and do not collide to produce new phonons with different momenta. Thus we can say that the phonons in a crystal constitute a kind of perfect gas. Moreover since any number of phonons can be in the same quantum state, and since there is no way to distinguish between these particles, the phonon gas is a *perfect boson gas*. If there existed small anharmonic terms in the classical energy of the crystal so that excitation of particular modes of the crystal would ultimately lead to the excitation of additional modes, we would say that the phonons interact with one another,† that is, that phonons can collide and "scatter," thus producing new phonons with different momenta.

Perfect Phonon Gas

Let us now explore the consequences of applying the methods of quantum statistics to the crystal. We first note that although there are 3N normal modes, there is no upper limit to the number of phonons. This is due to the fact that any phonon state can contain an arbitrary number of phonons.

† For an excellent discussion of phonon interactions in a harmonic crystal see L. Van Hove, Interactions of elastic waves in solids, in *Quantum Theory of Many Particle Systems* (L. Van Hove, N. M. Hugenholtz, and L. P. Howland, eds.), Benjamin, New York, 1961.

The number of phonons, $\mathcal{N}$, in the "gas" at equilibrium at a given temperature T and the given volume V of the crystal must be determined by the condition that the Helmholtz free energy, F, should be a minimum. Thus, we write

$$\left(\frac{\partial F}{\partial \mathcal{N}}\right)_{T,V} = 0. \tag{9.28}$$

Since $(\partial F/\partial \mathcal{N})_{T,V}$ is the chemical potential, μ, we conclude that the chemical potential of a phonon gas vanishes identically. It follows from Equation (8.32) that the average occupation numbers for phonons with energy ε_k are given by

$$n(\varepsilon_k) = \frac{1}{[\exp(\beta\varepsilon_k) - 1]}. \tag{9.29}$$

If we treat the phonon gas as a perfect gas and the crystal as isotropic, the number of quantum states in the momentum range $d\mathbf{p}$ is given by

$$\frac{4\pi V}{\mathbf{h}^3} p^2 \, dp = 4\pi V \mathrm{f}^2 \, d\mathrm{f}. \tag{9.30}$$

In order to compute the thermodynamic properties of an ideal phonon gas, it is necessary to determine the relation between the energy of a phonon, $\mathbf{h}\nu$, and its momentum, $\mathbf{h}\mathrm{f}$. In general, this depends on the interactions between the phonons as well as with the crystal lattice. Debye postulated that the velocity of propagation of a phonon is a constant, that is, that

$$\nu\lambda = c \quad \text{for} \quad \nu \leq \nu_{\mathrm{m}}. \tag{9.31a}$$

With the aid of the Einstein and de Broglie relations, this may also be written

$$\varepsilon = cp \qquad \text{for} \quad \varepsilon \leq \mathbf{h}\nu_{\mathrm{m}}, \tag{9.31b}$$

where ε is the energy of the phonon and p is its momentum. It will be seen in Chapter 10 that a similar relation is valid also for photons—the particles associated with electromagnetic waves.

The total number of quantum states is $3\mathbf{N}$, and this implies that

$$3\mathbf{N} = \frac{12\pi V}{c^3} \int_0^{\nu_{\mathrm{m}}} \nu^2 \, d\nu. \tag{9.32}$$

The factor $3/c^3$ is again equal to $(1/c_{\mathrm{l}}^3 + 2/c_{\mathrm{t}}^3)$. In the phonon picture we say that to longitudinal waves with velocity c_{l} there correspond "longitudinal phonons" with velocity c_{l}, and to transverse waves with velocity c_{t} there correspond "transverse phonons" with velocity c_{t}. The fact that Equation

(9.32) is identical with Equation (9.21) is a consequence of the wave–particle duality of all quantum-mechanical systems.

The internal energy, E, of the phonon gas can now be written

$$E = \frac{12\pi V}{c^3}\int_0^{\nu_m} \frac{(\mathbf{h}\nu)\nu^2\, d\nu}{\exp(\beta \mathbf{h}\nu) - 1}, \tag{9.33}$$

which is identical with Equation (9.23). The system partition function for $\mathcal{N}$ phonons, $\mathscr{Z}_{ph}$, is given by Equation (8.35) as

$$\ln \mathscr{Z}_{ph} = -\frac{12\pi V}{c^3}\int_0^{\nu_m} \ln[1 - \exp(-\beta \mathbf{h}\nu)]\nu^2\, d\nu$$

$$\ln \mathscr{Z}_{ph} = -\frac{9\mathbf{N}}{\nu_m^3}\int_0^{\nu_m} \ln[1 - \exp(-\beta \mathbf{h}\nu)]\nu^2\, d\nu. \tag{9.34a}$$

This can be integrated by parts to yield

$$\ln \mathscr{Z}_{ph} = \frac{E}{3\mathbf{k}T} - 3\mathbf{N}\ln[1 - \exp(-\beta \mathbf{h}\nu_m)]. \tag{9.34b}$$

The phonon pressure P_{ph} is given by

$$P_{ph} = \mathbf{k}T\frac{\partial \ln \mathscr{Z}_{ph}}{\partial V} = \mathbf{k}T\frac{\partial \ln \mathscr{Z}_{ph}}{\partial \nu_m}\frac{\partial \nu_m}{\partial V},$$

or, after some elementary manipulation,

$$P_{ph} = \frac{1}{3}\frac{E}{V}. \tag{9.35}$$

Equation (9.35) differs by a factor 2 from the usual result for the pressure of a perfect gas of particles whose number is fixed. The difference is due to the different energy–momentum relation in the two cases and will be considered further in Problem 9.13. The phonon pressure is exerted on the crystal faces and must be balanced by the molecular forces which keep the crystal together. However, if the crystal is subjected to a temperature increase, the phonon pressure increases and causes thermal expansion. The phonon pressure manifests itself most clearly in the coefficient of thermal expansion, α, defined by

$$\alpha = \left(\frac{\partial \ln V}{\partial T}\right)_P = -\left(\frac{\partial P}{\partial T}\right)_V \left(\frac{\partial \ln V}{\partial P}\right)_T. \tag{9.36a}$$

The last expression results from the application of the identity

$$\left(\frac{\partial x}{\partial y}\right)_z \left(\frac{\partial y}{\partial z}\right)_x \left(\frac{\partial z}{\partial x}\right)_y = -1. \tag{9.36b}$$

The factor $-(\partial \ln V/\partial P)_T$ is the coefficient of isothermal compressibility, κ, and we have

$$\alpha = \left(\frac{\partial P}{\partial T}\right)_V \kappa . \tag{9.36c}$$

At constant volume, only the phonon pressure changes with temperature, since the cohesive forces in the crystal depend, to a good approximation, on the volume alone. This confirms that the coefficient of thermal expansion constitutes a proper measure of the phonon pressure; it is given by

$$\alpha = \frac{\kappa c_v}{3V} . \tag{9.37}$$

Experimentally it is found that per mole

$$\alpha = \frac{\gamma \kappa c_v}{v_{\rm m}} , \tag{9.38}$$

where γ is a constant which varies from substance to substance; the latter is known as Grüneisen's constant.† Since κ is nearly independent of temperature, we find that c_v/α is also independent of temperature, the statement being known as Grüneisen's law.

For a crystal we may also obtain the relation

$$c_p - c_v = \frac{v_{\rm m}\,\alpha^2 T}{\kappa} = \frac{v_{\rm m}\,\alpha^2}{\kappa c_p^{\,2}}\, c_p^{\,2} T . \tag{9.39}$$

Experimental evidence suggests that the group

$$\Lambda = \frac{v_{\rm m}\,\alpha^2}{\kappa c_p^{\,2}}$$

remains nearly constant for most solids. Consequently we have

$$c_v = c_p - \Lambda c_p^{\,2} T . \tag{9.40}$$

This equation is known as the Nernst–Lindemann law and is usually employed to calculate the value of c_v from measured values of c_p, the values of Λ being determined from one set of data measured at a convenient temperature.

9.5. The Band Theory of Solids

So far, we have considered only the motion of the centers of mass of the atoms in the crystal lattice disregarding the modifications in electronic structure that the regularity of the array in a crystal imposes upon the atoms

† For numerical data see G. T. Furukawa and T. B. Douglas, in *American Institute of Physics Handbook*, 2nd ed., pp. 4–61 and 4–62, McGraw-Hill, New York, 1963.

in the solid. We may gain some understanding of these modifications by examining the fictitious process of varying the lattice spacing, d, between the atoms in a solid from infinity to that actually observed, d_0. At one extreme when $d \to \infty$, all atoms become independent of one another, and the energy levels of the electrons are those appropriate to an atom. As the spacing parameter d is allowed to decrease, the electrons of one atom come under a progressively stronger influence from the adjoining atoms. This influence is reinforced by the fact that the wave functions for electrons, even in pure atomic levels, admit a finite, if small, probability that the electrons themselves can be found at a separation of a few d from the nucleus. This places them close to another nucleus. The probability that an electron may exist at such distances is larger for the larger atoms increasing also for the less tightly bound electrons as they exist in one of their atomic levels. In this manner, the electronic levels become progressively modified by the presence of the adjoining nuclei with which they can interact. As a result, at values of $d = d_0$ observed in actual solid materials, the electronic levels have become considerably modified in comparison with their arrangement in an isolated atom, and it may no longer be true to say that all electrons associated with an atom remain localized in its structure. The electrons become mobile, and there is an equal probability of finding them near any atom in the lattice.

The energy levels available to the "delocalized" or *mobile* electrons cease to be discrete and form "bands" separated by definite "gaps." Each band contains a large number of discrete levels which are so closely spaced that they can be assumed to be distributed continuously within a band, one level corresponding to each atom in the crystal. In favorable cases the bands in the crystal may directly correspond to atomic levels, so that all levels in a particular band may be thought of as having developed from a particular atomic level, as suggested by Figure 9.5. However, this is not always the case.

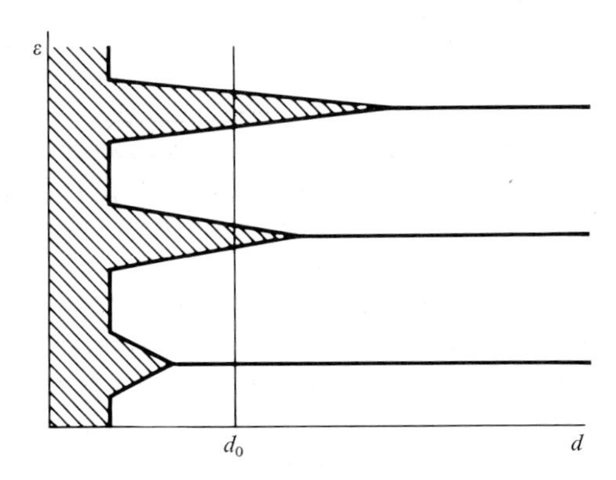

FIGURE 9.5. Schematic diagram of energy levels available to electrons as a function of the lattice spacing, d. The observed spacing is denoted by d_0.

It is seen that, normally, there are no levels available to the electrons between the bands, and the energy regions between them are often referred to as the "forbidden" regions or bands.

9.5.1. Metals, Insulators, and Semiconductors

The preceding considerations allow us to understand the differences between the various types of solids that occur in nature. For this purpose, we suppose now that the solid exists at a very low temperature and examine the distribution of electrons over the various bands. At absolute zero, the energy levels in the crystal are filled from the lowest energy state up, with two electrons to a level, until all electrons have been accommodated. The lowest energy levels are those in which the electrons are localized about particular atoms, and the electrons in them may be properly thought of as atomic electrons. The electrons with larger energies fill the bands in which the electrons are delocalized; they are referred to as the mobile or *conduction electrons*, since they and not the atomic electrons are active as carriers in the process of electric conduction. The energy of the highest filled level at absolute zero is called the *Fermi energy* and is denoted by μ_0. This nomenclature recalls that employed in the theory of the ideal fermion gas in which the energy of the uppermost filled level at absolute zero was equal to the chemical potential, $\mu\,(T = 0)$. However, in a crystal the relation between the Fermi energy, μ_0, and the potential at absolute zero, $\mu\,(T = 0)$, is not so simple.

The electrical and thermodynamic properties of the solid depend on the location of the uppermost filled level in relation to the bands discussed earlier. Suppose that the uppermost level is located within a partially filled band, with many adjacent levels remaining unfilled and available to the mobile electrons. In such circumstances, a small temperature change, or the application of an electric field, can easily excite the conduction electrons and force them into the unfilled levels inside the band. A solid of this type is a metal, since an electric current can now be easily passed through it. This is illustrated in Figure 9.6a where we have represented, for purpose of illustration only, a metal with three conduction bands, of which the uppermost is only partially filled at absolute zero.

If the highest filled level is at the top of a band, the electrons are not easily affected by temperature changes or external fields. In order to excite electrons into the next, unfilled, band and thus to cause electric conduction, it is necessary to overcome the forbidden gap which means that a relatively large amount of energy must be supplied to the crystal. There exist two types of materials of this kind: *intrinsic semiconductors* and *insulators*. The difference between the properties of these solids is only due to the magnitude of the

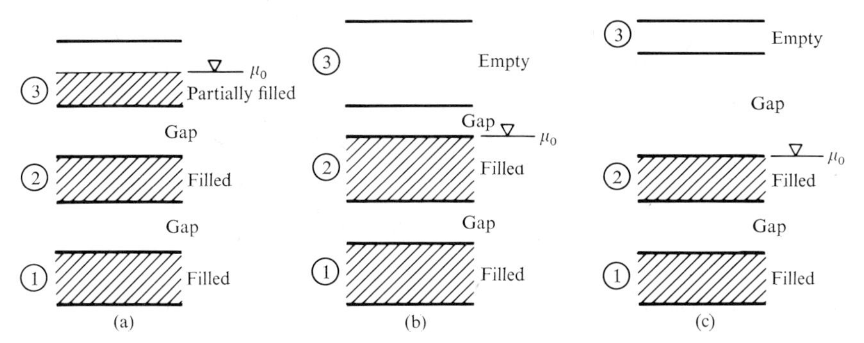

FIGURE 9.6. Schematic illustration of the band structure of solids; μ_o denotes the Fermi energy, that is, the energy of the highest filled level at 0 K; a, metal—band 3 is only partially filled; b, intrinsic semiconductor—there is a small gap between filled band 2 and empty band 3; c, insulator—there is a large gap between filled band 2 and empty band 3. [From D. Ter Haar, *Elements of Thermostatistics*, 2nd ed., p. 140, Holt, New York, 1966.]

energy gap which separates the last filled band from the next, unfilled, band. If the energy gap is small compared to $\mathbf{k}T$, an increase in temperature from absolute zero to T, say, causes a transfer of electrons to the next band, and the material becomes a conductor. It is customary to refer to any partially filled band at a given temperature T, either in a metal or in an intrinsic semiconductor, as a *conduction band.*

By introducing impurities† into a crystal it is possible to alter its band structure in several ways. The influence of the impurities on the band structure makes it possible to create two new kinds of materials: *donor and acceptor semiconductors.* The solid without impurity may be an insulator with a large energy gap between a filled band and a conduction band. The impurities produce new levels in the band structure which may lie in this energy gap. In N-type, or donor semiconductors, there is an impurity band which is filled at 0 K but which is close to a higher unfilled band of the solid. A small temperature increase results in electrons being excited into the unfilled band and the material becomes a conductor at these temperatures. In P-type, or acceptor semiconductors, at absolute zero there exists an unfilled impurity band that is close to a filled band. Now an increase in temperature causes a transfer of electrons into the adjacent impurity band, and the material starts to conduct. These situations are illustrated in Figure 9.7 in which we have sketched, for purposes of illustration, a simplified picture of the band structure of semiconductors. The impurity band is denoted by i, and it is seen that in its absence the material would display the properties of an electric insulator.

Examples of typical semiconducting materials are germanium, diamond, silicon, cadmium sulphide, and indium arsenide. As pure substances these

† Typically, the impurities amount to one part per 10^5.

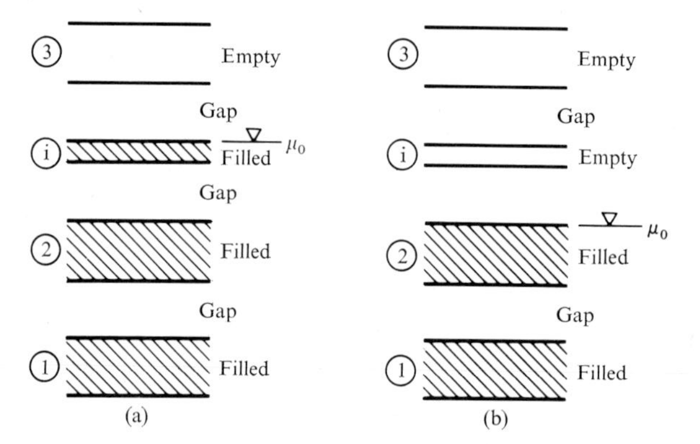

FIGURE 9.7. Donor and acceptor semiconductor band structure: a, donor semiconductor; band i is a filled impurity band which donates electrons to empty band 3; b, acceptor semiconductor; band i is an empty impurity band which accepts electrons from filled band 2.

materials are intrinsic semiconductors with energy gaps of the order of $\Delta\varepsilon/\mathbf{k} \approx 10^4$ K;† they become donor or acceptor semiconductors, depending on the nature of the impurity. Typical donor impurities are antimony, tellurium, and arsenic, which, present in amounts of the order of one part in 10^7, produce a donor semiconductor with an energy gap in the range from 100 to 1000 K. Acceptor impurities, such as boron or aluminum, produce energy gaps $\Delta\varepsilon/\mathbf{k} \approx 100$ to 1000 K.

The liberation of electrons into conduction bands turns the material into a good conductor of electricity. The delocalized electrons in the conduction bands are easily influenced by external electric fields, in contrast to the localized electrons in the valence bands which are not so easily influenced. As one might expect, the electrical resistance of a semiconductor decreases with increasing temperature because more electrons enter the conduction band. It is worth pointing out that the electrical resistance of a metal *increases* with temperature owing to collisions between the conduction electrons and the ions of the metal. Although such collisions take place in semiconductors, the filling of the conduction band dominates the electrical properties.

The increase in electrical conductivity which occurs when the electrons enter the conduction band in a semiconductor is accompanied by an increase in the thermal conductivity of the material. That is, the conduction electrons may easily carry heat, in the form of kinetic energy, from one part of the system to another. The details of the process are discussed in Section 12.5, where a more general treatment of thermal conductivity is given. Since the

† Materials regarded as insulators have energy gaps $\Delta\varepsilon/\mathbf{k} \approx 5$ to 10×10^4 K.

conduction of electricity and heat in a metal, or a semiconductor, is produced by the motion of free electrons, a good conductor of electricity must also be a good conductor of heat, as is well known.

9.5.2. Thermodynamic Properties of Solids

The band structure of a solid reveals itself in its thermodynamic properties as well as in its electrical properties. In particular, at low temperatures the specific heat of a solid is composed of two parts, one contribution deriving from the lattice vibrations, the other from the electronic motion. The former contribution, c_v^{lattice}, is given by Equation (9.24) as

$$c_v^{\text{lattice}} = 3\mathbf{R}\mathscr{D}\left(\frac{\mathbf{h}\nu_{\text{m}}}{\mathbf{k}T}\right).$$

The electronic contribution depends very sensitively on the structure of the solid. For insulators, at sufficiently low temperatures, there is effectively no contribution to the specific heat for temperatures much less than $\Delta\varepsilon/\mathbf{k}$, where $\Delta\varepsilon$ is the value of the energy gap between the last filled band and the next unfilled band. For temperatures less than $\Delta\varepsilon/\mathbf{k}$, the electrons are "frozen" in the filled band and make no contribution to the thermodynamic properties of the solid. We remind the reader that a similar situation exists with respect to the specific heat of polyatomic gases. At temperatures which are much smaller than the characteristic temperature associated with a particular kind of motion there is no contribution to the specific heats and the corresponding type of molecular motion is said to be "frozen." Thus for an *insulator*

$$c_v^{\text{electronic}} \approx 0 \qquad \text{at} \quad T \ll \Delta\varepsilon/\mathbf{k}.$$

In metals the situation is different, since in them there exist adjacent unfilled levels to which the electrons may be excited at any temperature. Thus, for them we expect an electronic contribution to the specific heat at all temperatures down to 0 K. If the electrons in the unfilled conduction band could be treated as a free fermion gas, we would expect to express the specific heat of the metal as

$$c_v^{\text{metal}} = 3\mathbf{R}\mathscr{D}\left(\frac{\Theta_{\text{D}}}{T}\right) + \frac{\pi^2}{2}\mathbf{R}\left(\frac{\mathbf{k}T}{\mu_0}\right), \tag{9.41}$$

as seen from Table 8.1, provided that the temperature were sufficiently low. Here μ_0 is the energy of the highest filled level at 0 K. This expression for the specific heat assumes that the electrons form a perfect gas with an energy given by

$$\varepsilon = \frac{1}{2\mathbf{m}_{\text{e}}}(p_x^2 + p_y^2 + p_z^2), \tag{9.42}$$

when $\mathbf{m}_e$ is the mass of an electron. If we consider the energy in momentum space, the surface of constant energy is a sphere of radius $(2\mathbf{m}_e\varepsilon)^{1/2}$. For the free-electron model, we can look upon the Fermi energy, μ_0, as defining a spherical surface in momentum space with radius

$$(2\mathbf{m}_e\mu_0)^{1/2} = (p_x{}^2 + p_y{}^2 + p_z{}^2)^{1/2}. \tag{9.43}$$

The occupied levels in momentum space are thus spherical surfaces whose radii are smaller than $(2\mathbf{m}_e\mu_0)^{1/2}$. The bounding sphere, defined by Equation (9.43), is referred to as the *Fermi surface* for the free electron model, as shown in Figure 9.8.

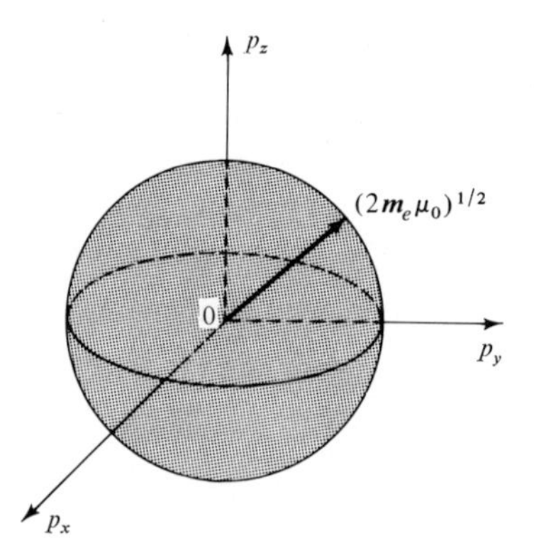

FIGURE 9.8. The Fermi surface at 0 K for a perfect electron gas.

The experimental data for the specific heats of metals at low temperatures can be fitted by an expression of the form

$$c_v^{\text{metal}} = 3\mathbf{R}\mathscr{D}\left(\frac{\Theta_D}{T}\right) + \gamma' T, \tag{9.44}$$

but the coefficient γ', the *electronic constant* listed in Table 9.1, is not equal to that given by expression (9.41). The explanation of the discrepancy lies in the fact that the electrons in the conduction bands are not really free but interact with each other, and, even more importantly, with the metallic ions of the lattice. As a result, the real Fermi surface of a metal is no longer spherical and becomes distorted. In particular, an electron whose wavelength is equal to an integral multiple of the lattice spacing of the solid will interact strongly with the ions of the lattice since its wavelength is appropriate for diffraction by the crystal lattice. In Figure 9.9 we illustrate a typical Fermi surface for a simple metal, copper. It is seen that the sphere is distorted at such values of the momenta p_x, p_z, p_y as are associated with wavelengths which are comparable with the lattice spacing.

TABLE 9.1[a]

The Ratio $m^*/\mathbf{m}_e$ of the Experimentally Determined Mass m^* to That of the Electron, $\mathbf{m}_e = 9.109558 \times 10^{-28}$ gr, for Various Metals (in Ascending Order of Atomic Number)

Metal	$\gamma' \times 10^4$ (cal/mol K^2)	$m^*/\mathbf{m}_e$
Beryllium, Be	0.54	0.46
Magnesium, Mg	3.25	1.33
Aluminum, Al	3.48	1.6
Iron, Fe	12	12
Nickel, Ni	17.4	28
Copper, Cu	1.78	1.5
Zinc, Zn	1.42	0.8
Silver, Ag	1.54–1.6	0.95–1.0
Platinum, Pt	16.0	13
Mercury, Hg	4.5–5.3	1.8–2.2

[a] J. G. Daunt, The electronic specific heat in metals, in *Progress in Low Temperature Physics*, Vol. I, (C. J. Gorter, ed.) Chapter XI, North-Holland Publ., Amsterdam, 1955.

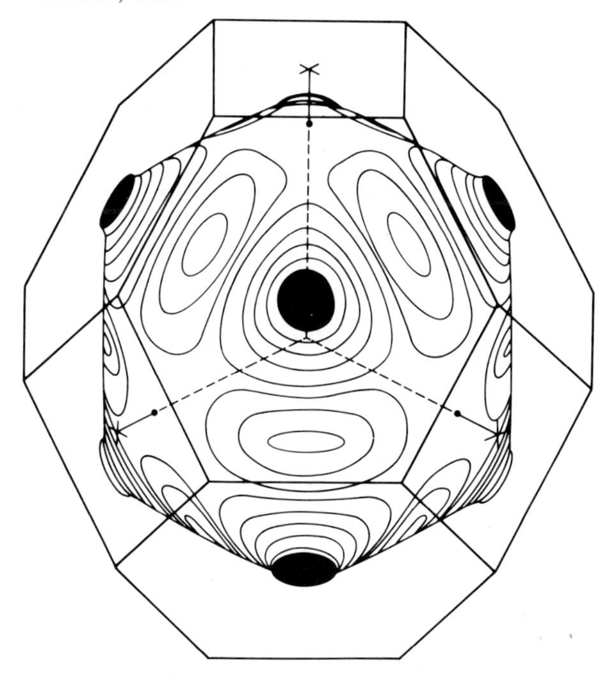

FIGURE 9.9. Fermi surface for copper. (From G. E. Smith, The anomalous skin effect, in *The Fermi Surface*, (W. A. Harrison and M. B. Webb, eds.), p. 192, Wiley, New York, 1960.)

At very low temperatures, usually below 4 K, the electronic contribution to the specific heat, which decreases as T, dominates the lattice contribution, which decreases as T^3. If experimental values of the quantity c_v/T for a metal are plotted against T^2 in this range, we obtain a straight line whose intercept at $T = 0$ is the electronic constant γ' and whose slope determines the Debye temperature Θ_D,

$$c_v/T = \gamma' + (77.93)(3\mathbf{R})\Theta_D^{-3}T^2. \tag{9.44a}$$

Here Equation (9.25) has been inserted for the lattice specific heat at low temperatures. A typical plot of this kind for Zn is shown in Figure 9.10.

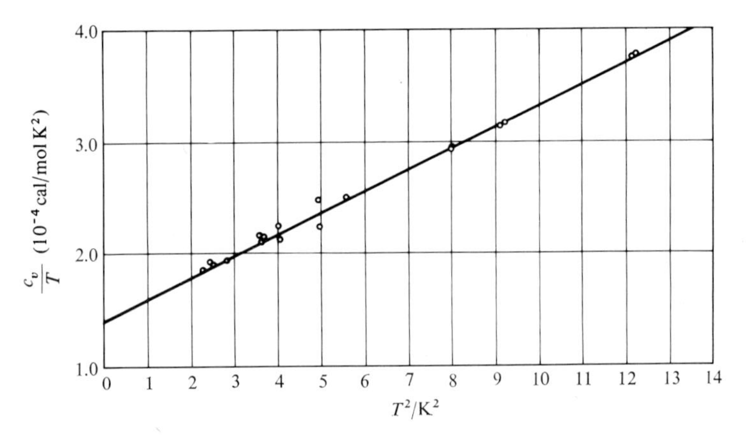

FIGURE 9.10. Plot of c_v/T against T^2 for zinc. (Measurements by A. A. Silvidi and J. G. Daunt; see J. G. Daunt, The electronic specific heat in metals, in *Progress in Low Temperature Physics*, Vol. I, (C. J. Gorter, ed.), Chapter XI, North-Holland Publ., Amsterdam, 1955).

The experimentally determined coefficient γ' in Equation (9.44) can be used to derive an "effective" mass, m^*, for the electrons in a metal. This is obtained by applying the free-electron theory of conduction electrons and by defining m^* through the formulas

$$\gamma' = \tfrac{1}{2}\pi^2\mathbf{R}\mathbf{k}/\mu_0{}^*$$

with

$$\mu_0{}^* = \frac{\mathbf{h}^2}{8m^*}\left(\frac{3\mathbf{N}}{\pi v_\mathrm{m}}\right)^{2/3}.$$

The ratio $m^*/\mathbf{m}_e$ obtained from experimental values of γ' for several metals can be found listed in Table 9.1. The Fermi surface of metals for which the ratio $m^*/\mathbf{m}_e$ differs appreciably from unity can become quite different from the spherical shape predicted by the free-electron theory. It is noteworthy that the process of electric conduction in silver ($m^*/\mathbf{m}_e = 0.95$–1.00) is quite close to that of a free-electron metal.

9.6. Thermionic Emission

The highest energy available to an electron in a metal at 0 K must be lower than that of a free electron outside it; otherwise electrons would escape from the metal. Instead, as is known, energy must be supplied to the electrons in order to remove them from the metal. For example, if the surface of the metal is illuminated, only light whose frequency exceeds a limiting value, ν_0—the photo-frequency of the metal—will liberate electrons from it. This is the photelectric effect.† The existence of this effect proves that at 0 K there is a finite energy gap, $W_0 = \mathbf{h}\nu_0$, called the *work function* of the metal, which extends between the energy of the highest filled electron state in the metal, μ_0, and the energy of an electron at rest outside the metal. The energy of the latter, at absolute zero, is $\mu_0 + W_0$. At sufficiently high temperatures a small fraction of the electrons have enough energy to escape from the crystal. This phenomenon is observed experimentally when a metal is heated in an electric field so that the emitted electrons can be collected on an anode at some distance from the metal. The experimentally determined current density, I, emitted per unit area of the metal surface satisfies the empirical equation

$$I = AT^2 \exp(-W/\mathbf{k}T), \tag{9.45}$$

where W is the energy difference between an electron with the Fermi energy at temperature T, and an electron at rest outside the metal.

The experimental result contained in Equation (9.45) can be understood by ignoring the band structure of the metal and by treating the electrons in the metal as a perfect fermion gas. We stipulate that the metal has been heated to a temperature T at which the Fermi energy and the work function are μ and W, respectively. The emitted current per unit area is equal to the number of electrons inside the metal that impinge on a unit area of the metal surface and have sufficient energy and momentum to escape from it. We imagine an infinite metal surface in the y, z plane and note that the electrons in the metal must overcome a potential barrier of height $W + \mu$ in the x direction. The situation is illustrated in Figure 9.11 in which the metal surface is at $x = 0$, and the interior of the metal is the region $x < 0$. An electron can escape

† For a full discussion of the photoelectric effect see F. K. Richtmeyer, E. H. Kennard, and T. Lauritsen, *Introduction to Modern Physics*, Chapter, 3, McGraw-Hill, New York, 1955.

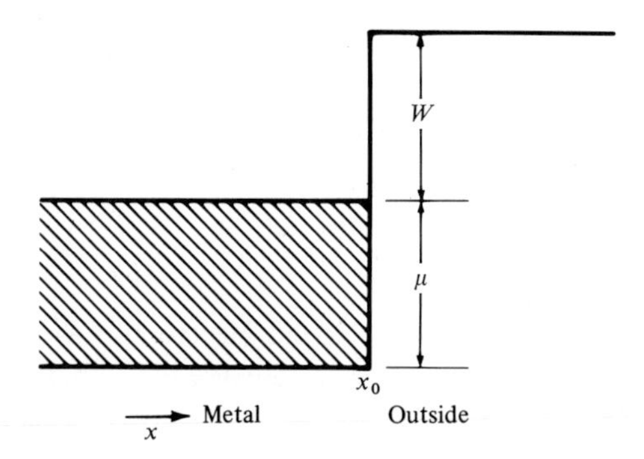

FIGURE 9.11. Schematic illustration of the electron energy inside and outside a metal. The quantity x denotes the distance from an origin fixed in the metal; x_0 is the location of the surface of the metal. An electron at rest outside the metal has an energy $\mu + W$.

from the metal surface at $x = 0$ if the x component of its momentum, p_x, is sufficiently high to overcome the barrier. The y and z components of the momentum are not affected by a potential which only varies in the x direction. As the x component of the momentum decreases by an amount $[2\mathbf{m}_e(W + \mu)]^{1/2}$, the electron escapes from the metal if

$$p_x \geq [2\mathbf{m}_e(W + \mu)]^{1/2}.$$

Naturally, this also implies that the total kinetic energy of an escaping electron must be greater than $W + \mu$.

To compute the number of electrons that strike a unit area of the metal surface in a time interval Δt, say, we consider first electrons with momentum $\mathbf{p}$ arriving at the surface at an angle θ with respect to x, and stipulate that $p_x > 0$ for them, Figure 9.12. If such an electron is to arrive within area ΔA in time Δt it must be contained in a cylinder centered on ΔA and having a length $(p/\mathbf{m}_e)\Delta t = (2\varepsilon/\mathbf{m}_e)^{1/2}\ \Delta t$, where ε is the energy of the electron. The

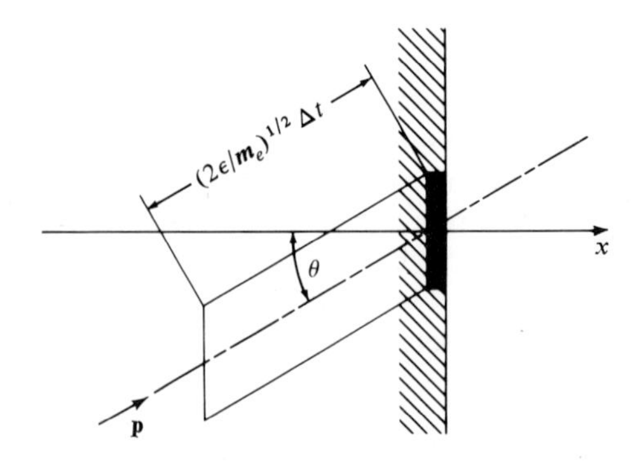

FIGURE 9.12. Electrons escaping from metal surface.

number of electrons contained in this cylinder can be taken to be equal to the volume of the cylinder multiplied into the number of such electrons per unit volume. Using Equation (8.25), we see that the number of electrons with energy ε which move in a direction interval $\langle \theta, \theta + d\theta \rangle$ and which collide with the surface per unit area and time is

$$\left(\frac{2\varepsilon}{\mathbf{m}_e}\right)^{1/2} \cos\theta \, \frac{4\pi}{\mathbf{h}^3} (2\mathbf{m}_e{}^3)^{1/2} \frac{\varepsilon^{1/2} \sin\theta \, d\theta}{\exp[\beta(\varepsilon - \mu)] + 1}. \tag{9.46}$$

If the electron is to escape, we must have

$$\varepsilon \geq W + \mu \tag{9.47a}$$

and

$$\frac{p_x{}^2}{2\mathbf{m}_e} = \frac{p^2}{2\mathbf{m}_e} \cos^2\theta = \varepsilon \cos^2\theta \geq W + \mu . \tag{9.47b}$$

The current density, I, produced is equal to the electronic charge, **e**, times the total number of electrons that escape per unit time and per unit area. Therefore

$$\begin{aligned}
I &= \mathbf{e}\left(\frac{2}{\mathbf{m}_e}\right)^{1/2} \frac{4\pi(2\mathbf{m}_e{}^3)^{1/2}}{\mathbf{h}^3} \int_{W+\mu}^{\infty} d\varepsilon \int_0^{\cos^{-1}[(W+\mu)/\varepsilon]^{1/2}} \frac{\varepsilon \sin\theta \cos\theta}{\exp[\beta(\varepsilon - \mu)] + 1} \, d\theta \\
&= \mathbf{e}\frac{4\pi \mathbf{m}_e}{\mathbf{h}^3} \int_{W+\mu}^{\infty} \frac{\varepsilon}{\exp[\beta(\varepsilon - \mu)] + 1} \left(1 - \frac{W + \mu}{\varepsilon}\right) d\varepsilon \\
&\approx \mathbf{e}\frac{4\pi \mathbf{m}_e}{\mathbf{h}^3} \int_{W+\mu}^{\infty} \exp[-\beta(\varepsilon - \mu)](\varepsilon - W - \mu) \, d\varepsilon \\
&= \mathbf{e}\frac{4\pi \mathbf{m}_e \mathbf{k}^2 T^2}{\mathbf{h}^3} \exp(-\beta W)
\end{aligned} \tag{9.48}$$

provided that $e^{\beta W} \gg 1$. This approximation is valid for most metals. Equation (9.48) is known as the *Richardson–Dushman* equation; it agrees in form with the experimentally obtained Equation (9.45), and predicts that

$$A = \frac{4\pi \mathbf{m}_e \mathbf{k}^2 \mathbf{e}}{\mathbf{h}^3} = 123 \text{ A/cm}^2 \text{ K}^2 . \tag{9.49}$$

The experimental verification of expression (9.49) for A is complicated in practice because of contamination and irregularities on the surface. Moreover, one cannot expect that Equation (9.49) will be exactly verified because the free-electron assumption is inadequate for metals whose Fermi surfaces are distorted appreciably.

PROBLEMS FOR CHAPTER 9

9.1. Compute the vibrational frequencies of a one-dimensional chain of harmonic oscillators as described in Section 9.3.1. Assume, however, that the ends of the chain are fixed (that is, $y_1 = y_{N+1} = 0$). Compare your results with those obtained when periodic boundary conditions are imposed. Are the frequencies the same for wavelengths which are small compared to the length of the chain? (*Hint:* Assume that the waves are of the form:

$$y_n = A \sin 2\pi[\nu t - (n-1)a/\lambda] + B \sin 2\pi[\nu t + (n-1)a/\lambda].)$$

9.2. The longitudinal and transverse sound velocities c_l and c_t in a solid are related to the shear modulus, μ, and the plate modulus, $\lambda + 2\mu$, in that $c_t = (\mu/\rho)^{1/2}$; $c_l = (\lambda + 2\mu/\rho)^{1/2}$, where ρ is the density. Making use of the data given below, calculate the Debye temperature, Θ_D, of aluminum, copper, and silver and compare the results with those in a standard table;† the latter are usually derived by fitting experimental data:

Metal	ρ (kg/m^3)	μ (N/m^2)	λ (N/m^2)
Aluminum, Al	2,700	2.4×10^{10}	6.1×10^{10}
Silver, Ag	10,400	2.7×10^{10}	8.6×10^{10}
Copper, Cu	8,940	4.6×10^{10}	13.1×10^{10}

9.3. Explain why the specific heat of diamond at room temperature is considerably below the Dulong–Petit value:

Specific Heat, c_p, of Diamond

T (K)	0	100	150	200	250	298.15	300	400
c_p (kcal/ K kmol)	0	0.0673	0.246	0.587	1.013	1.449	1.466	2.38

T (K)	500	600	700	800	900	1000	1100	1200
c_p (kcal/ K kmol)	3.14	3.79	4.29	4.66	4.90	5.03	5.10	5.16

9.4. The molar specific heat of aluminum varies with temperature according to the accompanying table. Determine the Debye temperature for this element:

T (K)	10	20	30	50	100	200	300	400	500	700	1000
c (kcal/kmol K)	0.01	0.05	0.20	0.91	3.12	5.16	5.82	6.12	6.42	7.31	7.00

† See, for example, Table XXXI in J. Kestin, *A Course in Thermodynamics*, Vol. II, Blaisdell, Waltham, Massachusetts, 1969. This lists $\Theta_D = 428$ K (Al), 226 K (Ag), 343 K (Cu).

9.5. Derive Equation (9.37) for a Debye solid.

9.6. Use thermodynamic arguments to derive Equation (9.39) in the text.

9.7. Show that the Debye frequency spectrum, Equation (9.20), may be obtained for an elastic solid in the shape of a rectangular parallelepiped, with either fixed boundaries or periodic boundary conditions.

9.8. The low-temperature specific heat of graphite is proportional to T^2. Show that this fact may be explained by assuming that graphite behaves like a two-dimensional harmonic crystal.

9.9. Compute $c_p - c_v$ for copper at $T/\Theta_D = \frac{1}{2}$ and 2. For copper, the Grüneisen constant $\gamma = 2$, and $\kappa = 7.5 \times 10^{-13}$ cm²/dyne, and $v = 11.8 \times 10^{-24}$ cm³/gram-atom.

9.10. Suppose that for $\varepsilon \ll \varepsilon_m$ the phonon energies of a certain system satisfy the relation $\varepsilon = A|\mathbf{p}|^n$, where A is a constant. Compute the low-temperature specific heat of such a system.

9.11. Calculate the molar specific heat, c_v, for copper, iron, and zinc at $T = 1$; 10; 100; and 1000 K.

Note: The Debye temperatures are: Cu, $\Theta_D = 343$ K; Fe, $\Theta_D = 467$ K; Zn, $\Theta_D = 310$ K.

9.12. Express the specific heat, c_v, of a Debye solid in the form of an expansion in terms of Θ_D/T. State the restriction on temperature for which this series is valid.

9.13. Show that the relation $PV = E/3$, valid for phonons, holds if the energy levels for a phonon are proportional to $V^{-1/3}$, where V is the volume of the system. Show, further, that the energy levels of a Debye solid are of this form.

9.14. Show that

$$\int_0^\infty [3N\mathbf{k} - c_v(T)]\,\mathrm{d}T = E(0),$$

where $E(0)$ is the energy of the solid at $T = 0$ K, if the specific heat of the solid, c_v, approaches $3N\mathbf{k}$ as $T \to \infty$.

SOLUTION: The integral may be written as

$$\int_0^\infty \frac{\mathrm{d}}{\mathrm{d}T}[3N\mathbf{k}T - E(T)]\,\mathrm{d}T,$$

from which the result follows.

9.15. Is it possible to formulate the description of an Einstein solid in terms of phonons? If so, what are the properties of Einstein phonons and in what way do they differ from Debye phonons?

9.16. Assume that the frequencies, ν_j, of vibration of a solid depend on the volume, V, in such a way that $\mathrm{d}\ln \nu_j(V)/\mathrm{d}\ln V = -\gamma$, where γ is independent of the frequency index j. Show that γ is the Grüneisen constant.

9.17. Calculate the thermionic current at $T = 1000$ K, from a surface whose work function is $W = 1.5$ eV.

9.18. Show that the thermionic current of electrons which obey Maxwell–Boltzmann statistics is given by $I_{MB} = A'T^{1/2}\exp(-\phi/\mathbf{k}T)$.

9.19. Suppose a quantity of a metal contains N atoms, then each of its energy bands can contain $2N$ electrons. Consider one such band and suppose that it contains $2N - \mathcal{N}$ electrons. Show that the contribution of these electrons to the thermodynamic properties of the metal is equivalent to that of $\mathcal{N}$ electrons which are distributed over levels with energies $-\varepsilon_i$, and for which the chemical potential is $-\mu$. Here, ε_i denotes the energy levels in the band, and μ is the chemical potential of the electrons present in the band.

SOLUTION: The thermodynamic properties of the $2N - \mathcal{N}$ electrons are determined by the relations

$$2N - \mathcal{N} = \sum_i \{\exp[\beta(\varepsilon_i - \mu)] + 1\}^{-1} \quad \text{and} \quad \tilde{E} = \sum_i \varepsilon_i\{\exp[\beta(\varepsilon - \mu)] + 1\}^{-1},$$

where $\tilde{E}$ denotes the contribution of the particular band to the total energy of the system. Using the identity

$$1 - \{\exp[\beta(\varepsilon_i - \mu)] + 1\}^{-1} = \{\exp[-\beta(\varepsilon_i - \mu)] + 1\}^{-1},$$

we may write these expressions in the form

$$\mathcal{N} = \sum_i \{\exp[\beta(-\varepsilon_i + \mu)] + 1\}^{-1}$$

and

$$\tilde{E} = \sum_i (-\varepsilon_i)\{\exp[\beta(-\varepsilon_i + \mu)] + 1\}^{-1} + \sum_i \varepsilon_i .$$

Here we have also used the fact that the total number of energy states in the band is $2N$, which we have expressed in the form $\sum_i 1 = 2N$. The sum, $\sum_i \varepsilon_i$, is a constant and may be ignored, since it does not affect the thermodynamic properties of the system. Thus, we see that $\mathcal{N}$ and $\tilde{E}$ are precisely the number of particles and energy, respectively, for a system of fermions with energy levels $-\varepsilon_i$, and chemical potential $-\mu$. This set of $\mathcal{N}$ "particles" is often described as a set of "holes," and we see that the thermodynamic contribution of an energy band may be discussed either in terms of particles or of "holes."

9.20. A semiconductor has an energy gap, ε_g, between a completely filled and an unfilled band at $T = 0$ K. Suppose that at a temperature $T_1 \neq 0$, $\mathcal{N}$ electrons have vacated the previously filled band and are in the conduction band. Show that the density, n, of conduction electrons is equal to the density, p, of holes in the previously filled band. Show, further, that

$$n = p = 2[(2\pi(\mathbf{m}_e m_h)^{1/2}\mathbf{k}T/\mathbf{h}^2]^{3/2}\exp(-\varepsilon_g/2\mathbf{k}T)$$

provided $\varepsilon_g - \mu \gg \mathbf{k}T$ and $\mu \gg \mathbf{k}T$; here m_h denotes the effective mass of the holes. Refer to Problem 9.19 for a discussion of "holes."

SOLUTION: If at $T = 0$ K the filled band contains $2N$ electrons, then at temperature $T_1 \neq 0$, the conduction band has $\mathcal{N}$ electrons, and the filled band has $2N - \mathcal{N}$ electrons. If we denote the energy levels of the filled band by ε_i and those of the conduction band by ε_j, we may write $2N = \sum_i 1$;

$$\mathcal{N} = \sum_j \{\exp[\beta(\varepsilon_j - \mu)] + 1\}^{-1};$$

and

$$2N - \mathcal{N} = \sum_i \{\exp[\beta(\varepsilon_i - \mu)] + 1\}^{-1}.$$

Using the results of Problem 9.19, we see that the number of holes in the filled band, $\mathcal{P}$, is equal to $\mathcal{N}$, and may be expressed as

$$\mathcal{P} = \mathcal{N} = \sum_i \{\exp[\beta(\mu - \varepsilon_i)] + 1\}^{-1}.$$

If we set the zero of energy at the top of the filled band, we may write $\varepsilon_i = -\mathbf{p}^2/2\mathbf{m}_h$; $\varepsilon_j = \varepsilon_g + \mathbf{p}^2/2\mathbf{m}_e$ and, using Equation (8.58a), we determine n and $\not{p}$ as

$$n = (2/\mathbf{h}^3) \int \mathbf{dp} \, \{\exp[\beta(\varepsilon_g + p^2/2\mathbf{m}_e - \mu)] + 1\}^{-1}$$

and

$$\not{p} = n = (2/\mathbf{h}^3) \int \mathbf{dp} \, \{\exp[\beta(\mu + p^2/2m_h)] + 1\}^{-1}.$$

Here the integrations extend over the momentum states in the conduction band and filled band, respectively. The result quoted for n may be obtained by observing that for $\varepsilon_g - \mu \gg \mathbf{k}T$ and $\mu \gg \mathbf{k}T$, μ is determined by $\mu/\mathbf{k}T = \varepsilon_g/2\mathbf{k}T + (\frac{3}{4}) \ln (m_h/\mathbf{m}_e)$.

9.21. Estimate the density of electrons in the conduction band of an intrinsic semiconductor at $T = 100$ K if $\varepsilon_g/\mathbf{k} = 10^4$ K, and for which both $\mathbf{m}_e$ and m_h are equal to the mass of an electron $\mathbf{m}_e = 9.107 \times 10^{-31}$ kg.

9.22. Using the electronic constant of zinc quoted in Table 9.1, evaluate the Fermi energy, μ_0, and the number of electrons, n, per cm³. The density of zinc is $\rho = 7$ gr/cm³.

9.23. Using Table 9.1, compute the specific heat of silver in the range from 0 to 30 K. The Debye temperature of silver is $\Theta_D = 226$ K. When does the contribution to c_v from the lattice vibrations dominate that stemming from the conduction electrons and vice versa?

LIST OF SYMBOLS FOR CHAPTER 9

Latin letters

A	Constant in Richardson–Dushman equation
$\mathbf{A}$	Amplitude of harmonic wave
a	Lattice spacing
c	Velocity of propagation of a wave

c_l Velocity of a longitudinal wave
c_t Velocity of a transverse wave
c_v Specific heat at constant volume
$\mathscr{D}$ Debye function
d Lattice spacing in a solid
E Energy
$\mathscr{E}$ Einstein function
$\mathbf{e}$ Electronic charge
F Helmholtz function
f Number of degrees of freedom
$\mathbf{f}$ Propagation vector
g Density of frequency distribution
$\mathbf{h}$ Planck's constant
I Electric current density
K Number of independent normal modes
k Force constant in harmonic solid
k' Number of single particle energy levels
$\mathbf{k}$ Boltzmann's constant
L Length
l Number of nodes in a standing wave
m Mass of a particle
$\mathbf{m}_e$ Electronic mass
m^* Effective mass
N Number of particles
$\mathbf{N}$ Avogadro's number
$\mathscr{N}$ Number of phonons
n Number of nodes in a standing wave
$\mathscr{n}$ Occupation number of phonons
P Pressure
$\mathbf{p}$ Linear momentum
$\mathbf{R}$ Perfect-gas constant
$\mathbf{r}$ Represents coordinates x, y, z
S Entropy
T Temperature
t Time
u Molar energy
V Volume
v Harmonic oscillator quantum number
v_m Molar volume
W Work function of electrons in a metal
W_0 Work function at $T = 0$ K
$\mathbf{Y}$ Displacement of an atom in harmonic solid
y Displacement of oscillator from its equilibrium position
Z_k Partition function for a single mode or particle
$\mathscr{Z}$ Partition function for an assembly of particles or modes
$\mathscr{Z}_{ph}$ Phonon partition function

Greek letters

α	Coefficient of thermal expansion
β	$(\mathbf{k}T)^{-1}$
γ	Grüneisen's constant
γ'	Electronic constant
ε	Single particle energy level
ε_{m}	Maximum energy level
Θ	Characteristic temperature of a Debye solid
Θ_{E}	Characteristic temperature of an Einstein solid
κ	Coefficient of isothermal compressibility
Λ	Constant in Nernst–Lindemann equation
λ	Wavelength
μ	Chemical potential
μ_0	Fermi energy of electrons in a solid
ν	Frequency
ν_{E}	Frequency of an Einstein solid
ν_{m}	Maximum frequency of a Debye solid
ν_0	Photo frequency of a metal
ξ	Ratio of vibrational to thermal energy $(\mathbf{k}T)$
ξ_{m}	Ratio of energy of oscillator with frequency ν_{m} to $\mathbf{k}T$
ϕ_{m}	Molar zero-point energy of a Debye solid

Subscript

i	Denotes impurity band
ph	Denotes phonon property

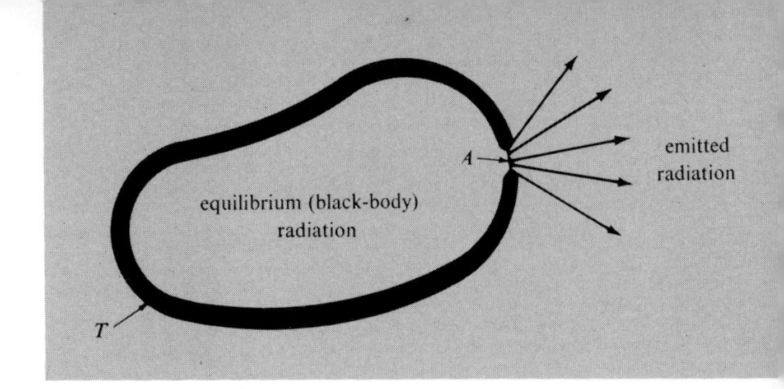

CHAPTER 10

RADIATION

10.1. A Descriptive Introduction

All radiation processes, such as the propagation of light, the emission of gamma or X rays, radiant heating, or the transmission of radio signals, are merely different facets of the same fundamental physical process. Although all such waves propagate with the speed of light

$$\mathbf{c} = (299{,}792.50 \pm 0.10)\ \text{km/sec}, \tag{10.1}$$

through empty space, the impression of variety in the different kinds of radiation is created by the immense range of wavelengths, λ, and frequencies, ν, encountered in nature. As in all wave motions, the two quantities are inversely proportional because they satisfy the dispersion relation†

$$\lambda\nu = \mathbf{c}. \tag{10.2}$$

The range of wavelengths and frequencies encountered in nature is illustrated in Figure 10.1. Cosmic rays have the highest frequencies (shortest wavelengths); they are followed by gamma rays and X rays. The range of radiation which can be detected by our senses is called thermal radiation and extends from $\lambda = 0.1\ \mu$ to $100\ \mu$ ($\nu = 3 \times 10^{15}$ 1/sec to 3×10^{12} 1/sec), the part between $\lambda = 0.36\ \mu$ to $0.76\ \mu$ ($\nu = 8.4 \times 10^{14}$ 1/sec to 3.9×10^{14} 1/sec) being in the visible range. The latter stretches from the violet to the red in increasing order of wavelengths.

† Some authors prefer to employ the *circular frequency* $\omega = 2\pi\nu$ and write $\omega\lambda = 2\pi\mathbf{c}$. Note that $\mathbf{h}\nu = \hbar\omega$.

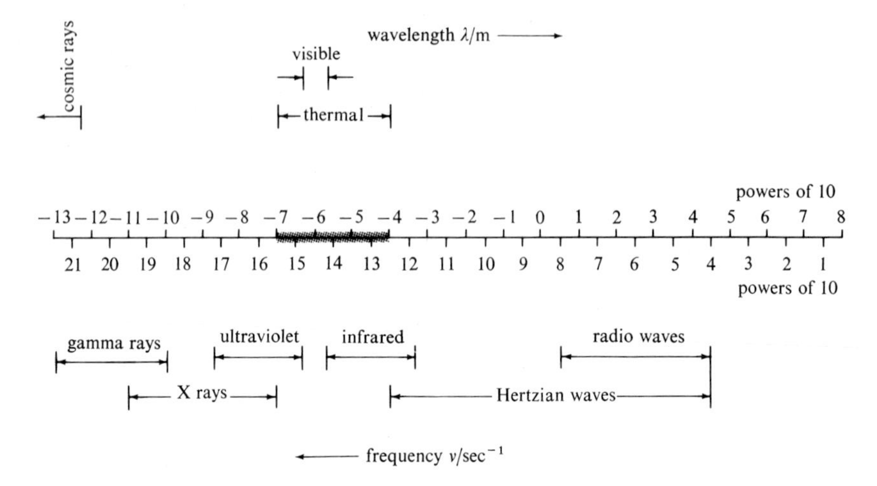

FIGURE 10.1. The spectrum of electromagnetic waves. (Note logarithmic scales: $\ln \lambda + \ln \nu = \ln \mathbf{c}$).

Radiation of very high frequency or very high energy, such as gamma rays, is associated with quantum transitions in the nucleus of an atom. X rays may be produced in atomic transitions by the rapid deceleration of charged particles. Most electronic transitions in atoms and molecules take place in the visible and ultraviolet region of the spectrum. Vibrational and rotational transitions in molecules produce radiation in the infrared. Microwave radiation with wavelengths of the order of a few centimeters can be created by molecular transitions, but longer wavelengths are usually produced by the acceleration of particles in antennas. We have discussed some of these processes in Section 3.15 on spectroscopy.

The common characteristic of all types of radiation is their ability to propagate in the absence of matter and so to transport energy over the empty spaces of the universe, as well as their electromagnetic origin. All such waves propagate in accordance with Maxwell's equations.

The wavelike nature has associated with it a particle-like character, and the latter is described by saying that radiation is the result of the motion of *photons*. The energy and momentum of a photon are

$$\varepsilon = \mathbf{h}\nu \tag{10.3}$$

and

$$\mathbf{p} = \mathbf{hf} \tag{10.4}$$

respectively. Here

$$|\mathbf{f}| = \frac{1}{\lambda} \tag{10.4a}$$

is the wave number, the direction of the wave vector $\mathbf{f}$ being collinear with that in which the wave itself is propagated, that is, with the direction of motion of the photon. Evidently, the rest mass, m_0, of a photon must vanish; otherwise its relativistic mass in motion

$$m = \frac{m_0}{[1 - (v/\mathbf{c})^2]^{1/2}}$$

would not remain finite, since for $v = \mathbf{c}$, m becomes infinite. This does not mean that its mass in motion or its momentum $|\mathbf{p}| = m\mathbf{c}$ also vanish. Equation (10.4a) shows that $|\mathbf{p}| = \mathbf{h}/\lambda$, or

$$|\mathbf{p}| = \frac{\mathbf{h}\nu}{\mathbf{c}} = \frac{\varepsilon}{\mathbf{c}};$$

the latter follows from the dispersion relation (10.2) and the quantum-mechanical equation (10.3). Even though the rest mass $m_0 = 0$, the effective mass is seen to be

$$m = \frac{|\mathbf{p}|}{\mathbf{c}} = \frac{\varepsilon}{\mathbf{c}^2} = \frac{\mathbf{h}\nu}{\mathbf{c}^2}. \tag{10.5}$$

Apart from the numerous, and obvious, practical applications, the study of radiation played a decisive part in the history of science. On the one hand, Lord Rayleigh's attempt to determine the spectral distribution of the specific energy, u, of radiation per unit volume in terms of frequency (or wavelength) demonstrated the intrinsic inadequacy of classical statistical thermodynamics in that it led to the so-called *ultraviolet catastrophe* (Section 10.3.6). On the other hand, M. Planck's formula (Section 10.3.4) successfully resolved this paradox, inducing him to initiate the development of quantum mechanics and so to pave the way for the explanation of many other paradoxes, notably those connected with the application of the principle of equipartition of energy in the calculation of the specific heats of gases (Section 6.5).

A cavity enclosed in a body of arbitrary shape whose walls are maintained at a constant temperature, T, Figure 10.2, contains radiation which can be treated as a thermodynamic system in equilibrium, and the laws of classical as well as statistical thermodynamics can be applied to it. In particular, the radiation in such a cavity can be regarded as a perfect gas of noninteracting photons which have the properties of bosons. Unlike a perfect gas, an adiabatic expansion or compression of the cavity which conserves the energy of the assembly of photons fails to conserve their number; the latter changes from temperature to temperature. Since all photons move with the same velocity $\mathbf{c}$, the equilibrium distribution of the number of photons per unit volume in a cavity at a given temperature T occurs by virtue of their having different frequencies. In a perfect gas, as we recall, the equilibrium distribution of

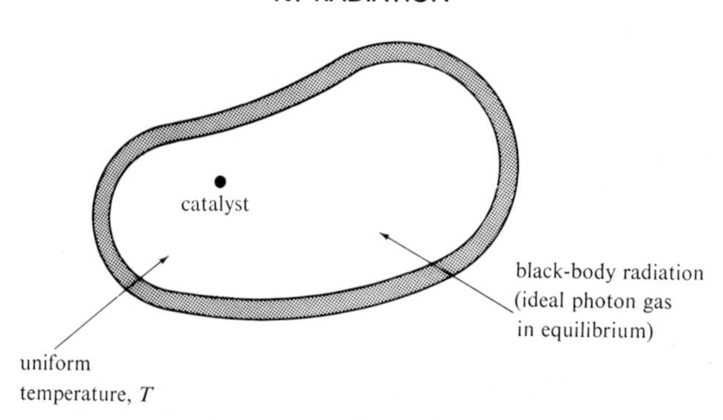

FIGURE 10.2. Black-body radiation as a thermodynamic system.

kinetic energy among the molecules resulted in a characteristic, Maxwellian *velocity distribution* which was maintained through an exchange of momentum in collisions. In our present case we must expect that equilibrium should manifest itself in the form of a characteristic distribution among *frequencies*. We shall study this distribution in Sections 10.3.2 and 10.3.4 where we shall call it the *Planck distribution* in honor of its discoverer.

From what has been said so far it is clear that the collisional mechanism which maintains a Maxwellian velocity distribution in a gas of molecules is not available in a gas of strictly noninteracting photons. This is provided by the walls of the cavity or by the introduction of a very small speck of dust into it, the speck of dust acting as a kind of catalyst. Generally speaking, the interaction of photons with matter, that is, their emission and absorption by it, is very complex and cannot be studied without clearly specifying the properties of the atoms constituting the solid walls of the cavity. Thus it may happen that the cavity becomes filled with photons of a single frequency or with photons which range over a narrow frequency band, that is, with a distribution which differs from that prevailing at equilibrium. We shall exclude such cases from our study assuming that either the walls or the speck of dust emit and reemit photons in a sufficiently random manner first to establish and then to maintain statistical equilibrium. In such circumstances it becomes unnecessary to specify in detail all the mechanisms which operate in the system, it being sufficient to assure that the distribution within it is that which corresponds to the wall temperature T. We can concentrate on the study of the equilibrium itself, since the fundamental postulates of statistical thermodynamics are now sufficient to reveal it to us. Furthermore, we can see clearly that the equilibrium state inside the cavity has become independent of the shape of the cavity or of the detailed nature of the walls surrounding it.

A photon gas which exists in an equilibrium state inside a cavity is said to contain *black-body radiation.* The reason for this term will be advanced later in this section.

Although intrinsically interesting, the study of black-body radiation would be of little practical importance if it could not be related to the radiation emitted by various material surfaces. The existence of such a connection can be perceived if we imagine that a small hole has been made in the body, as shown in Figure 10.3. The photons will now escape through area A of this

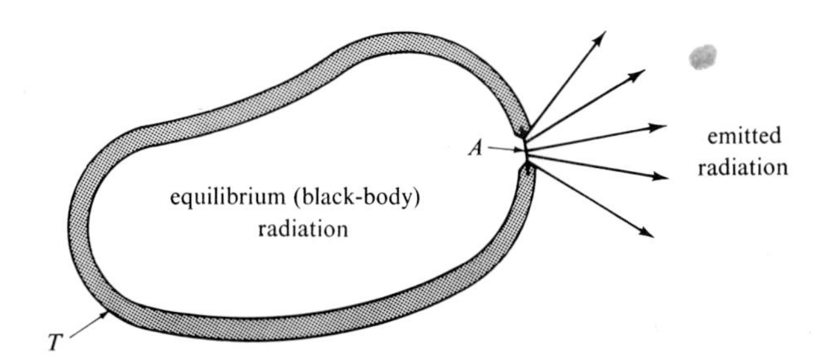

FIGURE 10.3. Emitter of black-body radiation.

hole and behave rather like the molecules of a perfect gas which are allowed to effuse through a pinhole. The existence of such a small opening has a negligible effect on the conditions of equilibrium inside the cavity, and yet, to an outside observer, the aperture will be hardly distinguishable from a small area A of a surface radiating energy into space. To be sure, the characteristics of the radiation from a real surface will differ from this black-body radiation. Nevertheless, the study of the latter forms a very good basis for the understanding of all radiation. In this, black-body radiation plays the same role with respect to that of an arbitrary surface as do the properties of perfect gases with respect to those of real gases.†

An external observer looking at area A will not see it as a black patch but will tend to ascribe to it that color which corresponds to the wavelength with the highest occupation number of photons. We shall see later (Section 10.3.4) that this shifts in the direction of *higher* frequencies (or lower wavelengths) as the temperature, T, is increased. In essence, this is what an operative observes when he estimates the temperature in a furnace by looking into it through a small spy-hole. Dull-red glow corresponds to about 1000 K.

† An extensive study of radiation processes can be found in E. M. Sparrow and R. D. Cess, *Radiation Heat Transfer*, Brooks/Cole Publ. Belmont, California, 1966. Astrophysical applications are given in V. V. Sobolev, *A Treatise on Radiative Transfer* (S. I. Gaposchkin, transl.), Van Nostrand, Princeton, New Jersey, 1963. Additional references are given on p. 439.

As the temperature increases, the dull-red changes to red and then to orange tending toward the blinding bright yellow of the sun whose "effective" surface temperature is about 5800 K. Thus, black-body radiation will appear dark to the eye only at very low temperatures.

The description "black-body radiation" has its origin in an observation which concerns the behavior of a cavity with a hole in the presence of *incident* radiation, Figure 10.4. It is easy to perceive that an incident beam entering

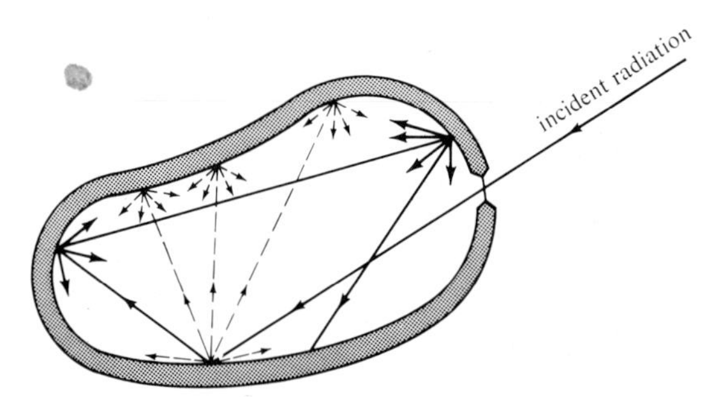

FIGURE 10.4. Perfect absorber of radiation.

the cavity through area A would be almost totally absorbed through repeated specular or diffuse reflections.† Thus, if the hole A did not radiate itself, it would appear black in an incident beam, since only a minute fraction of it would reach the eye of an external observer. Owing to its close connection with radiation from this kind of cavity, equilibrium radiation has been given the name black-body radiation.

As stated earlier, when photons interact with gases or certain solid surfaces (for example, fluorescent materials), they may cause the emission of photons whose frequencies are confined within narrow bands. In equilibrium, black-body radiation also consists of a discrete sequence of frequencies, ν, corresponding to the discrete nature of the eigenenergies $\mathbf{h}\nu$. However, these frequencies are so closely spaced, and span such a large range, that we can employ the usual Euler–Maclaurin approximation, replacing sums by integrals and pretending that the frequency spectrum ν extends over the range from 0 to ∞. This implies that the wavelengths $\lambda \approx \mathbf{c}/\nu_{\max}$ near and around that which corresponds to the most probable frequency, $\nu_{\max}$, must be very small compared with the linear dimensions of the area A and, *a fortiori*, of the cavity itself. In such circumstances, when convenient, we can describe the

† A reflection is said to be *specular* when each incident ray produces a single reflected ray so that the two are in a plane and form equal angles with the normal to the reflecting surface. In *diffuse* reflection, a single incident ray gives birth to a bundle of reflected rays pointing in all directions.

propagation of radiation by using the laws of geometrical optics, and may disregard any phase relations on condition that the observation times are large compared with the characteristic time $\nu_{\max}^{-1}$. Naturally, this excludes from further study the phenomena of diffraction and scattering.

The present chapter is devoted predominantly to a study of black-body radiation. However, in Section 10.8 we shall include a very general statement (Kirchhoff's law) concerning real, nonblack radiating surfaces. The last two sections (10.9 and 10.10) will be devoted to a brief analysis (due to Einstein) of the simplest case of interaction between photons and matter, namely that which occurs when a gas of very low density interacts with an equilibrium assembly of photons in a cavity. This will lead us to a qualitative understanding of the operation of a ruby laser.

10.2. Properties of Electromagnetic Radiation

In the classical theory of electromagnetism, electromagnetic radiation through empty space may be considered to be composed of plane-polarized, monochromatic waves. Each such wave is transverse and has a fixed frequency, ν, and wavelength, λ, both satisfying Equation (10.2). In other words, the direction of the wave oscillations are perpendicular to the direction of propagation of the wave. The former are oscillations of the electric and magnetic field vectors, **E** and **H**, which at any point along the wave are perpendicular to each other. For plane-polarized waves, the electric and magnetic field vectors remain in planes fixed in space throughout the entire length of the wave, as illustrated in Figure 10.5. For example a plane-polarized wave with frequency ν propagating in the z direction may be represented in the form

$$\mathbf{E}(\mathbf{r}, t) = E_0\,\mathbf{i}\cos 2\pi\nu(z/\mathbf{c} - t) \tag{10.6a}$$

$$\mathbf{H}(\mathbf{r}, t) = \frac{E_0}{\mu_0 \mathbf{c}}\,\mathbf{j}\cos 2\pi\nu(z/\mathbf{c} - t), \tag{10.6b}$$

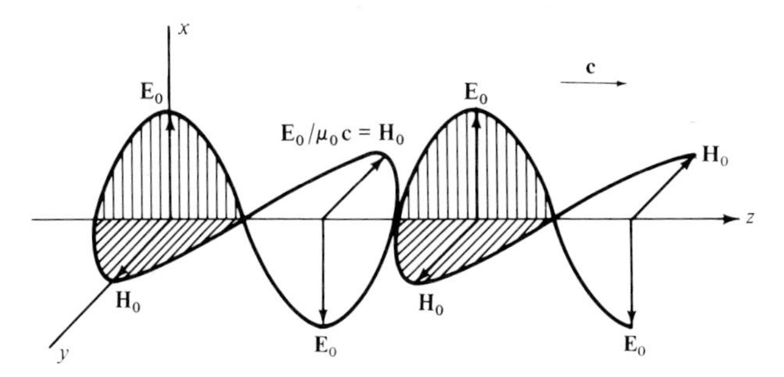

FIGURE 10.5. Plane-polarized electromagnetic wave.

where

$$\mu_0 = 4\pi \times 10^{-7} \text{ Wb/Am} \tag{10.6c}$$

is the *magnetic permeability* of empty space and where E_0 is a constant, $\mathbf{i}$ is a unit vector in the x direction and $\mathbf{j}$ is a unit vector in the y direction. Any plane-polarized wave of frequency v which propagates in the z direction and whose electric vector $\mathbf{E}$ is in a plane forming an angle with the x, z plane can be represented as a superposition of two waves located in the planes x, z and y, z respectively. The former is described by Equations (10.6a) and (10.6b), whereas the vectors $\mathbf{E}$ and $\mathbf{H}$ of the "perpendicular" wave are given by

$$\mathbf{E}_1(\mathbf{r}, t) = E_1 \mathbf{j} \cos 2\pi v(z/\mathbf{c} - t) \tag{10.7a}$$

$$\mathbf{H}_1(\mathbf{r}, t) = -\frac{E_1}{\mu_0 \mathbf{c}} \mathbf{i} \cos 2\pi v(z/\mathbf{c} - t), \tag{10.7b}$$

with suitably chosen values for E_0 and E_1. The electric vector of the combined wave is

$$\mathbf{E} = (E_0 \mathbf{i} + E_1 \mathbf{j}) \cos 2\pi v(z/\mathbf{c} - t), \tag{10.8a}$$

its magnetic vector being

$$\mathbf{H} = \frac{1}{\mu_0 \mathbf{c}} (E_0 \mathbf{j} - E_1 \mathbf{i}) \cos 2\pi v(z/\mathbf{c} - t). \tag{10.8b}$$

Since any electromagnetic plane wave can be so resolved, we say that there exist two degrees of freedom of polarization for it. Other types of polarization can be obtained by a superposition of the two plane-polarized waves represented by Equations (10.6) and (10.7). For example, elliptically polarized light may be represented as

$$\mathbf{E} = E_0 \mathbf{i} \cos 2\pi v(z/\mathbf{c} - t) + E_1 \mathbf{j} \cos[2\pi v(z/\mathbf{c} - t) + \phi] \tag{10.9a}$$

$$\mathbf{H} = \frac{E_0}{\mu_0 \mathbf{c}} \mathbf{j} \cos 2\pi v(z/\mathbf{c} - t) - \frac{E_1}{\mu_0 \mathbf{c}} \mathbf{i} \cos[2\pi v(z/\mathbf{c} - t) + \phi], \tag{10.9b}$$

where ϕ is a constant phase angle. For this type of radiation, the electric and magnetic field vectors at a fixed value of z or at a fixed instant, t, trace out ellipses. Circularly polarized radiation results if E_1 and E_0 are equal. *Unpolarized* radiation may be considered to be a superposition of many different polarized waves, all of which are out of phase with each other. That is, each wave has a phase, ϕ, whose value varies randomly from wave to wave. An arbitrary radiation field is composed of plane waves with varying frequencies, with various directions of propagation and with various directions of polarization. The magnetic field, **H**, and the electric field, **E**, of such radiation satisfy the general wave equations†

$$\frac{1}{\mathbf{c}^2}\frac{\partial^2 \mathbf{E}(\mathbf{r}, t)}{\partial t^2} = \left(\frac{\partial^2}{\partial x^2} + \frac{\partial^2}{\partial y^2} + \frac{\partial^2}{\partial z^2}\right)\mathbf{E}(\mathbf{r}, t) \tag{10.10a}$$

$$\frac{1}{\mathbf{c}^2}\frac{\partial^2 \mathbf{H}(\mathbf{r}, t)}{\partial t^2} = \left(\frac{\partial^2}{\partial x^2} + \frac{\partial^2}{\partial y^2} + \frac{\partial^2}{\partial z^2}\right)\mathbf{H}(\mathbf{r}, t). \tag{10.10b}$$

It can be immediately verified that the electromagnetic waves represented by Equations (10.6)–(10.9) are special solutions of Equations (10.10a) and (10.10b). The fact that the radiation field may be considered as a superposition of plane waves is a consequence of the linearity of Equations (10.10a) and (10.10b).

A final result that we shall need from classical electromagnetic theory concerns the pressure exerted by radiation on a material body. Electromagnetic radiation carries momentum and energy, and the reflection or absorption of an electromagnetic wave causes a change in momentum at the reflecting or absorbing surface, each resulting in a pressure, P, on the surface. Electromagnetic theory proves that this is given by

$$P = \tfrac{1}{3}u \tag{10.11}$$

where u is the energy of the radiation per unit volume.

The quantum-mechanical theory of electromagnetic radiation is based upon Planck's hypothesis and asserts that electromagnetic radiation can be regarded as a collection of photons whose energy and momentum are given by Equations (10.3) and (10.4). Corresponding to the two degrees of freedom for the polarization of an electromagnetic wave, we must associate with each plane wave exactly two photons. Since plane waves cannot be distinguished, we must assume that photons are also indistinguishable. We also know that there is no restriction on the number of waves—and, therefore, photons—with the same energy and momentum; this makes it clear that photons obey Bose–Einstein statistics. The photon gas is a *perfect* boson gas because photons never interact. This is a consequence of the linearity of the wave equations (10.10) and can be evidenced if we imagine that two photons of frequencies ν_1 and ν_2 meet at a given point, Figure 10.6. Using the wave mode of presentation, we assume that two plane waves of frequencies ν_1 and ν_2, respectively, have met at the same point. Since the incident waves satisfy Equations (10.10), and since superposition holds, the emergent wave is merely the sum of the two incident waves. In other words, the two photons pass

† The wave equations are obtained from Maxwell's field equations by retaining in them only the terms which vary as r^{-1} for large distances, r, measured from the charge-sources of the field. Terms which decay as r^{-2} or faster are systematically omitted from them.

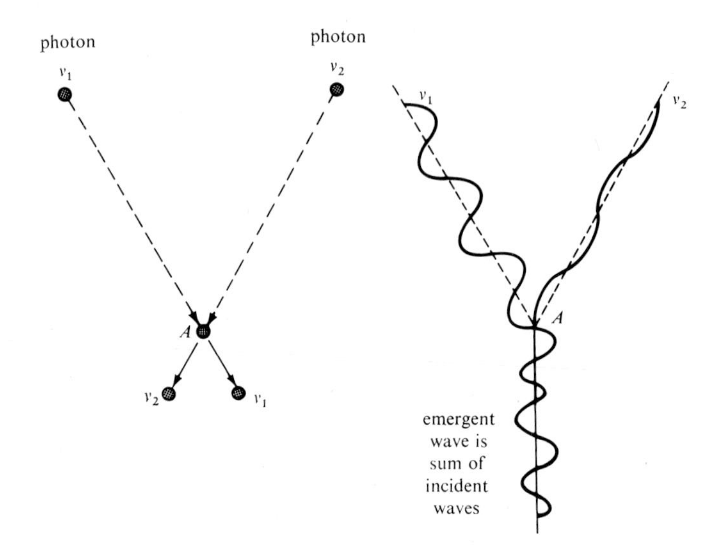

FIGURE 10.6. Two incident photons.

through the point of incidence completely unchanged, and we conclude that photons cannot collide and thereby change each other's energy or frequency.

We shall see later that the relation between the radiation pressure, P, and the energy per unit volume, u, quoted in Equation (10.11) from electromagnetic theory reappears in our statistical theory. This is not surprising because the statistical theory is based on a set of properties postulated for photons on the basis of Maxwell's equations.

10.3. The Photon Gas in Equilibrium

Summarizing the discussion of the preceding section, we reach the conclusion that electromagnetic radiation in equilibrium in an enclosed cavity, Figure 10.2, is equivalent to an assembly of noninteracting bosons each of which has an energy $\mathbf{h}\nu$ and a momentum $\mathbf{h}\nu/\mathbf{c}$. These bosons exist in two mutually perpendicular polarization states. It follows that the statistical description of the system under consideration must be based on the general principles which were developed in Chapter 8.

10.3.1. Number of Photons and Chemical Potential

Before we write down the principal equations of our theory, we must recall that equilibrium is established and maintained by the interactions between the photons and the atoms of a material body—the walls or the speck of dust. This means that the number N of photons present in the cavity cannot be imposed externally on the system, and we must assume that it is fixed by circumstances, that is by the volume V enclosed by and the temperature T maintained at the walls. Thus we may regard the walls of the cavity as a reservoir of photons which emits or absorbs them according to the requirements of thermodynamic equilibrium. It follows that for fixed values of V and T the only variation that can change the Helmholtz function, F, of the system is one with respect to the number of photons, N. Since the Helmholtz function must be a minimum, we may write that

$$\left(\frac{\partial F}{\partial N}\right)_{T,V} = 0 . \tag{10.12}$$

We recall that the derivative $(\partial F/\partial N)_{T,V}$ is equal to the chemical potential, μ, of the system, that is, to its specific Gibbs function. Hence we conclude that

$$\mu \equiv 0 , \tag{10.13}$$

and that the theory of photons is a special case of the theory of bosons, namely bosons whose chemical potential vanishes identically. All that remains to be done is to transcribe the equations of Chapter 8 with the substitutions†

$$\alpha = -\mu/\mathbf{k}T = 0 \qquad \text{and} \qquad \varepsilon_s = \mathbf{h}\nu_i .$$

10.3.2. Density of States

Equation (8.35) yields the partition function

$$\ln \mathscr{Z} = -\sum_i \ln[1 - \exp(-\mathbf{h}\nu_i/\mathbf{k}T)] . \tag{10.14}$$

The total number of photons, Equation (8.32), turns out to be

$$N = \sum_i [\exp(\mathbf{h}\nu_i/\mathbf{k}T) - 1]^{-1} , \tag{10.15}$$

whereas the distribution per energy level, that is among the frequencies, is governed by the relation

$$\bar{n}_i = \frac{1}{\exp(\mathbf{h}\nu_i/\mathbf{k}T) - 1} \tag{10.16}$$

transcribed from Equation (8.33). This is the famous *Planck distribution.*

The quantum states available to the photons are, in principle, determined by the geometry of the cavity. Each such quantum state corresponds to a standing electromagnetic wave, their number being finite. In accordance with

† The trivial change in the running index from s to i is made for convenience; it has no bearing on the argument.

our earlier remarks, we now take into account the fact that in volumes which are large compared with the wavelengths the energy levels are closely spaced. Consequently, it is possible to replace the actual discrete set by a continuous set in the same way as was done in Section 9.3.1 for the acoustic waves in a solid. This makes it necessary to calculate the density, $D(\nu)$, of states.

Retracing familiar steps, we assume, for the sake of simplicity, that we are faced with a cubic cavity of side L and impose periodic boundary conditions on the electromagnetic standing waves in the conviction that the exact boundary conditions in each particular case exert a negligible effect on the result. This means that

$$\mathbf{E}(x + L, y, z) = \mathbf{E}(x, y + L, z) = \mathbf{E}(x, y, z + L) = \mathbf{E}(x, y, z). \qquad (10.17)$$

Since a standing wave may be regarded as a superposition of two traveling waves propagating in opposite directions, we may apply the conditions (10.17) to the equivalent of Equation (10.7a) written for a wave whose vector, **f**, points in an arbitrary direction. This leads to the conclusion that

$$\mathrm{f}_x L = n_x, \qquad \mathrm{f}_y L = n_y, \qquad \mathrm{f}_z L = n_z, \qquad (10.18)$$

where n_x, n_y, n_z must be integers $(0, \pm 1, \pm 2, \ldots)$. Noting that $|\mathbf{f}| = \lambda^{-1}$, we conclude that

$$\frac{1}{\lambda^2} = \frac{1}{L^2}(n_x{}^2 + n_y{}^2 + n_z{}^2), \qquad (10.19)$$

or, that

$$\frac{\nu}{\mathbf{c}} = \frac{1}{L}(n_x{}^2 + n_y{}^2 + n_z{}^2)^{1/2}. \qquad (10.19a)$$

This equation is the exact equivalent of the earlier Equation (9.17). Continuing along the same lines, we can prove that

$$D(\nu) = 2V\frac{4\pi}{\mathbf{c}^3}\nu^2, \qquad (10.20)$$

the factor 2 having been appended to take into account the existence of two degrees of freedom of polarization.

10.3.3. Thermodynamic Properties

We are now in a position to evaluate the partition function explicitly. To this end we replace each term in Equation (10.14) by the corresponding generating function, multiply it by the density of states and integrate over the interval $\nu = 0$ to $\nu = \infty$. Thus,

$$\ln \mathscr{Z} \approx - \int_0^\infty D(\nu) \ln[1 - \exp(-\mathbf{h}\nu/\mathbf{k}T)] \, d\nu$$

$$= -\frac{8\pi V}{\mathbf{c}^3} \int_0^\infty \nu^2 \ln[1 - \exp(-\mathbf{h}\nu/\mathbf{k}T)] \, d\nu . \tag{10.21}$$

The explicit integral can be evaluated with the aid of the substitution

$$x = \frac{\mathbf{h}\nu}{\mathbf{k}T}, \tag{10.21a}$$

in which the dimensionless parameter x represents the ratio of a photon energy, $\mathbf{h}\nu$, to the reference energy, $\mathbf{k}T$. This leads us to the evaluation of the definite integral

$$\mathscr{I} = \left(\frac{\mathbf{k}T}{\mathbf{h}}\right)^3 \int_0^\infty x^2 \ln(1 - e^{-x}) \, dx .$$

Integrating once by parts, we find that

$$\mathscr{I} = -\frac{1}{3}\left(\frac{\mathbf{k}T}{\mathbf{h}}\right)^3 \int_0^\infty \frac{x^3 \, dx}{e^x - 1} = -\frac{\pi^4}{45}\left(\frac{\mathbf{k}T}{\mathbf{h}}\right)^3 . \tag{10.21b}$$

The last definite integral is a standard form† and has a value of $\pi^4/15$. Collecting terms, we can write for the Helmholtz function

$$F = -\mathbf{k}T \ln \mathscr{Z} = -\frac{8\pi^5 V(\mathbf{k}T)^4}{45(\mathbf{hc})^3} = -\frac{4\sigma}{3\mathbf{c}} V T^4, \tag{10.22}$$

where the abbreviation

$$\sigma = \frac{2\pi^5 \mathbf{k}^4}{15 \mathbf{c}^2 \mathbf{h}^3} \tag{10.22a}$$

has been introduced. The quantity

$$\sigma = (5.66961 \pm 0.00096) \times 10^{-8} \frac{W}{m^2 K^4} \qquad \left(\text{or} \times 10^{-5} \frac{\text{erg}}{\text{sec cm}^2 K^4}\right) \tag{10.22b}$$

is known as the *Stefan–Boltzmann constant*; it has been isolated in Equation (10.22) owing to its importance in the study of black-body radiation from surfaces, Equation (10.57b).

† See, for example, E. Jahnke, F. Emde, and F. Lösch, *Tables of Higher Transcendental Functions*, p. 37, McGraw-Hill, New York, 1960.

Employing by now standard and familiar formulas, we can easily derive the following expressions for black-body radiation in equilibrium:

$$E = \frac{4\sigma}{\mathbf{c}} VT^4 = -3F, \tag{10.23}$$

$$H = \frac{16\sigma}{3\mathbf{c}} VT^4, \tag{10.24}$$

$$c_v = \frac{16\sigma}{\mathbf{c}} T^3, \qquad \text{(per unit volume)} \tag{10.25}$$

$$c_p = \frac{64\sigma}{3\mathbf{c}} T^3,\dagger \qquad \text{(per unit volume)} \tag{10.26}$$

$$\gamma = \frac{c_p}{c_v} = \frac{4}{3}, \tag{10.27}$$

$$S = \frac{16\sigma}{3\mathbf{c}} VT^3, \tag{10.28}$$

$$P = \frac{4\sigma}{3\mathbf{c}} T^4, \tag{10.29}$$

$$PV = \frac{4\sigma}{3\mathbf{c}} VT^4 = \tfrac{1}{3}E, \tag{10.30}$$

$$G = F + PV \equiv 0. \tag{10.31}$$

It is interesting to continue this section with a derivation for the number of photons in an equilibrium system. This cannot be done conveniently with the aid of the partition function or by a direct evaluation of the sum in Equation (10.15). Instead, we turn to Equation (10.16) for Plank's distribution and write

$$\mathrm{d}N(\nu, T) = D(\nu)\bar{n}\,\mathrm{d}\nu = \frac{8\pi V}{\mathbf{c}^3} \frac{\nu^2\,\mathrm{d}\nu}{\exp(\mathbf{h}\nu/\mathbf{k}T) - 1} \tag{10.32}$$

for the number of photons whose frequencies are confined in the range $\mathrm{d}\nu$ about ν. Thus, the total number of photons per unit volume is

$$n = \frac{N(T)}{V} = 8\pi\left(\frac{\mathbf{k}T}{\mathbf{ch}}\right)^3 \int_0^\infty \frac{x^2\,\mathrm{d}x}{\mathrm{e}^x - 1},$$

† Here c_p is defined as $(\partial \mathrm{H}/\mathrm{d}T)_V$ and not as $(\partial H/\partial T)_P$.

with x from Equation (10.21a). The definite integral can be evaluated numerically† in terms of the Riemann zeta function of argument 3; it is equal to 2.404 .... Hence

$$n(T) \approx 0.244\left(\frac{2\pi \mathbf{k}T}{\mathbf{hc}}\right)^3. \tag{10.32a}$$

As the temperature T decreases, the photon density decreases with $n \to 0$ for $T \to 0$. Thus, the cavity becomes depleted of particles, and no condensation occurs as it would with bosons of finite rest mass and a nonzero chemical potential.

Reverting to the expression for entropy in Equation (10.28), we see that a photon gas enclosed in an adiabatic cylinder provided with a piston would undergo an isentropic process described by one of the equations

$$VT^3 = \text{const} \qquad \text{or} \qquad PV^{4/3} = \text{const}. \tag{10.33a, b}$$

10.3.4. Planck's Distribution Law

The most important result of the foregoing quantum-mechanical theory of black-body radiation is concerned with the spectral (monochromatic) distribution of energy in the cavity. This is contained in the expression

$$\mathrm{d}E(\nu, T) = \frac{8\pi \mathbf{h}\nu^3 V\,\mathrm{d}\nu}{\mathbf{c}^3[\exp(\mathbf{h}\nu/\mathbf{k}T) - 1]} \tag{10.34}$$

which is obtained simply by multiplying $\mathbf{h}\nu$ into $\mathrm{d}N(\nu, T)$ from Equation (10.32). The function multiplying $V\,\mathrm{d}\nu$ represents the *spectral density of specific energy* (energy per unit volume) which must be ascribed to photons of frequency ν. The need to introduce a spectral density is a consequence of the fact that we have approximated the discrete spectral distribution by a continuous one, and is quite analogous to the probability density discussed in Section 4.4 or to the usual density in a continuous distribution of masses. Denoting the spectral density of the specific energy by $u(\nu, T)$, we may write

$$\mathrm{d}E(\nu, T) = Vu(\nu, T)\,\mathrm{d}\nu$$

with

$$u(\nu, T) = \frac{8\pi \mathbf{h}\nu^3}{\mathbf{c}^3[\exp(\mathbf{h}\nu/\mathbf{k}T) - 1]}, \tag{10.35}$$

or

$$u(\nu, T) = \frac{c_1\nu^3}{\mathbf{c}^4[\exp(c_2\,\nu/\mathbf{c}T) - 1]}, \tag{10.35a}$$

† See, for example, E. Jahnke, F. Emde, and F. Lösch, *Tables of Higher Transcendental Functions*, p. 37, McGraw-Hill, New York, 1960.

where the *first radiation constant*

$$c_1 = 8\pi \mathbf{hc} \tag{10.35b}$$

and the *second radiation constant*

$$c_2 = \frac{\mathbf{hc}}{\mathbf{k}}, \tag{10.35c}$$

their numerical values being

$$c_1 = (4.992579 \pm 0.000038) \times 10^{-24}\,\mathrm{J\,m} \quad (\text{or} \quad \times 10^{-15}\,\mathrm{erg\,cm}) \tag{10.35d}$$

$$c_2 = (1.438833 \pm 0.000061) \times 10^{-2}\,\mathrm{m\,K} \quad (\text{or} \quad \times \mathrm{cm\,K}). \tag{10.35e}$$

Instead of discussing the function $u(\nu, T)$ itself, it is preferable to concentrate on one which is proportional to it and defined as

$$e_b(\nu, T) = \frac{\mathbf{c}u(\nu, T)}{4} = \frac{2\pi \mathbf{h}\nu^3}{\mathbf{c}^2[\exp(\mathbf{h}\nu/\mathbf{k}T) - 1]}$$

$$= \frac{c_1 \nu^3}{4\mathbf{c}^3[\exp(c_2 \nu/\mathbf{c}T) - 1]}. \tag{10.36}$$

The latter is known as *Planck's distribution law* or as the hemispherical *spectral* (monochromatic) *emissive power* of a source of black-body radiation; it will be introduced to the reader in the next section in which its relevance to the process of emission from surfaces will be demonstrated. Before we do this, it is advantageous to become familiar with its mathematical properties.

Equation (10.36) is frequently used in an alternative form in which ν has been eliminated in favor of λ with the aid of the dispersion relation

$$\lambda\nu = \mathbf{c} \qquad \text{with} \quad d\nu = -(\mathbf{c}/\lambda^2)\, d\lambda$$

leading to

$$e_b(\lambda, T) = \frac{c_1 \mathbf{c}}{4\lambda^5[\exp(c_2/\lambda T) - 1]}. \tag{10.36a}$$

When deriving this form, it must be realized that the function $e_b(\lambda, T)$ does not arise from $e_b(\nu, T)$ by the substitution of $\lambda = \mathbf{c}/\nu$ into Equation (10.36) as might appear at first sight. Since the energy carried in a given spectral band must be independent of the variable, λ or ν, selected for the representation, and since $d\nu > 0$ corresponds to $d\lambda < 0$ and *vice versa*, we must have

$$e_b(\lambda, T)\, d\lambda = -e_b(\nu, T)\, d\nu.$$

Substituting $d\nu = -(\mathbf{c}/\lambda^2)\, d\lambda$, we see that

$$e_b(\lambda, T)\, d\lambda = \frac{\mathbf{c}}{\lambda^2}\, e_b(\nu, T)\, d\lambda,$$

which shows that the two spectral densities are not identical. Thus we are led to Equation (10.36a). Some authors prefer to employ an alternative form of the first radiation constant by putting

$$C_1 = \frac{c_1 \mathbf{c}}{4}, \tag{10.37a}$$

so that Equation (10.36a) is written

$$e_b(\lambda, T) = \frac{C_1}{\lambda^5[\exp(c_2/\lambda T) - 1]} \tag{10.37b}$$

with

$$C_1 = (3.74184 \pm 0.00003) \times 10^{-16}\ \mathrm{J\ m^2/sec} \qquad (\text{or } \times 10^{-5}\ \mathrm{erg\ cm^2/sec}). \tag{10.37c}$$

The diagram in Figure 10.7 contains a plot of the spectral emissive power $e_b(\nu, T)$ from Equation (10.36) in terms of the frequency, ν, for several values of absolute temperature, T. The emissive power vanishes for $\nu \to 0$ as well

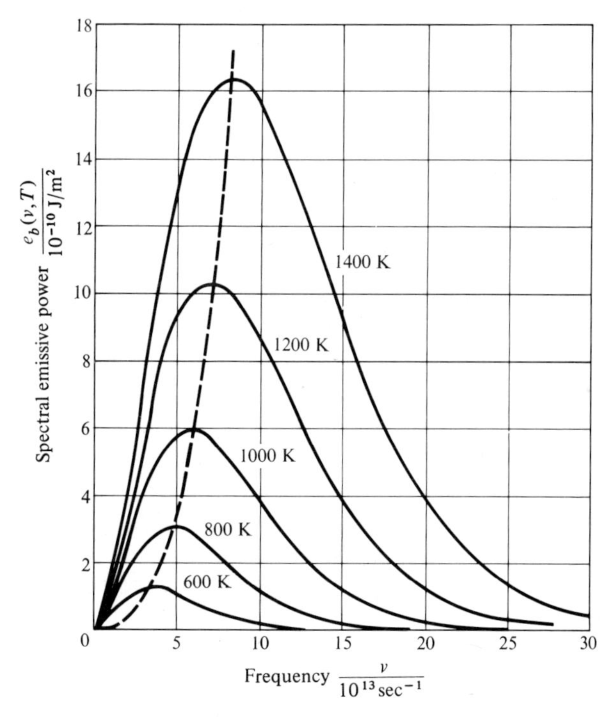

FIGURE 10.7. Spectral emissive power of black-body radiation according to Planck's distribution law, Equation (10.36).

as for $\nu \to \infty$, and it is easy to verify that the integral over the full range of frequencies is finite, as it should be. Consequently, each isotherm passes through a maximum which shifts in the direction of higher frequencies as the temperature is increased, that is, from blue toward yellow in the visible range.

10.3.5. Wien's Displacement Law

It is noteworthy that Equation (10.36) can be put in dimensionless form by regarding $\mathbf{h}^2\mathbf{c}^2 e_b/2\pi(\mathbf{k}T)^3$ as a function of $x = \mathbf{h}\nu/\mathbf{k}T$; the latter dimensionless argument was employed earlier to evaluate several related integrals. In this manner, the representation

$$\varepsilon = \frac{\mathbf{h}^2\mathbf{c}^2 e_b}{2\pi(\mathbf{k}T)^3} = \frac{x^3}{e^x - 1} \tag{10.38}$$

allows us to plot all isotherms in the form of a single curve. This form also allows us to recognize that the maximum in ε_b occurs for a specified value x_{max} of x. Solving the appropriate transcendental equation, we can find that

$$x_{max} = \frac{\mathbf{h}\nu_{max}}{\mathbf{k}T} = 2.821439, \tag{10.39a}$$

which proves that all maxima in Figure 10.7 lie on the cubic parabola

$$e_{b,\,max} = 0.6328707 \frac{2\pi\mathbf{h}}{\mathbf{c}^2} \nu_{max}^3 \tag{10.39b}$$

or

$$\frac{e_{b,\,max}}{\mathrm{J/m^2}} = (0.2931687 \pm 0.000022) \times 10^{-50} \frac{\nu_{max}^3}{\sec^{-3}}.$$

The cubic parabola has been indicated by the broken line in the diagram.

The fact that the maximum in the spectral emissive power of a black-body radiator shifts in the direction of increasing frequency is known as *Wien's displacement law*. This is the law which accounts for the change in color perceived by the human eye as the temperature of the radiator is increased, since the color (that is, average frequency) is judged by the frequency of the photons which are emitted in largest numbers.

The maximum in $e_b(\nu, T)$ itself is proportional to ν_{max}^3, as seen from Equation (10.39b), or, as is easy to verify, to T^3 according to the relation

$$e_{b,\,max}(\nu, T) = c_3\, T^3$$

with

$$c_3 = \frac{2\pi\mathbf{k}^3}{\mathbf{h}^2\mathbf{c}^2} \frac{x_{max}^3}{e^{x_{max}} - 1} = (0.0595609 \pm 0.0000085) \times 10^{-17} \frac{\mathrm{J}}{\mathrm{m^2K^3}}. \tag{10.40}$$

Since the function $e_b(\nu, T)$ is not identical with $e_b(\lambda, T)$, as remarked earlier, it is useful to record that in terms of wavelengths Wien's displacement law takes the form

$$y_{max} = \frac{\lambda_{max} T}{\text{cm K}} = 0.289788 \pm 0.000012 \tag{10.41}$$

with

$$e_{b,\,max}(\lambda, T) = C_3 T^5, \tag{10.41a}$$

where

$$C_3 = \frac{C_1}{y_{max}^5[\exp(c_2/y_{max}) - 1]} = (0.128647 \pm 0.0000029) \times 10^{-4} \frac{\text{J}}{\text{m}^3\,\text{K}^5}. \tag{10.41b}$$

Instead of plotting ε against x, it is more convenient to represent Planck's distribution law in reduced form by drawing

$$\frac{e_b(\nu, T)}{e_{b,\,max}(\nu, T)} \quad \text{in terms of} \quad x = \frac{\mathbf{h}\nu}{\mathbf{k}T},$$

as was done in the logarithmic diagram of Figure 10.8 which depicts the function

$$\varepsilon' = \frac{e_b(\nu, T)}{e_{b,\,max}(\nu, T)} = \frac{e^{x_{max}} - 1}{x_{max}^3}\,\frac{x^3}{e^x - 1} = 0.7035142\,\frac{x^3}{e^x - 1}. \tag{10.42}$$

Evidently, ε' is also a unique function of $\nu/\nu_{max} = x/x_{max}$ and this fact is emphasized by the auxiliary scale in the graph; the latter also contains a dimensional scale ν/T.

10.3.6. The Wien and Rayleigh–Jeans Approximations

For historical reasons, it is appropriate to derive two asymptotic approximations to Planck's distribution law from Equation (10.36). At very low frequencies with $\mathbf{h}\nu \ll \mathbf{k}T$ we can replace the exponential in the denominator by the first two terms of its series expansion. In this limit

$$e_b(\nu, T) = \frac{2\pi\nu^2\mathbf{k}T}{\mathbf{c}^2}, \tag{10.43}$$

and it is interesting to note that now Planck's constant, **h**, has disappeared from the relation. Equation (10.43) is called the *Rayleigh–Jeans formula* because it was derived by these two authors prior to Planck's discovery of quantum mechanics. It is the result of the assumption that each state participates equally in the distribution of energy, that is, by assigning an energy $\mathbf{k}T$ to each standing wave in accordance with the principle of equipartition. Attempts to integrate this density with respect to the full range of frequencies resulted in the "ultraviolet catastrophe" alluded to before, since extension of the sum to the high-frequency range causes the integral to diverge.

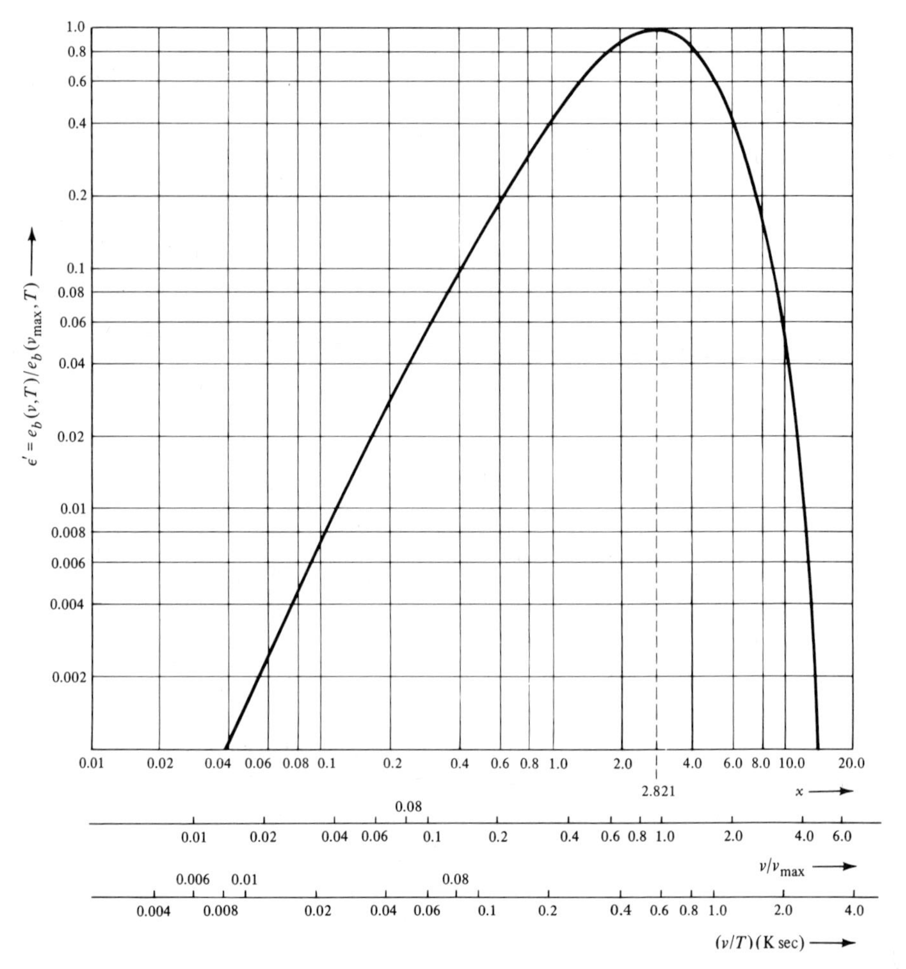

FIGURE 10.8. Planck's distribution law in reduced form.

In the opposite limit of $\mathbf{h}\nu \gg \mathbf{k}T$ we may neglect unity with respect to the exponential in Equation (10.36). In this manner we obtain

$$e_b(\nu, T) = \frac{2\pi\mathbf{h}\nu^3}{\mathbf{c}^2}\, e^{-\mathbf{h}\nu/\mathbf{k}T}, \tag{10.44}$$

the relation being known as *Wien's formula*, because W. Wien† first derived it on the basis of a few plausible, *ad hoc* assumptions.

The diagram in Figure 10.9 contains a comparison between Planck's exact equation and the two historical approximations which preceded it in time. It is remarkable how closely

† *Ann. Phys.* (*Leipzig*), **58** (1896), 662.

Wien's formula anticipated Planck's. At the time, the failure of the Rayleigh–Jeans derivation appeared incomprehensible and Planck's successful resolution of this paradox in 1900 lent credence to this, historically very first, success of the then new theory of quanta.

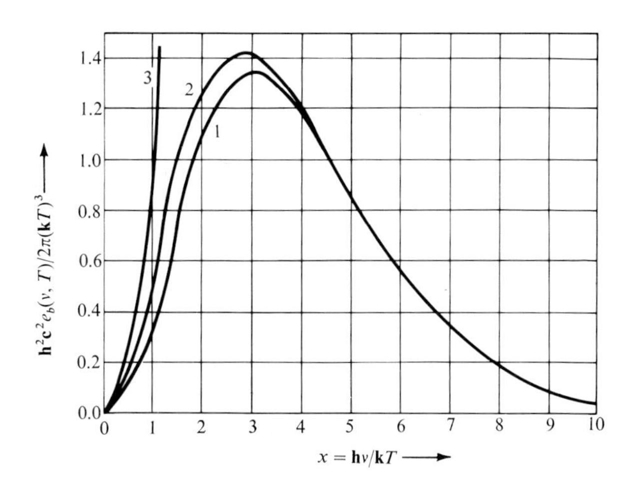

FIGURE 10.9. Plot of the function

$$\frac{\mathbf{h}^2\mathbf{c}^2}{2\pi(\mathbf{k}T)^3}\, e_b(\nu, T) = \frac{x^3}{e^x - 1}$$

in terms of the dimensionless parameter $x = \mathbf{h}\nu/\mathbf{k}T$ and its comparison with the Rayleigh–Jeans and Wien approximations. 1, Wien's formula, Equation (10.44); 2, Planck's formula, Equation (10.36); 3, the Rayleigh–Jeans formula, Equation (10.43). R. Kubo, H. Ichimura, T. Usui, N. Hashizume, *Statistical Mechanics*, p. 121, North-Holland Publ., Amsterdam, 1965.

10.4. Emission and Absorption of Black-Body Radiation

10.4.1. Radiation from a Surface

As far as practical applications are concerned, we are usually interested in the flow of energy from a solid body which can be assumed to exist at a uniform temperature T into a space surrounded by various bodies whose surface temperatures are different from each other and from T. We can schematize this problem into the one illustrated in Figure 10.10. We imagine here that the body A is kept at a uniform temperature T inside an enclosure whose inner surface S shows a temperature distribution $T'(\mathbf{r})$. Thus, in contrast with our previous considerations, the radiation filling the space cannot be said to be in equilibrium.

In such general circumstances there will be a net flow of energy to or from the body, and its temperature will change with time. Experience shows that

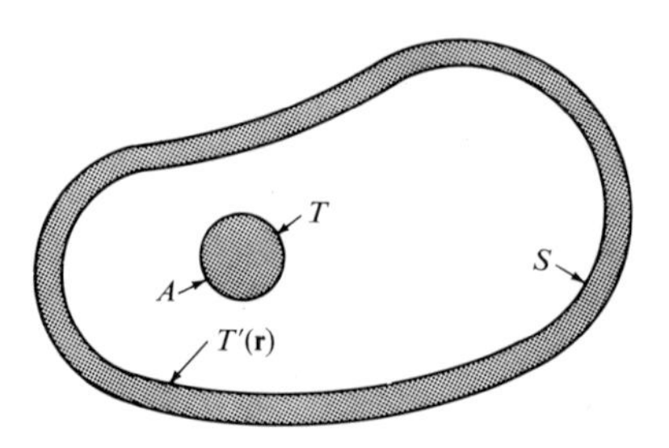

FIGURE 10.10. The practical problem of radiant heat exchange.

the rate of this net transfer of energy can be computed on the assumption that the body emits and absorbs energy *at every instant* as if its state were one of equilibrium. This assumption is equivalent to the principle of local state applied more generally in the study of irreversible processes in continuous systems formulated in Section 1.5.1; it will recur in Chapter 12 in our study of transport processes.

We know from the preceding sections that the nature of the radiation filling space is independent of the properties of body A and surface S when $T'(\mathbf{r}) = \text{const} = T$. In such circumstances its properties are those of black-body radiation. In the more general case, the absorption and emission of body A will become different. The simplest method of accounting for the differences between different materials is to study the departures of their properties from those of a black body.

Although, in principle, the absorption and emission of radiation from a material body must depend on its whole structure, it is found experimentally, particularly in the case of opaque solid bodies, that the departures from black-body emission and absorption can be characterized by considering its surface only. For this reason, we often say that *radiation is a surface phenomenon.*

A surface which most nearly retains the properties of an emitter and absorber of black-body radiation can be created by providing a small opening in a cavity, as illustrated earlier in Figure 10.4. Real surfaces deviate in varying degrees from the properties of such a black-body surface, the same surface showing different departures depending on temperature, on the range of wavelengths encountered in the process, on direction, and on polarization. For example, white enamel has properties which are close to those of a black body in the range of wavelengths from 3 to 8 μ (infrared), but quite different outside it.

We do not propose to enter into an exhaustive study of emission from and absorption by the immense variety of fluorescent and nonfluorescent opaque surfaces encountered in practice or of the radiation processes which occur in transparent media (gases, liquids, and certain solids). We shall restrict ourselves to the study of ideally black surfaces radiating into a homogeneous and isotropic empty space. This is due to the fact that the properties of such surfaces are directly related to those of radiation in equilibrium, and because they are taken as a background against which the properties of all other surfaces are understood. In doing this, it must be remembered that "black" in our context does not refer to the color of the surface as seen by the human eye but to its idealized radiative properties. In this way we hope to provide a firm basis for the reader's future studies.†

10.4.2. Geometric Relations

Before we can fully understand the radiation to and from black-body surfaces, we must establish several geometric concepts with which we can describe it. The need for doing this arises from the fact that we imagine that radiation energy is distributed continuously over the spectrum of frequencies or wavelengths. As already intimated, this allows us to resort to the same geometrical description as that used in ray optics. Thus we imagine that energy is propagated along straight geometric lines in a homogeneous and isotropic medium. An infinite number of them pass through any point in space and two rays moving in diametrically opposite directions behave as if they were entirely independent of one another. When necessary, we can imagine that each ray is monochromatic (that is, of fixed frequency or wavelength) and polarized in one particular direction.

According to this picture one ray has no energy associated with it, in the same way as a geometric line or surface in continuously distributed matter has no mass associated with it. To be able to set up energy balances, we must consider two elements of area dA_1 and dA_2, and we must realize that energy is localized in conical pencils of rays. More precisely, we refer to Figure 10.11 which shows two elements of surface, dA_1 (centered on P_1) and dA_2 (centered on P_2), both being oriented in an arbitrary fashion with respect to each other, and examine the rate at which energy radiated from surface dA_1 passes through dA_2. We denote the distance P_1P_2 by r, and assume

† More or less extensive treatments of the process of radiative heat transfer can be found in the following references (see also references quoted in the footnote on p. 421) M. Jakob, *Heat Transfer*, Vols. 1 and 2, Wiley, New York, 1949, 1957; E. R. G. Eckert and R. M. Drake, Jr., *Heat and Mass Transfer*, McGraw-Hill, New York, 1959; H. Gröber and S. Erk, *Fundamentals of Heat Transfer* (revised by U. Grigull; J. R. Moszynski, transl.), McGraw-Hill, New York, 1961; W. M. Rohsenow and H. Y. Choi, *Heat, Mass and Momentum Transfer*, Prentice-Hall, Englewood Cliffs, New Jersey, 1961; F. Kreith, *Principles of Heat Transfer*, 2nd ed., International Textbook, Scranton, Pennsylvania, 1965. An excellent, and historically very important, introduction into the physics of radiation is contained in M. Planck, *Theory of Heat* (H. L. Brose, transl.), Vol. 5 of "Introduction to Theoretical Physics," MacMillan, New York, 1932. See also, R. Siegel and J. R. Howell, *Thermal Radiation Heat Transfer*, 2 vols., NASA SP-164, Washington, D.C., 1968.

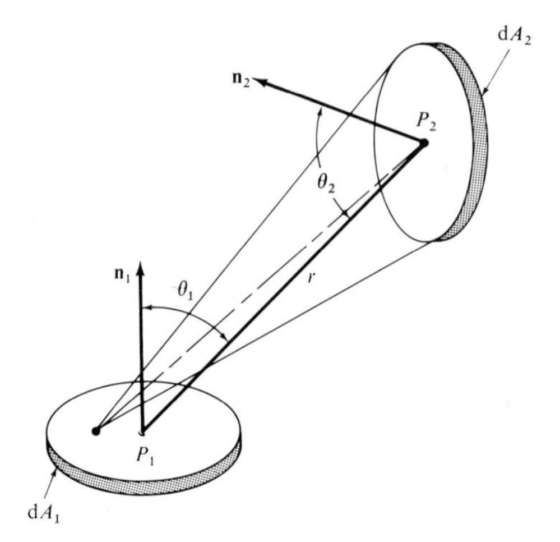

FIGURE 10.11. Radiation emitted by area dA_1 and intercepted by area dA_2.

that the angles formed by r and the respective normals, n_1 and n_2, are θ_1 and θ_2. The energy radiated is pictured contained in the infinity of cones based on all the points in dA_1 with dA_2 as a base. One such cone has been drawn in the diagram.

The magnitude of a cone in which energy can be thought to be localized is not measured by the base area but by its projection $dA_2 \cos\theta_2$ in a direction normal to the axis r, and it is clear that the amount of energy passing through dA_2 per unit time must be directly proportional to $dA_2 \cos\theta_2$ and inversely proportional to r^2. It is equally clear that the amount of energy emitted by dA_1 in the dircction P_1P_2 depends on its orientation with respect to r as measured by angle θ_1. For $\theta_1 = \pi/2$, no energy would reach dA_2; whereas for $\theta_1 = 0$, this would be highest. It is convenient to measure this orientation by $\cos\theta_1$ and to set the rate, $d\overline{\Phi}$, at which energy is emitted from dA_1 per unit time and intercepted by dA_2, proportional to it. Finally, the emitted power must also be proportional to dA_1, the product $dA_1 \cos\theta_1$ measuring the solid angle subtended from P_2 by dA_1 at the distance r. Thus we may put

$$d\overline{\Phi} = i\,\frac{dA_1\, dA_2 \cos\theta_1 \cos\theta_2}{r^2}. \tag{10.45}$$

The flux $\overline{\Phi}$, being an energy per unit time, is measured in units of

$$\{\overline{\Phi}\} = \frac{\text{J}}{\text{sec}}. \tag{10.45a}$$

The quantity i defined by Equation (10.45) is called the *intensity of radiation.* In principle, the intensity of radiation will depend on the physical nature of the emitting surface and on the orientation angle, θ_1. We shall see later that the intensity, i, is uniform and independent of orientation for a black-body emitter.

It is now convenient to introduce the flux, Φ which measures the rate of energy emission per unit area of emitting surface, so that

$$d\Phi = \frac{d\overline{\Phi}}{dA_1}, \tag{10.45b}$$

with

$$\{\Phi\} = \frac{\text{J}}{\text{m}^2\ \text{sec}}. \tag{10.45c}$$

Noting, further, that the solid angle $d\omega$ subtended by dA_2 at P_1 is

$$d\omega = \frac{dA_2 \cos\theta_2}{r^2},$$

we can rewrite the definition (10.45) as

$$i = \frac{1}{\cos\theta}\frac{d\Phi}{d\omega}. \tag{10.46}$$

Here, we have omitted the subscript 1 as unnecessary. Physically, and borrowing the terminology from optics, the intensity, i, describes the degree of brightness that an eye placed at P_2 would observe when looking on dA_1.

Equation (10.46) shows that the intensity of radiation, i, is proportional to the amount of energy, $d\Phi/d\omega$, per unit time (or power) which leaves a given surface in a prescribed direction per unit solid angle of a radiation cone. The intensity of radiation is measured in the same units as the flux because the solid angle, ω, is taken to be dimensionless. Nevertheless, the physical (or, more precisely, geometrical) characters of the two quantities are different.

In order to obtain the total flux of energy (energy per unit area and time) emitted into space by an element of surface, dA, it is necessary to integrate over all solid angles of half-space, as suggested in Figure 10.12. Since $d\omega = \sin\theta\, d\theta\, d\phi$, we find that

$$\Phi = \int_{\text{hemisph}} i\cos\theta\, d\omega = \int_0^{2\pi}\int_0^{\pi/2} i\sin\theta\cos\theta\, d\phi\, d\theta. \tag{10.47}$$

If, as will be the case with black-body surfaces, the intensity of radiation, i, is a constant, we obtain

$$\Phi = \pi i. \tag{10.47a}$$

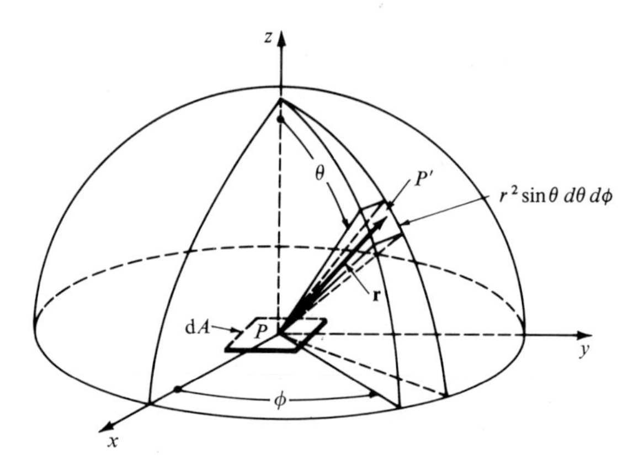

FIGURE 10.12. Radiation into half-space.

For a finite area A of uniform radiant surface, we would write

$$\overline{\Phi} = \int_A \Phi \, \mathrm{d}A = \int_A \int_0^{2\pi} \int_0^{\pi/2} i \sin\theta \cos\theta \, \mathrm{d}\theta \, \mathrm{d}\phi \, \mathrm{d}A. \qquad (10.48\text{a})$$

The quantity Φ is sometimes referred to as the *hemispherical* flux, and $\overline{\Phi}$ is the *hemispherical* power.

So far we have tacitly assumed that the optical rays carry the full spectrum of frequencies as well as the two directions of polarization It is clear, however, that the preceding geometrical quantities can be defined with respect to monochromatic or polarized rays. In this manner we can define a monochromatic intensity i_ν (measured in joules per square meter) or two monochromatic, polarized intensities $i_{\nu\perp}$ and $i_{\nu\parallel}$. Similarly, we may define polarized intensities which are resolved spectrally with respect to wavelength, $i_{\lambda\perp}$ and $i_{\lambda\parallel}$ (both measured in joules per cubic meter-second). Evidently,

$$\begin{aligned}\overline{\Phi} = &\int_0^\infty \int_A \int_0^{2\pi} \int_0^{\pi/2} i_{\nu\perp} \sin\theta \cos\theta \, \mathrm{d}\theta \, \mathrm{d}\phi \, \mathrm{d}A \, \mathrm{d}\nu \\ &+ \int_0^\infty \int_A \int_0^{2\pi} \int_0^{\pi/2} i_{\nu\parallel} \sin\theta \cos\theta \, \mathrm{d}\theta \, \mathrm{d}\phi \, \mathrm{d}A \, \mathrm{d}\nu. \qquad (10.48\text{b})\end{aligned}$$

If the intensities $i_{\nu\perp}$ and $i_{\nu\parallel}$ are equal and both independent of frequency, then

$$i = 2\int_0^\infty i_{\nu\perp} \, \mathrm{d}\nu = 2\int_0^\infty i_{\nu\parallel} \, \mathrm{d}\nu = \int_0^\infty i_\nu \, \mathrm{d}\nu. \qquad (10.48\text{c})$$

10.4.3. Detailed Balance in Black-Body Radiation and Lambert's Cosine Law

In order to specialize the preceding geometric quantities, $\overline{\Phi}$, Φ, and i, as well as their spectrally resolved counterparts to the case of black-body radiation, we shall use the subscript b for i, writing i_b, $i_{b\nu}$, $i_{b\nu\perp}$, or $i_{b\nu\parallel}$ as the case may be. As far as the flux of energy, Φ, radiated in a prescribed direction per unit area, time, and solid angle, is concerned, it is customary to use the symbol e instead and to employ the term *emissive power* for it. Thus, e_b denotes the *total* emissive power, $e_{b\nu}$ denotes the *spectral* or *monochromatic* (unpolarized) emissive power, whereas the same quantities for the two directions of polarization are denoted by $e_{b\nu\perp}$ and $e_{b\nu\parallel}$.

We now imagine that the two elementary surfaces from Figure 10.11 have been enclosed in a very large cavity of temperature T and brought to equilibrium with it. In this case, Figure 10.13, the emission of surface $\mathrm{d}A_1$ intercepted by $\mathrm{d}A_2$ can be written

$$\mathrm{d}\overline{\Phi}_{b1} = i_{b1} \frac{\mathrm{d}A_1\, \mathrm{d}A_2 \cos\theta_1 \cos\theta_2}{r^2}, \tag{10.49a}$$

whereas that emitted by surface $\mathrm{d}A_2$ and intercepted by $\mathrm{d}A_1$ is

$$\mathrm{d}\overline{\Phi}_{b2} = i_{b2} \frac{\mathrm{d}A_1\, \mathrm{d}A_2 \cos\theta_1 \cos\theta_2}{r^2}, \tag{10.49b}$$

both in accordance with Equation (10.45). At equilibrium the two quantities

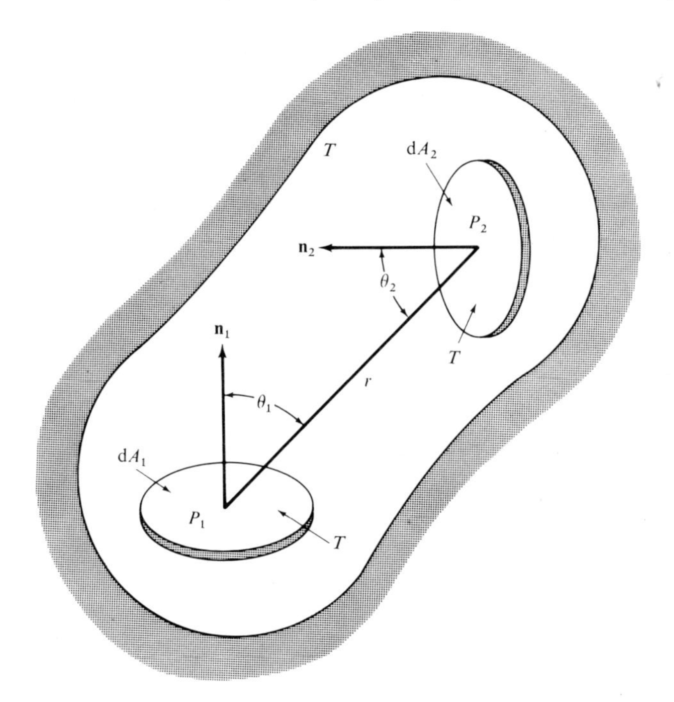

FIGURE 10.13. Two black-body surfaces in equilibrium.

must be equal because the balance is essentially independent of the envelop-ing surface; this proves that

$$i_{b1} = i_{b2}.$$

Since the mutual orientations of both radiating surfaces are quite arbitrary, we see immediately that the intensity of radiation from a black-body surface in equilibrium is constant:

$$i_b = \text{const}. \tag{10.50}$$

From the algebraic point of view, this result is a consequence of the symmetry of Equation (10.45) with respect to the subscripts 1 and 2. This was the real reason for the introduction of the factor $\cos\,\theta_1$ when Equation (10.45) was first written, and has rendered the quantity i defined by this equation independent of direction for black-body radiation *in equilibrium.*

We can understand the preceding result physically with the aid of the optical analogy. To this end we imagine an eye placed at P_2, looking towards dA_1 and scanning it along a meridian circle. If dA_1 were a black-body radiator, the eye would observe a circle of a certain brightness when $\theta = 0$. As θ is made to increase, an ellipse of progressively smaller apparent area would be observed. In spite of this, the diminishing area would appear to have a constant brightness.

Employing Equation (10.46) we deduce that

$$\frac{de_b}{d\omega} = i_b \cos\theta. \tag{10.51}$$

This equation shows that the black-body emissive power per unit solid angle in a given direction θ with respect to the normal is proportional to the cosine of the angle θ *when the surface is in equilibrium.* Equation (10.51) is known in the literature on radiation as *Lambert's cosine law.* The polar diagram of Figure 10.14 illustrates the variation of i_b and $de_b/d\omega$ for a radiating black-body surface dA.

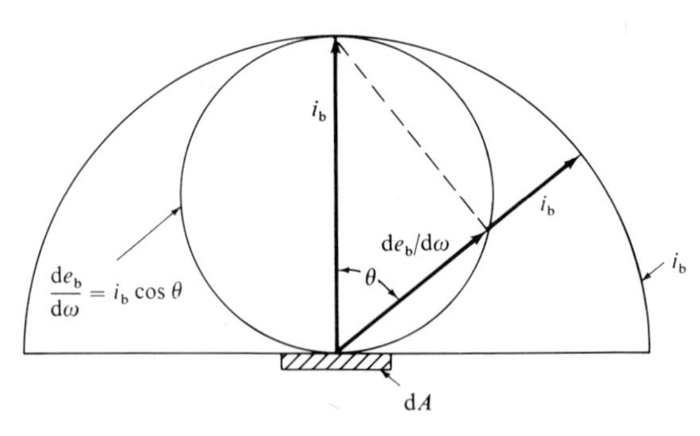

FIGURE 10.14. Lambert's cosine law.

Carrying the preceding human-eye simile a step further, we can imagine a human face scanning area dA_1 along the meridian plane. Although the brightness appears constant to the eye, the feeling of hotness which is proportional to $de_b/d\omega$ would change from a maximum at $\theta = 0$ to practically nothing as $\theta \to \pi/2$. This is the reason why we instinctively tend to look toward incandescent surfaces at glancing angles.

Reverting to the system of Figure 10.13, we can now imagine that one of the surfaces has been covered, successively, with a variety of monochromatic filters or polarizers. This will have the effect of filtering out equally the incident, full black-body radiation as well as the emitted radiation. It is felt that in such circumstances equilibrium should be preserved, otherwise, and in violation of the Second law, one of the surfaces would spontaneously increase or decrease its temperature. In this manner we prove that in equilibrium the emitted and the absorbed fluxes of radiation must be *balanced in detail.* In other words, balance must exist not only with respect to the total energy, but also with respect to every wavelength and direction of polarization *separately*. Furthermore, this means that

$$\frac{de_{b\nu}}{d\omega} = i_{b\nu} \cos\theta\,, \tag{10.52}$$

regardless of the presence or absence of polarization. In any case, in black-body radiation

$$i_{b\nu\perp} = i_{b\nu\parallel} \tag{10.52a}$$

owing to isotropy.

The essential assumption in the practical applications of the present theory is the statement that all *the preceding relations continue to apply even if the surrounding envelope is not at the temperature T of the surface*, provided only that the latter is truly black. The surface opening in a cavity constitutes a very close approximation to such conditions.

Applying the principle of detailed balance to Equation (10.48a), we can write

$$e_{b\nu} = \pi i_{b\nu} \tag{10.53a}$$

for spectrally resolved quantities (with or without polarization) and

$$e_b = \pi i_b \tag{10.53b}$$

for total (hemispheric) quantities. Thus, for black-body radiation the spectral as well as the total emissive power differ from the spectral or total intensity of radiation, respectively, merely by a factor of π.

The only remaining problem is to relate the spectral black-body emissive power, $e_{b\nu}$, to the spectral density of specific energy, $u(\nu, T)$, of black body radiation.

10.5. Relation between Black-Body Emissive Power and Spectral Density of Specific Energy

The problem formulated at the end of the last section is best solved by abandoning the ray-optics mode of description in favor of the representation in terms of photons. This induces us to discuss the mechanics of photons effusing through an opening dA made in the walls of a cavity filled with equilibrium radiation, Figure 10.15.

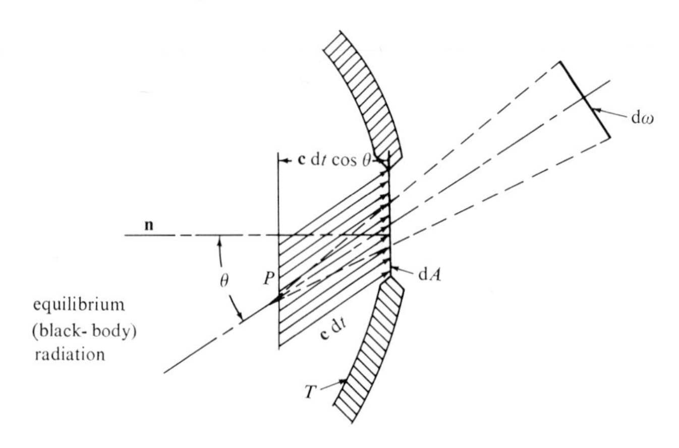

FIGURE 10.15. Effusion of photons.

The fraction of photons which escape in a prescribed direction (defined by angle θ shown in the figure and an angle ϕ which is not shown) is equal to the ratio $d\omega/4\pi$ of a solid angle $d\omega$ drawn about this direction and the total solid angle 4π of space.† Moreover, the photons which leave the cavity through the aperture dA in time dt from an interior region are contained in a cylinder of height $\mathbf{c}\cos\theta\,dt$ whose volume is $\mathbf{c}\cos\theta\,dt\,dA$. This cylinder encompasses an amount of energy

$$u(\nu, T)\mathbf{c}\cos\theta\,dt\,dA\,,$$

where $u(\nu, T)$ is the spectral density of specific energy per unit volume first given in Equation (10.35). Of these, only the fraction $d\omega/4\pi$ can escape in the form of emitted radiation through the radiation cone shown in the sketch. It follows that the monochromatic density of energy, $de_b/d\omega$, emitted in a prescribed direction per unit area, time and solid angle is

$$\frac{de_{b\nu}}{d\omega} = \frac{\mathbf{c}u(\nu, T)}{4\pi}\cos\theta\,. \tag{10.54}$$

† Not of half-space, because at an *interior* point P photons move in *all* directions.

This quantity turns out to be proportional to $\cos\theta$, except for a factor which is independent of the angle θ. Thus, the microscopic argument leads us again to *Lambert's cosine law*. It is useful to notice that the present derivation, like the one given in Section 10.4.3, hinges on the assumption that the photons have an equilibrium distribution and merely represents a consequence of this fact. The same assumption was made in Section 10.4.3, except that there it was stated in the equivalent, macroscopic mode of expression.

Integrating over the half-space, in the same manner as in Equation (10.47a), we see that

$$e_{b\nu} \equiv e_b(\nu, T) = \frac{\mathbf{c}u(\nu, T)}{4}. \tag{10.55}$$

This is the required link between the properties of equilibrium radiation and the emission of an ideal black surface. The function $e_b(\nu, T)$ was discussed in detail in Section 10.3.4 and was quoted explicitly as Equation (10.36).

Referring to Equations (10.52) and (10.35) we can see that

$$i_{b\nu} = \frac{\mathbf{c}u(\nu, T)}{4\pi} = \frac{2\mathbf{h}\nu^3}{\mathbf{c}^2[\exp(\mathbf{h}\nu/\mathbf{k}T) - 1]}, \tag{10.56}$$

confirming that the intensity of black-body radiation is independent of direction.

10.6. The Stefan–Boltzmann Law

The total black-body emissive power, e_b, is found by integration over frequencies, and is

$$e_b(T) = \pi \int_0^\infty i_{b\nu}\, d\nu = \frac{2\pi\mathbf{h}}{\mathbf{c}^2} \int_0^\infty \frac{\nu^3}{\exp(\mathbf{h}\nu/\mathbf{k}T) - 1}\, d\nu\,. \tag{10.57a}$$

The definite integral has been evaluated in Equation (10.21b), and we may record the result as

$$e_b(T) = \frac{2\pi^5\mathbf{k}^4}{15\mathbf{c}^2\mathbf{h}^3}\, T^4 = \sigma T^4, \tag{10.57b}$$

where the *Stefan–Boltzmann constant*, σ, is identical with the one defined in Equation (10.22a); its numerical value was given as Equation (10.22b). Equation (10.56) embodies the *law of Stefan and Boltzmann* which asserts that the total (hemispherical) emissive power of a black-body surface is proportional to the fourth power of its absolute temperature.

10.7. The International Practical Temperature Scale of 1968

The equations for the total black-body emissive power, Equation (10.57b), as well as Equation (10.36) for the spectral emissive power have been derived from first principles in terms of the thermodynamic temperature, T. The inclusion of the appropriate numerical value of the Boltzmann constant, $\mathbf{k}$, assures us that the temperature of the triple point of water has the conventional value of $T_3 = 273.16$ K, as explained in Section 6.3. In other words, if a measurement is made of $e_b(T)$, $e_b(\nu, T)$ or of some other convenient quantity derived from them, it would be possible to determine the temperature of the radiator (provided that its surface is sufficiently close to being ideally black!) on the internationally accepted *thermodynamic* scale directly. In this manner, a black body can be used as a primary thermometer, at least in principle.

In practice, it is found that such direct, *absolute* measurements are difficult to perform. Instead, relative measurements are made with the aid of suitably calibrated and properly designed pyrometers. The international practical temperature scale of 1968, now in force, defines the scale above the freezing point of gold, $T_{68}(\mathrm{Au}) = 1337.58$ K, with reference to Equation (10.37b) by means of the formula

$$\frac{e_{b\lambda}(T_{68})}{e_{b\lambda}[T_{68}(\mathrm{Au})]} = \frac{\exp[c_2/\lambda T_{68}(\mathrm{Au})] - 1}{\exp(c_2/\lambda T_{68}) - 1}, \tag{10.58}$$

with $c_2 = 1.4388 \times 10^{-2}$ m K which is very close to the best value given as Equation (10.36e). Thus, a measurement of the ratio of the spectral emissive power at an unknown temperature, T_{68}, with respect to that at the gold point allows us to determine that temperature with respect to the temperature $T_{68}(\mathrm{Au})$. At the present time it is estimated that the adopted value for $T_{68}(\mathrm{Au})$ is burdened with an uncertainty of ± 0.2 K in relation to the thermodynamic scale of temperatures.

10.8. Kirchhoff's Law

In this last section concerned with distributed radiation, we slightly widen our scope and consider opaque emitters and absorbers of radiation which are not necessarily black. This will lead us to a general relation between the properties of such surfaces which describe their behavior with respect to absorption and emission. This is known as *Kirchhoff's law*.

In the ensuing derivation we shall attempt to give a slightly more general form than is customary.† The essential idea of the proof is the recognition of the fact that any opaque nonfluorescent‡ radiator *can* be brought into a state of equilibrium with a black-body radiator in an envelope, the resulting

† See M. Planck, *Theory of Heat* (H. L. Brose, transl.), Vol. 5 of "Introduction to Theoretical Physics," p. 189, Macmillan, New York, and L. D. Landau and E. M. Lifshitz, *Statistical Physics* (E. and R. F. Peierls, transl.), p.176, Addison-Wesley, Reading, Massachusetts, 1958.

‡ These restrictions are introduced to keep the exposition simple; it is possible to derive more general forms of Kirchhoff's law for transparent and fluorescent bodies too.

temperature of both systems being the same. Since photons do not interact, equilibrium must be established by means of a detailed balance between emission and absorption, and this must lead to a relation between them. The central, additional assumption of this section is the statement that this relation, being a property of the material or surface, is the same in the presence of any other field of radiation.

In order to characterize the emission from a real surface, it is common to introduce a coefficient ε, called *emittance*, which relates the intensity of the body's radiation to that of a black body. In view of the need for detailed balance, we write

$$i_{\nu\|} = \varepsilon_{\nu\|}(\theta) i_{b\nu} \qquad \text{and} \qquad i_{\nu\perp} = \varepsilon_{\nu\perp}(\theta) i_{b\nu} \tag{10.59}$$

for the two degrees of polarization, and designate the coefficients as the directional, monochromatic polarized emittances of the surface. Since the spectral intensity of radiation depends, generally speaking, on the angle of emission and polarization, the same must be true of the emittances, though this is not the case for a black-body emitter. Thus, the amount of monochromatic energy per unit solid angle, area, and time radiated in a given direction θ with respect to the normal is

$$\frac{d\Phi}{d\omega} = [\varepsilon_{\nu\|}(\theta) + \varepsilon_{\nu\perp}(\theta)] i_{b\nu} \cos\theta \,. \tag{10.60}$$

We denote the intensity of radiation impinging on a surface by the general symbol h and assume that a fraction α of it is absorbed. The coefficient α is given the generic name of *absorptance*. For example, in the case of a black-body surface the absorptance

$$\alpha_b \equiv 1 \tag{10.61}$$

for every spatial direction, direction of polarization, and frequency. In any other case, the amount of monochromatic energy per unit solid angle, area, and time absorbed is denoted by $d\Psi/d\omega$ and given by

$$\frac{d\Psi}{d\omega} = [\alpha_{\nu\|}(\theta) h_{\nu\|}(\theta) + \alpha_{\nu\perp}(\theta) h_{\nu\perp}(\theta)] \cos\theta \,, \tag{10.62}$$

where $\alpha_{\nu\|}$ and $\alpha_{\nu\perp}$ are the directional, monochromatic, polarized absorptances.

In order to derive Kirchhoff's law, we first suppose that the fraction of impinging radiation which is not absorbed is reflected in a specular way without change in polarization. For equilibrium with black-body radiation to be possible, the principle of detailed balance requires that the absorbed and emitted radiation must be equal when $h_{\nu\|} = h_{\nu\perp} = i_{b\nu}$. Hence

$$\varepsilon_{\nu\|}(\theta) i_{b\nu} \cos\theta = \alpha_{\nu\|}(\theta) i_{b\nu} \cos\theta$$

and

$$\varepsilon_{\nu\perp}(\theta) i_{b\nu} \cos\theta = \alpha_{\nu\perp}(\theta) i_{b\nu} \cos\theta,$$

proving that

$$\varepsilon_{\nu\|}(\theta) = \alpha_{\nu\|}(\theta) \qquad \text{and} \qquad \varepsilon_{\nu\perp}(\theta) = \alpha_{\nu}(\theta). \tag{10.63}$$

This is our first form of Kirchhoff's law; it states that the directional, monochromatic, polarized *emittance* is *equal to* its corresponding *absorptance* for surfaces which reflect specularly.

In cases when the reflection is diffuse, the balance equation must be written in a form of an integral over all directions and both directions of polarization, since the latter does, generally, change upon reflection. This leads to

$$\int_0^{2\pi} \int_0^{\pi/2} [\varepsilon_{\nu\|}(\theta) + \varepsilon_{\nu\perp}(\theta)] \sin\theta \cos\theta \, d\theta \, d\phi$$

$$= \int_0^{2\pi} \int_0^{\pi/2} [\alpha_{\nu\|}(\theta) + \alpha_{\nu\perp}(\theta)] \sin\theta \cos\theta \, d\theta \, d\phi. \tag{10.64}$$

In most cases of practical importance the emittances and absorptances can be taken to be independent of direction and polarization. Then,

$$\Phi_\nu = \int_0^{2\pi} \int_0^{\pi/2} i_\nu \sin\theta \cos\theta \, d\theta \, d\phi = \varepsilon_\nu e_{b\nu},$$

whereas the amount absorbed is

$$\Psi_\nu = \int_0^{2\pi} \int_0^{\pi/2} i_{b\nu} \sin\theta \cos\theta \, d\theta \, d\phi = \alpha_\nu e_{b\nu},$$

leading to

$$\varepsilon_\nu = \alpha_\nu. \tag{10.65}$$

The coefficients ε_ν and α_ν are called monochromatic, hemispherical emittance and absorptance, respectively.

Since for an ideal black-body surface $\alpha = 1$ for all wavelengths and directions of polarization, we find that $\varepsilon = 1$ for it as well. This proves that at a given temperature an ideal *black-body* surface has *the highest emissive power* of all possible surfaces, monochromatically or hemispherically, and for every possible direction of polarization. Being a perfect absorber, it is also the best emitter. It is this property which makes the study of black-body radiation so important.

10.9. Radiative Atomic Transitions. The Einstein Coefficients

The major part of this chapter has been devoted to the study of interactions between matter and photons in circumstances in which the changes in the energy levels of the material particles could be disregarded. Section 3.15, devoted to the study of molecular spectra, contained a short discussion of interactions between photons and gaseous molecules in equilibrium without paying attention to the mechanism which maintained it. In the present section we propose to study a very simple example, first investigated by A. Einstein, of such a mechanism. Thus it becomes necessary to examine the rates at which transitions occur. This cannot be done on the basis of equilibrium arguments alone and additional, heuristic principles will have to be invoked.

The simplest case of interaction between photons and matter occurs in a space filled with black-body radiation and the molecules of a gas of very low density. In such circumstances the molecules become distinguishable, and collisions between them can be disregarded. Moreover, it may be assumed that molecular transitions (electronic, vibrational, or rotational) occur solely by the emission or absorption of photons.

We consider two quantum states at the energy levels ε_1 and ε_2, with $\varepsilon_2 > \varepsilon_1$. In equilibrium, the occupation numbers n_1 and n_2 for these states are related by

$$\frac{n_2}{n_1} = \exp[-(\varepsilon_2 - \varepsilon_1)/\mathbf{k}T)]. \tag{10.66}$$

We imagine that the transition $1 \to 2$ occurs through the absorption of a *single* photon of frequency

$$\nu_{21} = \frac{\varepsilon_2 - \varepsilon_1}{\mathbf{h}}, \tag{10.67}$$

and assume, with Einstein, that the rate at which molecules make the transition from state 1 to state 2 is proportional to the number of molecules in state 1 as well as to the density of photons at the frequency ν_{21}, that is to $u(\nu_{21}, T)/\mathbf{h}\nu_{21}$. This can be written as

$$R_{12} = B_{12}\, n_1 u(\nu_{21}, T), \tag{10.68a}$$

where the factor of proportionality B_{12} contains the factor $(\mathbf{h}\nu_{21})^{-1}$ in it; it is called the *coefficient of induced absorption.*

There exist two processes whereby photons are emitted during the reverse transition $2 \to 1$. The first is the process of *spontaneous emission* of a photon by an excited molecule which occurs even in the absence of a radiation field. It is assumed that this occurs at a rate

$$R'_{21} = A_{21} n_2, \tag{10.68b}$$

where A_{21} is the *coefficient of spontaneous emission.* The second process, called *induced emission,* is thought of as occurring through a collision with a photon which is not absorbed simultaneously. In analogy with Equation (10.64a), we stipulate that the rate at which this occurs is

$$R''_{21} = B_{21} n_2 u(\nu_{21}, T), \tag{10.68c}$$

the coefficient of proportionality, B_{21}, being called the coefficient of *induced emission.*

In equilibrium the forward and reverse rates must balance, and this establishes that

$$u(\nu_{21}, T) = \frac{A_{21} n_2}{B_{12} n_1 - B_{21} n_2} = \frac{A_{21} \exp(-\mathbf{h}\nu_{21}/\mathbf{k}T)}{B_{12} - B_{21} \exp(-\mathbf{h}\nu_{21}/\mathbf{k}T)}. \tag{10.68d}$$

Substituting the expression for $u(\nu_{21}, T)$ from Equation (10.35) we prove that

$$B_{12} = B_{21}, \tag{10.69a}$$

and that

$$\frac{A_{21}}{B_{21}} = \frac{8\pi \mathbf{h} \nu_{21}^3}{\mathbf{c}^3}. \tag{10.69b}$$

The three coefficients, A_{21}, B_{12}, and B_{21}, are referred to as the *Einstein coefficients*. It is clear from Equations (10.69a) and (10.69b) that one of these coefficients is sufficient to determine the other two. The preceding relations find their principal application in atomic spectroscopy and in the general theory of the interaction of radiation with matter. In the succeeding section we discuss another interesting application, namely the theory of the *laser*.

10.10. The Laser

The phenomenon of stimulated emission of radiation has led to the development of the *laser*† and similar devices for the amplification of electromagnetic radiation. A typical laser, the ruby laser, is illustrated in Figure 10.16. The laser contains a cylindrical ruby crystal with reflecting surfaces at each end, the material containing chromium ions as impurities. The distribution of these ions among their quantum states forms the basis for the operation of the laser.

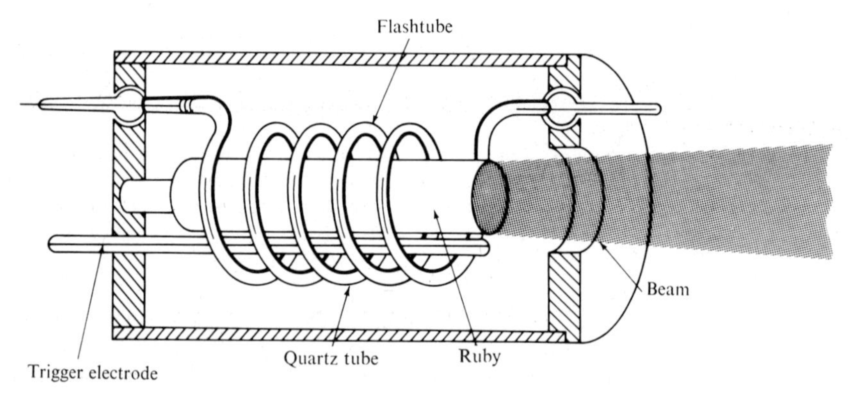

FIGURE 10.16. The ruby laser. [From B. A. Lengyel, *Introduction to Laser Physics*,, p. 50, Wiley, New York, 1966.]

† This is an acronym for "light amplification by the stimulated emission of radiation."

The quartz tube surrounding the crystal contains xenon at a pressure of 150 mm of mercury. It is flashed by discharging a capacitor bank through it, so that it provides about 2000 J of radiant energy to the crystal over the interval of a few milliseconds. This radiation serves to excite a large number of chromium ions in the crystal into a higher energy state so that there are more ions in this excited state than in the ground state. Such a situation is referred to as a *population inversion*. Suppose now that some of the excited ions return to their ground state by the spontaneous emission of photons. Owing to reflections from the ruby crystal, these photons remain within the crystal and serve to stimulate more atomic transitions between the ground state and the excited state. As a result, a nonequilibrium state is created in which the energy density of radiation at frequency ν, $u(\nu, t)$, changes with time, ν being determined by the energy difference, $\mathbf{h}\nu$, between the ground and excited states.

We may give a simple description of this process if we regard the chromium ions as independent and suppose that the only way that photons of frequency ν enter or leave the radiation field is by emission or absorption by the ions. In such circumstances the process may be described entirely in terms of the Einstein coefficients defined in the previous section. It is clear that the monochromatic density of specific energy, $u(\nu, T)$, changes with time through the following processes:

a. spontaneous emission from the excited state increases the energy of the radiation field at a rate

$$\mathbf{h}\nu A_{21} n_2(t),$$

since each single spontaneous emission contributes an energy $\mathbf{h}\nu$;

b. induced emissions increase this energy at a rate

$$\mathbf{h}\nu B_{21} n_2(t) u(\nu, T);$$

c. induced absorptions deplete the energy of the radiation field at the rate

$$\mathbf{h}\nu B_{12}\, n_1(t) u(\nu, t).$$

Summing up all these influences, we can easily write an equation for the rate of change $\mathrm{d}u(\nu, t)/\mathrm{d}t$ of the specific energy of the field in the form

$$\frac{\mathrm{d}u(\nu, t)}{\mathrm{d}t} = \mathbf{h}\nu[A_{21} n_2(t) + B_{21}\, n_2(t) u(\nu, t) - B_{12}\, n_1(t) u(\nu, t)].$$

Using the relations between the Einstein coefficients, Equations (10.69a) and (10.69b), we can transform this to

$$\frac{\mathrm{d}u(\nu, t)}{\mathrm{d}t} = A_{21}\mathbf{h}\nu\left[n_2(t) + \frac{\mathbf{c}^3}{8\pi\mathbf{h}\nu^3}[n_2(t) - n_1(t)]u(\nu, t)\right].$$

This equation assures an increase in the energy density at frequency ν with time as long as $n_2(t) > n_1(t)$, that is, as long as the population inversion is maintained. At the beginning of the process, $n_2 \gg n_1$, so that $u(\nu, t)$ increases exponentially as

$$u(\nu, t) \cong u(\nu, 0)\exp\left[\frac{A_{21}\mathbf{c}^3}{8\pi\nu^2} n_2(0)t\right] + \frac{8\pi\mathbf{h}\nu^3}{\mathbf{c}^3}\left\{\exp\left[\frac{A_{21}\mathbf{c}^3}{8\pi\nu^2} n_2(0)t\right] - 1\right\}.$$

However, with the passage of time the variations in $n_1(t)$ and $n_2(t)$ must be accounted for, and the relations become complicated. As long as the population inversion exists, there will be an amplification of radiation until, eventually, the number of ions in the excited state becomes so depleted by emission that the inequality $n_2(t) > n_1(t)$ is no longer satisfied. The radiation in the ruby becomes so intense that it cannot be contained in the crystal any longer and a sharp burst of radiation emerges from it. The reflecting surfaces at the ends of the crystal collimate the beam of light, since the radiation in the crystal is reflected so many times from the end surfaces that light other than that traveling precisely in the normal direction is eliminated by destructive interference. Moreover, the waves are all in phase, since destructive interference results in the elimination of any phase differences among the waves. The burst, typically, emits radiation of a wavelength $\lambda = 6943$ Å, a beam width of about 1 mrad and a frequency width $\Delta\nu/\nu \approx 10^{-10}$, corresponding to a frequency width of 200 kc or a wavelength spread of 0.1 Å.

Although the most commonly used laser is the ruby laser described above, there are many different types presently available. Some of these use other ion impurities in crystals such as samarium in a calcium fluoride crystal, which produces laser light at 7083 Å. Other lasers use liquid or gaseous materials as substances in which the required population inversion can be produced. Among these are the helium–neon laser in which excited states of both He and Ne atoms are involved in the laser action, to produce light at a wavelength of 1.2 μ.

The use of stimulated emission to amplify electromagnetic radiation is not confined to visible light. The first instruments that employed the laser principle were in fact *masers* in which radiation in the microwave region was amplified. Other regions of the spectrum have since been explored, leading to the production of the *iraser* (infrared) and, of course, the laser.

PROBLEMS FOR CHAPTER 10

10.1. The electric field, **E**, and the magnetic intensity, **H**, of electromagnetic radiation in a region free of charges satisfy the relations

$$\operatorname{div}\mathbf{E} = \frac{\partial E_x}{\partial x} + \frac{\partial E_y}{\partial y} + \frac{\partial E_z}{\partial z} = 0 \quad \text{and} \quad \operatorname{div}\mathbf{H} = 0.$$

Show that in these circumstances, the plane-wave solutions of the wave equation,

Equations (10.10a) and (10.10b), represent transverse waves, that is, that $\mathbf{E} \cdot \mathbf{f} = \mathbf{H} \cdot \mathbf{f} = 0$, where $\mathbf{f}$ is the propagation vector.

SOLUTION: Let us suppose that $\mathbf{E} = \mathbf{E}_0 \cos 2\pi(\nu t - z/\lambda)$, where $\mathbf{E}_0$ is a constant vector; the condition div $\mathbf{E} = 0$ leads to $\mathbf{E}_{0z} = \mathbf{E}_0 \cdot \hat{\mathbf{z}} = 0$. The wave propagates in the z direction and the electric vector points in a perpendicular direction. A more general wave propagating in the z direction can be synthesized as a superposition of such plane waves, each of which is transverse. Here $\hat{\mathbf{z}}$ is a unit vector in the z direction.

10.2. Compute the specific energy and monochromatic emissive power for a hypothetical two-dimensional photon gas. (*Hint:* The relation between $u(\nu, T)$ and $e_b(\nu, T)$ is not the same in two and three dimensions.)

10.3. Determine the specific energy, $u(T)$, of a hypothetical photon gas, assuming that its constituents obey Fermi–Dirac statistics and that their total number is not fixed but depends upon temperature in such a way as to render the chemical potential equal to zero.

10.4.† Planck's formula, Equation (10.35), gives an exact expression for the dependence of the energy density of black-body radiation, u, on the *variables* ν and T (the *only* variables in the problem) and on the universal physical *constants* $\mathbf{c}$, $\mathbf{k}$, and $\mathbf{h}$. This equation was derived in Chapter 10 on the basis of a physically verified model and by the use of quantum statistics.

Suppose that we attempt to analyze the problem (as did Lord Rayleigh and J. H. Jeans) without any knowledge of quantum mechanics. In such circumstances, we would conclude that the energy density, u, depended only on the *two* constants $\mathbf{c}$ and $\mathbf{k}$, apart from the variables ν and T.

(a) Employing a dimensional argument, show that the relation would *have* to be of the form

$$u(\nu, T) = A\nu^2 \mathbf{k}T/\mathbf{c}^3\,, \tag{1}$$

[which is essentially the same as the Rayleigh–Jeans equation (10.43), originally derived with the aid of *classical* statistics]. Here A denotes a constant.

(b) Prove that this result is physically unacceptable.

SOLUTION: (a) We can employ four fundamental dimensions E, L, T, t (energy, length, temperature, time), and note the derived dimensions:

u	ν	$\mathbf{k}$	T	$\mathbf{c}$
$\mathrm{EL}^{-3}\mathrm{t}$	t^{-1}	ET^{-1}	T	Lt^{-1}.

The monomial $\nu^a \mathbf{k}^b T^c \mathbf{c}^d$ must have the dimension of u, giving four linear algebraic equations for the four exponents a, b, c, d. Hence, the unique solution is: $a = 2$, $b = 1$, $c = 1$, $d = -3$, and Equation (1) follows. Note that $\Pi = u\mathbf{c}^3/\nu^2 \mathbf{k}T$ is the only dimensionless group which can be formed with our two variables and two constants, except for the trivial addition of arbitrary powers Π^α of Π.

(b) It is evident that $\int_0^\infty \nu^2 \, d\nu$ is divergent ("ultraviolet catastrophe").

† Adapted from A. Sommerfeld, *Thermodynamics and Statistical Mechanics* (J. Kestin, transl.), p. 141, Academic Press, New York, 1956.

The preceding dimensional argument proves very succinctly that the problem of black-body radiation cannot be solved with the aid of any model, however complex, if the body of physics employed (classical statistics) admits the existence of only two relevant universal physical constants. The nature of the supplementary universal constant required in the theory is investigated in Problem 10.5.

10.5.† Investigate the nature of the additional universal physical constant, **a**, required to solve the problem of black-body radiation, bearing in mind that the total black-body emissive power must be proportional to T^4 (Stefan's law, known to us from classical thermodynamics, Problem 1.7). In particular, determine the relation between **a** and Planck's constant **h**.

SOLUTION: The introduction of one more parameter, without a change in the number of fundamental units, must lead to the appearance of one additional dimensionless group, Π_1, (pi-theorem). Without any loss in generality, we may postulate that

$$\Pi_1 = \mathbf{a}\nu T^n/\mathbf{k} \qquad \text{with} \quad u(\nu, T) = \frac{\nu^2 \mathbf{k} T}{\mathbf{c}^3} \mathrm{f}(\Pi_1),$$

where $\mathrm{f}(\Pi_1)$ is an undetermined (but universal) function of the dimensionless variable Π_1. Integrating over all frequencies, we have

$$\int_0^\infty u(\nu, T)\, \mathrm{d}\nu = \frac{\mathbf{k}^4 T^{1-3n}}{\mathbf{a}^3 \mathbf{c}^3} \int_0^\infty \Pi_1{}^2 \mathrm{f}(\Pi_1)\, \mathrm{d}\Pi_1,$$

where the last integral represents a universal constant. It follows that $1 - 3n = 4$, yielding $n = -1$. Since $\Pi_1 = \mathbf{a}\nu/\mathbf{k}T$ must be dimensionless, we find that the dimension of **a** is Et which is the same as that of Planck's constant. Thus, we may choose $\mathbf{a} \equiv \mathbf{h}$, and conclude that

$$u(\nu, T) = \frac{\nu^2 \mathbf{k} T}{\mathbf{c}^3} \mathrm{f}\left(\frac{\mathbf{h}\nu}{\mathbf{k}T}\right).$$

Thus, dimensional analysis prescribes a universal relation between the groups $u\mathbf{c}^3/\nu^2 \mathbf{k} T$ and $\mathbf{h}\nu/\mathbf{k}T$. The detailed form of this relation must, evidently, be obtained on the basis of additional physical information. This task was achieved by Planck when he first introduced the concept of quanta.

10.6. Assume that the sun emits black-body radiation at $T = 5800$ K. Calculate the fraction of energy that lies in the visible range of the spectrum, as well as the two fractions on both sides of it.

10.7. Obtain Planck's distribution law assuming that the radiation field of a black body can be looked upon as composed of quantum-mechanical harmonic oscillators whose number $\mathrm{d}n = 8\pi\nu^2/\mathbf{c}^3$ in the range $\mathrm{d}\nu$ about ν.

10.8. At what temperature, T, is the radiation pressure of a black body equal to $P = 1$ atm?

† See footnote on p. 455.

10.9. Determine the average number of photons of black-body radiation at temperature T which occupy a particular quantum state. At what frequency ν' is $\bar{n} \gtrless 1$. Are Maxwell–Boltzmann statistics valid for $\nu > \nu'$ or $\nu < \nu'$? Give reasons for your answers.

10.10. The surface temperature of the sun is $T = 5800$ K. Calculate the wavelength, λ, for which the monochromatic emitted power $e_b(\lambda)$ is a maximum. To what color does this correspond? What is the radiation pressure, P, of a black body at $T = 5800$ K compared with that in the terrestrial atmosphere?

10.11. The diameter of the sun is $D = 1.4 \times 10^9$ m and its distance from the earth is $R = 1.5 \times 10^{11}$ m; the diameter of the earth is $D' = 1.27 \times 10^7$ m. Assume that the sun is an emitter of black-body radiation ($T = 5800$ K) and suppose that the sun and earth are perfect absorbers of radiant energy. Determine the temperature, T', which the earth would assume in steady state.

10.12. Calculate the maximum value of the spectral emissive power, e, of a surface at $T = 1000$ K if its emittance $\varepsilon = 0.8$.

10.13. Two infinite parallel plates possess black-body surfaces and are maintained at temperatures T_1 and T_2, respectively. Calculate the rate $q_{12}(\nu)$ at which energy at frequency ν is transferred from one surface to the other per unit area and time. Calculate the total rate, q_{12}, of energy transfer.

SOLUTION: The spectral emissive powers are $e_b(\nu, T_1)$ and $e_b(\nu, T_2)$, respectively. Since the emittance and absorptance of black-body surfaces are equal to unity, the rate at which energy is transferred per unit area *to* the plate at T_2 (or *from* the plate at T_1) is $q_{12}(\nu) = e_b(\nu, T_1) - e_b(\nu, T_2)$. Integrating over all frequencies, we obtain $q_{12} = \sigma(T_1^4 - T_2^4)$.

10.14. (a) Revert to Problem 10.13, and assume that the two plates have emittances ε_1 and ε_2, equal for every wavelength, respectively (so-called grey bodies). Show that

$$q_{12} = \frac{\sigma(T_2^4 - T_1^4)}{(1/\varepsilon_1) + (1/\varepsilon_2) - 1}.$$

(b) Solve the same problem for two concentric cylindrical and two concentric spherical surfaces. (*Hint:* Instead of following a ray as it is absorbed and reflected an infinite number of times, set up a balance equation. Denote the total quantities radiated by the surfaces by H_1 and H_2, so that $q_{12} = H_1 - H_2$. The energy leaving surface 1 is $H_1 = e_1 + (1 - \varepsilon_1)H_2$; that leaving surface 2 is $H_2 = e_2 + (1 - \varepsilon_2)H_1$, with $e_1 = \varepsilon_1 e_{b1}$, $e_2 = \varepsilon_2 e_{b2}$.)

Note: See references at end of Problem 10.15.

10.15. (a) Show that the rate (per unit area and time) at which energy is exchanged between two black-body surfaces A_1 and A_2 is

$$q_{12} = \int_0^\infty d\nu \int_{A_1} dA_1 \int_{A_2} dA_2 (i_b(\nu, T_1) - i_b(\nu, T_2)) \cos\theta_1 \cos\theta_2 / r^2.$$

(b) Give an explicit result for two parallel disks of radius R_1 and R_2, which are separated by a distance D.

(c) Generalize to the case of grey surfaces. (*Hint:* The integral

$$J = \int_{A_1} dx_1 \int dy_1 \int_{A_2} dx_2 \int dy_2 \frac{D^2}{[(x_1 - x_2)^2 + (y_1 - y_2)^2 + D^2]^2},$$

where $\cos\theta_1 = \cos\theta_2 = D/[(x_1 - x_2)^2 + (y_1 - y_2)^2 + D^2]^{1/2}$. For disks, this integral reduces to

$$J = \int_0^{R_1} dr_1 \int_0^{R_2} dr_2 \int_0^{2\pi} d\theta_1 \int_0^{2\pi} d\theta_2 \frac{D^2 r_1 r_2}{[r_1{}^2 + r_2{}^2 - 2r_1 r_2 \cos(\theta_1 - \theta_2) + D^2]^2}.$$

Hence

$$q_{12} = \tfrac{1}{2}\pi\sigma(T_1{}^4 - T_2{}^4)\{R_1{}^2 + R_2{}^2 + D^2 - [(R_1{}^2 + R_2{}^2 + D^2)^2 - 4R_1{}^2R_2{}^2]^{1/2}\}.)$$

Note: The factor multiplying $\sigma(T_1{}^4 - T_2{}^4)$ is known as the *shape or angle factor.* For further details, see E. R. G. Eckert and R. M. Drake, Jr., *Heat and Mass Transfer*, McGraw-Hill, New York, 1959; or E. M. Sparrow and R. D. Cess, *Radiation Heat Transfer*, Brooks/Cole Publ., Belmont, California, 1966.

10.16. A thin polished copper plate ($\varepsilon = 0.03$) is inserted as a radiation shield between two parallel black-body surfaces maintained at temperatures of $T_1 = 800$ K and $T_2 = 300$ K, respectively.

(a) Determine the temperature of the copper plate and the fractional reduction in the rate of heat transfer due to the presence of the shield.

(b) What would the fractional reduction be if the shield became oxidized and if its emittance increased to $\varepsilon' = 0.6$?

10.17. Compute the black-body emissive power per unit solid angle, $de_b/d\omega$, in directions forming an angle $\theta = 0$; 30; 45; and 60° with the normal to the hole of a cavity maintained at $T = 400$ K. What is the total energy radiated per unit time and unit area of the opening?

10.18. Compare the relation $P = u/3$ valid for photons with the relation $P = 2u/3$ valid for a perfect gas. What is the source of the difference? (*Hint:* Show that the energy of a photon in a container of volume V is proportional to $V^{-1/3}$.)

10.19. What are the differences between the thermodynamic properties of a gas of *phonons* and a gas of *photons*? What is the difference between a phonon and a photon?

10.20. Derive the form of Wien's displacement law given by Equation (10.41).

10.21. Assume that the process of induced emission has been neglected in the theory of radiative atomic transitions. Prove that the expression for $u(\nu, T)$ which balances the transition rates at equilibrium must be of the Rayleigh–Jeans form.

LIST OF SYMBOLS FOR CHAPTER 10

Latin letters

A	Area; coefficient of spontaneous emission
B	Coefficient of induced absorption or emission
C	Radiation constant; ratio of maximum value of the spectral emissive power to fifth power of the temperature
c	Radiation constant; ratio of the maximum value of the spectral emissive power to the cube of the temperature; specific heat
$\mathbf{c}$	Velocity of light
D	Density of states
E	Energy
$\mathbf{E}$	Electric field
e	Spectral emissive power
F	Helmholtz function
$\mathbf{f}$	Propagation vector of an electromagnetic wave or photon
G	Gibbs function
H	Enthalpy
$\mathbf{H}$	Magnetic intensity
h	Intensity of radiation impinging on a surface
$\mathbf{h}$	Planck's constant
$\mathcal{I}$	Definite integral appearing in the photon partition function
i	Intensity of radiation
$\mathbf{i}$	Unit vector in x direction
$\mathbf{j}$	Unit vector in y direction
$\mathbf{k}$	Boltzmann's constant
L	Length
m	Mass
m_0	Rest mass
N	Number of photons
n	Number of modes in a standing wave; number of photons per unit volume; occupation number for photons or atoms in an energy level
$\mathbf{n}$	Unit vector normal to a surface
P	Pressure; point in space
$\mathbf{p}$	Momentum
R	Transition rate
r	Distance between two points
$\mathbf{r}$	Represents coordinates x, y, z
S	Entropy; surface
T	Temperature
t	Time
u	Specific energy of radiation
V	Volume
$\mathbf{v}$	Represents velocity coordinates v_x, v_y, v_z

x	Ratio of photon energy to thermal energy
y	Product of photon wavelength and temperature
$\mathscr{Z}$	Partition function for a collection of photons

Greek letters

α	Absorptance; also $(= -\mu/\mathbf{k}T)$
β	$(\mathbf{k}T)^{-1}$
γ	Ratio of specific heats
ε	Single-particle energy level; emittance; dimensionless representation of emissive power
ε'	Ratio of spectral emissive power to its maximum value
θ	Angle
λ	Wavelength
μ	Chemical potential
ν	Frequency
σ	Stefan–Boltzmann constant
Φ	Hemispherical flux; flux for finite solid angle
ϕ	Phase angle
Ψ	Energy absorbed per unit area and time
ω	Solid angle

Superscripts

$\bar{}$	Refers to average value

Subscripts

b	Denotes properties of black body
max	Denotes maximum value
ν, λ	Denote spectrally resolved properties
$\parallel, \perp$	Denote states of polarization
1,2,..., 5	Subscripts distinguishing various radiation constants defined in Section 10.3

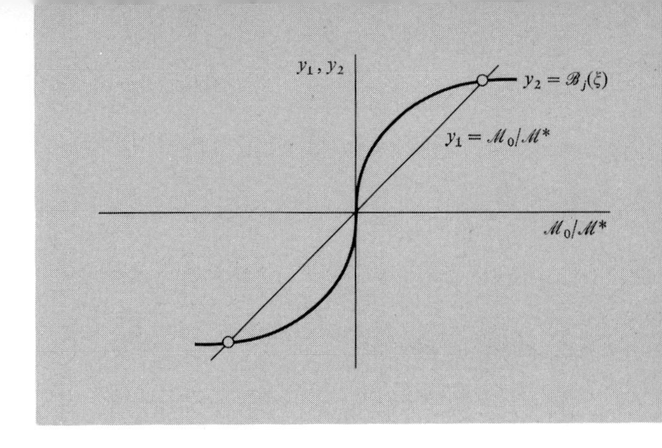

CHAPTER 11

MAGNETIC PROPERTIES

11.1. Introduction

The atoms or molecules of substances which interact with externally imposed magnetic fields themselves develop magnetic moments. Every electron orbiting about the nucleus can be regarded as an elementary current loop which produces a magnetic dipole. This creates a magnetic field which is additive to that provided by the intrinsic magnetic moment of the electron itself. In the absence of an external magnetic field all such internal fields cancel each other owing to the randomness of their distribution. The application of an external field affects the frequency of orbiting without changing the radius, and, by Lenz's rule, the resulting change in the dipole moment weakens the applied field. The material is then said to possess *diamagnetic* properties. However, in most materials, this diamagnetic weakening of external fields is so small as to be negligible, and the material appears to be inert. In alkali halides (for example NaCl, LiBr) the diamagnetic effect can become of some importance. In superconducting metals the induced field is so large that it completely cancels any applied field (Meissner effect).

In *paramagnetic* materials the effects of orbiting electrons together with the effect of the electron spin do not cancel. The atoms behave like small magnetic dipoles which tend to align with an applied, external field and to reinforce it. Thus, the total magnetic induction

$$\mathbf{B} = \mathbf{B}_0 + \mathbf{B}' \qquad (\{\mathbf{B}\} = \mathrm{Wb/m^2} = \mathrm{N/A\ m}) \tag{11.1}$$

consists of the sum of the applied induction, $\mathbf{B}_0$, and the internally induced induction, $\mathbf{B}'$.

In isotropic bodies the three vectors, $\mathbf{B}$, $\mathbf{B}'$, and $\mathbf{B}_0$ are parallel, and the additional vector $\mathbf{B}'$ provides a measure of the magnetic moment of the aligned, elementary dipoles. Instead of using the induced vector, $\mathbf{B}'$, and the applied induction, $\mathbf{B}_0$, for the description of the resulting magnetic field, it is customary to employ two different vectors: the magnetic field intensity, $\mathbf{H}$, and the magnetization, $\mathbf{M}$, per unit volume. The latter differ by a multiplicative constant, and are related to the former by the equations†

$$\mathbf{B} = \mathbf{B}_0 + \mathbf{B}' = \mu_0(\mathbf{H} + \mathbf{M}) \qquad (\{\mathbf{H}\} = \{\mathbf{M}\} = \mathrm{A/m}) \tag{11.2}$$

and

$$\mathbf{B}_0 = \mu_0 \mathbf{H}. \tag{11.3}$$

Here

$$\mu_0 = 4\pi \times 10^{-7}\ \mathrm{Wb/A\ m} \tag{11.4}$$

is the *magnetic permeability* of empty space.

In isotropic paramagnetic substances the resultant vector $\mathbf{B}$ is proportional to the applied vector $\mathbf{B}_0$ provided that the applied field is not very large. The coefficient of proportionality, κ_m, is known as the *permeability* of the substance. Thus for homogeneous and isotropic paramagnetic substances, we can write

$$\mathbf{B} = \kappa_m \mathbf{B}_0 = \mu_0 \kappa_m \mathbf{H}, \tag{11.5a}$$

or

$$\mathbf{M} = \chi_m \mathbf{H} \qquad (\{\chi_m\} = 1) \tag{11.5b}$$

with

$$\chi_m = \kappa_m - 1. \tag{11.5c}$$

The quantity χ_m is called the *magnetic susceptibility* of the substance. The magnetic susceptibility of most paramagnetic materials depends on their temperature and density; Equation (11.5b) together with the P, v, T relation for the substance can be regarded as constituting the thermal equation of state of the system. The latter now contains *three* independent thermodynamic variables.

Below a certain temperature, known as the *Curie* (or Néel) *temperature*, the properties of some paramagnetic crystals change in that the proportionality expressed in Equation (11.5b) no longer holds. At such temperatures, groups of atoms form so-called Weiss domains inside which the elementary dipoles become magnetically aligned, the direction of the resultant magnetic

† We employ the rationalized MKSA system of units.

moment varying from domain to domain. An external field causes the alignment of magnetic moments of rather large magnitude, and this results in a very large susceptibility. The material has become *ferromagnetic*. The process of magnetization in a ferromagnetic substance is not reversible, the reverse process of demagnetization forming a hysteresis loop with the forward process.

Above the Curie temperature, the susceptibility of a paramagnetic crystal may be represented by the Curie–Weiss law (see Section 11.7)

$$\chi_m = \frac{C}{T - T_C}, \tag{11.6a}$$

where T_C is the Curie temperature, and C is the *Curie constant*, both being characteristics of a given paramagnetic substance. At high temperatures, the approximation

$$\chi_m = \frac{C}{T}, \tag{11.6b}$$

known as *Curie's law* (see Section 11.4), proves to be adequate.

As usual, the purpose of a microscopic theory of magnetism is to derive statistically the equations of state of different materials on the basis of suitably postulated models. In this chapter, we shall concentrate on ideal paramagnetic materials whose properties can be derived rigorously from first principles. The treatment of ferromagnetic properties and the phase transition across the Curie temperature cannot be treated in full generality, and we shall be satisfied with a more heuristic discussion of these topics.

11.2. Fundamental Equation of Paramagnetic System

In order to derive the fundamental equation of a paramagnetic system it is first necessary to establish an expression for the work, dW°, of reversibly changing the magnetization of a substance. This is usually supplied by the electric current in the winding of a solenoid, so that in addition to affecting the magnetic field in the material body, we also affect the magnetic field in the surrounding space. Consequently, it is convenient to include the latter in the thermodynamic system under consideration. To fix our ideas, we begin the discussion with a consideration of the magnetostatic field brought into being by a current in a vacuum.

11.2.1. Magnetostatic Field in a Vacuum

The magnetostatic field created by a current in a vacuum is described by the vector of *magnetic induction*, $\mathbf{B}_0$, which plays the same part as a field intensity in spite of the fact that vector $\mathbf{H}$ is called the *magnetic field strength* or *intensity*. In a vacuum the difference is of no importance, the two vectors being proportional to each other:

$$\mathbf{B}_0 = \mu_0 \mathbf{H}; \tag{11.7}$$

here μ_0 is the magnetic *permeability constant*; μ_0, Equation (11.4), is a universal physical constant, but we refrain from using a boldface symbol for it in this section to avoid confusion with a vector which will appear later.

A uniform magnetic field $\mathbf{B}_0$ is established with a good degree of approximation in a toroidal body shown in Figure 11.1a when the linear dimensions of its cross section, A, are small compared with the mean radius of the torus.

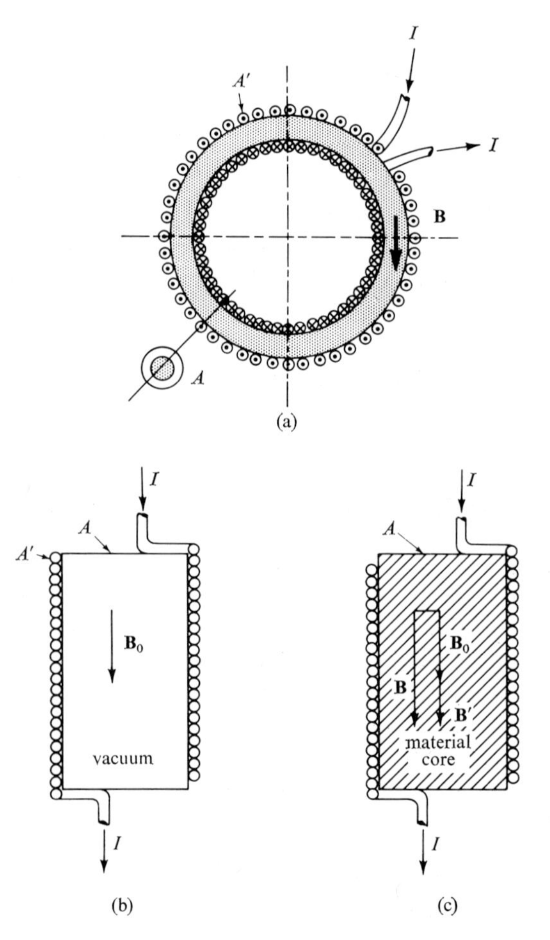

FIGURE 11.1. Infinite solenoid or Rowland ring: a, Rowland ring; b, solenoid in empty space; c, solenoid with material core.

The field can be created by passing a d.c. current, I, through the spirally wound conductor. Such an arrangement is known as a *Rowland ring*. Another possibility is to produce a closely wound, "infinite" solenoid.

We now wish to examine the work of *reversibly* changing the uniform magnetic field $\mathbf{B}_0$ by $d\mathbf{B}_0$. However, the winding which passes the current possesses a finite resistance which makes this impossible. Consequently, we shall disregard this resistance and ignore the emf required to overcome it. In this manner we extrapolate to the case when the wire cross section, A', of the winding tends to infinity. A good, practical approximation to such conditions can be achieved at very low temperatures when many metals become *superconducting*, their resistivity decreasing to virtually zero.

The work of changing this field in a vacuum is governed by *Faraday's law of magnetic induction* which leads to the conclusion that the inductance of the coil creates in it a "back-emf" whose value depends on the number of flux linkages Φ. In a vacuum, the number of flux linkages Φ_0 in a length l of solenoid or torus is measured by the product of the flux $\Phi_B = \mathbf{B}_0 A$ of vector $\mathbf{B}_0$ through the cross-sectional area A of the torus or solenoid into the number, N, of windings in that length. Hence,

$$\Phi_0 = N\Phi_B = NB_0 A. \tag{11.7a}$$

According to Ampère's law,

$$|\mathbf{B}_0| = \begin{cases} 0 & \text{outside the solenoid} \\ \mu_0 In & \text{inside the solenoid}, \end{cases} \tag{11.7b}$$

where n is the number of turns per unit length of solenoid. This vector is parallel to the axis of the solenoid as shown in the sketch. The back-emf is given by the equation

$$\mathscr{E}_0 = -\frac{d\Phi_0}{dt} \qquad (\{\mathscr{E}_0\} = \mathrm{V}) \tag{11.8}$$

and depends only on the rate of change of the number of flux linkages with time. This emf acts in a direction which forces the current, I, to perform work *on* the system when Φ_0 increases, it being noted that no work is performed ($\mathscr{E}_0 = 0$) to maintain a steady state. The external source must provide an opposite emf, and

$$\frac{dW^\circ}{dt} = -(-\mathscr{E}_0 I) = -I\frac{d\Phi_0}{dt}. \tag{11.9a}$$

This equation is equivalent to

$$dW^\circ = -I\,d\Phi_0. \tag{11.9b}$$

Since

$$\Phi_0 = nlB_0 A = \mu_0 n^2 lAI,$$

we may also write

$$dW^\circ = -\mu_0 n^2 lAI \, dI,$$

or, per unit volume,

$$dw^\circ = \frac{dW^\circ}{Al} = -\mu_0 n^2 I \, dI = -\mathbf{H}_0 \cdot d\mathbf{B}_0. \tag{11.10}$$

Since the magnetic field vanishes outside the solenoid, the expression for work can be put in the form

$$dW^\circ = -\int_{\text{all space}} (\mathbf{H}_0 \cdot d\mathbf{B}_0) \, dV. \tag{11.11}$$

The preceding derivation was related to a particularly simple geometrical arrangement leading to a uniform magnetic field. Nevertheless, the final expression contained in Equation (11.11) is not restricted to this case only. A more general derivation starting with J. C. Maxwell's field equation would lead to exactly the same result.† On this assurance, we shall accept Equation (11.11), as well as its consequences, as representing the work performed in an arbitrary reversible process of changing a continuous magnetic field in a vacuum.

In view of the relation in Equation (11.7), we can integrate the expression for work to yield

$$W^\circ = -\int_0^{\mathbf{H}} \int_{\text{all space}} [d(\tfrac{1}{2}\mu_0 \mathbf{H}^2)] \, dV$$

$$= -\int_{\text{all space}} (\tfrac{1}{2}\mu_0 \mathbf{H}^2) \, dV, \tag{11.11a}$$

and so to obtain the work of magnetizing empty space when a field of intensity $\mathbf{H}(x, y, z)$ is created in it. Since, obviously, the process of magnetizing empty space is adiabatic, the term

$$u_m = \tfrac{1}{2}\mu_0 \mathbf{H}^2 \tag{11.11b}$$

represents the local energy of the magnetic field per unit volume.

† V. Heine, The thermodynamics of bodies in static electromagnetic field, *Proc. Cambridge Phil. Soc.* **2** (1956), 546.

11.2.2. Magnetostatic Field in a Material Body

We now imagine that the interior of the Rowland ring of Figure 11.1 has been filled with a paramagnetic substance. As already mentioned, the body will produce an induction $\mathbf{B}'$ which must be added to $\mathbf{B}_0$. The vector of magnetic intensity, $\mathbf{H}$, is so defined that it remains unaffected by the presence of the material system and retains the same value as in a vacuum, as shown in Figure 11.1b; its value is

$$|\mathbf{H}| = \begin{cases} 0 & \text{outside the solenoid} \\ nI & \text{inside the solenoid}. \end{cases} \tag{11.12}$$

The magnetic flux, Φ, is determined by the total magnetic induction $\mathbf{B} = \mathbf{B}_0 + \mathbf{B}'$, where $\mathbf{B}_0 = \mu_0 \mathbf{H}$. Hence, we have

$$\Phi = nlAB.$$

Since Faraday's law of magnetic induction remains valid, we may write

$$\mathrm{d}W^\circ = -(lA)H\,\mathrm{d}B,$$

for the solenoid, asserting, once again, that in the general case the equation would remain unaltered. This means that the work of reversibly magnetizing a paramagnetic system is

$$\mathrm{d}W^\circ = -\int_{\text{all space}} (\mathbf{H}\cdot\mathrm{d}\mathbf{B})\,\mathrm{d}V. \tag{11.13}$$

It is customary to eliminate the vector of magnetic induction, $\mathbf{B}$, in favor of the magnetization, $\mathbf{M}$, with the aid of the definition in Equation (11.2). In this manner Equation (11.13) can be transformed to

$$\mathrm{d}W^\circ = -\int_{\text{all space}} [\mathrm{d}(\tfrac{1}{2}\mu_0 \mathbf{H}^2)]\,\mathrm{d}V - \int_{\text{all space}} (\mu_0 \mathbf{H}\cdot\mathrm{d}\mathbf{M})\,\mathrm{d}V. \tag{11.14}$$

The first term on the right-hand side of this equation represents the change in the magnetic energy of empty space, Equations (11.11a) and (11.11b). The second term is the expression for the reversible work performed on a paramagnetic substance with magnetization $\mathbf{M}$. The integral extends only over the volume, V, of the paramagnetic material, since outside the material $\mathbf{M} = 0$. For a uniform external field $\mathbf{H}$,

$$\int_{\text{all space}} (\mu_0 \mathbf{H}\cdot\mathrm{d}\mathbf{M})\,\mathrm{d}V = \mu_0 \mathbf{H}\cdot\mathrm{d}\mathscr{M}, \tag{11.15}$$

where the symbol $\mathscr{M}$ has been introduced for the integral

$$\mathscr{M} = \int_V \mathbf{M}\,\mathrm{d}V. \tag{11.15a}$$

11.2.3. The Gibbs Equation

The Gibbs equation for the process of magnetizing a paramagnetic system can now be written as

$$dE = T\,dS - dW_0^\circ + \int_{\text{all space}} (\mathbf{H} \cdot d\mathbf{B})\,dV, \tag{11.16}$$

where

$$dW_0^\circ = \sum_r Y_r\,dz_r$$

represents the work performed reversibly by other than magnetic processes. For example, in a paramagnetic gas this may represent the work, $P\,dV$, of compression or expansion.

Taking into account Equations (11.14) and (11.15), we may also write

$$d(E - U_m) = dE_m = T\,dS - \sum_r Y_r\,dz_r + \mu_0 \mathbf{H} \cdot d\mathcal{M}, \tag{11.17}$$

if the external magnetic field is uniform. Here

$$dU_m = \int_{\text{all space}} [d(\tfrac{1}{2}\mu_0 \mathbf{H}^2)]\,dV, \tag{11.18a}$$

so that

$$U_m = \int_{\text{all space}} (\tfrac{1}{2}\mu_0 \mathbf{H}^2)\,dV. \tag{11.18b}$$

The symbol $E_m = E - U_m$ represents the energy of the paramagnetic system *exclusive* of the energy of the accompanying field in empty space.

Equation (11.17) is the required Gibbs equation for a paramagnetic system in a uniform field. In this form it implies that the magnetic internal energy, E_m, is conceived as a function of entropy, S, the deformation coordinates, z_r, of the nonmagnetic work, W_0°, and of the magnetization $\mathcal{M}$ of the paramagnetic system when placed in a uniform external field, $\mathbf{H}$. In the statistical analysis, it is more convenient to employ the Legendre transformation of Equation (11.17) which results from an interchange in the roles played by the pairs of variables T, S and $\mathbf{H}$, $\mathcal{M}$. Introducing the *magnetic Helmholtz function*

$$F_m = E_m - TS - \mu_0 \mathbf{H} \cdot \mathcal{M}, \tag{11.19}$$

we deduce that

$$dF_m = -S\,dT - \sum_r Y_r\,dz_r - \mu_0 \mathcal{M} \cdot d\mathbf{H}, \tag{11.20}$$

implying that F_m is treated as a function of temperature, T, of the deformation variables, z_r, and of the externally applied field intensity, **H**. It is noted that, in contrast with previous cases, the set of independent variables includes an additional intensive parameter, **H**, apart from temperature, T.

11.3. The Mechanical Model

A perfect paramagnetic substance is imagined as a collection of atoms, molecules, or ions each of which is endowed with an angular momentum as well as a magnetic dipole. The angular momentum consists of two terms: the angular momentum contributed by the orbital motion of the electrons and the angular momentum associated with their spins. The magnetic dipole is created by the fact that the electrons are charged. As stated earlier, the fact that the electrons perform an orbital motion is equivalent to the existence of an elementary current loop which, in turn, is accompanied by a magnetic dipole. Thus, each element of a paramagnetic substance is pictured as a rotor of angular momentum **J** (denoted by **p** in Section 3.10) to which there corresponds a definite value of the magnetic moment, $\boldsymbol{\mu}$.

We shall refrain from specifying whether the macroscopic system is liquid, solid, or gaseous imagining only that the atoms, molecules, or ions perform the appropriate, additional microscopic motions. In the case of a perfect paramagnetic substance, we stipulate that these additional motions (translations, rotations, vibrations) are not coupled to the orbital motion of the electrons, the former being entirely independent of the latter. Evidently, we shall assume that the elementary dipoles are coupled to any external magnetic field that may be imposed on the system. However, we shall imagine that the magnetic field does not affect the other motions. Finally, we shall assume that the microscopic elements of the system are independent (do not interact with *each other*) and distinguishable.

The quantum-mechanical characteristics of the orbital motion are the same as those discussed generally in Section 3.10. In particular, it is recalled that the state of a rotor is described by two quantum numbers, j and m. In contrast with the case treated earlier, the quantum number j can now assume half-integer values ($0, \frac{1}{2}, 1, 1\frac{1}{2}, \ldots$), owing to the contribution that the electron spin makes to the motion.

The angular momentum of a particular quantum state is given by

$$\mathbf{J}^2 = j(j+1)\hbar^2, \tag{11.21}$$

and is independent of the second quantum number, m. Thus the kinetic energy is

$$\varepsilon_j = \frac{j(j+1)}{2I}\hbar^2, \tag{11.21a}$$

also regardless of the value of m. A rotor of given $\mathbf{J}$, ε_j, and j has a degeneracy

$$g = 2j + 1, \tag{11.22}$$

each particular state being characterized by the second quantum number, m, which can assume values ranging from $-j$ to $+j$, successive values differing by unity exactly.

The magnetic moment, $\boldsymbol{\mu}$, associated with the orbital motion must be proportional to the angular momentum, $\mathbf{J}$. Thus we may put

$$\boldsymbol{\mu} = -g_j \frac{\boldsymbol{\mu}_B}{\hbar} \mathbf{J} \tag{11.23}$$

where

$$\boldsymbol{\mu}_B = \frac{\mathbf{e}\hbar}{2\mathbf{m}_e \mathbf{c}} = (9.274096 \pm 0.000065) \times 10^{-24} \mathrm{J/T}, \dagger \tag{11.24}$$

known as the *Bohr magneton*, is simply a convenient reference unit formed with the electronic charge, $\mathbf{e}$, Planck's constant, $\hbar$, the mass of an electron, $\mathbf{m}_e$, and the speed of light, $\mathbf{c}$; g_j, known as the *Landé g-factor* is a characteristic, tabulated‡ coefficient of proportionality which differs from substance to substance. If we neglect the effects of any induced, internal fields§, the magnetic contribution to the energy of an element of the system may be thought of as arising from the interaction of the magnetic dipole, $\boldsymbol{\mu}$, with the applied magnetic field of intensity $\mathbf{H}$. This interaction energy is

$$\varepsilon_{mj} = -\boldsymbol{\mu} \cdot \mu_0 \mathbf{H}. \tag{11.25}$$

In our analysis it is sufficient to allow for a uniform intensity H_z, assumed directed parallel to the z axis of a Cartesian system of coordinates. Hence

$$\varepsilon_{mj} = \frac{g_j \boldsymbol{\mu}_B J_z H_z}{\hbar}.$$

Recalling from Section 3.10 that the projection J_z of angular momentum is quantized according to the equation

$$J_z = m\hbar, \tag{11.26}$$

we can see that the magnetic contribution to the energy of an element is also quantized, and that it is equal to

$$\varepsilon_{mj} = g_j \boldsymbol{\mu}_B \mu_0 H_z m. \tag{11.27}$$

† 1 T (tesla) = 10^4 G (gauss) = 1 Wb/m^2.

‡ See, for example, E. U. Condon and G. H. Shortley, *The Theory of Atomic Spectra*, Table 1^{16}, p. 382, Cambridge Univ. Press, London and New York, 1959.

§ These will be accounted for in Section 11.7.

According to our assumptions of independence, the total energy of an element is equal to the sum of three terms: the energy ε_i which arises from translational, rotational, and vibrational motions and which is characterized by an appropriate set of quantum numbers that need not be specified in detail; the energy ε_j of the orbital motion, and the magnetic energy, ε_{mj}. Thus the total energy is

$$\varepsilon_{imj} = \varepsilon_i + \varepsilon_j + \varepsilon_{mj}. \tag{11.28}$$

The last two terms are given by Equations (11.21a) and (11.27). The energy levels of a system with $j = 5/2$ are shown schematically in Figure 11.2 on the

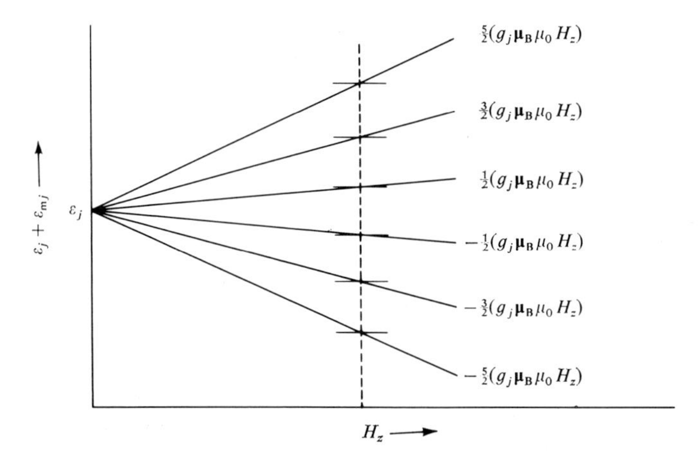

FIGURE 11.2. The variation of the energy level $\varepsilon_j + \varepsilon_{mj}$ with the externally applied electric field, H_z (Zeeman effect).

assumption that $\varepsilon_i = 0$ for convenience. The associated values of the second quantum number are now

$$m = -\tfrac{5}{2}, \quad -\tfrac{3}{2}, \quad -\tfrac{1}{2}, \quad +\tfrac{1}{2}, \quad +\tfrac{3}{2}, \quad +\tfrac{5}{2};$$

the level being six-fold degenerate. It is seen that for a given value of ε_i, the energy level ε_j from Equation (11.21a) has become split by the magnetic field into six separate energy levels for each value of ε_i. The degeneracy of the combined level $\varepsilon_j + \varepsilon_{mj}$ is now $g = 1$, and the separation between these levels varies linearly with the intensity H_z of the applied field. This splitting of a single orbital-rotational level is referred to as the *Zeeman effect*.

Ordinarily, the spacing between the Zeeman levels is considerably smaller than the spacing between successive electronic levels, ε_j, in the atom. Typically, the spacing $g_j \mu_B \mu_0 H_z$ between two Zeeman levels corresponds to a few kelvins, whereas the spacing between two electronic levels may reach an order of several hundred kelvins. It follows that at low or ordinary temperatures, the higher electronic levels remain unexcited and only the ground electronic state of the system needs to be taken into account. The diagram in Figure 11.3 illustrates the spacing between the electronic levels of the trivalent ion Sm^{3+} which appears in the paramagnetic salt $(Sm)_2(SO_4)_3$. The diagram shows a comparison between this spacing and the reference quantity $\mathbf{k}T$ with $T = 300$ K, thus emphasizing the wide discrepancy which exists between the Zeeman and electronic levels.

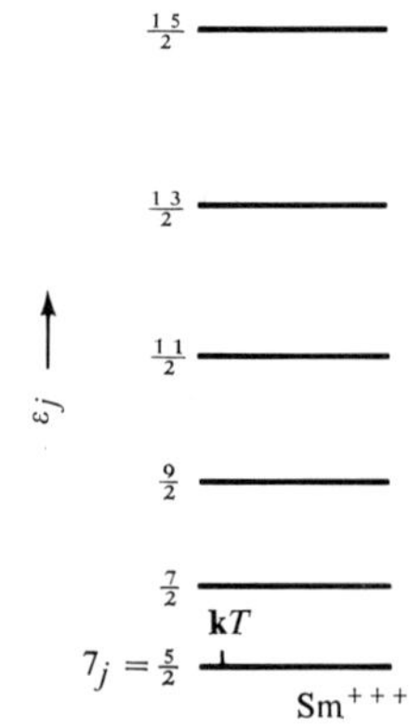

FIGURE 11.3. Spacing of the electronic level of Sm^{3+} compared to $\mathbf{k}T$ at $T = 300$ K. [From J. H. Van Vleck, *The Theory of Electric and Magnetic Susceptibilities*, p. 246. Oxford (Clarendon), London, 1932.]

11.4. The Partition Function

The macroscopic magnetization of a perfect paramagnetic substance is equal to the sum of the *average* values of the z component of the magnetic moments of its N identical constituents. If a single dipole has an energy ε_{imj} given by Equation (11.28), the probability of finding a particle at that energy level is described by the Boltzmann factor

$$\exp[-(\varepsilon_i + \varepsilon_j + \varepsilon_{mj})/\mathbf{k}T].$$

Since the associated z component of the magnetic moment is $-g_j \boldsymbol{\mu}_B m$, Equations (11.23) and (11.26), the ensemble average for one dipole is

$$\langle \mu_z \rangle = \frac{\sum_i \sum_j \sum_{m=-j}^{j} (-g_j \boldsymbol{\mu}_B m) \exp[-(\varepsilon_i + \varepsilon_j + \varepsilon_{mj})/\mathbf{k}T]}{\sum_i \sum_j \sum_{m=-j}^{j} \exp[-(\varepsilon_i + \varepsilon_j + \varepsilon_{mj})/\mathbf{k}T]}. \quad (11.29)$$

The denominator of Equation (11.29)

$$Z = \sum_i \sum_j \sum_{m=-j}^{j} \exp[-(\varepsilon_i + \varepsilon_j + \varepsilon_{mj})/\mathbf{k}T)], \tag{11.30}$$

represents the *single-particle* partition function, the system partition function being

$$\mathscr{Z} = Z^N. \tag{11.30a}$$

In a perfect paramagnetic system, the energy ε_i is unaffected by the orbiting motion or by the magnetic field. This means that every value of ε_i must be associated with all possible values of ε_j and ε_{mj} leading to the appearance of a factor $\exp(-\varepsilon_i/\mathbf{k}T)$ in front of a set of factors $\exp[-(\varepsilon_j + \varepsilon_{mj})/\mathbf{k}T]$ in which j and m range over all admissible values. The latter can, in turn, be factored out, showing that the particle partition function Z contains the factor

$$Z_0 = \sum_i \exp(-\varepsilon_i/\mathbf{k}T). \tag{11.30b}$$

This proves that the particle partition function can be written in the form of the product

$$Z = Z_0 Z_m, \tag{11.30c}$$

where

$$Z_{\mathrm{m}} = \sum_j \sum_{m=-j}^{j} \exp[-(\varepsilon_j + \varepsilon_{mj})/\mathbf{k}T] \tag{11.30d}$$

represents the *magnetic* single-particle partition function. Hence the system partition function can be factored, giving

$$\mathscr{Z} = \mathscr{Z}_0 \mathscr{Z}_m \quad \text{with} \quad \mathscr{Z}_0 = Z_0{}^N \quad \text{and} \quad \mathscr{Z}_m = Z_m{}^N. \tag{11.31}$$

It is worth noticing that

$$\sum_j \exp(-\varepsilon_j/\mathbf{k}T)$$

does *not* appear as an independent factor, because ε_{mj} is related to ε_j through the index j. However, a factor Z_0 does appear in the numerator of Equation (11.29) for the same reason that it appeared in its denominator.

The sum-of-states Z from Equation (11.30) has been expressed here as a function of the r deformation coordinates, z_r, of nonmagnetic work since

$$\varepsilon_i = \varepsilon_i(z_1, \ldots, z_r), \qquad \text{and hence} \qquad Z_0 = Z_0(z_1, \ldots, z_r).$$

The partition function contains, further, the field intensity, H_z, because

$$\varepsilon_{mj} = \varepsilon_{mj}(H_z), \qquad \text{and hence} \qquad Z_{\mathrm{m}} = Z_{\mathrm{m}}(H_z).$$

Finally, it contains the temperature, T, through the factor $\mathbf{k}T$ which appears in Z_0 as well as in Z_{m}. In all past examples, the partition function contained only temperature and a set of *extensive* deformation parameters, and the reader may have expected that the magnetization, $\mathscr{M}$, should have appeared as a variable in Z_m instead of the intensity H_z. It is, however, clear that the energy levels ε_{mj} from Equation (11.27) are more conveniently expressed in terms of the magnetic field intensity, and this suggests that we must search for the proper thermodynamic potential which ought to be associated with the present partition function

$$\mathscr{Z}(T, z_1, \ldots, z_r, H_z, N) = [Z_0(T, z_1, \ldots, z_r)Z_m(T, H_z)]^N.$$

A simple enumeration of the independent variables in the above expression shows that they are identical with those appearing in the Gibbs equation (11.20), thus making it plausible that the identification with the magnetic Helmholtz function, F_m, through the equation

$$F_m = -\mathbf{k}T \ln \mathscr{Z} = -\mathbf{k}T \ln \mathscr{Z}_0 - \mathbf{k}T \ln \mathscr{Z}_m \tag{11.32}$$

should be appropriate. It is easy to verify that this is, indeed, the case. First, we notice that

$$\frac{\partial}{\partial z_r}(-\mathbf{k}T \ln \mathscr{Z}) = \frac{\partial}{\partial z_r}(-\mathbf{k}T \ln \mathscr{Z}_0) = -Y_r(T, z_1, \ldots, z_r), \tag{11.32a}$$

where the functions Y_r are, obviously, the intensive parameters appearing in the expression dW_0° for the nonmagnetic work. Similarly we can verify that the magnetization $\mathscr{M}$, which is the average value of the magnetic dipole moment of the system in the direction of the applied magnetic field is given by

$$\mathscr{M} = N\langle \mu_z \rangle = N \frac{\sum_j \sum_{m=-j}^{j} (-g_j \mu_B m)\exp[-(\varepsilon_j + \varepsilon_{mj})/\mathbf{k}T]}{Z_m}, \tag{11.32b}$$

which follows from Equation (11.29) after the cancellation of the common factor Z_0.† Secondly, by taking the proper derivatives, we can verify that

$$\begin{aligned} \mathscr{M} = N\langle \mu_z \rangle &= -\frac{1}{\mu_0}\frac{\partial}{\partial H_z}(-\mathbf{k}T \ln \mathscr{Z}) \\ &= -\frac{1}{\mu_0}\frac{\partial}{\partial H_z}(-\mathbf{k}T \ln \mathscr{Z}_m). \end{aligned} \tag{11.32c}$$

Finally, it is equally easy to obtain an expression for the energy, E_m, and so to show that the entropy, S, is equal to the partial derivative of $\mathbf{k}T \ln \mathscr{Z}$ with respect to temperature, T, as implied in the Gibbs equation (11.20).

† We use boldface $\boldsymbol{\mathscr{M}}$ to denote an arbitrary magnetization vector, and $\mathscr{M}$, its component in the direction of an applied magnetic field. For perfect paramagnetic systems, the components of the vector $\boldsymbol{\mathscr{M}}$ perpendicular to the field vanish, and Equation (11.5b) is satisfied.

The fact that the partition function $\mathscr{Z}$ separates into two independent factors leads to two additive terms in the corresponding expression for the magnetic Helmholtz function. In turn, this signifies that each of the two terms will define a separate thermal equation of state: one relates the temperature, T, to the intensities, Y_r, and the deformation parameters, z_r, regardless of the values assumed by the magnetic parameters $\mathscr{M}$ and H_z. The second equation relates T to $\mathscr{M}$ and H_z in an equation which does not contain the previous variables. Consequently, the present theory is incapable of encompassing the phenomenon of magnetostriction which results from the coupling between the magnetic field and the stress tensor.

In continuing our analysis, we may now center attention on the magnetic partition function, $\mathscr{Z}_{\mathrm{m}}$, and on the derivation of the relation $\mathrm{F}(H_z, \mathscr{M}, T)$, realizing that contributions from the partition function $\mathscr{Z}_0$ are, essentially, additive and known from our previous studies.

At ordinary temperatures only the electronic ground state contributes significantly to ε_j, and we may use the approximation

$$
\begin{aligned}
Z_m &= \mathrm{e}^{-\varepsilon_j/\mathbf{k}T} \sum_{m=-j}^{j} \exp(-\varepsilon_{mj}/\mathbf{k}T) \\
&= \mathrm{e}^{-\varepsilon_j/\mathbf{k}T} \sum_{m=-j}^{j} \exp[(-g_j \boldsymbol{\mu}_{\mathrm{B}} \mu_0 H_z m)/\mathbf{k}T]. \qquad (11.33)
\end{aligned}
$$

Here j denotes the first quantum number for the electronic ground state—the only one taken into consideration.

The sum which appears in Equation (11.33) can be evaluated explicitly, since it represents a simple geometric progression. In this manner we obtain

$$
\begin{aligned}
Z_m &= \mathrm{e}^{-\varepsilon_j/\mathbf{k}T} \frac{\exp(-g_j \boldsymbol{\mu}_{\mathrm{B}} \mu_0 j H_z/\mathbf{k}T) - \exp(+g_j \boldsymbol{\mu}_{\mathrm{B}} \mu_0 (j+1) H_z/\mathbf{k}T)}{1 - \exp(g_j \boldsymbol{\mu}_{\mathrm{B}} \mu_0 H_z/\mathbf{k}T)} \\
&= \mathrm{e}^{-\varepsilon_j/\mathbf{k}T} \frac{\sinh[(j+\frac{1}{2}) g_j \boldsymbol{\mu}_{\mathrm{B}} \mu_0 H_z/\mathbf{k}T]}{\sinh[\frac{1}{2} g_j \boldsymbol{\mu}_{\mathrm{B}} \mu_0 H_z/\mathbf{k}T]}. \qquad (11.34)
\end{aligned}
$$

11.5. Magnetization

The magnetization of a perfect paramagnetic substance consisting of N identical particles is

$$
\mathscr{M} = \frac{\mathbf{k}T}{\mu_0} \left(\frac{\partial \ln \mathscr{Z}_{\mathrm{m}}}{\partial H_z} \right)_T = N g_j \boldsymbol{\mu}_{\mathrm{B}} j \mathscr{B}_j(x), \qquad (11.35)
$$

where

$$\mathscr{B}_j(x) = \frac{1}{j}\{(j+\tfrac{1}{2})\coth[(j+\tfrac{1}{2})x] - \tfrac{1}{2}\coth(\tfrac{1}{2}x)\} \tag{11.35a}$$

is the so-called Brillouin function, and where the argument

$$x = g_j \mu_B \mu_0 H_z/\mathbf{k}T. \tag{11.35b}$$

The diagram in Figure 11.4 shows a plot of the Brillouin function in

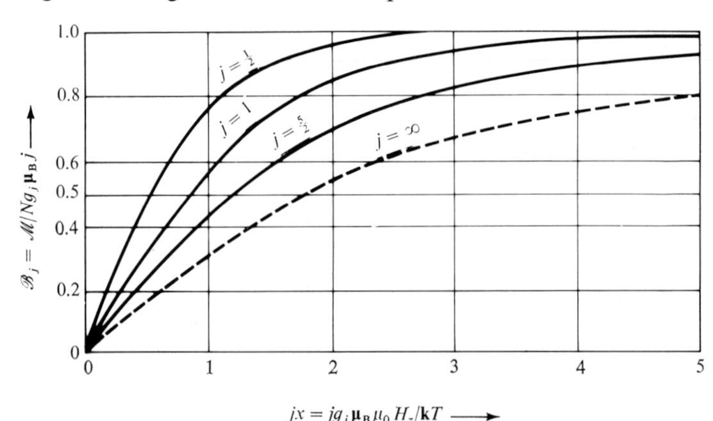

FIGURE 11.4. Plot of the Brillouin function $\mathscr{B}_j(x)$ from Equation (11.35a) in terms of the argument $jx = jg_j \mu_B \mu_0 H_z/\mathbf{k}T$ for several values of the quantum number, j. The broken line represents Langevin's function from Equation (11.37a). [From D. Ter Haar, *Elements of Thermostatistics*, 2nd ed., p. 87, Holt, New York, 1966.]

terms of its argument x with j as a parameter. At very low temperatures, or at high values of the applied magnetic field ($x \gg 1$), that is when

$$g_j \mu_B \mu_0 H_z \gg \mathbf{k}T,$$

the hyperbolic cotangents tend to unity, and so does the Brillouin function. In this limit,

$$\mathscr{M}_\infty = Ng_j \mu_B j \qquad (x \to \infty). \tag{11.36a}$$

In the opposite limit ($x \ll 1$), that is, for low magnetic field intensities or for high temperatures, we can use the approximation

$$\coth x \approx \frac{1}{x} + \frac{x}{3},$$

and the Brillouin function simplifies to

$$\mathscr{B}_j(x) \approx \frac{j+1}{3} x,$$

so that

$$\mathscr{M} = \frac{Nj(j+1)(g_j \mu_B)^2 \mu_0 H_z}{3\mathbf{k}T}$$

assumes the form of Curie's law

$$\mathscr{M} = \frac{C' H_z}{T} \qquad (\{C\} = \text{Km}^3)$$

with

$$C' = \frac{Nj(j+1)(g_j \mu_B)^2 \mu_0}{3\mathbf{k}}, \tag{11.36b}$$

The constant C' is related to Curie's constant appearing in Equation (11.6b) by the relation

$$C' = VC \tag{11.36c}$$

The preceding two limits can be interpreted physically as follows. At low temperatures or at high values of magnetic field intensity ($x \to \infty$) the dipoles are as closely aligned with the field as quantum mechanics allows them to be, and the system becomes "saturated." At low temperatures, the dipoles are in their lowest energy state with $m = -j$; it is recalled from Figure 3.12 that this is the closest possible alignment with the z axis. A similar alignment occurs at high values of H_z owing to the magnetic interaction. In the opposite limit ($x \to 0$, that is low H_z or high T) no such alignment is possible, the distribution of the momentum vectors is more nearly random, and the magnetic moment becomes proportional to the field intensity and inversely proportional to the temperature.

Table 11.1 and Figure 11.5 contain a comparison of this simple theory

TABLE 11.1

Comparison between Calculated and Measured Values of Curie's Constant for Several Paramagnetic Salts[a]

Paramagnetic salt	M gram-ionic mass of crystal	$j(j+1)$	Measured $C' \times 10^3$ (m^3 K/ kg ion)	Calculated $C' \times 10^3$ (m^3 K/ kg ion)
$CrK(SO_4)_2 \cdot 12H_2O$	499	3.75	23.12	23.62
$Fe(NH_4)(SO_4)_2 \cdot 12H_2O$	482	8.75	55.17	55.04
$Gd_2(SO_4)_3 \cdot 8H_2O$	373	15.75	98.02	98.90

[a] Data from American Institute of Physics Handbook, 2nd ed., pp. 5–219, McGraw Hill, New York, 1963.

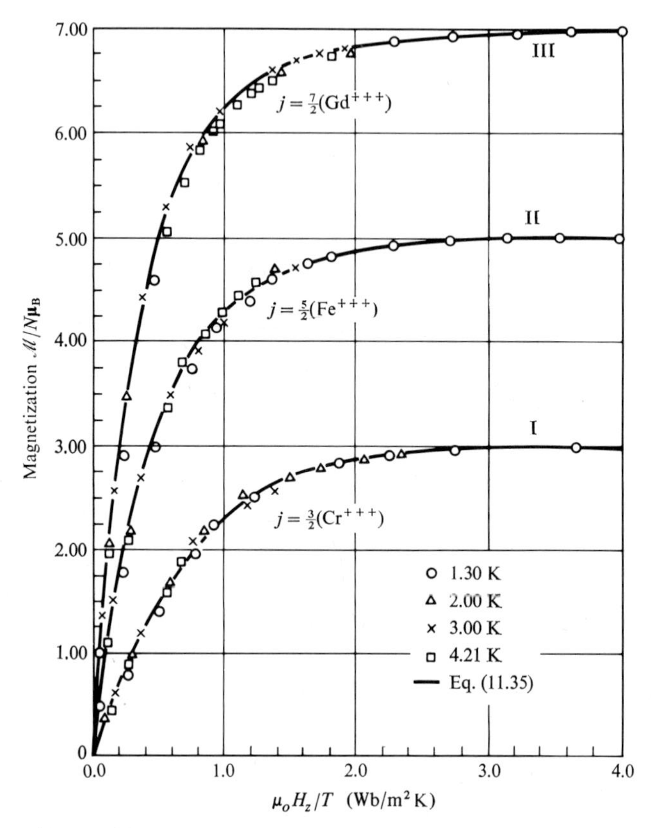

FIGURE 11.5. Magnetization per magnetic dipole expressed in Bohr magnetons and plotted in terms of the ratio $\mu_0 H_z/T$. Comparison between Equation (11.35) and measurements due to W. E. Henry [*Phys. Rev.* **88** (1952), 561]. I, chromium potassium alum, $j = \frac{3}{2}$; II, iron ammonium alum, $j = \frac{5}{2}$; III, gadolinium sulfate, $j = \frac{7}{2}$.

with experiment. The remarkable agreement between the calculated and measured values of Curie's constant in the table shows that many substances possess the properties of perfect paramagnetic systems. The conclusion is reinforced by the data contained in the figure in which a comparison is made between measurements and Equation (11.35).

11.6. Classical Limit

The first theory of paramagnetism was formulated in 1905 by P. Langevin. At that time, of course, Langevin could only employ a classical argument, since the quantum-mechanical theory of rotational motion and spin was yet to come. It is useful to retrieve Langevin's formula by passing to the limit $j \to \infty$, $\mathbf{h} \to 0$ with $j\mathbf{h}$ finite from the quantally derived Equations (11.35) and (11.35a). Noting that

$$\mu = g_j \frac{\mathbf{eh}j}{4\pi \mathbf{m_e c}},$$

we conclude that $\mu = \text{const}$ as $j \to \infty$ and $\mathbf{h} \to 0$. Hence the argument from Equation (11.35b)

$$x = \frac{\mu_0 \mathbf{eh} g_j H_z}{4\pi \mathbf{m_e ck}T} \to 0 \qquad \text{as} \quad \mathbf{h} \to 0,$$

and we can replace the second term in Equation (11.35a) by

$$\coth x \approx \frac{1}{x} + \frac{x}{3}.$$

Hence it can be seen that

$$\mathscr{B}_j(x) \to \coth(\mu_0 \mu H_z/\mathbf{k}T) - \frac{\mathbf{k}T}{\mu_0 \mu H_z}$$

The expression

$$\mathscr{L}(\alpha) = \coth \alpha - \frac{1}{\alpha} \tag{11.37a}$$

of the argument

$$\alpha = \frac{\mu_0 \mu H_z}{\mathbf{k}T} \tag{11.37b}$$

is known as *Langevin's function*, and Langevin's formula can be given the form

$$\mathscr{M} = N\mu\mathscr{L}\left(\frac{\mu_0 \mu H_z}{\mathbf{k}T}\right). \tag{11.38}$$

The diagram in Figure 11.4 shows a comparison between the single Langevin function and the more exact family of Brillouin functions.

11.7. Ferromagnetism

Ferromagnetism is the result of strong mutual internal interactions between the elementary dipoles as well as external interactions with the imposed field. Unfortunately, even in the presence of the simplest kinds of interaction, the evaluation of the partition function presents us with a challenging problem which still awaits solution. In 1944, L. Onsager† succeeded in deriving a partition function for the hypothetical case in which the dipoles are assumed to be arranged in a two-dimensional lattice and are allowed to interact with their "nearest neighbors," the interaction being restricted to the components of the magnetic moments that are normal to the plane of the lattice. This simplified form of interaction was first proposed by D. Ising‡ ("Ising model").

Owing to these difficulties, it is necessary to proceed heuristically, and in this connection the "model" proposed by P. Weiss§ best serves the purpose. Weiss postulated that in ferromagnetic materials there exist small regions or *domains* in which the elementary dipoles are aligned, the direction of the alignment changing from domain to domain. Thus there arises an internal, molecular field, H_i, and the assumption is made that its strength is proportional to the magnetization, $\mathscr{M}$, present in it. The intensity of this internal field exceeds that of the external field by many orders of magnitude, and we write

$$H_\mathrm{i} = \frac{\alpha}{\mu_0}\mathscr{M}. \tag{11.39}$$

In the absence of an external field, the system appears to be nonmagnetic because the direction of the molecular field varies randomly from one Weiss domain to another. The moment exerted by an external field, H, on the domains is quite different from that exerted on individual, elementary dipoles, and ferromagnetism arises because whole domains can align with the external field. However, the principal effect is due to irreversible, abrupt turns performed by the dipoles at the boundary of domains and to the wall displacements connected with them. For this reason, a typical ferromagnetic magnetization–demagnetization process results in a dissipative hysteresis loop.

According to this picture, the energy of an atom is

$$\varepsilon_i = (g_j \boldsymbol{\mu}_\mathrm{B} m)(\mu_0 H + \alpha\mathscr{M}). \tag{11.40}$$

The expression in Equation (11.40) is identical with that for a perfect paramagnetic material in a field $H + \alpha\mathscr{M}/\mu_0$. Consequently, the magnetization is

$$\mathscr{M} = N g_j \boldsymbol{\mu}_\mathrm{B} j \mathscr{B}_j[g_j\boldsymbol{\mu}_\mathrm{B}(\mu_0 H + \alpha\mathscr{M})/\mathbf{k}T], \tag{11.41}$$

as seen from Equation (11.35). This is a very complicated, implicit equation

† *Phys. Rev.* **65** (1944), 117. See also G. H. Wannier, *Statistical Physics*, Wiley, New York, 1966.

‡ *Z. Phys.* **31** (1925), 253. For the history of this model, see S. Brush, *Rev. Mod. Phys.* **39** (1967), 883.

§ *J. Phys. (Paris)* **6** (1907), 667.

for the magnetization, and the obvious path to take is to resort to a graphical method of solution.

We begin by considering the simplest case where the external field, H, is absent. This leads us to the equation

$$\mathscr{M}_0 = Ng_j\mu_B j\,\mathscr{B}_j(\alpha g_j \mu_B \mathscr{M}_0/\mathbf{k}T),$$

which can also be written

$$\frac{\mathscr{M}_0}{\mathscr{M}^*} = \mathscr{B}_j\left(\frac{\alpha N(g_j\mu_B)^2 j}{\mathbf{k}T}\frac{\mathscr{M}_0}{\mathscr{M}^*}\right), \tag{11.42}$$

with

$$\mathscr{M}^* = Ng_j\mu_B\, j. \tag{11.42a}$$

The diagram in Figure 11.6 contains a plot of the curves

$$y_1 = \frac{\mathscr{M}_0}{\mathscr{M}^*} \qquad \text{and} \qquad y_2 = \mathscr{B}_j(\xi),$$

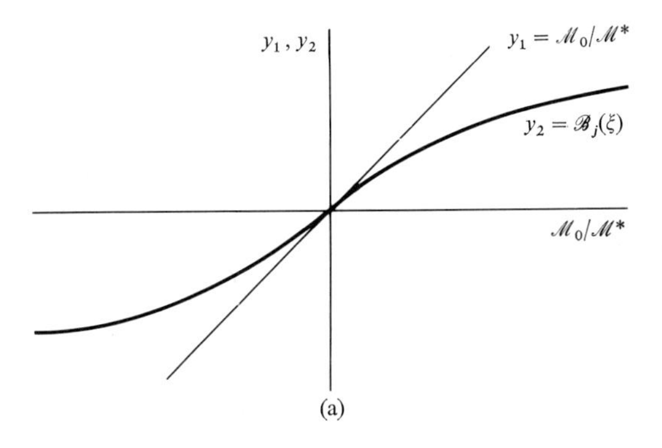

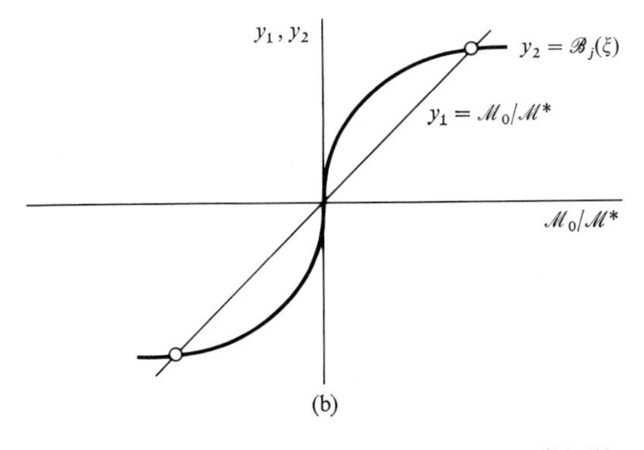

FIGURE 11.6. Graphical solution of Equation (11.42).

with

$$\xi = \frac{\alpha N(g_j \boldsymbol{\mu}_B)^2 j}{\mathbf{k}T} \frac{\mathcal{M}_0}{\mathcal{M}^*}. \tag{11.42b}$$

Each intersection of the two curves represents a possible solution of Equation (11.42). There always exists the obvious solution $\mathcal{M}_0/\mathcal{M}^* = 0$, but nonzero solutions may also exist. The conditions for the existence of such solutions is that the slope of y_2 at the origin should exceed unity, or that

$$\left[\frac{\mathrm{d}}{\mathrm{d}(\mathcal{M}_0/\mathcal{M}^*)} \mathcal{B}_j(\xi)\right]_{\xi=0} \geq 1. \tag{11.42c}$$

Since the Brillouin function is symmetric with respect to the origin, there exist two nonzero solutions when condition (11.42c) is satisfied, as seen in the sketch. It is easy to verify that the explicit form of condition (11.42c) is

$$\frac{j(j+1)\alpha N}{3\mathbf{k}T} (\boldsymbol{\mu}_B g_j)^2 \geq 1, \tag{11.42d}$$

or

$$T \leq T_C \tag{11.42e}$$

with

$$T_C = \frac{j(j+1)\alpha N(\boldsymbol{\mu}_B g_j)^2}{3\mathbf{k}}. \tag{11.42f}$$

The preceding argument leads to the conclusion that there exists a critical temperature T_C (Curie point) which separates states with $\mathcal{M}_0 = 0$ $(T > T_C)$ from states with additional nonzero values for $\mathcal{M}_0(T < T_C)$; the latter are equal but opposite in sign. It can be shown that the two nonzero solutions correspond to lower values of the Helmholtz free energy than that with $\mathcal{M}_0 = 0$; hence the nonzero solutions represent the more stable states. The existence of a magnetic moment which remains after the externally applied magnetic field is reduced to zero represents an essential property of a ferromagnetic material. In this manner the theory of Weiss's domains provides us with a description of the transition from paramagnetic behavior at $T > T_C$ to ferromagnetic behavior of $T < T_C$.

The diagram in Figure 11.7 gives an idea of the variation of the residual magnetization, $\mathcal{M}_0$, with temperature in relation to the Curie point T_C. At low temperatures, a saturation value, $\mathcal{M}^*$, is quickly reached.

In order to complete our discussion of the molecular field theory, we need to consider the case where there is a nonzero applied field. Here again, the behavior of the system can be deduced by graphical methods. However, for small magnetic fields, H, and for temperatures above the Curie point T_C,

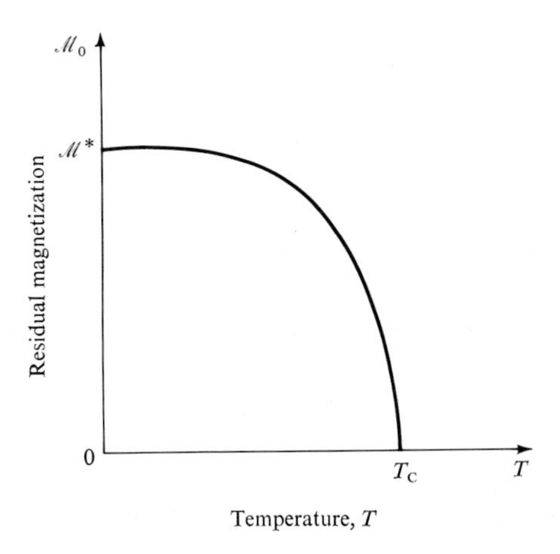

FIGURE 11.7. Residual magnetization, $\mathscr{M}_0$.

we can find a closed-form expression for the magnetization. We expand the Brillouin function as

$$\mathscr{B}_j(g_j \boldsymbol{\mu}_B(\mu_0 H + \alpha\mathscr{M})/\mathbf{k}T) \approx \tfrac{1}{3}(j+1)g_j \boldsymbol{\mu}_B(\mu_0 H + \alpha\mathscr{M})/\mathbf{k}T, \quad (11.43)$$

which is acceptable if H and $\mathscr{M}$ are both small. Hence

$$\mathscr{M} = \frac{\mathscr{M}^*}{3}(j+1)g_j \boldsymbol{\mu}_B(\mu_0 H + \alpha\mathscr{M})/\mathbf{k}T \quad (11.43a)$$

or

$$\mathscr{M} = \frac{C_W H}{T - T_C} \quad (11.43b)$$

with

$$C_W = \frac{\mathscr{M}^*(j+1)g_j \boldsymbol{\mu}_B \mu_0}{3\mathbf{k}} = \frac{N(j+1)j}{3\mathbf{k}}(g_j \boldsymbol{\mu}_B)^2 \mu_0, \quad (11.43c)$$

which is the Curie–Weiss law for the magnetization of a paramagnetic substance near its Curie temperature.

We remark in conclusion that there exists a strong analogy between Weiss's theory for a ferromagnet and the van der Waals equation for gases. It can be shown, in fact, that Weiss's theory results from a microscopic model in which the magnetic moments interact at long range. The existence of a weak, long-range interaction is also a characteristic feature of the microscopic model for the van der Waals gas.†

† For more complete details, see M. Kac, Mathematical mechanisms of phase transitions, in *Brandeis University Summer Institute in Theoretical Physics* (M. Chretien, E. P. Gross, and S. Deser, eds.), Vol. I, Gordon and Breach, New York, 1966.

PROBLEMS FOR CHAPTER 11

11.1 Compute the energy per unit volume contained in a long, evacuated solenoid in which a magnetic induction of $B = 1\ \text{Wb/m}^2$ is maintained. Calculate the change in this specific energy if the interior of the solenoid is filled with a material whose magnetic susceptibility $\chi = 2 \times 10^{-4}$, and in which the total magnetic induction is maintained at $B = 1\ \text{Wb/m}^2$.

11.2. Compute the separation, in degrees kelvin, between adjacent Zeeman levels for an ion with angular momentum quantum number $j = 2$ and Landé factor $g_j = 1$ when placed in a magnetic field $H = (2\pi)^{-1} \times 10^7$ A/m.

11.3. Obtain an expression for the energy, E_m, of a perfect paramagnetic system in terms of the partition function $\mathscr{Z}$ defined in Equation (11.31). Using the expressions for the intensive parameters Y_r and the magnetization $\mathscr{M}$ given in Equations (11.32a) and (11.32c), show that the quantity

$$\frac{1}{T}(\mathrm{d}E_m + \sum Y_r\,\mathrm{d}z_r - \mu_0 \mathbf{H} \cdot \mathrm{d}\mathscr{M})$$

is an exact differential. Hence, derive expression for the entropy, S, of a magnetic system. Verify the Gibbs relation, Equation (11.20).

11.4 Compute the first two terms in the series expansion of $\mathscr{B}_j(x)$ in powers of x.

11.5. Show that $\mathscr{B}_{1/2}(x) = \tanh(x/2)$.

11.6. Compute the contribution to the specific heat at constant magnetic field, $c_{\mathbf{H}}$, stemming from the interaction of magnetic dipoles with the magnetic field in a perfect paramagnetic substance. How does this term depend on temperature, T, and on the magnetic field, $\mathbf{H}$?

11.7 Compute the energy, E, specific heat at constant magnetization, $c_{\mathscr{M}}$, and entropy, S, for a classical system of independent paramagnetic molecules and for a quantum system of molecules in an electronic quantum state whose total angular momentum quantum number is j and whose Landé factor is g_j.

11.8. Gadolinium sulphate, $Gd_2(SO_4)_3 \cdot 8H_2O$, has a molecular weight $M = 747$ gr/gmol, and a density $\rho = 3.01\ \text{gr/cm}^3$. The salt is in an electronic state labeled by $j = 7/2$ and $g_j = 2$. Compute the magnetization per mole at $T = 4$ K in a magnetic field $B = 16\ \text{Wb/m}^2$. Compute also the magnetic susceptibility per mole, χ.

11.9. Calculate the magnetic susceptibility per mole, χ, at 300 K for the complex $[Mn(H_2O)_6^{2+}][SiF_6^{2-}]$ for which $j = \frac{5}{2}$, $g_j = 2$, $\rho = 1.9\ \text{gr/cm}^3$ and $M = 305$ gr/gmol.

SOLUTION: The magnetic susceptibility χ is defined by Equation (11.5b) as $\mathscr{M}/V = \chi\mathbf{H}$. At high temperature this relation is satisfied by a perfect paramagnetic substance with the molar value of χ given by Equation (11.36c) as $\chi = (\mathbf{N}/V)j(j+1)(g_j\mu_B)^2\mu_0/3\mathbf{k}T$. In order to evaluate χ numerically we need only to remember that one mole of this salt occupies a volume $V = (305\ \text{gr/gmol})/(1.9\ \text{gr/cm}^3)$.

11.10. Consider a paramagnetic crystal in an angular momentum state labeled by j and g_j, when placed in a magnetic field. What is the heat, Q, rejected by the crystal in a reversible process during which the temperature is kept constant but in which the field is decreased?

11.11. Oxygen is a paramagnetic gas whose susceptibility per mole at $T = 293$ K and $P = 1$ atm is $\chi = 1.80 \times 10^{-6}$. Compute the Curie constant. What is the value of the Landé factor g_j if $j = 1$?

11.12. Given that the Landé g_j-factors for chromium potassium alum,

$$CrK(SO_4)_2 \cdot 12H_2O$$

iron ammonium alum, $Fe(NH_4)(SO_4)_2 \cdot 12H_2O$; and gadolinium sulphate, $Gd_2(SO_4)_3 \cdot 8H_2O$ are each equal to $g_j = 2$, verify the calculated values of the Curie constant given in Table 11.1.

11.13. Show that the spontaneous magnetization of a Weiss ferromagnet at a temperature below, but close to, the Curie temperature, T_C, is proportional to $(T_C - T)^{1/2}$.

SOLUTION: The spontaneous magnetization may be expressed with the aid of Equation (11.42) as $\mathcal{M}_0/\mathcal{M}^* = \mathcal{B}_j(3T_C\mathcal{M}_0/(j+1)T\mathcal{M}^*)$. Near the Curie point $\mathcal{M}_0/\mathcal{M}^*$ is small, and we may expand the Brillouin function as

$$\frac{\mathcal{M}_0}{\mathcal{M}^*} = (j+1)\left[\frac{T_C}{(j+1)T}\frac{\mathcal{M}_0}{\mathcal{M}^*} - \frac{(j^2+j+\frac{1}{2})}{18}\frac{27T_C{}^3\mathcal{M}_0{}^3}{(j+1)^3T^3\mathcal{M}^{*3}}\right].$$

One solution is, of course, $\mathcal{M}_0/\mathcal{M}^* = 0$. However, below T_C, another solution is possible; this is given by

$$\left|\frac{\mathcal{M}_0}{\mathcal{M}^*}\right| = \left(\frac{T_C - T}{T_C}\right)^{1/2}\left[\frac{2(j+1)^2}{3(j^2+j+\frac{1}{2})}\right]^{1/2}$$

for T close to T_C.

11.14. The classical theory of perfect paramagnetic ions in a magnetic field regards the ions as particles with a constant magnetic moment $\boldsymbol{\mu}$. The particles are free to rotate about a fixed point. The interaction energy of the ion with the field is $-\mu_0\boldsymbol{\mu}\cdot\mathbf{H}$. Using classical statistical mechanics, compute the magnetization of a collection of N such ions and show that your answer is the classical limit of the quantum formula in Equation (11.38). In order to obtain the classical limit, it is necessary to put $j \to \infty$, $\mathbf{h} \to 0$, but to postulate that the product $j\mathbf{h}$ must remain finite. (Explain why this must be so.)

11.15. Compute the magnetization of a classical perfect paramagnetic system at high and low temperatures and magnetic fields. Provide a physical explanation for these properties. What is the value of the Curie constant for this system?

11.16. The Curie temperature for iron is $T_C^{Fe} = 1043$ K and for nickel its value is $T_C^{Ni} = 631$ K. Assume that $g_j = 1$ and $j = 1$ for these atoms and compute the internal field constant, α, in the Weiss theory for one mole of each substance.

11.17. Compute the paramagnetic susceptibility, χ, for a system of electrons at $T = 0$ K when placed in a weak field $B(\mu_0 \gg \mu_B B$, where μ_0 is the Fermi energy).

SOLUTION: The energy of an electron in a magnetic field B is $p^2/2\mathbf{m}_e \pm \mu_B B$; the + (−) sign referring to those situations where the electron spin is aligned to be parallel (antiparallel) with the direction of the magnetic field. At $T = 0$ K, the number of electrons whose spin is parallel to the field may be obtained from Equations (8.40b) and (8.57), with the energy per particle given by $\varepsilon = \mathbf{h}^2 s^2/8\mathbf{m}_e V^{2/3} + \mu_B B$. Hence

$$N_+ = \frac{\pi}{2}\int_0^{s^*} s^2\,ds = \frac{\pi}{6}s^{*3},$$

where

$$s^* = \left[\frac{8V^{2/3}\mathbf{m}_e}{\mathbf{h}^2}(\mu_0 - \mu_B B)\right]^{1/2}$$

or

$$N_+ = (4\pi V/3\mathbf{h}^3)[2\mathbf{m}_e(\mu_0 - \mu_B B)]^{3/2}.$$

Similarly, the number of electrons whose spin is aligned to be antiparallel with the field is

$$N_- = (4\pi V/3\mathbf{h}^2)[2\mathbf{m}_e(\mu_0 + \mu_B B)]^{3/2}.$$

The magnetization $\mathcal{M}$ is

$$\mathcal{M} = -\mu_B(N_+ - N_-) = +(4\pi V/3\mathbf{h}^2)\mu_B\{[2\mathbf{m}_e(\mu_0 + \mu_B B)]^{3/2} - [2\mathbf{m}_e(\mu_0 - \mu_B B)]^{3/2}\}.$$

When $\mu_0 \gg \mu_B B$, we have

$$\mathcal{M} \approx \tfrac{3}{2}(\mu_B{}^2 N/\mu_0)B.$$

LIST OF SYMBOLS FOR CHAPTER 11

Latin letters

A	Area
$\mathcal{B}$	Brillouin function
$\mathbf{B}$	Magnetic induction vector
C	Curie constant
C'	Product of Curie constant and volume
C_W	Curie-Weiss constant
$\mathbf{c}$	Speed of light
E	Energy
$\mathcal{E}$	Electromotive force
$\mathbf{e}$	Electronic charge
F	Helmholtz function
F_m	Magnetic Helmholtz function
g	Degeneracy factor
g_J	Landé g-factor
$\mathbf{H}$	Magnetic intensity
$\mathbf{h}$	Planck's constant

I	Electric current
$\mathbf{J}$	Angular momentum
j	Angular momentum quantum number
$\mathbf{k}$	Boltzmann's constant
$\mathscr{L}$	Langevin function
l	Length of a solenoid
M	Magnetization per unit volume
$\mathscr{M}$	Magnetization
$\mathscr{M}$	Component of magnetization in the direction of an applied magnetic field
m	Magnetic quantum number
$\mathbf{m}_e$	Electronic mass
N	Number of windings in a solenoid; number of particles
$\mathbf{N}$	Avogadro's number
n	Number of windings per unit length
P	Pressure
S	Entropy
T	Temperature
T_C	Curie temperature
t	Time
U_m	Energy of a magnetic field in a vacuum
u_m	Specific energy of a magnetic field in a vacuum
V	Volume
v	Specific volume
W°	Reversible work
Y	Intensive property
Z	Single-particle partition function
Z_m	Magnetic partition function for a single particle
$\mathscr{Z}$	Partition function for an assembly of particles
z	Deformation variable

Greek letters

α	Ratio of magnetic energy to $\mathbf{k}T$; constant of proportionality in internal field equation
β	$(\mathbf{k}T)^{-1}$
ε	Energy of a single particle, or single magnetic dipole
κ_m	Permeability
μ_0	Magnetic permeability
$\boldsymbol{\mu}$	Magnetic dipole moment
$\boldsymbol{\mu}_B$	Bohr magneton
Φ	Magnetic flux
χ_m	Magnetic susceptibility

Subscripts

m	Denotes properties of a magnetic system
0	Denotes properties of a magnetic field in a vacuum

Special symbols

$\langle\ \rangle$	Denotes average value

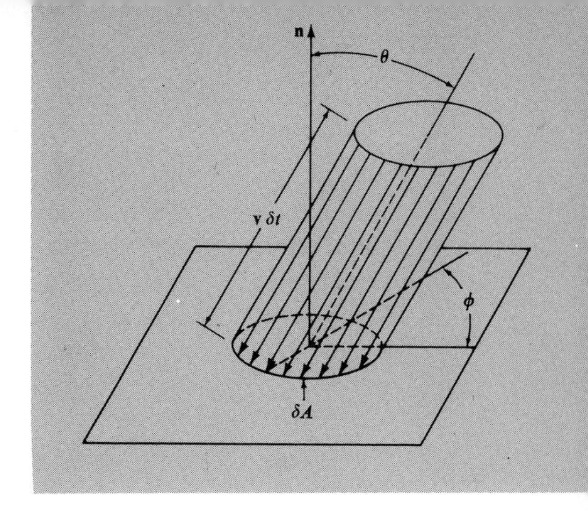

CHAPTER 12

KINETIC THEORY OF GASES

12.1. Prefatory Remarks

Perhaps the most remarkable feature of equilibrium statistical mechanics is the circumstance that all thermodynamic properties of a system can be obtained from the evaluation of the partition function. In turn, the partition function can be calculated if the energy levels available to the system are known. However, our discussion of equilibrium systems never touched upon the fact that the equilibrium properties of a system may also be interpreted as averages over the dynamical quantities of the system. As an example, let us consider a gas in equilibrium. At any instant, molecules collide with each other, with the wall, and move freely between collisions. The collisions with the wall are responsible for the pressure, and the intermolecular collisions are responsible for the deviation from ideality in the gas. All these events are accounted for in the partition function, of course, but this occurs in a way that does not refer to their dynamical character. The dynamic picture of equilibrium suggests that any system should display fluctuations about the equilibrium state. For example, the force exerted by a gas on a unit area of its container cannot remain constant in time, but ought to fluctuate about a mean value—the equilibrium pressure. We conclude that only the average behavior of a system in equilibrium is described by the partition function, and that its fluctuations are disregarded, as stressed on several previous occasions in this book.

The situation becomes even more complex when the system is in a non-equilibrium state. In principle, such states must be described with the aid of a solution of the complete, time-dependent Schrödinger equation. However, not all nonequilibrium states are so far removed from equilibrium that a full treatment of this kind becomes mandatory. If a system is in a late stage of its approach to equilibrium, it may display almost all the properties that we normally associate with equilibrium. Nevertheless, we cannot study it by means of the partition function, and we must develop an alternative theory based on the *dynamical* properties of the system.

In the case of gases, even equilibrium states were, historically, first studied from the dynamical point of view. This was pioneered by J. J. Waterston, J. Herapath, R. Clausius, J. C. Maxwell, and L. Boltzmann in an era which preceded J. W. Gibbs's introduction of the method of ensembles. In the course of their researches, these authors extended the theory to include nonequilibrium states as well, and in doing so attained a level of excellence which has never been surpassed.

The branch of physics that bridges the gap between equilibrium statistical mechanics and the Schrödinger equation for a gas is referred to as the *kinetic theory of gases*, and we shall devote to it this and the next chapter of this book. When we consider nonequilibrium states, we shall be primarily concerned with *transport* processes in the gas, that is, with the transfer of particles, momentum, or energy from one part of the system to another. The transfer of particles is associated with the phenomenon of diffusion; the transfer of momentum is responsible for viscosity; the flow of energy gives rise to thermal conduction. We now proceed to develop theories for these processes, starting with the simplest mean-free-path arguments and culminating in a discussion based on the famous *Boltzmann* transport *equation.*

12.2. Some Elementary Ideas

We begin by considering a *dilute* monatomic gas whose molecules interact through the intermediary of short-range forces and assume that quantum-mechanical effects are unimportant. This allows us to use classical mechanics. In order to describe the state of the gas, we introduce a *distribution function*, $f(\mathbf{r}, \mathbf{v}, t)$. This is so defined that $f(\mathbf{r}, \mathbf{v}, t)\, d\mathbf{r}\, d\mathbf{v}$ represents the instantaneous number of molecules which are contained in the small volume $d\mathbf{r}$ centered about the point $\mathbf{r}$ in the container and whose velocity is confined to the range from $\mathbf{v}$ to $\mathbf{v} + d\mathbf{v}$ in momentum space. Accordingly, the function $f(\mathbf{r}, \mathbf{v}, t)$ must satisfy the equation

$$\int_V d\mathbf{r} \int f(\mathbf{r}, \mathbf{v}, t)\, d\mathbf{v} = N, \tag{12.1}$$

where N is the number of (monatomic) molecules in the gas.

There are two distinct types of distribution functions to which the preceding definition may be applied. We may define $f(\mathbf{r}, \mathbf{v}, t)$ as the distribution function for the molecules in a *particular* container. Alternatively, we may look upon it as the *average* distribution function for an ensemble of macroscopically identical systems. This distinction will turn out to be a crucial one later, for we shall draw conclusions that can only be correct for the average of an *ensemble* of similar systems. It will turn out that such conclusions cannot be rigorously true when applied to a particular system. Failure to make this simple distinction in the past has led to many difficulties in the development of the kinetic theory of gases. For the moment, we assume that the distribution function $f(\mathbf{r}, \mathbf{v}, t)$ is interpreted as the average over an ensemble. The specification of the distribution function for a particular system requires a knowledge of the position and velocity of each molecule in the gas; this information is usually unavailable.

In order to construct $f(\mathbf{r}, \mathbf{v}, t)$ as an ensemble average we consider a collection of $\mathcal{N}$ systems which have been prepared so that at some instant of time the macroscopic features of all of them are identical. For example, we might prepare an ensemble in which each member has all its molecules compressed into a particular region of the container. Of course, there exists an enormous number of ways to realize the preceding macrostate, each particular realization corresponding to a particular microstate of the system. In order to describe the temporal behavior of this ensemble, we define the average density of microstates in the ensemble in the $6N$-dimensional Γ-space of the system, $\rho(\mathbf{r}_1, \mathbf{p}_1, \ldots, \mathbf{r}_N, \mathbf{p}_N, t)$, as was done in Section 5.4, by the probability density

$$\rho(\mathbf{r}_1, \mathbf{p}_1, \ldots, \mathbf{r}_N, \mathbf{p}_N, t) = \frac{\mathcal{N}_i}{\Delta\mathcal{V}}.$$

Here $\mathcal{N}_i$ is the number of systems whose representative points in the ensemble are confined at instant t to a volume

$$\Delta\mathcal{V} = \Delta\mathbf{r}_1\,\Delta\mathbf{p}_1 \cdots \Delta\mathbf{r}_N\,\Delta\mathbf{p}_N,$$

in Γ-space, and we take ρ to be a symmetric function of the N particles. The temporal behavior of ρ is governed by the Liouville equation

$$\frac{\partial\rho}{\partial t} + \sum_i \left(\dot{\mathbf{r}}_i \frac{\partial\rho}{\partial\mathbf{r}_i} + \dot{\mathbf{p}}_i \frac{\partial\rho}{\partial\mathbf{p}_i}\right) = \frac{d\rho}{dt} = 0. \tag{12.2}$$

It is recalled that $\rho\,\Delta\mathcal{V}$ represents the number of systems of N particles whose microstates are located in $\Delta\mathcal{V}$, whereas $f(\mathbf{r}, \mathbf{v}, t)\,d\mathbf{r}\,d\mathbf{v}$ represents the number of particles in a single system whose microstates are located in the phase-space volume $d\mathbf{r}\,d\mathbf{v}$ regardless of the location of the remaining $N-1$ particles. Since $d\mathbf{r}\,d\mathbf{v}$ is a subspace of $\Delta\mathcal{V}$, the two functions must be related in that $f(\mathbf{r}, \mathbf{v}, t)$ represents an integral of $\rho(\mathbf{r}_1, \mathbf{p}_1, \ldots, \mathbf{r}_N, \mathbf{p}_N)$ taken over all coordinates and momenta except for one set, say, $\mathbf{r}_1, \mathbf{p}_1$. In other words, the ensemble-average probability of finding one molecule at $\mathbf{r}$ in the small volume $d\mathbf{r}$ and with momentum $\mathbf{p} = m\mathbf{v}$ in the range $d\mathbf{p}$ ($= m\,d\mathbf{v}$) is obtained from ρ by

$$f(\mathbf{r}, \mathbf{v}, t)\, d\mathbf{r}\, d\mathbf{v} = \frac{N}{\mathcal{N}}\, d\mathbf{r}(m\, d\mathbf{v}) \int d\mathbf{r}_2 \int d\mathbf{p}_2 \cdots \int d\mathbf{r}_N \int d\mathbf{p}_N \times \rho(\mathbf{r}, \mathbf{p}, \mathbf{r}_2, \mathbf{p}_2, \ldots, \mathbf{r}_N, \mathbf{p}_N, t), \tag{12.3a}$$

where N is the number of particles in each system of the ensemble. In the same way as was done in Section 5.4.2, we normalize ρ to satisfy the equation

$$\int d\mathbf{r}_1 \int d\mathbf{p}_1 \cdots \int d\mathbf{r}_N \int d\mathbf{p}_N\, \rho(\mathbf{r}_1, \mathbf{p}_1, \ldots, \mathbf{r}_N, \mathbf{p}_N, t) = \mathcal{N}. \tag{12.3b}$$

Consequently,

$$f(\mathbf{r}, \mathbf{v}, t) = \frac{mN}{\mathcal{N}} \int d\mathbf{r}_2 \int d\mathbf{p}_2 \cdots \int d\mathbf{r}_N \int d\mathbf{p}_N\, \rho(\mathbf{r}, \mathbf{p}, \ldots, \mathbf{r}_N, \mathbf{p}_N, t). \tag{12.4}$$

The properties of the gas may be expressed in terms of $f(\mathbf{r}, \mathbf{v}, t)$ as average values. The number density $n(\mathbf{r}, t)$, that is, the number of particles per unit volume of the gas at a point $\mathbf{r}$ in the container at time t, is found by integrating $f(\mathbf{r}, \mathbf{v}, t)$ over all velocities,

$$n(\mathbf{r}, t) = \int d\mathbf{v}\, f(\mathbf{r}, \mathbf{v}, t). \tag{12.5a}$$

The average velocity of molecules at a point $\mathbf{r}$ at time t, $\mathbf{u}(\mathbf{r}, t)$, is

$$u(\mathbf{r}, t) = \frac{\int d\mathbf{v}\, \mathbf{v} f(\mathbf{r}, \mathbf{v}, t)}{\int d\mathbf{v}\, f(\mathbf{r}, \mathbf{v}, t)} = \frac{1}{n} \int d\mathbf{v}\, \mathbf{v} f(\mathbf{r}, \mathbf{v}, t). \tag{12.5b}$$

The average value of any function of velocity, $g(\mathbf{v})$, is expressed by

$$\langle g(\mathbf{r}, t) \rangle = \frac{1}{n} \int d\mathbf{v}\, g(v) f(\mathbf{r}, \mathbf{v}, t). \tag{12.5c}$$

If the distribution function $f(\mathbf{r}, \mathbf{v}, t)$ is known, as it is in equilibrium (Section 6.6), then all preceding properties of the gas are obtained by integration. However, if the system is not in equilibrium, $f(\mathbf{r}, \mathbf{v}, t)$ must be obtained by solving an equation for it. In the next chapter we shall introduce the Boltzmann transport equation which is appropriate for a dilute gas. In this chapter, we propose to develop the kinetic theory as far as possible without having to specify $f(\mathbf{r}, \mathbf{v}, t)$. It will turn out that it is possible to proceed quite far without knowing $f(\mathbf{r}, \mathbf{v}, t)$, and that the equilibrium form for $f(\mathbf{r}, \mathbf{v}, t)$ is very useful even for nonequilibrium calculations if we are prepared to accept an additional hypothesis—that of *local equilibrium*.

12.3. A Dynamical Derivation of the Perfect-Gas Law

As our first example of the methods of kinetic theory, we derive the perfect-gas law

$$Pv_{\mathrm{m}} = \mathbf{R}T$$

by studying the collisions of molecules with the walls of the container. We assume that the gas is composed of molecules which do not collide with each other, but only with the walls of the container. We assume, further, that the molecules make perfectly elastic collisions with the walls, Figure 12.1.

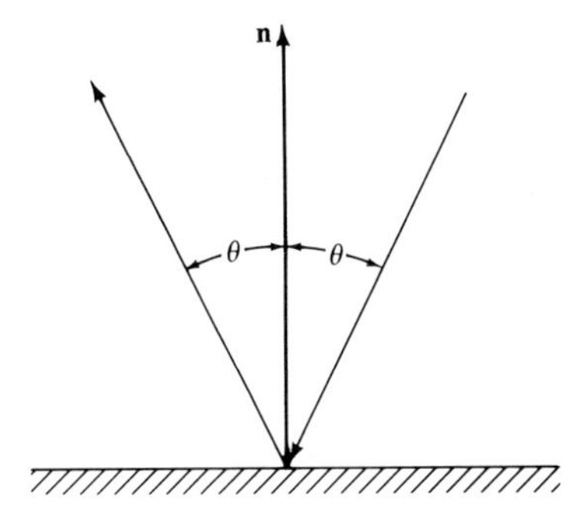

FIGURE 12.1. A perfectly elastic collision of a molecule with a wall.

The pressure is defined as the momentum transferred by the molecules to a unit area of the wall per unit time. In order to compute this quantity, we center attention on a small flat area, δA, of the wall and evaluate the momentum transferred in time δt.

First, consider the contribution to the pressure made by molecules traveling with a velocity $\mathbf{v}$. We suppose that $\mathbf{v}$ makes an angle θ with the normal, $\mathbf{n}$, to the wall, as illustrated in Figure 12.1. All such molecules will make a collision with the area δA during time δt if they lie inside the small cylinder shown in Figure 12.2. The cylinder is based on δA; its side, parallel to $\mathbf{v}$, has a length equal to $v\,\delta t$.† If $f(\mathbf{r}, \mathbf{v}, t)$ denotes the distribution function for the molecules, the number of molecules, n_v, with velocity $\mathbf{v}$ in the cylinder under consideration is given by

$$n_v = \int_{\delta V} d\mathbf{r}\, f(\mathbf{r}, \mathbf{v}, t), \tag{12.6}$$

† There is a strong similarity between many of the computations in this section and those given in Chapter 10. There the ideal gas was a photon gas.

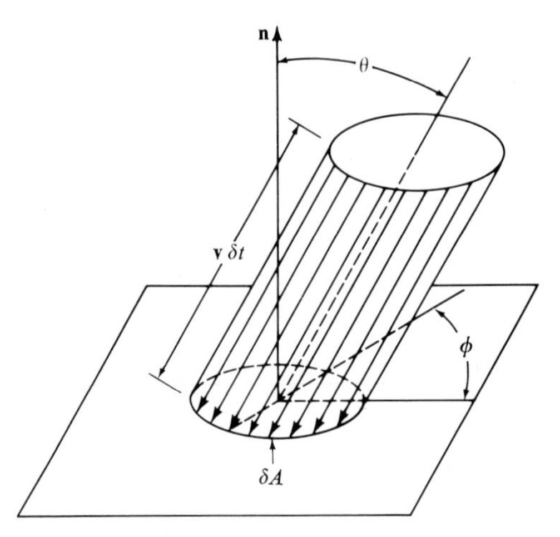

FIGURE 12.2. Momentum transferred by molecules which move with a velocity **v**.

where δV is the volume $\delta A\,\delta t\,v\cos\theta$ of the small cylinder based on δA. Since the collisions of the molecules with the wall are elastic, the momentum transferred to the wall during *each collision* is

$$2m\,|\mathbf{v}|\cos\theta\,.$$

This is the magnitude of the difference between the momentum of the molecule after colliding with the wall and that before the collision. The total momentum, p_v, transferred to the area δA of wall by the molecules which move with velocity **v**, is

$$p_v = 2m\,|\mathbf{v}|\cos\theta \int_{\delta V} d\mathbf{r}\,f(\mathbf{r}, \mathbf{v}, t)\,.$$

The total momentum, p, transferred to the wall by molecules of all possible velocities is obtained by integrating over $|\mathbf{v}|$, θ, and ϕ, where ϕ measures the orientation of the cylinder, as shown in Figure 12.2. Thus

$$p = \int_0^\infty dv\,v^2 \int_0^{\pi/2} d\theta\,\sin\theta \int_0^{2\pi} d\phi\,2mv\cos\theta \int_{\delta V} d\mathbf{r}\,f(\mathbf{r}, \mathbf{v}, t)\,. \quad (12.7)$$

The integration extends over $0 \leq \theta \leq \pi/2$, since for angles greater than $\pi/2$ the molecules travel away from the wall, instead of toward it. By contrast, ϕ varies from 0 to 2π.

This is as far as we can proceed without specifying $f(\mathbf{r}, \mathbf{v}, t)$. Suppose now that the gas is spatially homogeneous, that is, that $f(\mathbf{r}, \mathbf{v}, t)$ is independent of $\mathbf{r}$. This is a further assumption about the distribution function that is certainly not always satisfied. However, if the gas is spatially homogeneous, we can write

$$f(\mathbf{r}, \mathbf{v}, t) = n f(\mathbf{v}, t) \qquad \text{(spatially homogeneous gas)}, \qquad (12.8)$$

where $n = N/V$ is the number density. With this assumption Equation (12.7) becomes

$$p = n\,\delta A\,\delta t \int_0^\infty dv \int_0^{\pi/2} d\theta \int_0^{2\pi} d\phi\, v^2 \sin\theta\, 2mv\cos\theta(v\cos\theta)\, f(\mathbf{v}, t). \quad (12.9)$$

Finally, to compute the equilibrium pressure, let us suppose that $f(\mathbf{v}, t)$ is the Maxwell–Boltzmann distribution

$$f(\mathbf{v}, t) = \left(\frac{m}{2\pi \mathbf{k} T}\right)^{3/2} \exp\left(-\frac{mv^2}{2\mathbf{k}T}\right), \qquad (12.10a)$$

known to us from Section 6.6. The pressure, P, that is the momentum transferred per unit area and time is now easily seen to be

$$P = n\left(\frac{m}{2\mathbf{k}T\pi}\right)^{3/2} 2m \int_0^\infty dv \int_0^{\pi/2} d\theta \int_0^{2\pi} d\phi$$
$$\times\, v^4 \sin\theta \cos^2\theta \exp(-mv^2/2\mathbf{k}T) = n\mathbf{k}T. \qquad (12.10b)$$

The preceding calculation, which has led us to the perfect-gas law, was based upon the following assumptions:

1. The molecules do not collide with each other. We have assumed that all molecules heading toward the wall will impinge on it. If the molecules collided with each other, this assumption would be incorrect.
2. The gas is spatially homogeneous. The molecules are evenly distributed throughout the container. This assumption cannot be true at all times since this would imply that any molecule leaving a small volume is instantly replaced by an identical molecule. This is an assumption that can only be accepted for an ensemble average, but not for a particular system.
3. The gas is in equilibrium. This assumption specifies the distribution function completely, and is responsible for our ability to derive the perfect-gas equation of state.

An almost identical calculation allows us to compute the *number* of particles that strike the area δA in a time δt. It is easily seen that this number is obtained by the following steps:

a. The number of particles, n_v', with a given velocity $\mathbf{v}$ that collide with δA in time δt is

$$n_v' = \int_{\delta V} d\mathbf{r}\, f(\mathbf{r}, \mathbf{v}, t). \qquad (12.11)$$

b. The total number of particles n' is obtained by integrating Equation (12.11) over all velocities $\mathbf{v}$ and over $0 \leq \theta \leq \pi/2$ together with $0 \leq \phi \leq 2\pi$:

$$n' = \int_0^\infty \mathrm{d}v \int_0^{\pi/2} \mathrm{d}\theta \int_0^{2\pi} \mathrm{d}\phi\, v^2 \sin\theta \int_{\delta V} \mathrm{d}\mathbf{r} f(\mathbf{r}, \mathbf{v}, t). \qquad (12.12)$$

If the gas is partially homogeneous, we may use Equation (12.8) to obtain

$$n' = n\,\delta t\,\delta A \int_0^\infty \mathrm{d}v \int_0^{\pi/2} \mathrm{d}\theta \int_0^{2\pi} \mathrm{d}\phi\, v^2 \sin\theta v \cos\theta f(\mathbf{v}, t). \qquad (12.13)$$

Finally, in equilibrium, the number of particles that strike area δA in time δt, denoted by $a\,\delta A\,\delta t$, is given by

$$\begin{aligned} a\,\delta A\,\delta t &= n\,\delta A\,\delta t \left(\frac{m}{2\pi\mathbf{k}T}\right)^{3/2} \int_0^\infty \mathrm{d}v \int_0^{\pi/2} \mathrm{d}\theta \int_0^{2\pi} \mathrm{d}\phi \\ &\quad \times v^3 \sin\theta \cos\theta \exp\left(-\frac{mv^2}{2\mathbf{k}T}\right) \\ &= n\,\delta A\,\delta t \left(\frac{\mathbf{k}T}{2m\pi}\right)^{1/2}. \end{aligned} \qquad (12.14)$$

This number can be put into a more familiar form if we utilize the fact that the average speed, u, in equilibrium is given by

$$\begin{aligned} u = \langle|\mathbf{v}|\rangle &= \left(\frac{m}{2\pi\mathbf{k}T}\right)^{3/2} \int_0^\infty \mathrm{d}v \int_0^{\pi} \mathrm{d}\theta \int_0^{2\pi} \mathrm{d}\phi\, v^3 \sin\theta \exp\left(-\frac{mv^2}{2\mathbf{k}T}\right) \\ &= \left(\frac{8\mathbf{k}T}{\pi m}\right)^{1/2} \end{aligned} \qquad (12.15)$$

Thus,

$$a = \tfrac{1}{4}nu. \qquad (12.16)$$

Suppose now that the small area δA is actually a hole in the wall of the container. Then a represents the number of particles that escape from the container per unit time through unit area. This is the phenomenon of *effusion* (also called the *Knudsen effect*), and the quantity a is referred to as the *rate of effusion.*

It is now useful to compute the average momentum transferred to the wall per collision. This is most easily evaluated for equilibrium systems, for which we obtain

$$\frac{n\mathbf{k}T}{\frac{1}{4}nu} = \frac{4\mathbf{k}T}{u} = (2\pi m\mathbf{k}T)^{1/2} = \tfrac{1}{2}\pi m u. \qquad (12.17)$$

Since mu represents the average momentum per particle, Equation (12.17) states that, on the average, each particle that collides with the wall transfers $\frac{1}{2}\pi$ times its momentum to the wall.

12.4. The Mean Free Path

So far we have considered some of the properties of a gas whose molecules collide only with the walls of the container. Now we extend our discussions to include molecules that collide with each other as well as with the walls. Such molecules will be thought of as interacting through an intermolecular potential whose range of interaction, σ, is short compared to the average distance between the molecules in the gas, whereas the molecules themselves will be imagined as spheres of diameter σ. Since molecules now collide with each other, they cannot, usually, move an arbitrarily large distance without colliding with another molecule. One of the most important quantities which characterize such a gas is called the *mean free path*, that is, the distance that a molecule covers, on the average, between two successive collisions.

In order to discuss collisions it is necessary to introduce the concept of a *collision cylinder*. Suppose that a particular molecule in the container moves with velocity $\mathbf{v}$ and that a molecule with velocity $\mathbf{v}_1$ collides with it in time δt. To describe this collision, we choose a coordinate system and place its origin at the center of the molecule so that the z axis is drawn in the direction of the relative velocity vector $\mathbf{v}_1 - \mathbf{v}$. Evidently, the molecules can collide with each other only if the distance between their centers is smaller than or equal to σ. This means that the center of the molecule with velocity $\mathbf{v}_1$ must, at time t, lie inside the "cylinder" shown sketched in Figure 12.3, if a collision is to take place in the succeeding interval of time, δt. The height of this *collision cylinder* is $|\mathbf{v}_1 - \mathbf{v}|\, \delta t$ and its volume is $\pi\sigma^2|\mathbf{v}_1 - \mathbf{v}|\, \delta t$.

We may use the collision cylinder to compute the number of binary collisions that take place in a small volume of gas in time δt between molecules with velocity $\mathbf{v}$ and $\mathbf{v}_1$, respectively. In a small volume $\delta\mathbf{r}$ in the gas there exist $f(\mathbf{r}, \mathbf{v}, t)\, \delta\mathbf{r}\, \delta\mathbf{v}$ molecules with velocity $\mathbf{v}$ in the range $\delta\mathbf{v}$. Imagine that to each of these molecules there is attached a collision cylinder, appropriate for collisions with molecules of velocity $\mathbf{v}_1$ and within a time interval δt. The number of such cylinders is then equal to $f(\mathbf{r}, \mathbf{v}, t)\, \delta\mathbf{r}\, \delta\mathbf{v}$. The total volume occupied by the collision cylinders is the number of cylinders times the volume per cylinder, which is

$$f(\mathbf{r}, \mathbf{v}, t)\, \delta\mathbf{r}\, \delta\mathbf{v}\, \pi\sigma^2\, |\mathbf{v} - \mathbf{v}_1|\, \delta t\,. \tag{12.18}$$

In order to obtain the number of $(\mathbf{v}, \mathbf{v}_1)$ collisions, we must compute the number of molecules with velocity $\mathbf{v}_1$ that are present in the collision cylinders

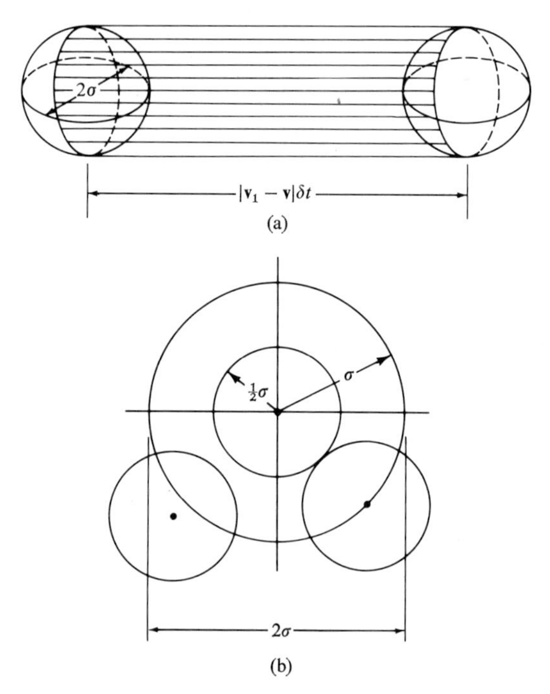

FIGURE 12.3. a, Illustrates the collision cylinder; b, illustrates the fact that the particles begin to interact if their centers are within a distance σ.

at the beginning of the time interval. If there are more than one $\mathbf{v}_1$-molecules in a particular collision cylinder, it is necessary to determine which one of them actually leads to a collision. However, this alone does not exhaust the difficulties, since it is possible that either the $\mathbf{v}$-molecule or the $\mathbf{v}_1$-molecule may collide with a molecule moving with a different velocity at an earlier instant. To complicate matters, the collision cylinders can become extremely long, extending beyond the boundaries of the volume $\delta\mathbf{r}$ if the magnitude of the relative velocity $|\mathbf{v} - \mathbf{v}_1|$ is chosen sufficiently large; in extreme cases, they may reach beyond the container. It should now be obvious that to find our way through all these complications it would be necessary to know the exact positions and velocities of all molecules in the gas at the instant t and to follow the motion of each molecule during the time interval δt. Clearly, it is impossible to do this in such detail. However, we recall now that we are dealing with an ensemble of systems and that we merely ought to compute the *average* number of $(\mathbf{v}, \mathbf{v}_1)$ collisions. To achieve this, we assume (a) that the gas is sufficiently dilute for the collision cylinders to contain at most one molecule with velocity $\mathbf{v}_1$ and such molecules lead to $(\mathbf{v}, \mathbf{v}_1)$ collisions; and (b) that the number of molecules with velocity $\mathbf{v}_1$ present in the $(\mathbf{v}, \mathbf{v}_1)$

collision cylinders at the instant t is equal to the number of $\mathbf{v}_1$-molecules per unit volume multiplied into the total volume of the $(\mathbf{v}, \mathbf{v}_1)$ cylinders. Combining these two assumptions we arrive at the statement that *every* $\mathbf{v}_1$-molecule in the cylinder at the start of the interval actually leads to a $(\mathbf{v}, \mathbf{v}_1)$ collision, and that

the number of collisions occurring within a time interval δt between molecules with velocity $\mathbf{v}$ in the range $\delta\mathbf{v}$ and molecules with velocity $\mathbf{v}_1$ in range $\delta\mathbf{v}_1$ in a volume $\delta\mathbf{r}$ of gas centered about a point $\mathbf{r}$ is given by the expression

$$f(\mathbf{r}, \mathbf{v}, t) f(\mathbf{r}, \mathbf{v}_1, t) |\mathbf{v} - \mathbf{v}_1| \pi\sigma^2 \, \delta\mathbf{v} \, \delta\mathbf{v}_1 \, \delta\mathbf{r} \, \delta t. \tag{12.19}$$

Equation (12.19) is not exact but, rather, a statement about the average behavior of an ensemble of similar systems. This equation forms the foundation for our work in kinetic theory and, as such, is quite important. We shall refer to the assumptions (a) and (b), as well as to Equation (12.19) which expresses them mathematically, as the *Stosszahlansatz*, "the collision-number assumption." This term was first used by P. and T. Ehrenfest, and the English-language literature on kinetic theory still refers to Equation (12.19) by this German word. We shall see later that it plays a vital role in the theory of irreversibility and in the derivation of Boltzmann's transport equation.

If we use the *Stosszahlansatz*, we can compute the following quantities:

a. The *total* number of collisions suffered by molecules of velocity $\mathbf{v}$ in $\delta\mathbf{r}$ in time δt is given by

$$f(\mathbf{r}, \mathbf{v}, t)\pi\sigma^2 \, \delta\mathbf{r} \, \delta t \int d\mathbf{v}_1 f(\mathbf{r}, \mathbf{v}_1, t) |\mathbf{v} - \mathbf{v}_1|. \tag{12.20}$$

b. The total number of collisions suffered by molecules of all velocities in $\delta\mathbf{r}$ in time δt is obtained by integrating Equation (12.20) over velocity $\mathbf{v}$, and is

$$\delta\mathbf{r} \, \delta t \pi\sigma^2 \int d\mathbf{v} \int d\mathbf{v}_1 f(\mathbf{r}, \mathbf{v}, t) f(\mathbf{r}, \mathbf{v}_1, t) |\mathbf{v} - \mathbf{v}_1|. \tag{12.21}$$

These are the most general statements for the preceding quantities expressed in terms of the single-particle distribution function $f(\mathbf{r}, \mathbf{v}, t)$. Let us now assume that the gas is spatially homogeneous, so that the single-particle distribution functions do not depend upon $\mathbf{r}$. This means that we may use Equation (12.8) to write:

a. The number of collisions in the gas per unit volume per unit time at instant t that involve molecules with velocity $\mathbf{v}$ is

$$n^2 f(\mathbf{v}, t)\pi\sigma^2 \int d\mathbf{v}_1 f(\mathbf{v}_1, t) |\mathbf{v} - \mathbf{v}_1|. \tag{12.22}$$

b. The total number of collisions suffered by molecules of all velocities per unit volume and per unit time at instant t is

$$n^2\pi\sigma^2 \int d\mathbf{v} \int d\mathbf{v}_1 f(\mathbf{v}, t) f(\mathbf{v}_1, t) |\mathbf{v} - \mathbf{v}_1| . \tag{12.23}$$

The average number of collisions that a molecule suffers per unit time, or the collision frequency, ν, is evaluated by dividing the total number of collisions per unit volume per unit time, Equation (12.23), by the number of molecules per unit volume, n. This yields

$$\nu = n\pi\sigma^2 \int d\mathbf{v} \int d\mathbf{v}_1 f(\mathbf{v}, t) f(\mathbf{v}_1, t) |\mathbf{v} - \mathbf{v}_1| . \tag{12.24}$$

The average time *between* collisions for a particle is ν^{-1}, and if u is the average speed of the molecules, the average distance traveled between collisions, or the mean free path, l, becomes equal to

$$l = \frac{u}{\nu} = \frac{\int d\mathbf{v} |\mathbf{v}| f(\mathbf{v}, t)}{n\pi\sigma^2 \int d\mathbf{v} \int d\mathbf{v}_1 f(\mathbf{v}, t) f(\mathbf{v}_1, t) |\mathbf{v} - \mathbf{v}_1|} . \tag{12.25}$$

The mean free path is inversely proportional to the density and to the cross-sectional area of the interaction sphere. The velocity integrals in Equation (12.25) form a ratio of the average speed to the average relative speed.

In equilibrium, we may completely determine the mean free path by evaluating the integral

$$\int d\mathbf{v} \int d\mathbf{v}_1 |\mathbf{v} - \mathbf{v}_1| f(\mathbf{v}) f(\mathbf{v}_1) = \left(\frac{m}{2\pi\mathbf{k}T}\right)^3 \int d\mathbf{v} \int d\mathbf{v}_1 |\mathbf{v} - \mathbf{v}_1| \times \exp\left[-\frac{m}{2\mathbf{k}T}(v_1{}^2 + v^2)\right] . \tag{12.26}$$

To do this, we change the variables from $\mathbf{v}$, and $\mathbf{v}_1$ to the velocity of the center of mass $\mathbf{V} = \frac{1}{2}(\mathbf{v}_1 + \mathbf{v})$ and the relative velocity $\boldsymbol{\rho} = \mathbf{v}_1 - \mathbf{v}$,

$$\mathbf{v} = \mathbf{V} - \tfrac{1}{2}\boldsymbol{\rho} \tag{12.27a}$$

and

$$\mathbf{v}_1 = \mathbf{V} + \tfrac{1}{2}\boldsymbol{\rho} . \tag{12.27b}$$

Moreover,

$$v^2 + v_1{}^2 = 2V^2 + \tfrac{1}{2}\rho^2 \tag{12.27c}$$

and

$$d\mathbf{v}\, d\mathbf{v}_1 = d\mathbf{V}\, d\boldsymbol{\rho} . \tag{12.27d}$$

yielding

$$\int d\mathbf{v} \int d\mathbf{v}_1 \, |\mathbf{v} - \mathbf{v}_1| f(\mathbf{v}) f(\mathbf{v}_1)$$

$$= \left(\frac{m}{2\pi \mathbf{k}T}\right)^3 \int_0^\infty dV \int_0^\pi d\theta_V \int_0^{2\pi} d\phi_V \int_0^\infty d\rho \int_0^\pi d\theta_\rho \int_0^{2\pi} d\phi_\rho$$

$$\times \exp\left(-\frac{mV^2}{\mathbf{k}T} - \frac{m\rho^2}{4\mathbf{k}T}\right) V^2 \rho^3 \sin\theta_V \sin\theta_\rho$$

$$= \left(\frac{16\mathbf{k}T}{\pi m}\right)^{1/2} = \sqrt{2}\, u \,. \tag{12.28}$$

Here, as usual, the angles θ_ρ, ϕ_ρ, θ_V and ϕ_V are drawn in the respective phase spaces. Finally, we combine Equations (12.25) and (12.28) to obtain

$$l_{\text{eq}} = \frac{1}{\sqrt{2}\, n\pi\sigma^2} \tag{12.29}$$

for the mean free path in equilibrium.

A typical molecular diameter, say that for oxygen, is $\sigma = 4 \times 10^{-8}$ cm. Thus, in oxygen at $T = 300$ K and atmospheric pressure $n = P/\mathbf{k}T \approx 2.4 \times 10^{19}$ molecules/cm^3. Hence $l_{\text{eq}} \approx 10^{-5}$ cm. The collision frequency, $\nu \approx 6 \times 10^9$ collisions/sec, if we put $m = 5 \times 10^{-23}$ gr. The average speed is $u \approx 5 \times 10^4$ cm/sec.

12.5. The Mean Free Path and Transport Properties

One of the most useful applications of the mean free path concept occurs in the theory of transport processes in systems in which there exists a gradient of a quantity such as the temperature, the average velocity, or the density. The existence of such gradients causes a transfer of energy, momentum, or particles, respectively, from one region of the gas to another. The elementary theory of such *transport processes* rests on three assumptions:

a. The gas is dense enough for the mean free path to be small compared to a characteristic dimension of the container. This means that the particles collide with each other much more often than they collide with the walls of the container.

b. The gas is sufficiently dilute for the mean free path to be much greater than the molecular diameter. Thus, our considerations in the previous section remain applicable, and the mean free path is determined by binary collisions.

c. The temperature, average velocity, and density do not vary appreciably over distances of the order of a mean free path.

12.5.1. Thermal Conduction

The conduction of heat occurs within a system when the temperature distribution in it is not uniform. If the system is a gas or a liquid, temperature differences create differences in density which, in turn, introduce changes in the gravitational forces throughout the fluid. Thus the fluid begins to move on a macroscopic scale initiating the process of *natural convection*. The macroscopic motion causes an additional transfer of energy which may outweigh by far that due to the molecular motion. In what follows, we shall assume that the process of natural convection has been suppressed in the gas, and that only conduction occurs in it. In practice, this can be achieved by reducing the thickness of the layer of gas in the direction of the gravitational force, or by making sure that the denser, and therefore cooler, layers lie below the warmer and lighter layers.†

In order to simplify our ideas, we first imagine a layer of gas, shown in Figure 12.4, which is small enough to exclude convection and yet many orders of magnitude larger than the mean free path. We further imagine that the temperature is maintained constant at T_1 and $T_2 > T_1$ along two (real or imaginary) semi-infinite planes separated by a distance δ which is also several orders of magnitude larger than the mean free path. Thus, for the moment, we exclude the possibility that heat may be conducted in any direction except that of the z axis placed at right angles to the planes. Finally, we postulate that the two-dimensional temperature field, also sketched in Figure 12.4, has attained a steady state, so that the temperature at any point is independent of time.

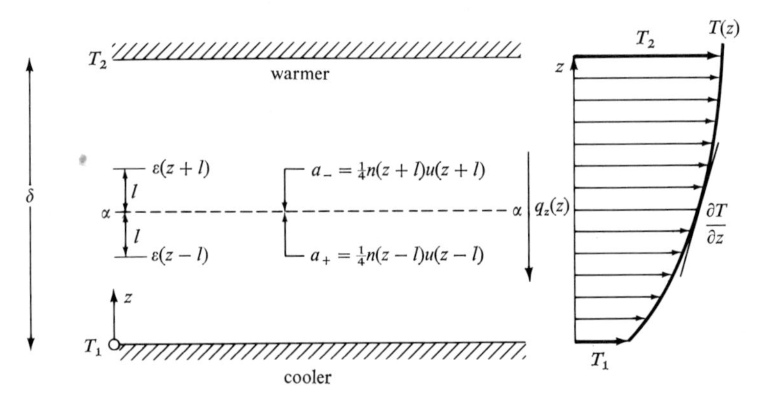

FIGURE 12.4. Steady-state heat conduction.

† This stratification may, however, become unstable if the temperature difference or the thickness of the layer are increased beyond certain limits. Once again convection sets in, and, once again, we assume that this is not the case.

We center our attention on an imaginary plane α–α and on its neighborhood of thickness δ. It should be realized that the distances l in the sketch have been enormously exaggerated, since l is several orders of magnitude smaller than δ. On a microscopic scale, we shall observe that molecules, carrying kinetic energy with them, cross the plane α–α in both directions. In the steady state under consideration, the upward directed flux, a_-, of molecules exactly balances that directed downward, a_+. The same is not true about the flux of energy because the molecules moving downward carry with them a larger average energy than those traveling in the opposite direction. This creates a net flow of energy across plane α–α.

The conditions described so far exist in a fluid regardless of its density. In our present case, that is in a dilute gas, we assume, heuristically, that the molecules crossing plane α–α carry with them the average energy that is characteristic of the temperature which prevails at a distance l from the plane. In other words, the molecules that cross the plane are assumed to be in *local equilibrium* with the molecules separated by exactly one mean free path. This is the distance at which, on the average, the molecule has undergone its last collision, thus, again on the average, acquiring the average velocity which corresponds to the appropriate temperature.

Referring to Figure 12.4, we deduce that the number of molecules crossing the plane at z from above is

$$a_-(z) = \tfrac{1}{4}n(z + l)u(z + l), \tag{12.30a}$$

where n and u denote the average number density and the average velocity prevailing at level $z + l$. The opposite flux is

$$a_+(z) = \tfrac{1}{4}n(z - l)u(z - l). \tag{12.30b}$$

In steady state, the two molecular fluxes balance, and we record that

$$a_+ = a_- \equiv a[= n(z)u(z)] \qquad \text{(steady state)}. \tag{12.31}$$

The average energy, $q_-(z)$, transferred per unit area and unit time from above the plane α–α is equal to the molecular flux multiplied into the average energy, ε, carried by each molecule. The latter is the energy $\varepsilon(z + l)$ appropriate to a location at $z + l$. Hence

$$q_-(z) = a\varepsilon(z + l), \tag{12.32a}$$

and, analogously,

$$q_+(z) = a\varepsilon(z - l). \tag{12.32b}$$

The net amount of energy transferred per unit area and time in the positive z direction is, thus,

$$
\begin{aligned}
q_z(z) &= a[\varepsilon(z - l) - \varepsilon(z + l)] \\
&= a\left\{\left[\varepsilon(z) - l\,\frac{\partial\varepsilon(z)}{\partial z} + \frac{1}{2}\,l^2\,\frac{\partial^2\varepsilon(z)}{\partial z^2} + \cdots\right]\right. \\
&\qquad \left. - \left[\varepsilon(z) + l\,\frac{\partial\varepsilon(z)}{\partial z} + \frac{1}{2}\,l^2\,\frac{\partial^2\varepsilon(z)}{\partial z^2} + \cdots\right]\right\} \\
&= -2al\,\frac{\partial\varepsilon(z)}{\partial z} + O\left(\frac{\partial^3\varepsilon(z)}{\partial z^3}\right).
\end{aligned}
\tag{12.33}
$$

Neglecting the terms of the third and higher orders in the gradient of the energy, we obtain an expression for the net flow (the *heat flux*) in the form,

$$q_z = -\lambda\,\frac{\partial T(z)}{\partial z} \tag{12.34a}$$

where

$$\lambda = 2al\,\frac{\partial\varepsilon}{\partial T}. \tag{12.34b}$$

This approximation presupposes that the energy gradient and the temperature gradient are small, so that neither the energy nor the temperature change appreciably over a distance comparable to the mean free path. This condition is satisfied with great accuracy under most experimental conditions and breaks down only at pressures lower than a few microns of mercury. However, when this is the case, the continuum approximation breaks down too.

Equation (12.34a) states that the heat flux, q_z, is proportional to the component $\partial T/\partial z$ of the temperature gradient parallel to it, and is known as *Fourier's law* of thermal conduction. It is easy to see that the preceding argument can be applied to each of the three perpendicular directions of a most general temperature field. It follows that the transport of energy through a small area is, generally speaking, described by a *heat-flux vector*, $\mathbf{q}(q_x, q_y, q_z)$, whose components are separately proportional to the local temperature-gradient vector, grad T. In an isotropic medium, the coefficient of proportionality, λ, is the same in all directions, and we may write Fourier's law in its most general form

$$\mathbf{q}(x, y, z) = -\lambda \operatorname{grad} T. \tag{12.34c}$$

In these circumstances, the vector $\mathbf{q}$ is collinear with the gradient of temperature and opposed to it in direction, indicating that the transport of heat occurs in the direction of decreasing temperatures, since λ is obviously positive. This is in accordance with the Second law of thermodynamics.

Experiments show that Fourier's law, Equation (12.34c), applies under much wider conditions than the preceding derivation would lead us to believe.

First, it applies to a fluid of any density and molecular structure, as well as to a solid. Secondly, it applies instantaneously to a nonsteady temperature field. In fact, the theory of thermal conduction in solids, liquids, and gases is firmly based on the acceptance of its validity† as suggested by J. B. J. Fourier in 1807.‡ The same cannot be said about the factor of proportionality, λ, which is known as the (coefficient of) *thermal conductivity*, and is given in the form of the theoretical equation (12.34b). This is valid, and even so with reservations, only for dilute monatomic gases. Thus, the preceding argument should not be regarded as a derivation of Fourier's law, but rather as an elementary derivation of an expression for the thermal conductivity of a monatomic gas at low density.

It is easy to verify that the units of thermal conductivity are

$$\{\lambda\} = \frac{\text{kcal}}{\text{m sec K}}, \quad \frac{\text{kJ}}{\text{m sec K}}, \quad \text{or} \quad \frac{\text{Btu}}{\text{ft h R}}.$$

In order to obtain an explicit expression for the thermal conductivity in a dilute monatomic gas, we make the additional assumption that all quantities which appear in Equation (12.34b) may be computed with the aid of the Maxwell–Boltzmann distribution function. Noting that $\partial\varepsilon/\partial T = mc_v$, substituting the expression for l_{eq} from Equation (12.29), and utilizing Equation (12.15) for u in the expression for a from Equation (12.31), we calculate§

$$\lambda = 2aml_{\text{eq}}c_v = \tfrac{1}{2}nul_{\text{eq}}mc_v = \frac{umc_v}{2\sqrt{2}\,\pi\sigma^2} \tag{12.35}$$

and

$$\lambda = \frac{c_v}{2\sqrt{2}\,\pi\sigma^2}\left(\frac{8m\mathbf{k}T}{\pi}\right)^{1/2}. \tag{12.36}$$

This result asserts that the thermal conductivity of a gas of moderate density is a unique function of temperature, increasing as $T^{1/2}$. More specifically, the thermal conductivity is independent of density (or pressure), contrary to expectation. Nevertheless, this characteristic of thermal conductivity is easily explained physically, for if the density of the gas is doubled, the molecular flux, a, will double, but the mean free path will be halved simultaneously. Experiments show that as long as the density is small, the thermal conductivity

† For an exhaustive study of solutions of Fourier's equation which is based on Fourier's law, see H. S. Carslaw and J. C. Jaeger, *Conduction of Heat in Solids*, Oxford Univ. Press, (Clarendon), London and New York, 1959.

‡ *Théorie Analytique de la Chaleur*, Paris, 1822. English translation, *The Analytical Theory of Heat* (A. Freeman, transl.), Dover, New York, 1955.

§ It should be noted that ε is the energy carried by a molecule, so that $\mathrm{d}\varepsilon/\mathrm{d}T$ is the specific heat per molecule or per m units of mass. Hence the specific heat per unit mass $c_v = (1/m)(\mathrm{d}\varepsilon/\mathrm{d}T)$.

is essentially independent of it. However, the increase with temperature, even for $c_v = \text{const}$, occurs faster than our theoretical result would suggest. A more rigorous theory due to S. Chapman and D. Enskog† and based on Boltzmann's equation (Section 13.4) supplies us with a correction to this zero-order temperature dependence by taking into account that molecules interact with each other through an appropriately chosen force potential. The results for most monatomic gases can be brought to a reasonable agreement with this improved theory by accepting for them the validity of the Lennard-Jones potential‡ from Equation (2.5b). Regardless of the assumed form of the potential, the thermal conductivity of a monatomic gas is given by the following, first-order approximation

$$\lambda = \frac{25}{32} \frac{(\pi m \mathbf{k} T)^{1/2} c_v}{\pi \sigma^2 \Omega^{(2,2)*}(\mathbf{k}T/\varepsilon)}, \tag{12.36a}$$

where $\Omega^{(2,2)*}$ is a tabulated function of the argument $\mathbf{k}T/\varepsilon$ ("collision integral") which depends on the form of intermolecular force potential. For a rigid-sphere molecule, $\Omega^{(2,2)*} \equiv 1$.

The diagram in Figure 12.5 represents the experimental results obtained

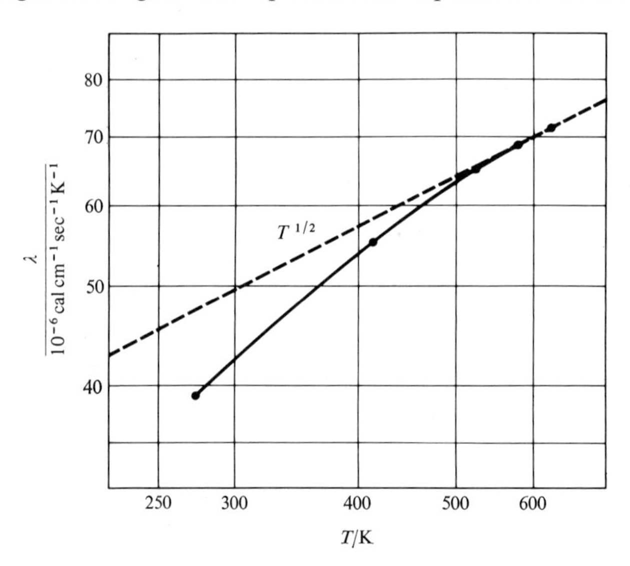

FIGURE 12.5. The thermal conductivity of argon, as measured by F. G. Keyes and R. G. Vines, in a logarithmic plot.

† See S. Chapman and T. G. Cowling, *The Mathematical Theory of Non-Uniform Gases*, Cambridge Univ. Press, 3rd ed., London and New York, 1970; or J. O. Hirschfelder, C. F. Curtiss, and R. B. Bird, *Molecular Theory of Gases and Liquids*, Wiley, New York, 1964.

‡ Or a more general $6 - n$ potential, where n is adjustable.

by F. G. Keyes and R. G. Vines† for argon in a logarithmic plot, comparing it with a line proportional to $T^{1/2}$. It is seen that at high temperatures, the trend of the experimental points seems to approach that stipulated in Equation (12.36).

The representation of experimental results can be improved by defining an "effective" hard-sphere molecular cross section

$$\sigma_{\text{eff}} = \left[\frac{c_v}{2\sqrt{2\pi}\lambda}\left(\frac{8\mathbf{k}mT}{\pi}\right)^{1/2}\right]^{1/2}, \tag{12.37}$$

and assuming that it depends on temperature. The effective molecular diameter, σ_{eff}, has been plotted as a function of temperature in Figure 12.6 which shows that it decreases with increasing temperature, varying by about 10–15% over a span of 100 K or so. This is understood if it is realized that the average velocity of the molecules in a gas increases with temperature allowing molecules further to penetrate into their respective fields of forces. The simple theory interprets this process as a decrease in the molecular diameter.

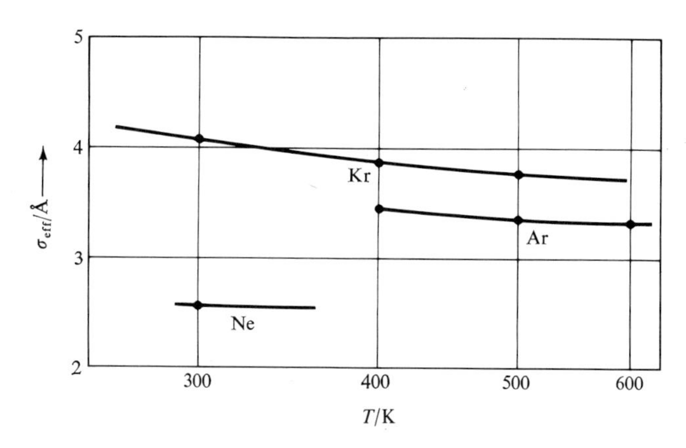

FIGURE 12.6. Effective molecular diameter for several monatomic gases (computed from their thermal conductivity).

12.5.2. The Assumption of Local Equilibrium

It is useful now to interrupt for a moment our study of transport phenomena and to devote a few remarks to a discussion of the assumption that the equations for the transport coefficient, such as Equation (12.34b) for thermal conductivity, can be brought to a closed form by evaluating all averages with reference to the Maxwell–Boltzmann distribution which is strictly valid for equilibrium states only. When this is done, we say that the *assumption of local equilibrium* has been made. At first, this assumption appears to be

† *J. Heat Transfer* **87**, (1965), 177.

inconsistent with the fact that a system in which transport phenomena occur is in a nonequilibrium state. Nevertheless, this assumption leads to good agreement with experiment if applied judiciously. Physically, it is implied that over distances of the order of several mean free paths, still measured in units of 10^{-5} cm in a normal system, the enormous frequency of collisions establishes a velocity distribution which is very closely, even if not exactly, Maxwellian. This *local* Maxwellian distribution corresponds to the local temperature, and nothing in our reasoning precludes it from varying over distances which are large compared with a mean free path and yet very small compared with a characteristic, macroscopic dimension.

An equivalent assumption known under the name of the *principle of local state* (Section 1.5.1) is made in the macroscopic theory of transport processes.† The preceding discussion convinces us that this assumption is justified when temperature changes over one mean free path remain small compared with the local temperature, when changes in macroscopic velocities over the same distance are small compared with the microscopic, molecular velocities, and when the density is large enough to assure a very large collision frequency in times comparable with those of interest in the study of the process. Such conditions are indeed satisfied in a preponderant majority of problems of practical importance.

12.5.3. Viscosity

A continuous distribution of velocity gives rise to a transport of momentum in complete analogy with the transfer of energy which results from a continuous distribution of temperature. When a fluid is made to move on a macroscopic scale, it is found that a true steady state can be maintained in it only if its velocity is small. As the velocity in a given fluid is increased beyond a certain limit, the fluid develops an instability which manifests itself in macroscopic oscillations causing the flow to become unsteady in a random fashion, every macroscopic velocity fluctuating locally about a mean value. When this is the case, we say that the flow has become *turbulent*. A truly steady flow is termed *laminar*, and in what follows we shall assume that turbulence has been inhibited. Viewed microscopically, the motion of molecules, even in laminar flow, is not steady. However, the unsteadiness occurs on a time scale as well as on a length scale which are of the order of a mean free path or of the inverse of the collision frequency, respectively. The scale of turbulent flow is many orders of magnitude larger than this, whereas the frequency is many orders of magnitude smaller.

When a fluid moves with a local macroscopic velocity, $\mathbf{U}(U_x, U_y, U_z)$,

† See also J. Kestin, *A Course in Thermodynamics*, Vol. I, p. 211, and Vol. II, p. 386, Blaisdell, Waltham, Massachusetts, 1966 and 1969.

it is clear that the average microscopic velocity $\mathbf{v}(v_x, v_y, v_z)$ can no longer vanish over a small element of volume $\delta\mathbf{r}$, but must average to $\mathbf{U}(U_x, U_y, U_z)$, so that

$$u_x = \langle v_x \rangle = U_x \qquad \text{and so on for } y \text{ and } z. \tag{12.38}$$

Thus, a transport of momentum $m\mathbf{v}$ is set up because molecules arriving at some point from a distance l carry with them, on the average, that velocity which was acquired during the last collision. This transport of momentum, which gives rise to internal friction and is responsible for the viscosity of a gas (or a liquid, for that matter), can be described in very much the same terms as those used in Section 12.5.1 for the transport of energy.

For simplicity, we first consider a two-dimensional flow field extending over a distance $\delta \gg l$, Figure 12.7, in which $U_z = U_y = 0$, but in which U_x

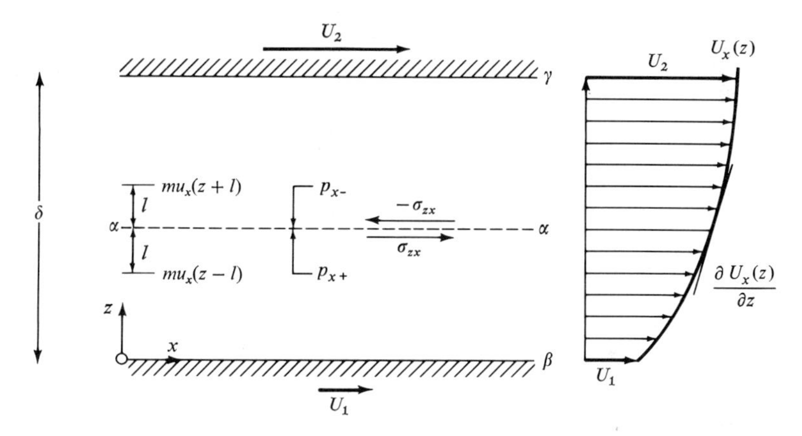

FIGURE 12.7. Two-dimensional shear flow.

varies with the coordinate z only. This means that in the planes β at $z = 0$ and γ at $z = \delta$, which may be real or fictitious, the macroscopic velocity U_x is uniform, varying, say, from U_1 at $z = 0$ to U_2 at $z = \delta$. If we now center attention on the imaginary plane α–α, we conclude that there must develop in it a stress (force per unit area) σ_{zx} tending to slow down the faster fluid and to accelerate its slower layers. This force results from the fact that the molecules which arrive from below move with a velocity whose x component is smaller than that prevailing at z. Similarly, the reverse flux of molecules carries with it a momentum whose x component is larger than that prevailing at α–α. Thus, the flow of molecules in the z direction tends to change the values of the x component of the average velocity and momentum above and below the plane. This change in momentum is equivalent to a force (or stress if measured per unit area) in the x direction.

Retracing our steps from Section 12.5.1, we now write down an expression for the flux of x momentum

$$p_{x+} = amu_x(z - l) \qquad \text{and} \qquad p_{x-} = amu_x(z + l), \tag{12.39}$$

where a is the molecular flux from Equation (12.31), and u_x denotes the x component of the average molecular velocity. The net flow, or change in momentum, that is the shearing stress, is, thus,

$$\begin{aligned} \sigma_{zx} = p_{x+} - p_{x-} &= ma[u_x(z - l) - u_x(z + l)] \\ &= -2mal \frac{\partial u_x(z)}{\partial z} + O\left(\frac{\partial^3 u_x(z)}{\partial z^3}\right). \end{aligned} \tag{12.40}$$

In a gas in equilibrium, the average velocity component in any direction is zero. However, if the gas participates in a macroscopic motion in one direction (x in this case), the macroscopic velocity component [$U_x(z)$ in this case] becomes equal to the new average, as already stated. It follows from Equation (12.38) that

$$u_x(z) = U_x(z) \qquad \text{and} \qquad \frac{\partial u_x(z)}{\partial z} = \frac{\partial U_x(z)}{\partial z}, \tag{12.40a}$$

and, hence,

$$\sigma_{zx} = -2mal \frac{\partial U_x(z)}{\partial z} \tag{12.40b}$$

or

$$\sigma_{zx} = -\eta \frac{\partial U_x(z)}{\partial z}, \tag{12.41a}$$

with

$$\eta = 2alm = \tfrac{1}{2} nulm. \tag{12.41b}$$

Equation (12.41a) expresses *Newton's law of friction* for a fluid in motion which states that the shear stress is proportional to the gradient of the macroscopic velocity taken in a normal direction. The factor of proportionality, η, is known as the (absolute or shear) *viscosity* of the fluid, and is measured in units of

$$\{\eta\} = \text{Poise} \equiv \frac{\text{gr}}{\text{cm sec}} \qquad \text{or} \qquad \frac{\text{lbf sec}}{\text{ft}^2}$$

Like Fourier's law, Equation (12.34a), Newton's law of friction enjoys much wider applicability than our derivation would warrant, in that it holds in steady as well as in unsteady motions of a wide class of gases and liquids known as *Newtonian fluids*. Newton's law of friction can also be generalized to a three-dimensional velocity field, but the attendant mathematical complications exceed the scope of this course. These are due to the fact that we must now relate a tensor (stress) to the gradient of a vector (velocity).†

As far as our present interest is concerned, we regard our theory as a means of deriving the theoretical formula (12.41b) for the viscosity of a dilute gas. Applying the principle of local equilibrium, we can easily derive the explicit expression

$$\eta = \frac{(m\mathbf{k}T)^{1/2}}{\pi^{3/2}\sigma^2}. \tag{12.42}$$

We can now repeat much of the discussion which followed Equation (12.36) in Section 12.5.1. In particular, we notice that the viscosity of a dilute gas is independent of density (and pressure) for the same reasons as its thermal conductivity. Once again experiment shows that the viscosity of a gas increases faster than $T^{1/2}$ tending to this type of behavior at higher temperatures. This is illustrated in the diagram of Figure 12.8 in which we have plotted the ratio

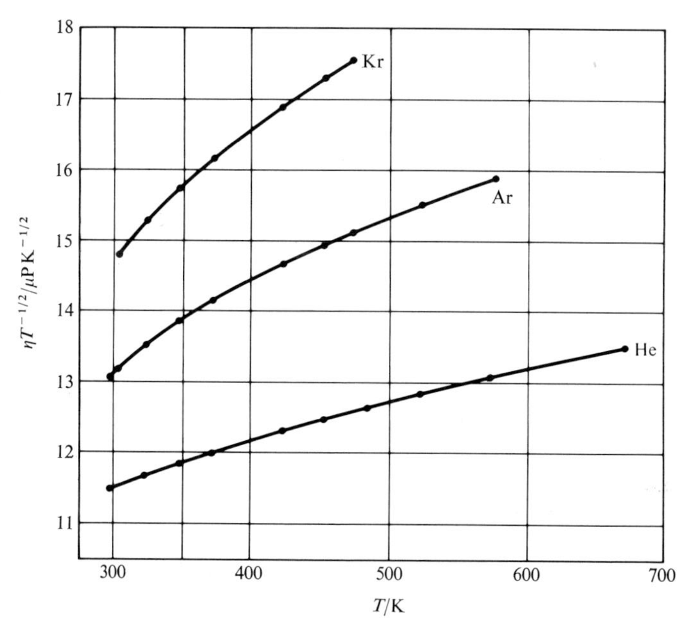

FIGURE 12.8. The viscosity of several monatomic gases as measured by R. DiPippo and J. Kestin.

† See, for example, J. Kestin, *A Course in Thermodynamics*, Vol. II, Chapter 24, Blaisdell, Waltham, Massachusetts, 1969.

$\eta/T^{1/2}$ for several monatomic gases as measured by R. DiPippo and J. Kestin.† It is also useful to know that the Chapman–Enskog theory mentioned earlier in connection with thermal conductivity succeeds in improving agreement with experiment if a suitable force-potential can be chosen. Often, this turns out to be the Lennard-Jones potential or the more general $6\text{-}n$ potential. The first-order approximation to viscosity, that is, the analog of Equation (12.36a), is now

$$\eta = \frac{5}{16} \frac{(\pi m \mathbf{k} T)^{1/2}}{\pi \sigma^2 \Omega^{(2,\,2)*}(\mathbf{k}T/\varepsilon)} \tag{12.42a}$$

in which $\Omega^{(2/2)*}(\mathbf{k}T/\varepsilon)$ is the same collision integral as for thermal conductivity, Equation (12.36a).

The improved theory,‡ though, strictly speaking, derived for dilute monatomic gases, represents the viscosity of more complex molecules as well, provided that they are not polar. This is in marked contrast with thermal conductivity which is strongly affected by the internal motions in polyatomic molecules. The contribution of such motions to the transport of momentum is negligible, being very considerable for the transport of energy.

It is once again possible to define a temperature-dependent effective molecular diameter, σ'_{eff}. However, it turns out that the quantity computed in this way differs from the analogous quantity evaluated with the aid of thermal conductivity. This points to the inadequacy of the elementary theory and of the *ad hoc* assumption that postulating a temperature-dependent molecular diameter is sufficient to improve it.

A comparison between Equations (12.41b) and (12.36) shows that our simple theory suggests that the transport coefficients η and λ must be related to each other at any given temperature, since the mean free path, l, can be eliminated between them. In this manner, we find that the ratio

$$\frac{\eta c_v}{\lambda} = 1 \tag{12.43}$$

is a universal constant. The more exact theory due to Chapman and Enskog, Equations (12.36a) and (12.42a), leads to a similar relation

$$\frac{\eta c_v}{\lambda} = \frac{1}{C} \qquad (C = 2.5), \tag{12.43a}$$

† The viscosity of seven gases up to 500°C and its statistical interpretation, in *Proceedings 4th Symposium on Thermophysical Properties* (J. R. Moszynski, ed.), p. 304, Amer. Soc. Mech. Engineers, New York, 1968.

‡ S. Chapman and T. G. Cowling, *The Mathematical Theory of Non-Uniform Gases*, 3rd ed., Cambridge Univ. Press, London and New York, 1970; or J. O. Hirschfelder, C. F. Curtiss, and R. B. Bird, *Molecular Theory of Gases and Liquids*, Wiley, New York, 1964. See also, H. J. M. Hanley, ed., *Transport Phenomena in Fluids*, Dekker, New York, 1969.

where the constant C is, to a first approximation, the same for all gases;† in fact it varies slightly from gas to gas. This relation is satisfied with a very high degree of accuracy for monatomic gases, as seen from Table 12.1.

Owing to strong molecular attractions between the molecules of a fluid and those in a solid wall, the relative macroscopic velocity of a fluid at the wall is identical with that of the wall (no-slip condition). This phenomenon is of great importance in fluid mechanics, but it should be realized that it is a result of molecular interactions and not of a transport of momentum.

TABLE 12.1

The Constant C in Equation (12.43a) for Several Monatomic Gases[a]

Gas	Temperature T (K)	C	Gas	Temperature T (K)	C
A	90.1	2.493	Ne	90.18	2.507
	194.6	2.527		194.6	2.508
	273.1	2.508		273.15	2.536
	298.1	2.503		298.15	2.492
	373.1	2.515		373.15	2.516
	491.0	2.489		491.15	2.485
	579.0	2.482		579.05	2.460
Kr	273.15	2.508			
	373.15	2.497			
	491.15	2.488			
	579.05	2.488			

[a] Data from H. J. M. Hanley, ed., *Transport Phenomena in Fluids*, Dekker, New York, 1969.

To conclude this section, it is useful to inform the reader that the viscosity of a liquid decreases with increasing temperature whereas, as we know, that of a gas increases. This is due to the fact that in a liquid the dominant mechanism for momentum transport is the direct interaction of molecules on the two sides of an element of plane via the intermolecular force potential. By contrast, in a gas, as assumed here at the outset, the dominant mechanism is the actual flow of molecules from one side of the elementary plane to the other.

† Since for monatomic gases $\gamma = c_p/c_v = 1.667$ and is constant, the dimensionless group, known as the Prandtl number

$$\mathrm{Pr} = \frac{\eta c_p}{\lambda} = \frac{\gamma}{C} = 0.667,$$

is also a constant. This fact is of great importance in the study of heat transfer, because the Prandtl number of most gases, including polyatomic ones, is of the order of 0.7.

12.5.4. Internal Degrees of Freedom and the Eucken Factor

So far, and strictly speaking, we have studied the transport properties of monatomic gases only. Whereas the equations for viscosity provide us with a workable first approximation even for polyatomic gases, those for thermal conductivity do not for the reasons mentioned in the preceding section, that is, owing to the fact that the internal degrees of freedom in polyatomic molecules contribute significantly to the transport of energy.

In 1913 A. Eucken† advanced a very simple, intuitive argument which allowed him to extend the validity of Equation (12.43a) to include polyatomic gases. In order to understand the gist of this generalization, it is useful first to examine the reasons which increase the value of the factor C from $C = 1$ in Equation (12.43) to $C = 2.5$ in Equation (12.43a). This is intimately connected with the mechanical details of the process of molecular collisions. The elementary theory given in Section 12.5.2 ignored all such details and worked with velocity averages directly. By contrast, the more exact Chapman–Enskog theory, too complex in its structure for this book, is based on a full study of all orbits which one *monatomic* molecule can follow about another in a binary collision. The difference between the two approaches can be clarified, albeit only qualitatively, with the aid of Figure 12.9 in which we imagine

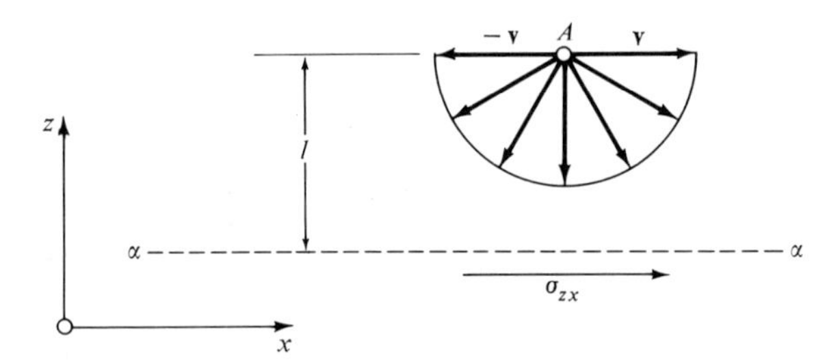

FIGURE 12.9. Difference between the transport of energy and momentum.

a group of particles, represented by the single point A, approaching the control plane α–α from a distance l with the same velocity $|\mathbf{v}|$ but in different directions. Each of these molecules carries with it a fixed amount of kinetic energy of translation, fully accounted for in Equation (12.43a). Furthermore, if the molecule is polyatomic, it transports a supplementary quantity of kinetic energy due to rotation and vibration which is left out of account in the Chapman–Enskog theory. It may be assumed that the internal energy is uncorrelated with a molecule's kinetic energy. By contrast, the amount of momentum contributing to the stress σ_{zx}, that is its x component, varies

† *Physik. Z.* **14** (1913), 324.

with the angle of approach from a negative value through zero to a positive value. This is also fully accounted for in Equation (12.43a), and it is noted that the internal degrees of freedom contribute little to the transport of momentum. All this means that the elementary theory underestimates the transport of translational energy expressed as the thermal conductivity, λ, in comparison with the transport of translational momentum embodied in the viscosity, η. For this reason, the ratio in Equation (12.43) decreases by a factor C.

According to these arguments, we first separate the transport of translational energy from that due to internal motions, and replace c_v by

$$c_v = c_v^{\mathrm{tr}} + c_v{}^{\mathrm{i}}, \tag{12.44a}$$

remembering that

$$c_v^{\mathrm{tr}} = \frac{3}{2}\frac{\mathbf{R}}{M} = \frac{3}{2}(c_p - c_v). \tag{12.44b}$$

Since the internal motions are uncorrelated with the translational velocity, the factor C should only be applied to c_v^{tr} and not to $c_v{}^{\mathrm{i}}$. Hence, we write

$$\lambda = \tfrac{1}{2}nulm(Cc_v^{\mathrm{tr}} + c_v{}^{\mathrm{i}})$$

$$\eta = \tfrac{1}{2}nulm,$$

leading to

$$\frac{\lambda}{\eta c_v} = \frac{Cc_v^{\mathrm{tr}} + c_v{}^{\mathrm{i}}}{c_v}.$$

Introducing $C = 2.5$, $\gamma = c_p/c_v$ and $c_v^{\mathrm{tr}} = \frac{3}{2}c_v(\gamma - 1)$ from Equation (12.44b), we obtain

$$\frac{\eta c_v}{\lambda} = \frac{1}{f} \qquad \text{where} \quad f = \tfrac{1}{4}(9\gamma - 5). \tag{12.45a, b}$$

The constant f in this equation is known as the *Eucken factor*. With $\gamma = \frac{5}{3}$ for a monatomic gas, we retrieve the previous result, that is, we see that $f \equiv C$.

Table 12.2 contains a comparison between measured results and those implied by Equations (12.45a, b). In contrast with monatomic gases, this comparison convinces us that further improvements in the theory are needed. In recent times, Eucken's theory has been reexamined in terms of the more exact Chapman–Enskog theory by E. A. Mason and his collaborators.†

† E. A. Mason and L. Monchick, *J. Chem. Phys.* **36** (1962), 1622. L. Monchick, A. N. G. Pereira, and E. A. Mason, *J. Chem. Phys.* **42** (1965), 3241.

TABLE 12.2

Comparison between Measurements and Equations (12.45a,b)[a]

Temperature T (K)	Nitrogen, N_2: $f_1 = \frac{\lambda}{\eta c_v}$	Nitrogen, N_2: $f_2 = \frac{9\gamma - 5}{4}$	Carbon dioxide, CO_2: $f_1 = \frac{\lambda}{\eta c_v}$	Carbon dioxide, CO_2: $f_2 = \frac{9\gamma - 5}{4}$
100	1.85	1.90	—	—
200	1.89	1.90	1.74	1.78
300	1.97	1.90	1.69	1.65
400	2.02	1.89	1.70	1.57
500	2.04	1.88	1.75	1.52
600	2.01	1.86	1.81	1.48
800	1.94	1.81	—	—
1000	1.85	1.77	—	—

[a] Data courtesy of Professor E. A. Mason.

12.5.5. Self-Diffusion

As a third application of the mean free path theory of transport coefficients, we consider the flow of molecules in a region in which there exists a density gradient. Such a situation is most easily realized by preparing a mixture of gases whose concentration is not uniform throughout the container. As an example, we shall consider a mixture of two gases, one of which is a radioactive isotope of the other, postulating that the gases are identical in all other respects and that the difference in the molecular masses may be neglected.

Suppose that the gas as a whole is at a constant temperature but that the density $n_R(z)$ of the radioactive isotope is not uniform in the z direction. Referring to Figure 12.10, we again consider a hypothetical plane α-α at a height z in the gas and assume that the density of the radioactive molecules increases in the z direction. The number of molecules crossing a unit area per unit time is

$$n_+ = \tfrac{1}{4} u n_R(z + l), \qquad \text{and} \qquad n_- = \tfrac{1}{4} u n_R(z - l). \tag{12.46a, b}$$

The net flow of the molecule in the upward direction, J_z, is

$$J_z = \tfrac{1}{4} u [n_R(z - l) - n_R(z + l)] = -\tfrac{1}{2} u l \frac{\partial n_R(z)}{\partial z} + \cdots. \tag{12.46c}$$

Neglecting the higher-order derivatives, we obtain an expression for the particle flow. The latter is known as *Fick's law*,

$$J_z = -\mathscr{D} \frac{\partial n_R(z)}{\partial z} \tag{12.47a}$$

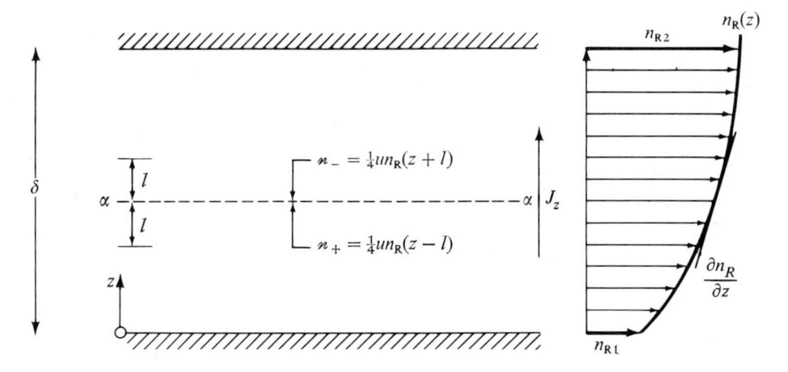

FIGURE 12.10. Self-diffusion.

with

$$\mathscr{D} = \tfrac{1}{2}ul. \tag{12.47b}$$

The constant $\mathscr{D}$ is known as the *coefficient of diffusion*,† and Fick's law, like Fourier's and Newton's, is now well established for gases as well as for liquids.

If the density of the entire system is $n_t = n_R + n_0$, where n_0 is the density of the nonradioactive isotope, the mean free path of the radioactive molecule becomes,

$$l = \frac{1}{\sqrt{2}\,\pi\sigma^2 n_t} \tag{12.48a}$$

so that $\mathscr{D}$ may be expressed as

$$\mathscr{D} = \frac{1}{n_t\pi\sigma^2}\left(\frac{\mathbf{k}T}{\pi m}\right)^{1/2}. \tag{12.48b}$$

Unlike thermal conductivity and viscosity, the coefficient of self-diffusion does depend upon the density, varying inversely with it. We can apply the perfect-gas law $n_t = P/\mathbf{k}T$ to eliminate the density in favor of the pressure, and obtain

$$\mathscr{D} = \frac{(\mathbf{k}T)^{3/2}}{\pi\sigma^2(\pi m)^{1/2}P}. \tag{12.49}$$

It is seen that the diffusion coefficient varies as $T^{3/2}$ at constant pressure. The first approximation to the Chapman–Enskog expression for the coefficient

† Here self-diffusion is more accurate since we are considering the flow of a radioactive gas in an otherwise identical gas. If the gases were chemically different, we would speak of binary diffusion.

of self-diffusion is

$$\mathscr{D} = \frac{3(\pi \mathbf{k}T)^{1/2}}{8\pi m^{1/2} n_t \sigma^2 \Omega^{(1,1)*}(\mathbf{k}T/\varepsilon)}, \tag{12.50}$$

where $\Omega^{(1,1)*}(\mathbf{k}T/\varepsilon)$ is a different collision integral from that which appeared in the expressions for thermal conductivity and viscosity, Equations (12.36a) and (12.42a).

The problem of binary or multicomponent diffusion is too complex to be discussed here.

PROBLEMS FOR CHAPTER 12

12.1. List *all* the assumptions that enter into the derivation of the perfect-gas equation of state from kinetic arguments, as presented in Section 12.3.

12.2. Compute the number of collisions per unit time per unit volume in a two-component gas mixture in which the number density of species A is n_A and that of species B is n_B. Apply the expression to the computation of the number of collisions per unit time and volume of an equimolar mixture of H_2 and I_2 at $T = 300$ K and $P = 1$ atm. The diameter $\sigma_1 = 2.18$ Å for H_2 and $\sigma_2 = 3.74$ Å for I_2. What are the rates of the H_2- H_2 and $I_2 - I_2$ collisions?

12.3. Calculate the mean free path, l, in nitrogen at $T = 300$ K and $P = 10^{-9}$ mm Hg (very "hard" vacuum). The diameter of N_2 is $\sigma = 3.80$ Å.

12.4. (a) Compute the number of hydrogen molecules in a gas ($m = 3.34 \times 10^{-27}$ kg) which impinge in 1 sec on an area A = 1 cm² of the walls of the vessel containing the gas at $T = 0$ C when its number density is $n = 2.0 \times 10^{19}$ molecules/cm³.

(b) Derive a general expression for the number of molecules, a', striking the walls of the container per unit time and area with a velocity greater than c_0. Show that

$$a' = n(\mathbf{k}T/2\pi m)^{1/2}(mc_0^2/2\mathbf{k}T + 1)\exp(-mc_0^2/2\mathbf{k}T)$$

and then compute this number for hydrogen under the conditions stated in (a) if $c_0 = 12{,}000$ m/sec.

12.5. Consider a hypothetical two-dimensional gas in which the molecules are constrained to move in a plane, and whose density is given. Show that the number striking a unit length of the boundary per unit time is $n\langle v\rangle/\pi$, where $\langle v\rangle$ is the average molecular speed. Assume that the molecules are in equilibrium and compute $\langle v\rangle$. (*Hint:* The Maxwell–Boltzmann distribution function in two dimensions is $(\beta m/2\pi)\exp(-\beta m v^2/2)$.)

12.6. Show that, on the average, a molecule travels a distance $2l/3$ between an imaginary plane in the gas and the point of its last collision before crossing the plane.

12.7. Prove that the probability, $\mathscr{P}(L)$, that a molecule in a gas travels a distance L without suffering a collision is $\mathscr{P}(L) = l^{-1} \exp(-L/l)$, where l is the mean free path of a molecule in the gas.

SOLUTION: The probability that a molecule will suffer a collision in moving a small distance δL in the gas is $\delta L/l$. Therefore, we may write $\mathscr{P}(L + \delta L) = \mathscr{P}(L)(1 - \delta L/l)$ for sufficiently small δL. Expanding in powers of δL and keeping only the terms of order δL, we obtain the differential equation $d\mathscr{P}(L)/dL = -l^{-1}\mathscr{P}(L)$ whose solution is $\ln \mathscr{P}(L) = -L/l +$ constant. The constant is determined by normalizing $\mathscr{P}(L)$.

12.8. A spherical bulb whose radius $R = 10$ cm is maintained at a temperature $T = 300$ K except for an area $A = 1$ cm² on the surface which is kept at a very low temperature. The bulb contains water vapor, initially at a pressure of $P_1 = 10$ mm Hg. Assuming that every water molecule striking the cold area condenses and adheres to the surface, compute the time needed for the pressure to decrease to $P_2 = 10^{-4}$ mm Hg.

12.9. The resistance to the conduction of electricity in a metal or in a slightly ionized gas is due to the collisions of the charged particles carrying the electrical current with the atoms or the ions of the surrounding medium. Consider a dilute gas composed of charged particles of mass m and charge $\mathbf{e}$ with number density n in a system in which the charged particles travel on the average a time τ between collisions with neutral atoms. A weak electric field, E_z, acting in the z direction is imposed on this gas. Show that the number of charged particles crossing a unit area perpendicular to the z direction per unit time, $j_z/\mathbf{e}$, is given by $j_z = \sigma E_z$, where $\sigma = n\mathbf{e}^2\tau/m$.

SOLUTION: Consider an imaginary plane placed at right angles to the z direction in the gas. The average number of charged particles crossing a unit area of the plane per unit time is $n\langle v_z \rangle$, where $\langle v_z \rangle$ is the average value of the z component of the velocity of the charged particles. If there were no electric field, $\langle v_z \rangle = 0$, and there would be no conduction of electricity. If, however, there is an electric field, $\langle v_z \rangle$ may be approximated as follows. The velocity of a free charged particle at time t in an electric field in the z direction is given by $v_z(t) = (\mathbf{e}E_z/m)t + v_z(0)$, where $v_z(0)$ is the velocity of the particle at $t = 0$. If the charged particle moves in a gas whose particles collide, on the average, at time intervals τ, we may assume that the average velocity $\langle v_z \rangle$, in the z direction is $(\mathbf{e}E_z/m)\tau$. This supposes that the field is weak and that the collisions between the charged particles and the atoms in the gas are not affected by the field, so that the average velocity of the charged particles immediately after a collision is zero. Using this expression for $\langle v_z \rangle$, we obtain $j_z/\mathbf{e} = (n\mathbf{e}E_z\tau)/m$; j_z is the electric current density.

12.10. Compute the pressure on the walls of a cylinder which contains black-body radiation at temperature T. Consider that the photons make elastic collisions with the walls, and compute the momentum transferred by the photons to a unit area of the wall in unit time. The momentum of a photon with frequency ν is $\mathbf{h}\nu/\mathbf{c}$.

SOLUTION: Let **f**, defined by Equation (10.4), be the propagation vector of a photon. We consider these photons in which **f** makes an angle with respect to the interior normal to the wall. The momentum transferred to the wall by such photons is $2\mathbf{h}\,|\mathbf{f}|\,\cos\theta = (2\mathbf{h}\nu/\mathbf{c})\,\cos\theta$. The number of photons traveling with **f** directed into a solid angle $d\omega = \sin\theta\, d\theta\, d\phi$ is $[n(\nu, T)/4\pi]\sin\theta\, d\theta\, d\phi$, where $n(\nu, T)$ is the density of photons with frequency ν at temperature T. The pressure exerted by the photons on the wall is

$$P = \int_0^\infty \int_0^{\pi/2} \int_0^{2\pi} [n(\nu, T)/4\pi](\mathbf{c}\cos\theta)(2\mathbf{h}\nu/\mathbf{c})\cos\theta \sin\theta\, d\theta\, d\phi\, d\nu$$

$$= \tfrac{1}{3}\int_0^\infty n(\nu, T)\mathbf{h}\nu\, d\nu = \tfrac{1}{3}u(T),$$

where $u(T)$ is the specific energy of the radiation field.

12.11. Consider molecules in a container in equilibrium at temperature T. A small hole is made in the wall of the container and molecules may effuse into a vacuum.

(a) What is the velocity distribution of the molecules that escape through the hole? Assume that the hole does not affect the Maxwell–Boltzmann distribution within the container.

(b) Calculate the rate of energy loss suffered by the vessel, as well as the rate of change of its temperature. Assume that the walls of the vessel are adiabatic.

(c) Calculate the rate of loss of momentum suffered by the container. This quantity represents the *thrust*, or effective force, exerted on the container by the escaping molecules.

12.12. A vertical cylinder is fitted with a frictionless piston of mass M which is acted upon by gravity (see the figure). The cylinder contains a sphere of mass $m \ll M$ moving up and down in a vertical direction with a velocity c and reflecting perfectly from the piston as well as from the bottom of the cylinder. Neglect the influence of gravity on the motion of the sphere and assume that the law of reflection of the sphere from the piston is unaffected by the latter's motion.

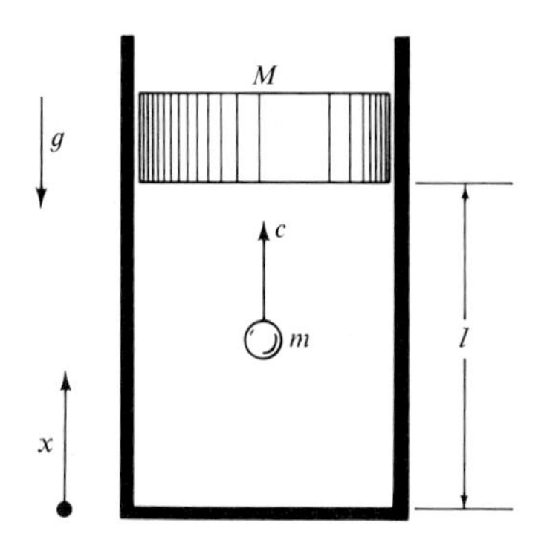

(a) Write the condition of equilibrium for the piston and compare it with the perfect-gas equation, ignoring the dimensions of the sphere.

(b) Perform the same calculation for a sphere of radius r. Which gas law is analogous to this result?

(c) The piston is now being withdrawn very slowly with a velocity $\mathscr{V} \ll c$. Show that the loss in the kinetic energy of the sphere, $\frac{1}{2}m(c'^2 - c^2)$, during a time interval Δt is $-P\,\Delta V$, where c' is the velocity of the sphere after impact and $\Delta V = A\,\Delta x$ is the volume displaced by the piston as it travels a distance Δx in time Δt.

Are the preceding analogies fortuitous?

12.13. A gas is contained in a cylinder provided with a frictionless piston. The piston moves with a velocity v_0 which is small compared to the average velocity of the gas molecules. Evaluate the change, ΔE, in the energy of the gas in a time Δt produced by molecules colliding with the piston, and show that $\Delta E = P\,\Delta V$, where ΔV is the change of the volume of the cylinder in time Δt. You may neglect any effects due to the presence of walls of the cylinder adjacent to the face of the piston.

12.14. Calculate the collision frequency in nitrogen (a) at $T = 300$ K and $P = 1$ atm; (b) at $T = 300$ K and $P = 10$ mm Hg.

12.15. A well-insulated vessel is separated into two compartments, each of volume V, with the aid of a thin, but rigid, diathermal partition. Initially the first compartment (denoted by subscript 1) contains a perfect gas which is in equilibrium at a pressure P_0 and a temperature T, while the second compartment is evacuated. At the instant of time $t = t_0$, a small hole in the partition is uncovered. The area A of the hole is microscopically large.

(a) Show that at any time $t \geq t_0$ the pressure P_1 in the first compartment is

$$P_1(t) = \tfrac{1}{2}P_0\left\{1 + \exp\left[-\frac{2A}{V}\left(\frac{\mathbf{k}T}{2\pi m}\right)^{1/2}(t - t_0)\right]\right\}.$$

(b) Compute the time required for P_1 to drop to three-quarters of its initial value when the gas is helium ($m = 6.64 \times 10^{-27}$ kg) at a temperature $T = 300$ K; the volume of each compartment is $V = 1$ m^3, and the area of the hole is $A = 0.01$ cm^2.

(c) Demonstrate that the average energy ΔE transferred per molecule is $\Delta E = 2\mathbf{k}T$. Why is this value greater than $\frac{3}{2}\mathbf{k}T$?

12.16. By treating each species in a mixture of two gases as independent of each other, prove Dalton's law of partial pressures:

$$P_{\text{tot}} = P_1 + P_2 = \mathbf{k}T(N_1/V + N_2/V)$$

where V is the volume of the system and N_i is the number of particle of the corresponding species.

12.17. Equation (12.23) may be used to evaluate the collision frequency, $\nu(\mathbf{v})$, of molecules whose velocity is $\mathbf{v}$. Thus,

$$\nu(\mathbf{v}) = n\pi\sigma^2 \int d\mathbf{v}_1 |\mathbf{v} - \mathbf{v}_1|\, f(\mathbf{v}_1)\,.$$

Show that in equilibrium, we have

$$\nu(\mathbf{v}) = n\sigma^2(2\pi/\beta m)^{1/2}[e^{-x^2} + (2x + x^{-1})\mathrm{erf}(x)],$$

where $x = v(m\beta/2)^{1/2}$ and

$$\mathrm{erf}(x) = \int_0^x \exp(-y^2)\,dy$$

is the error function. Evaluate $[\nu(\mathbf{v})/\nu]$ for molecules traveling with the average thermal velocity $u = (8/\pi m\beta)^{1/2}$.

Note: $\mathrm{erf}(2/\sqrt{\pi}) = 0.79$.

12.18. List *all* assumptions that enter into the derivation of Equation (12.29) for the mean free path.

12.19. The mean free path of the molecules of a certain gas at $T = 298$ K is $l = 2.63 \times 10^{-5}$ mm; the radius of each molecule is $r = 2.56 \times 10^{-10}$ m. Compute the number of collisions made by a typical particle in moving a distance $L = 1$ m as well as the pressure, P, of the gas.

12.20. How does the mean free path, l, in nitrogen at $T = 273$ K and $P = 1$ atm compare with the average distance, r, between the molecules at the same temperature and pressure? Assume that the molecules are spheres of diameter $\sigma = 3.80$ Å.

12.21. A spherical satellite d ft in diameter moves through the earth's atmosphere with a speed $\mathscr{V}$ ft/sec at an altitude where the number density is n molecules/ft^3. How many molecules strike the satellite in one second? Derive an expression for the drag experienced by the satellite, assuming that all molecules which strike the sphere adhere to it.

12.22. Estimate the magnitude of the viscosity, η, and the thermal conductivity, λ, of argon at $T = 293$ K and $P = 1$ atm. The atomic weight of argon is $M = 40$ gr/gmol; argon freezes into a solid packed structure with a density $\rho = 1.65$ gr/cm^3. You may assume that the atoms are hard spheres. The measured values are $\eta = 22.3 \times 10^{-5}$ gr/cm sec, and $\lambda = 0.0387 \times 10^{-3}$ cal/cm sec K.

12.23. A molecule of methane (molecular weight 16) may be considered to be a sphere whose volume is about five times the volume of an argon atom. Compute the ratio of the viscosities of methane and argon at a given temperature, T. The chemical formula for methane is CH_4, and the molecule is not linear. Further compute the ratio of the thermal conductivities of these two gases making use of the Eucken factor introduced in Section 12.5.4.

12.24. Suppose that the molecules of a gas repel each other according to the force law $F(r) = dr^{-s}$. Using dimensional arguments, deduce the power law which describes the dependence of the thermal conductivity upon the temperature.

SOLUTION: We start with the effective diameter of the molecule defined by Equation (12.37). This quantity depends on the mass of the molecules, m, the average velocity, u, and the constant, d, in the force law. If we write $\sigma_{eff} \sim m^a u^b d^c$ and adjust a, b, and c so that both sides have the dimensions of a length, we obtain $\sigma_{eff} \sim (d/mu^2)^{1/(s-1)}$. Since $u \sim T^{1/2}$, we find that $\sigma_{eff} \sim T^{-1/(s-1)}$ and $\lambda \sim T^{\omega}$ where $\omega = [\frac{1}{2} + 2/(s-1)]$.

12.25. Given that the standard density of air is $\rho = 1.29 \times 10^{-3}$ gr/cm³, that $\langle v \rangle = 4.6 \times 10^4$ cm/sec and that the thermal conductivity is $\lambda = 0.0548 \times 10^{-3}$ cal/cm sec K, estimate the viscosity, η, of air and compare your rough result with the measured value $\eta = 18.19 \times 10^{-5}$ gr/cm sec, at $P = 1$ atm and $T = 298$ K.

12.26. Compute the Eucken correction to the thermal conductivity for a diatomic gas whose specific heat $c_v = \frac{5}{2}\mathbf{R}$.

12.27. A tube length $L = 50$ cm and diameter $D = 5$ cm contains methane. Half of the molecules contain the radioactive carbon isotope, ^{14}C. At time $t = 0$ there exists a linear concentration gradient in the tube, the methane at the extreme left consisting of 100% radioactive molecules. Determine the initial rate of diffusion of the radioactive isotope across a plane drawn through the center of the tube. The state is one of $P = 1$ atm and $T = 300$ K. The viscosity of the gas is $\eta = 11 \times 10^{-5}$ P.

LIST OF SYMBOLS FOR CHAPTER 12

Latin Letters

A	Area
$\mathscr{a}$	Rate of effusion
a	Number of molecule crossing unit area of a plane per unit time
C	Eucken factor for a monatomic gas
c_v	Specific heat at constant volume
c_p	Specific heat at constant pressure
$\mathscr{D}$	Coefficient of self diffusion
f	Single-particle distribution function; Eucken factor
J	Net flux of molecules across a unit area of a plane
$\mathbf{k}$	Boltzmann's constant
l	Mean free path
m	Mass of a molecule
N	Number of particles
$\mathscr{N}$	Number of systems in an ensemble
n	Number of particles per unit volume
$\mathscr{n}$	Flux of particles across unit area of a plane
P	Pressure
$\mathbf{p}$	Momentum vector
$\mathbf{q}$	Heat flux
$\mathbf{R}$	Perfect-gas constant
$\mathbf{r}$	Represents coordinates x, y, z
T	Temperature

t	Time
$\mathbf{U}$	Macroscopic fluid velocity
$\mathbf{u}$	Average molecular velocity
V	Volume
$\mathbf{V}$	Center-of-mass velocity
$\mathscr{V}$	Volume in Γ-space
v_m	Molar volume
$\mathbf{v}$	Represents velocity components v_x, v_y, v_z

Greek letters

β	$(\mathbf{k}T)^{-1}$; denotes a plane in space
Γ	$6N$-dimensional phase space
γ	Ratio of specific heats
δ	Denotes a distance
ε	Energy per particle; energy parameter in Lennard-Jones 6–n potential
η	Viscosity
θ	Denotes an angle
λ	Thermal conductivity
ν	Collision frequency
ρ	Density of systems in $6N$-dimensional phase space
$\boldsymbol{\rho}$	Relative velocity of two particles
σ	Molecular diameter
σ_{xy}	Element of stress tensor
ϕ	Denotes an angle
Ω	Tabulated collision integral

Superscripts

i	Denotes properties due to internal degrees of freedom
tr	Denotes properties due to translational degrees of freedom

Subscripts

eq	Denotes equilibrium properties
R	Denotes radioactive molecules
(1, 1)*	Refers to collision integral for self-diffusion
(2, 2)*	Refers to collision integrals for thermal conductivity and viscosity

Special symbols

$\langle\ \rangle$	Denotes average value

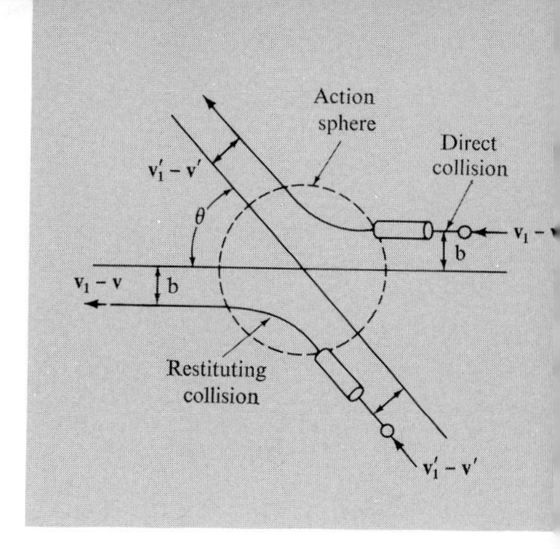

CHAPTER 13

THE BOLTZMANN EQUATION

13.1. Introduction

The elementary, mean-free-path theory of transport processes of gases at low density outlined in Chapter 12 has proved modestly successful when its results were compared with experiment. This can be taken as an indication that the underlying physical ideas are in accord with reality. Nevertheless, the theory went astray in important numerical details, such as its failure to assign a value close to 2.5 for the ratio $\lambda/\eta c_v$ in a monatomic gas, or its inability to lead to the correct temperature dependence of thermal conductivity and viscosity insofar as both deviate from the predicted proportionality to $T^{1/2}$. It thus appears that an improved accurate calculation of the transport properties must be based on a more systematic development of the foundations of kinetic theory, that is, in particular, on a more accurate knowledge of the time-dependent distribution functions which describe the nonequilibrium states of a gas. This is achieved by the formulation of Boltzmann's equation for the single-particle distribution function.

13.2. The Rate Equation

Before we outline the derivation of the integro-differential equation which must be satisfied by the single-particle distribution function, $f(\mathbf{r}, \mathbf{v}, t)$, it is useful to restate the assumptions which will circumscribe its validity. They are essentially the same as those discussed in Section 12.2 in the simpler context, and the only difference now is the greater rigor in mathematically interpreting our physical concepts.

We assume that the density of the gas is low. More precisely, we assume that the density is low enough for the molecules to spend most of the time in traveling freely *between* collisions. This time is assumed to be much larger than that spent on undergoing collisions with other molecules or with the walls of the container. On the other hand, we assume that the density is sufficiently high to ensure that the intermolecular collisions are more frequent than those with the walls of the container. In terms of the mean free path, l, the theory stipulates that the latter is large compared with the molecular diameter, σ, and yet small compared with a characteristic linear dimension of the vessel, $V^{1/3}$. Thus, the mean free path must satisfy the double inequality

$$\sigma \ll l \ll V^{1/3}. \tag{13.1}$$

It must also be remembered that internal motions are disregarded† thus limiting our considerations to *monatomic* gases. In practice, the theory is used at pressures varying from about 0.5 atm to 5 or even 10 atm and its application is not always confined to monatomic gases. Evidently, all such extensions must be carefully tested against experimental data, and the expected precision of its predictions must be compromised as the real state differs more and more from that specified here.

The single-particle distribution function, $f(\mathbf{r}, \mathbf{v}, t)$, represents the average density of particles at a point $\mathbf{r}$ with velocity $\mathbf{v}$ at time t, the average being taken over an ensemble of macroscopically identical systems. As intimated in the preceding chapter, the distribution function $f(\mathbf{r}, \mathbf{v}, t)$ is obtained from the ensemble distribution function $\rho(\mathbf{r}_1, \mathbf{p}_1, \mathbf{r}_2, \mathbf{p}_2, \ldots, \mathbf{r}_N, \mathbf{p}_N, t)$ by integrating over the positions and momenta of all but one of the particles. Since the time variation of ρ is implied in the Liouville equation (5.21), it follows that the time variation of $f(\mathbf{r}, \mathbf{v}, t)$ may be obtained by similarly integrating the Liouville equation. This approach, although correct, would involve us in rather complex procedures before a useful equation for $f(\mathbf{r}, \mathbf{v}, t)$ could be derived; it is essential to adopt it for the discussion of the behavior of gases at high densities.‡ Instead of proceeding in this direction, we shall follow a more intuitive method which was invented by L. Boltzmann§ in his original derivation, and simply state that the same equation is obtained by means of the rigorous methods which start with the Liouville equation.

† C. S. Wang Chang, G. E. Uhlenbeck, and J. de Boer, The heat conductivity and viscosity of polyatomic gases, in *Studies in Statistical Mechanics*, Vol. II (G. E. Uhlenbeck and J. de Boer, eds.), North-Holland Publ., Amsterdam, 1964.

‡ See articles by E. G. D. Cohen, in *Transport Phenomena in Fluids* (H. J. M. Hanley, ed.), Dekker, New York, 1969.

§ L. Boltzmann, *Lectures on Gas Theory* (S. Brush, transl.), Univ. of California Press, Berkeley, California, 1964.

We begin the derivation of the Boltzmann equation by supposing that there are no external forces acting on the molecules, so that the molecules change their velocities only by colliding with each other or with the walls of the vessel. Let us focus our attention on a region of volume $d\mathbf{r}$ about the point $\mathbf{r}$ in the vessel and consider molecules moving with velocity $\mathbf{v}$ in the range $d\mathbf{v}$. If there were no collisions in the gas, we could easily compute the number of particles entering and leaving the six-dimensional volume element $d\mathbf{r}\, d\mathbf{v}$ about the point $(\mathbf{r}, \mathbf{v})$ in a small time interval δt. This would be done by a method similar to that used in Section 5.4.3 in the derivation of the Liouville equation. To this end we consider the six-dimensional volume element $d\mathbf{r}\, d\mathbf{v}$, a two-dimensional projection of which is sketched in Figure 13.1.

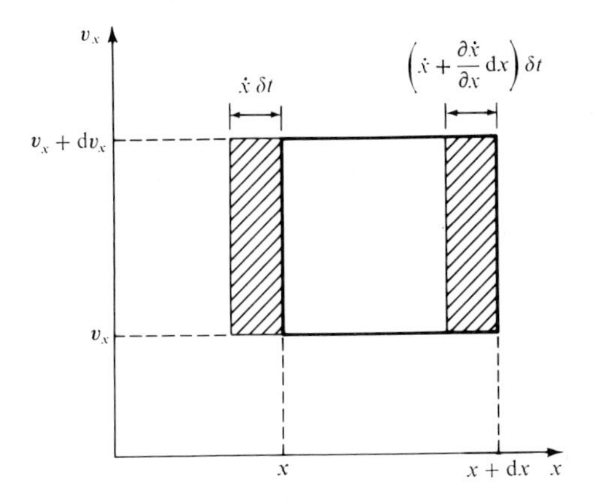

FIGURE 13.1. Projection of the volume element $d\mathbf{r}\, d\mathbf{v}$ onto the plane v_x, x. The shaded areas represent regions from which particles enter and leave the volume element in time δt.

The number of particles in the region at time t is given by $f(\mathbf{r}, \mathbf{v}, t)\, d\mathbf{r}\, d\mathbf{v}$. The number of particles entering the region through the face x in the time interval δt is equal to the number of particles in the shaded area at x, and is given by

$$f(x, y, z, v_x, v_y, v_z, t)\dot{x}\, \delta t\, dy\, dz\, dv_x\, dv_y\, dv_z .$$

The number of particles leaving the region through the face at $x + dx$ is equal to the number of particles in the shaded region at $x + dx$, that is to

$$f(x + dx, y, z, v_x, v_y, v_z)\left(\dot{x} + \frac{\partial \dot{x}}{\partial x} dx\right) \delta t\, dy\, dz\, dv_x\, dv_y\, dv_z .$$

The net number of particles entering the region in time δt through the faces at x and at $x + \mathrm{d}x$ is thus

$$
\begin{aligned}
& f(x, y, z, v_x, v_y, v_z, t)\dot{x}\,\mathrm{d}y\,\mathrm{d}z\,\mathrm{d}v_x\,\mathrm{d}v_y\,\mathrm{d}v_z\,\delta t \\
& -f(x, y, z, v_x, v_y, v_z, t)\dot{x}\,\mathrm{d}y\,\mathrm{d}z\,\mathrm{d}v_x\,\mathrm{d}v_y\,\mathrm{d}v_z\,\delta t \\
& -\dot{x}\frac{\partial f}{\partial x}(x, y, z, v_x, v_y, v_z, t)\,\mathrm{d}x\,\mathrm{d}y\,\mathrm{d}z\,\mathrm{d}v_x\,\mathrm{d}v_y\,\mathrm{d}v_z\,\delta t \\
& -\frac{\partial \dot{x}}{\partial x}f(x, y, z, v_x, v_y, v_z, t)\,\mathrm{d}x\,\mathrm{d}y\,\mathrm{d}z\,\mathrm{d}v_x\,\mathrm{d}v_y\,\mathrm{d}v_z\,\delta t \\
& \qquad = -\frac{\partial}{\partial x}[\dot{x}f(x, y, z, v_x, v_y, v_z)]\,\mathrm{d}\mathbf{r}\,\mathrm{d}\mathbf{v}\,\delta t. \qquad (13.2)
\end{aligned}
$$

The total change in the number of particles in the region $\mathrm{d}\mathbf{r}\,\mathrm{d}\mathbf{v}$ in time δt (still assuming a complete absence of collisions) is

$$[f(\mathbf{r}, \mathbf{v}, t + \delta t) - f(\mathbf{r}, \mathbf{v}, t)]\,\mathrm{d}\mathbf{r}\,\mathrm{d}\mathbf{v},$$

and is equal to the net number of particles entering the region. For a small interval δt, this is

$$
\begin{aligned}
& -\left\{\frac{\partial}{\partial x}[\dot{x}f(\mathbf{r}, \mathbf{v}, t)] + \frac{\partial}{\partial y}[\dot{y}f(\mathbf{r}, \mathbf{v}, t)] + \frac{\partial}{\partial z}[\dot{z}f(\mathbf{r}, \mathbf{v}, t)]\right. \\
& \qquad \left. + \frac{\partial}{\partial v_x}[\dot{v}_x f(\mathbf{r}, \mathbf{v}, t)] + \frac{\partial}{\partial v_y}[\dot{v}_y f(\mathbf{r}, \mathbf{v}, t)] + \frac{\partial}{\partial v_z}[\dot{v}_z f(\mathbf{r}, \mathbf{v}, t)]\right\}\mathrm{d}\mathbf{r}\,\mathrm{d}\mathbf{v}\,\delta t.
\end{aligned}
$$

Finally, we can write

$$\frac{\partial f(\mathbf{r}, \mathbf{v}, t)}{\partial t}\,\mathrm{d}\mathbf{r}\,\mathrm{d}\mathbf{v}\,\delta t = -\mathbf{v}\cdot\nabla_r f(\mathbf{r}, \mathbf{v}, t)\,\mathrm{d}\mathbf{r}\,\mathrm{d}\mathbf{v}\,\delta t. \qquad (13.3)$$

The gradient operator ∇_r is a vector whose components in the x, y, and z directions, $(\partial/\partial x, \partial/\partial y, \partial/\partial z)$, represent the spatial rate of variation of the scalar function which it modifies. In arriving at Equation (13.3) we have made use of the fact that

$$\frac{\mathrm{d}\mathbf{v}}{\mathrm{d}t} = 0 \qquad (13.4\text{a})$$

and that

$$\frac{\partial v_x}{\partial x} = \frac{\partial v_y}{\partial y} = \frac{\partial v_z}{\partial z} = 0. \qquad (13.4\text{b})$$

These express the assumptions that there are no external forces acting on the molecules, and that x and v_x and so on are independent of each other in our considerations.

In order to allow for collisions, it is necessary to modify Equation (13.3) and to include their effects on the distribution function. Evidently, the effect of intermolecular collisions is to modify the count of molecules entering and leaving the region. This count is affected in two ways. Some molecules which have been counted as entering or leaving will not do so, and some which have been omitted from the count will either enter or leave the element **dr dv** as a result of a collision undergone by them. We may summarize this remark by writing that the time rate of change of the number of particles satisfies the equation

$$\frac{\partial f(\mathbf{r}, \mathbf{v}, t)}{\partial t} \, d\mathbf{r} \, d\mathbf{v} \, \delta t = -\mathbf{v} \cdot \nabla_r f(\mathbf{r}, \mathbf{v}, t) \, d\mathbf{r} \, d\mathbf{v} \, \delta t + \left(\frac{\partial f}{\partial t}\right)_c d\mathbf{r} \, d\mathbf{v} \, \delta t, \quad (13.5)$$

where

$$\left(\frac{\partial f}{\partial t}\right)_c d\mathbf{r} \, d\mathbf{v} \, \delta t$$

is the net change in the number of particles in the region brought about by collisions. We may further decompose $(\partial f/\partial t)_c$ into two parts

$$\left(\frac{\partial f}{\partial t}\right)_c d\mathbf{r} \, d\mathbf{v} \, \delta t = (J_+ - J_-) \, d\mathbf{r} \, d\mathbf{v} \, \delta t, \quad (13.6)$$

in which

$J_+ \, d\mathbf{r} \, d\mathbf{v} \, \delta t$ is the number of particles that enter the region in time δt through the agency of collisions and

$J_- \, d\mathbf{r} \, d\mathbf{v} \, \delta t$ is the number of molecules that leave the region in time δt owing to collisions.

The resulting equation for $f(\mathbf{r}, \mathbf{v}, t)$, namely

$$\frac{\partial f(\mathbf{r}, \mathbf{v}, t)}{\partial t} + \mathbf{v} \cdot \nabla_r f(\mathbf{r}, \mathbf{v}, t) = J_+ - J_-, \quad (13.7)$$

is exact but not very useful until an explicit representation of the molecular collisional fluxes, J_+ and J_-, can be found. In order to derive the required expressions, we must study the mechanics of a collision in detail. Accordingly, we temporarily interrupt our main derivation and turn to the theory of binary collisions.

13.3. The Dynamics of a Binary Collision

Since we are studying transport processes in dilute gases, we confine our attention to the effects of *binary* collisions and ignore the contributions to $\partial f(\mathbf{r}, \mathbf{v}, t)/\partial t$ that arise from collisions involving three or more particles simultaneously. This leads us to a discussion of elastic collisions between two particles of equal mass, m. As in our previous work on transport theory, we apply classical mechanics.

Suppose that two molecules moving with velocities $\mathbf{v}_1$ and $\mathbf{v}_2$ collide, and change them to $\mathbf{v}_1'$ and $\mathbf{v}_2'$, respectively, after the collision. The laws of conservation of momentum and energy can be written

$$\mathbf{v}_1 + \mathbf{v}_2 = \mathbf{v}_1' + \mathbf{v}_2', \tag{13.8a}$$

$$v_1^2 + v_2^2 = v_1'^2 + v_2'^2, \tag{13.8b}$$

because the masses are equal. If we introduce the center of mass velocity $\mathbf{V}$ and the relative velocity $\mathbf{v}$ by the definitions

$$\mathbf{V} = \tfrac{1}{2}(\mathbf{v}_1 + \mathbf{v}_2), \tag{13.9a}$$

and

$$\mathbf{v} = \mathbf{v}_1 - \mathbf{v}_2, \tag{13.9b}$$

we may contract the conservation Equations (13.8a) and (13.8b) to

$$\mathbf{V} = \mathbf{V}' \tag{13.10a}$$

and

$$v^2 = v'^2, \tag{13.10b}$$

respectively. Equations (13.10a, b) state that the velocity of the center of mass is not affected by the collision, and that the relative velocity may be rotated in the course of the collision, but that its magnitude cannot be changed by it. It follows that a specification of $\mathbf{V}$, $\mathbf{v}$, and the angles θ, and ϕ of $\mathbf{v}'$ with respect to $\mathbf{v}$ is sufficient completely to describe the collision. A collision has been represented graphically in the form of the hodograph of Figure 13.2 which illustrates the relations between the velocity vectors, $\mathbf{v}_1$, $\mathbf{v}_2$, $\mathbf{v}_1'$, $\mathbf{v}_2'$, $\mathbf{V} = \mathbf{V}'$, $\mathbf{v}$, and $\mathbf{v}'$. Here θ is the angle between $\mathbf{v}$ and $\mathbf{v}'$, and ϕ denotes the angle of orientation of the plane containing $\mathbf{v}$ and $\mathbf{v}'$ with respect to a reference plane fixed in space and containing $\mathbf{v}$. A different representation of the collision, that in an absolute frame of reference ("laboratory coordinates"), is shown in Figure 13.3a. Here, two particles, 1 and 2, approach each other with velocities $\mathbf{v}_1$ and $\mathbf{v}_2$, respectively. The particles are assumed to interact through their central fields of forces which cause them to follow the

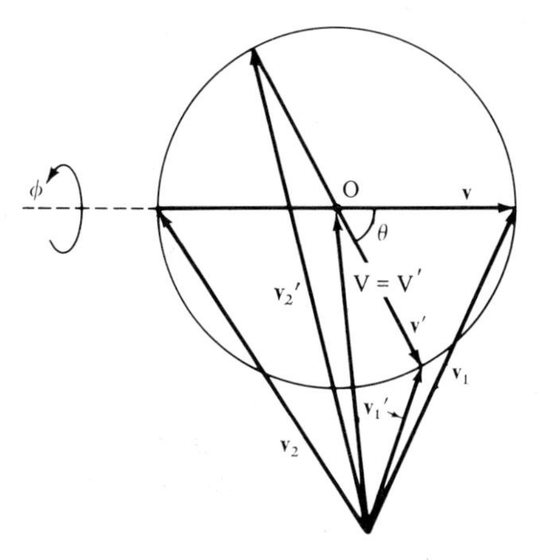

FIGURE 13.2. Hodograph of a collision drawn in accordance with conservation of momentum and energy. [From K. Huang, *Statistical Mechanics*, p. 60. Wiley, New York, 1963.]

curved paths s_1 and s_2. From points 1′ and 2′ onwards, the particles follow straight-line trajectories with the velocities $\mathbf{v}_1'$ and $\mathbf{v}_2'$ because the forces of interaction have decreased to vanishingly small values. Yet another representation of the process of binary collision is shown in Figure 13.3b. Here, the frame of reference is fixed at the center of mass of the two colliding particles. In this system, we may focus our attention on one of the molecules since in this system the position and velocity of each particle are opposite with respect to the center of mass. The collision now appears as a scattering of each particle by a fictitious force located at the center of mass, point O in Figure 13.3b. One

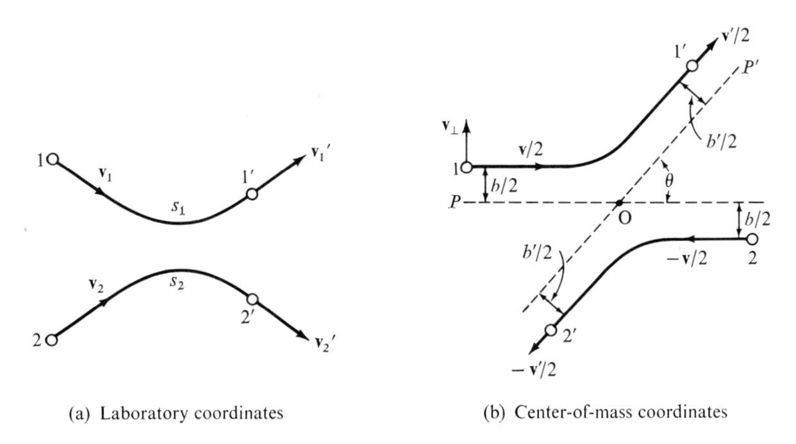

FIGURE 13.3. Two representations of a binary collision. Note that $b = b'$. [From K. Huang, *Statistical Mechanics*, p. 61. Wiley, New York, 1963.]

of the molecules approaches point O with the velocity $\mathbf{v}/2$ before the collision. The perpendicular distance between the two lines of approach is called the *impact parameter*, b. After the collision, the molecule moves away from O with the velocity $\mathbf{v}'/2$, and it is noted that the perpendicular distance, b', between the two lines of escape is also equal to the impact parameter, b, since the angular momentum about the point O must be the same before and after the collision. In other words, we must have

$$|\mathbf{v}|\, b = |\mathbf{v}'|\, b', \tag{13.11}$$

and, since $|\mathbf{v}'| = |\mathbf{v}|$, it follows that $b = b'$. The scattering angle θ depends on the impact parameter b, and on the magnitude of the relative velocity $|\mathbf{v}|$ in a way that is determined by the potential of interaction between the particles. This means that the intermolecular force potential must be known in order explicitly to compute θ from assumed values of b and $|\mathbf{v}|$. It is important to note that the scattering angle, θ, depends on the sign of the parameter $b/2$ as measured from the collision axis, OP. For negative values of $b/2$, as illustrated in Figure 13.4, the scattering angle is $-\theta$, because the angular

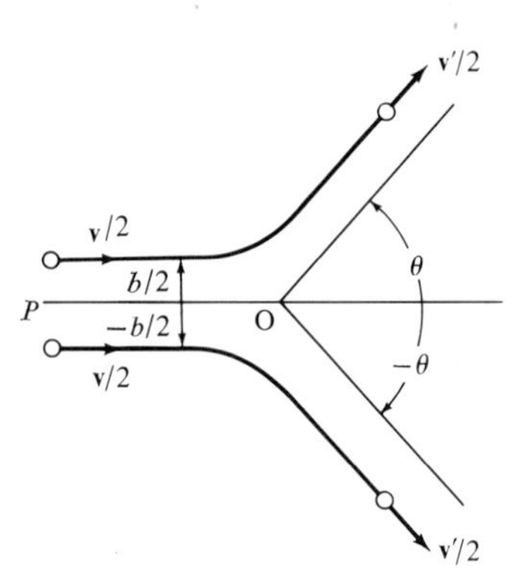

FIGURE 13.4. The dependence of the scattering angle, θ, on the sign of the impact parameter, as measured from the collision axis OP.

momentum about the center of mass at O must be conserved throughout the collision.

We may express the relation between $\mathbf{v}$ and $\mathbf{v}'$ in terms of the scattering angle, θ. Returning to Figure 13.3b, we suppose that $\mathbf{v}$ points in the x direction, so that

$$\mathbf{v} = v\mathbf{i},$$

where $\mathbf{i}$ is the corresponding unit vector. In this coordinate system $\mathbf{v}'$ may be written as

$$\mathbf{v}' = \mathbf{v} \cos \theta + \mathbf{v}_\perp \sin \theta, \tag{13.12}$$

where $\mathbf{v}_\perp$ is a vector of magnitude $|\mathbf{v}|$ but directed at right angles to it and turned counterclockwise in the figure. Using this equation for $\mathbf{v}'$ and the fact that $\mathbf{V}$ remains constant during the collision, we may express the final velocities $\mathbf{v}_1'$ and $\mathbf{v}_2'$ by the forms

$$\mathbf{v}_1' = \tfrac{1}{2}[\mathbf{v}_1 + \mathbf{v}_2 + \mathbf{v} \cos \theta + \mathbf{v}_\perp \sin \theta]; \tag{13.13a}$$

$$\mathbf{v}_2' = \tfrac{1}{2}[\mathbf{v}_1 + \mathbf{v}_2 - \mathbf{v} \cos \theta - \mathbf{v}_\perp \sin \theta]. \tag{13.13b}$$

Having thus obtained the final velocities $\mathbf{v}_1'$ and $\mathbf{v}_2'$ in terms of the initial velocities $\mathbf{v}_1$ and $\mathbf{v}_2$, we now find it useful to examine the inverse of the preceding problem because we shall need its solution later. Accordingly, we imagine that the two prescribed velocities $\mathbf{v}_1$ and $\mathbf{v}_2$ (regardless of their points of application, but treated vectorially), instead of describing the motion *before* a collision should describe it after a collision. A related motion of this kind, Figure 13.5, is called a *restituting collision*, and the problem is to find the

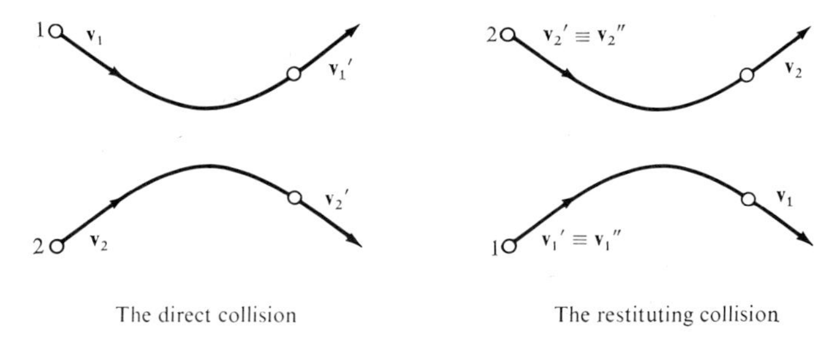

FIGURE 13.5. The restituting collision is obtained by allowing molecules with velocities $\mathbf{v}_2'$ and $\mathbf{v}_1'$ to collide with the negative of the same scattering angle as in the direct collision.

initial velocities, $\mathbf{v}_1''$ and $\mathbf{v}_2''$, that might produce it, it being clear that the positions of the particles must now be interchanged, as shown in the figure.

The simplest way to solve the problem is, first, to allow the direct collision to take place, so that the velocities $\mathbf{v}_1'$ and $\mathbf{v}_2'$ result. Secondly, starting with the latter, and interchanging their positions, we examine a collision for which $\mathbf{v}_1'$ and $\mathbf{v}_2'$ constitute the initial velocities, and for which the impact parameter b and the angle θ are equal in magnitude (but now opposite in sign) to those in the direct collision. We claim that this is a restituting collision, that is, that $\mathbf{v}_1'' = \mathbf{v}_1'$ and $\mathbf{v}_2'' = \mathbf{v}_2'$. The same motion is shown represented in the center-of-mass system in Figure 13.6.

To describe the velocities resulting from the restituting collision, we may

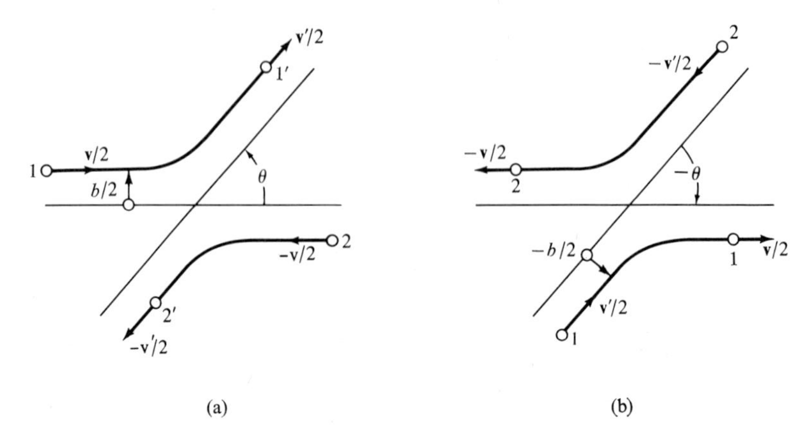

FIGURE 13.6. The direct collision (a) and the restituting collision (b), compared in the center of mass system. The impact parameter of the direct collision is b and the scattering angle is θ; the corresponding quantities for the restituting collision are $-b$, and $-\theta$.

use Equations (13.13a, b) if we replace $\mathbf{v}_1$ and $\mathbf{v}_2$ by $\mathbf{v}_1'$ and $\mathbf{v}_2'$, respectively, as well as θ by $-\theta$. Then, noticing that

$$\mathbf{v}_1' + \mathbf{v}_2' = \mathbf{v}_1 + \mathbf{v}_2$$

and

$$\mathbf{v}_\perp' = -\mathbf{v}\sin\theta + \mathbf{v}_\perp \cos\theta,$$

we obtain for the final velocities:

$$\begin{aligned}
&\tfrac{1}{2}[\mathbf{v}_1' + \mathbf{v}_2' + \mathbf{v}'\cos\theta - \mathbf{v}_\perp'\sin\theta] \\
&\quad = \tfrac{1}{2}[\mathbf{v}_1 + \mathbf{v}_2 + (\mathbf{v}\cos\theta + \mathbf{v}_\perp\sin\theta)\cos\theta - (-\mathbf{v}\sin\theta + \mathbf{v}_\perp\cos\theta)\sin\theta] \\
&\quad = \tfrac{1}{2}[\mathbf{v}_1 + \mathbf{v}_2 + \mathbf{v}] = \mathbf{v}_1
\end{aligned}$$

and

$$\begin{aligned}
&\tfrac{1}{2}[\mathbf{v}_1' + \mathbf{v}_2' - \mathbf{v}'\cos\theta + \mathbf{v}_\perp'\sin\theta] \\
&\quad = \tfrac{1}{2}[\mathbf{v}_1 + \mathbf{v}_2 - (\mathbf{v}\cos\theta + \mathbf{v}_\perp\sin\theta)\cos\theta + (-\mathbf{v}\sin\theta + \mathbf{v}_\perp\cos\theta)\sin\theta] \\
&\quad = \tfrac{1}{2}[\mathbf{v}_1 + \mathbf{v}_2 - \mathbf{v}] = \mathbf{v}_2 .
\end{aligned} \tag{13.14}$$

Thus, the restituting collision is indeed seen to produce the prescribed end velocities $\mathbf{v}_1$ and $\mathbf{v}_2$.

It is important to realize that the *restituting* collision is *not* the same as the *inverse* collision, which is obtained when molecules collide with velocities $-\mathbf{v}_1'$ and $-\mathbf{v}_2'$. For central potentials, this collision leads to molecules with final velocities $-\mathbf{v}_1$ and $-\mathbf{v}_2$, not $\mathbf{v}_1$ and $\mathbf{v}_2$ as in the restituting collision. The difference between the restituting collision and the inverse collision is illustrated in Figure 13.7, which should be compared with Figure 13.5.

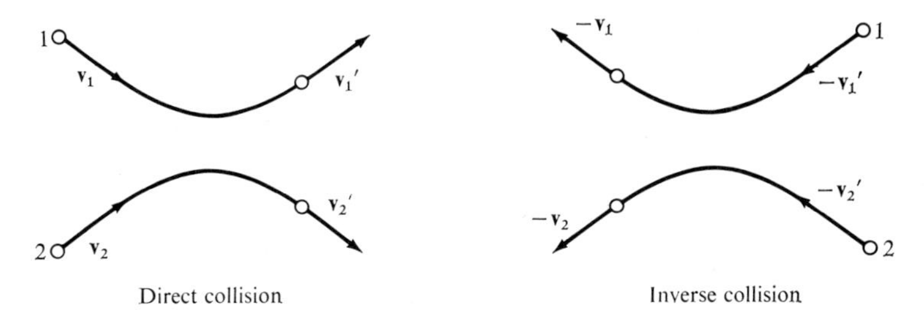

FIGURE 13.7. The direct and inverse collision compared.

In our further work in this chapter we shall have occasion to consider collisions between molecules whose velocities may lie in the ranges $d\mathbf{v}_1$ and $d\mathbf{v}_2$ about $\mathbf{v}_1$ and $\mathbf{v}_2$, respectively. The velocities after collision, and consequently, the restituting velocities, will then be confined to some ranges, say, $d\mathbf{v}_1'$ and $d\mathbf{v}_2'$. We assert that

$$d\mathbf{v}_1\, d\mathbf{v}_2 = d\mathbf{v}_1'\, d\mathbf{v}_2', \tag{13.15}$$

that is, that the Jacobian of the transformation from the colliding velocities to the restituting velocities is identically equal to unity. First, we consider the transformation from the velocities $\mathbf{v}_1$ and $\mathbf{v}_2$ to the center of mass and relative velocities given by Equations (13.9a) and (13.9b). Thus,

$$d\mathbf{v}_1\, d\mathbf{v}_2 = d\mathbf{v}\, d\mathbf{V}, \tag{13.16a}$$

as may easily be verified by explicitly writing out the Jacobian determinant. Since the center of mass velocity does not change during the collision, it follows that

$$d\mathbf{V} = d\mathbf{V}' \tag{13.16b}$$

where $\mathbf{V}'$ is the velocity of the center of mass after the collision. Moreover, since $\mathbf{v}'$ is obtained from $\mathbf{v}$ by a rotation, the magnitudes of $\mathbf{v}$ and $\mathbf{v}'$ are equal, and it follows that†

$$d\mathbf{v} = d\mathbf{v}'. \tag{13.16c}$$

Combining Equations (13.16a)–(13.16c), we obtain

$$d\mathbf{v}_1\, d\mathbf{v}_2 = d\mathbf{v}\, d\mathbf{V} = d\mathbf{v}'\, d\mathbf{V}'; \tag{13.17a}$$

but

$$d\mathbf{v}'\, d\mathbf{V}' = d\mathbf{v}_1'\, d\mathbf{v}_2', \tag{13.17b}$$

and this leads us to Equation (13.15).

† For a simple proof that volume elements remain unchanged under rotations, see E. Hille, *Analysis*, Vol. II, p. 428, Blaisdell, Waltham, Massachusetts, 1966.

13.4. The Boltzmann Equation

We are now in a position to compute the molecular fluxes J_+ and J_-; we recall that they represent the rates at which molecules are scattered into and out of a region $d\mathbf{r}\, d\mathbf{v}$ by collisions,† and that we are restricted to cases when binary collisions alone are responsible for the variation of the distribution function $f(\mathbf{r}, \mathbf{v}, t)$ with time. We begin with the simpler quantity J_-, and remind the reader that $J_-\, d\mathbf{r}\, d\mathbf{v}\, \delta t$ represents the number of molecules that leave the μ-space element $d\mathbf{r}\, d\mathbf{v}$ in time δt through collisions. We now center attention on collisions between molecules which move with the velocity $\mathbf{v}$ in $d\mathbf{r}$ and molecules which move with a different velocity $\mathbf{v}_1$ so that the impact parameter of the $(\mathbf{v}, \mathbf{v}_1)$ collision falls in the range from b to $b + db$, the azimuthal angle of the collision being confined to the range ϕ to $\phi d + \phi$ about a plane fixed in space and containing the relative velocity $\mathbf{v}_1 - \mathbf{v}$. This situation is seen illustrated in Figures 13.8a, b. The so-called action sphere which appears in Figure 13.8b is centered at one particle and has a radius which is equal to the range of the forces between two particles. Thus, we assume that a collision takes place if the centers of the two molecules are located inside this sphere.

The number of molecules moving with a velocity $\mathbf{v}$ in the region $d\mathbf{r}$ is $f(\mathbf{r}, \mathbf{v}, t)\, d\mathbf{r}\, d\mathbf{v}$, and the total volume of the collision cylinders for the $(\mathbf{v}, \mathbf{v}_1)$ collisions is

$$f(\mathbf{r}, \mathbf{v}, t)\, d\mathbf{r}\, d\mathbf{v}\, b\, db\, d\phi\, |\mathbf{v} - \mathbf{v}_1|\, \delta t.$$

In order to determine the number of $(\mathbf{v}, \mathbf{v}_1)$ collisions we must first compute the number of $\mathbf{v}_1$-molecules in the collision cylinders, and then determine how many of them are actually involved in a $(\mathbf{v}, \mathbf{v}_1)$ collision. In the same way as in the preceding chapter, we assume the validity of the *Stosszahlansatz.* This assumption consists of two parts: (a) the collision cylinders contain at most one molecule with velocity $\mathbf{v}_1$, and each such molecule is involved in a $(\mathbf{v}, \mathbf{v}_1)$ collision, and (b) the number of $\mathbf{v}_1$-molecules in the collision cylinders is equal to the number of molecules with velocity $\mathbf{v}_1$ per unit volume multiplied by the total volume of the collision cylinders. With these assumptions, the number of collisions under consideration becomes

$$f(\mathbf{r}, \mathbf{v}, t)\, d\mathbf{v}\, f(\mathbf{r}, \mathbf{v}_1, t)\, d\mathbf{v}_1\, d\mathbf{r}\, |\mathbf{v} - \mathbf{v}_1|\, b\, db\, d\phi\, \delta t.$$

The preceding expression contains an error in that we supposed that the $\mathbf{v}_1$-molecules are located at $\mathbf{r}$, whereas in reality they are distributed over different points in the collision cylinders. In order to make sure that this

† As always, $d\mathbf{r}$ is supposed to be large enough to contain many molecules but small compared to a macroscopic volume. Here $d\mathbf{r}$ is stipulated to be smaller than a volume of the order of the cube of the mean free path l^3. Now $\mathbf{v}$ denotes the velocity of one particle.

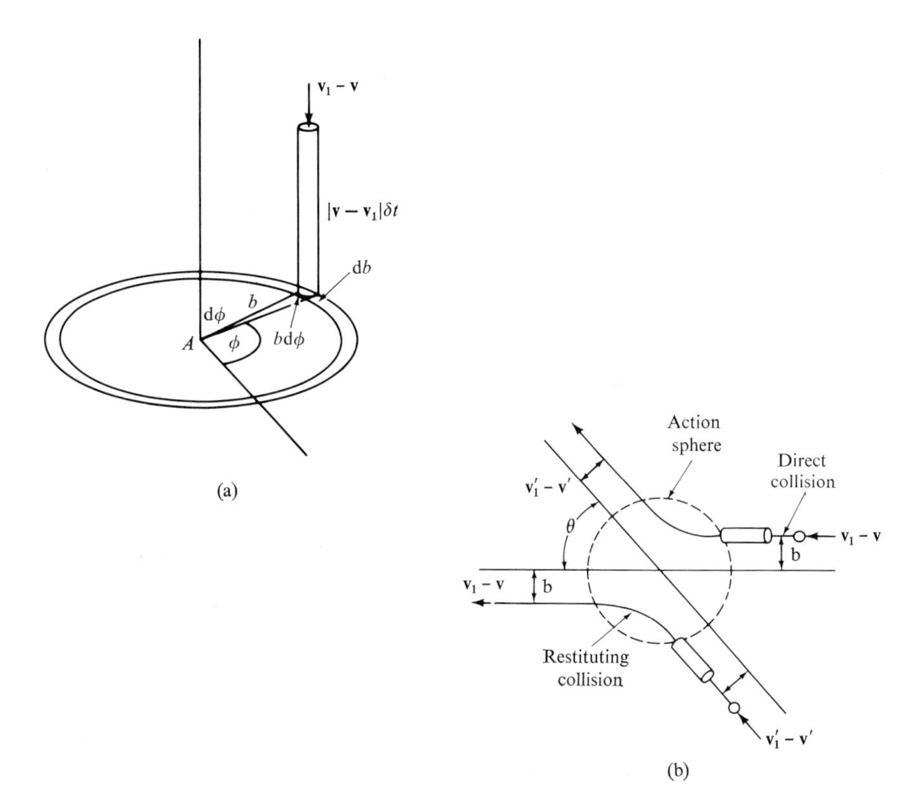

FIGURE 13.8. The collision cylinders for the computation of J_+ and J_-. In Figure 13.8a we illustrate the collision cylinder for a collision in which b is in the range b to $b + \mathrm{d}b$ and ϕ in the range ϕ to $\phi + \mathrm{d}\phi$. We represent the location of the cylinders with respect to the center of one particle. [Figure 13.8b from G. E. Uhlenbeck and G. W. Ford, *Lectures in Statistical Mechanics,* p. 78, Amer. Math. Soc., Providence, Rhode Island, 1963.]

error is small we must further assume that, in time δt, at least on the average, the colliding $\mathbf{v}_1$-molecules travel a distance which is much smaller than the mean free path. Furthermore, we must stipulate that the distribution function $f(\mathbf{r}, \mathbf{v}, t)$ does not change appreciably over the distance covered by a molecule in time δt.

The molecular flux J_- is obtained by integration over the velocity space, and this leads to the expression

$$J_- \,\mathrm{d}\mathbf{r}\,\mathrm{d}\mathbf{v}\,\delta t = \left[\int \mathrm{d}\mathbf{v}_1 \int_0^\sigma \mathrm{d}b \int_0^{2\pi} \mathrm{d}\phi\, b\,|\mathbf{v} - \mathbf{v}_1| f(\mathbf{r}, \mathbf{v}, t) f(\mathbf{r}, \mathbf{v}_1, t)\right] \mathrm{d}\mathbf{r}\,\mathrm{d}\mathbf{v}\,\delta t, \tag{13.18}$$

where σ denotes the range of the forces.

We now turn to the computation of the flux J_+, that is of the number of molecules in volume $d\mathbf{r}$ that *acquire* the velocity $\mathbf{v}$ through collision in time δt. It is clear that we must now count the restituting collisions. The restituting velocities $\mathbf{v}'$ and $\mathbf{v}_1'$ are given in terms of $\mathbf{v}$, $\mathbf{v}_1$, and b in Equations (13.13a) and (13.13b). Each restituting collision of this kind injects a molecule of velocity $\mathbf{v}$ into the region $d\mathbf{r}$. The number of such molecules is

$$f(r, \mathbf{v}', t) f(r, \mathbf{v}_1', t)\, |\mathbf{v}_1' - \mathbf{v}'|\, b\, db\, d\phi\, d\mathbf{v}'\, d\mathbf{v}_1'\, d\mathbf{r}\, \delta t.$$

Now, to obtain J_+ we make use of the fact that $d\mathbf{v}'\, d\mathbf{v}_1' = d\mathbf{v}\, d\mathbf{v}_1$, as seen from Equation (13.17) and put $|\mathbf{v}_1' - \mathbf{v}'| = |\mathbf{v}_1 - \mathbf{v}|$, Equation (13.10b). The total number of restituting collisions is found by integrating over b, ϕ, and all *final* velocities of the second particle in the restituting collision. Hence we obtain

$$J_+\, d\mathbf{r}\, d\mathbf{v}\, \delta t = \left[\int d\mathbf{v}_1 \int_0^\sigma db \int_0^{2\pi} d\phi\, b\, |\mathbf{v}_1 - \mathbf{v}|\, f(\mathbf{r}, \mathbf{v}', t) f(\mathbf{r}, \mathbf{v}_1', t)\right] d\mathbf{r}\, d\mathbf{v}\, \delta t. \tag{13.19}$$

Combining Equation (13.7) with Equations (13.18) and (13.19), we are led to the *Boltzmann equation*

$$\frac{\partial f(\mathbf{r}, \mathbf{v}, t)}{\partial t} + \mathbf{v} \cdot \nabla_r f(\mathbf{r}, \mathbf{v}, t) = \int d\mathbf{v}_1 \int_0^\sigma db \int_0^{2\pi} d\phi\, b\, |\mathbf{v}_1 - \mathbf{v}|\, [f' f_1' - f f_1]. \tag{13.20}$$

Here, we have made the following changes in notation:

$$\begin{aligned} f_1(\mathbf{r}, \mathbf{v}, t) &\to f, & f_1(\mathbf{r}, \mathbf{v}', t) &\to f', \\ f_1(\mathbf{r}, \mathbf{v}_1, t) &\to f_1, & f_1(\mathbf{r}, \mathbf{v}_1', t) &\to f_1'. \end{aligned} \tag{13.21}$$

In order to evaluate the distribution function $f(\mathbf{r}, \mathbf{v}, t)$ for a dilute gas, it is necessary to solve the nonlinear, integro-differential Boltzmann equation (13.20).

13.5. Concluding Remarks

L. Boltzmann's equation (13.20) constitutes the basis of a number of interesting, practical as well as theoretical, developments, all of which exceed the scope of this course. First, it is useful to know that S. Chapman† and, independently, D. Enskog‡ developed a systematic method of solving Boltz-

† S. Chapman, *Phil. Trans. A*, **211** (1912), 433; **216** (1915), 279. Also *Mem. Manchester Lit. Philos. Soc.* **66** (1922), 66.

‡ D. Enskog, *Kinetische Theorie der Vorgänge in mässig verdünnten Gasen.* Inaug. Diss., Uppsala, 1917. Also, *Arkiv. Mat. Astro. Fys.* **16** (1921), No. 16, and *Kungl. Svenska Vetenskapsakademiens Handlingar*, **63** (1922), No. 4.

mann's equation.† As a result, they derived the exact expressions for the transport coefficients quoted earlier as Equations (12.36a) and (12.42a). By taking progressively higher moments, it is possible to derive the basic equations of fluid mechanics (conservation of mass, momentum, and energy, that is, the so-called Navier–Stokes equations) as low-order, general approximations to the Boltzmann equation,‡ and so to justify their use, strictly speaking, for monatomic gases of moderate density. Finally, it is worth mentioning that Boltzmann's equation forms the basis of modern studies of the dynamics of the motion of rarefied gases.§

13.6. Approach to Equilibrium. The *H*-Theorem

The specific applications of Boltzmann's equation enumerated in the preceding section lead to results which agree well with experiments; this establishes our confidence in it. Nevertheless, it must be realized that the equation contains a bold generalization in the form of the *Stosszahlansatz* and this makes it necessary to investigate whether its intrinsic character is compatible with the most general principles of equilibrium thermodynamics. Boltzmann himself studied this question and gave a partial answer in the form of his celebrated *H-theorem.*

More precisely, the first question that should be answered is whether Boltzmann's equation necessarily implies that a gas approaches the Maxwellian distribution, Equation (6.19a), whenever the external boundary conditions are so arranged as to secure the onset of equilibrium after the lapse of a sufficiently long interval of time. Whereas it is easy to verify that the Maxwellian distribution itself satisfies Boltzmann's equation identically, the answer to the wider question is not so simple. In other words, does Equation (13.20) really indicate an *irreversible* approach to a *homogeneous* state of equilibrium from *any* initial state, provided that suitable boundary conditions are imposed? We shall see later that an affirmative answer can be given in the case of an adiabatically isolated system which is prevented from interacting with its surroundings from a certain instant onward.

The second problem concerns the second part of the Second law of

† See also S. Chapman and T. G. Cowling, *The Mathematical Theory of Non-Uniform Gases*, 3rd ed., Cambridge Univ. Press, London and New York, 1970; or J. O. Hirschfelder, C. F. Curtiss, and R. B. Bird. *Molecular Theory of Gases and Liquids*, Wiley, New York, 1964.

‡ See the last two references cited and A. Sommerfeld, *Thermodynamics and Statistical Mechanics* (J. Kestin, transl.), Chapter V, Section 43, Academic Press, New York, 1956.

§ W. G. Vincenti and C. H. Kruger, Jr., *Introduction to Physical Gas Dynamics*, Wiley, New York, 1965. J. W. Bond, Jr., K. M. Watson, and J. A. Welch, Jr., *Atomic Theory of Gas Dynamics*, Addison-Wesley, Reading, Massachusetts, 1965; J. F. Clarke and M. McChesney, *The Dynamics of Real Gases*, Butterworth, London, 1964.

thermodynamics and poses the question as to whether such an approach to equilibrium occurs with a monotonically increasing entropy and ends in a state of maximum entropy when the system is confined within a rigid, adiabatic boundary. Having posed this second question, we immediately realize that Boltzmann's equation deals with a succession of nonequilibrium states for which even a clear *definition* of entropy does not exist.

Since Equation (13.20) is of the first order in time, the analysis may be formulated as an initial-value problem resulting in a statement about the long-time behavior of the distribution function, $f(\mathbf{r}, \mathbf{v}, t)$, when its value $f(\mathbf{r}, \mathbf{v}, 0)$ has been prescribed. Leaving aside the mathematical problem of the *existence* of solutions of his equation for $t \to \infty$,† Boltzmann defined a function $H(\mathbf{r}, t)$, known simply as Boltzmann's H, in the form of the integral

$$H(\mathbf{r}, t) = -\mathbf{k} \int f(\mathbf{r}, \mathbf{v}, t) \ln f(\mathbf{r}, \mathbf{v}, t) \, d\mathbf{v} \tag{13.22}$$

which plays a role similar to that expected of entropy. It is easy to verify that rendering f equal to the Maxwellian distribution, Equation (6.19a), reduces H to Equation (6.27c), which is an expression for the entropy of a monatomic gas with the correction for indistinguishability omitted from it.

In order to derive a differential equation for H, we multiply Equation (13.20) by $1 + \ln f$ and integrate it over all velocities. In this manner, we obtain that

$$\frac{\partial H(\mathbf{r}, t)}{\partial t} + \nabla \cdot \mathbf{S}(\mathbf{r}, t) = -\mathbf{k} \int d\mathbf{v} \int d\mathbf{v}_1 \int db \int d\phi \, b \, |\mathbf{v} - \mathbf{v}_1| (1 + \ln f)(f' f_1' - f f_1), \tag{13.23}$$

where

$$\mathbf{S}(\mathbf{r}, t) = -\mathbf{k} \int \mathbf{v} f \ln f \, d\mathbf{v} \tag{13.23a}$$

and

$$\nabla \cdot \mathbf{S} = \frac{\partial S_x}{\partial x} + \frac{\partial S_y}{\partial y} + \frac{\partial S_z}{\partial z}. \tag{13.23b}$$

The vector $\mathbf{S}$, being an integral of the integrand of H multiplied into the velocity $\mathbf{v}$, represents the local *flux* of the quantity H. Consequently, the left-hand side of Equation (13.23) becomes identical with the left-hand side of

† T. Carleman, *Problèms Mathématiques dans la Théorie Cinétique des Gaz*, Vol. 2, Publ. scientifiques de l'Institut Mittag-Leffler, Uppsala, 1957.

Equation (1.48), and all that remains to be done to show that Boltzmann's equation is compatible with the second part of the Second law, is to prove that the right-hand side of Equation (13.23) is nonnegative. By interchanging the integration variables $\mathbf{v}$ and $\mathbf{v}_1$ in the integral on the right-hand side of Equation (13.23), we obtain the identity

$$\int d\mathbf{v} \int d\mathbf{v}_1 \int db \int d\phi b\,|\mathbf{v}-\mathbf{v}_1|(1+\ln f)[f'f_1'-ff_1]$$
$$= \int d\mathbf{v}_1 \int d\mathbf{v} \int db \int d\phi\, b\,|\mathbf{v}-\mathbf{v}_1|(1+\ln f_1)[f'f_1'-ff_1], \quad (13.23\text{c})$$

which allows us to write Equation (13.23) in the form

$$\frac{\partial H(\mathbf{r},t)}{\partial t} + \nabla\cdot\mathbf{S}(\mathbf{r},t)$$
$$= -\frac{\mathbf{k}}{2}\int d\mathbf{v} \int d\mathbf{v}_1 \int db \int d\phi\, b\,|\mathbf{v}-\mathbf{v}_1|(2+\ln f+\ln f_1)(f'f_1'-ff_1). \quad (13.24)$$

If we now insert the relation between the direct and restituting velocities, Equations (13.11) and (13.15), we obtain the identity

$$\int d\mathbf{v}_1 \int d\mathbf{v} \int db \int d\phi\, b\,|\mathbf{v}-\mathbf{v}_1|(2+\ln ff_1)(f'f_1'-ff_1)$$
$$= \int d\mathbf{v}' \int d\mathbf{v}_1' \int db \int d\phi\, b\,|\mathbf{v}'-\mathbf{v}_1'|(2+\ln ff_1)(f'f_1'-ff_1). \quad (13.25\text{a})$$

Interchanging the parts played by $\mathbf{v}$ and $\mathbf{v}_1$ on the one hand, and by $\mathbf{v}'$ and $\mathbf{v}_1'$ on the other, we may also write

$$\int d\mathbf{v}_1 \int d\mathbf{v} \int db \int d\phi\, b\,|\mathbf{v}-\mathbf{v}_1|(2+\ln ff_1)(f'f_1'-ff_1)$$
$$= \int d\mathbf{v}_1 \int d\mathbf{v} \int db \int d\phi\, b\,|\mathbf{v}-\mathbf{v}_1|(2+\ln f'f_1')(ff_1-f'f_1') \quad (13.25\text{b})$$

for the restituting collision. Finally, combining Equations (13.24) and (13.25b), we obtain

$$\frac{\partial H(\mathbf{r},t)}{\partial t} + \nabla\cdot\mathbf{S}(\mathbf{r},t)$$
$$= -\frac{\mathbf{k}}{4}\int d\mathbf{v} \int d\mathbf{v}_1 \int db \int d\phi\, b\,|\mathbf{v}-\mathbf{v}_1| \ln\left(\frac{ff_1}{f'f_1'}\right)(f'f_1'-ff_1). \quad (13.26)$$

Regardless of the values of f, f_1, f', and f_1', the quantity†

$$\ln\left(\frac{ff_1}{f'f_1'}\right)(f'f_1' - ff_1) \begin{cases} = 0 & \text{if } f'f_1' = ff_1 \\ < 0 & \text{otherwise}. \end{cases} \tag{13.27}$$

This proves that

$$\dot{\sigma} = \frac{\partial H}{\partial t} + \nabla \cdot \mathbf{S} \geq 0, \tag{13.28}$$

as required by the Second law.

We now turn our attention to the manner in which the system approaches equilibrium. If at instant $t = 0$ the system were surrounded by a rigid, adiabatic boundary, the flux $\mathbf{S}$ of H through the boundary would cease because the velocity $\mathbf{v}$ would vanish identically on the bounding surface. Thus, regardless of the distribution $f(\mathbf{r}, \mathbf{v}, 0)$ prevailing at that instant, the H-content of the system would change only through internal collisions, satisfying the relation

$$\frac{\mathrm{d}\mathscr{H}(t)}{\mathrm{d}t} \geq 0 \tag{13.29}$$

with

$$\mathscr{H}(t) = \int_V H(\mathbf{r}, t)\, \mathrm{d}\mathbf{r}, \tag{13.29a}$$

as it evolved. This follows immediately from the observation that

$$\frac{\mathrm{d}\mathscr{H}(t)}{\mathrm{d}t} = \int_V \frac{\partial H(\mathbf{r}, t)}{\partial t}\, \mathrm{d}\mathbf{r},$$

in the presence of a rigid boundary enclosing a fixed volume, and that the latter integrand is nonnegative in view of Equation (13.29). Thus, our main question reduces itself to an investigation whether $\mathscr{H}(t)$ must reach a maximum in the circumstances. This can, indeed, be proved, as discussed in Problem 13.5. Moreover, it can be shown that a maximum is reached for a unique form of the distribution function, namely the Maxwellian distribution

$$f(\mathbf{r}, \mathbf{v}, t) = \frac{N}{V}\left(\frac{m}{2\pi \mathbf{k} T}\right)^{3/2} \exp\left(-\frac{mv^2}{2\mathbf{k}T}\right). \tag{13.30}$$

Since $\mathscr{H}(t)$ is monotonic [see Equation (13.29)], the system reaches its maximum through a trajectory along which $\mathscr{H}$ increases monotonically with time.

† Note that the equality $f'f_1' = ff_1$ implies the existence of a Maxwellian distribution (see Problem 13.1).

Finally, once this maximum has been reached, no further change is possible because the collision integral has vanished causing f to remain constant.

The reader should note that the preceding argument carefully avoids the statement that H or $\mathscr{H}$ represent the entropy of the gas, except in the limiting case when equilibrium has been reached. The function H does however serve the purpose which the second part of the Second law of thermodynamics requires of a nonequilibrium entropy, namely, it discriminates between nonequilibrium states which are accessible from a given nonequilibrium state and those which are not accessible, in the course of the system's approach to equilibrium.

Boltzmann's H-theorem gave rise to a number of objections.† The most important objection, the so-called reversibility objection, sees a paradox in the fact that an equation rooted in essentially reversible laws of motion‡ nevertheless implies that a mechanical system must evolve in a uniquely defined direction—toward equilibrium. The resolution of this particular objection turns on the fact that Boltzmann's equation is not merely a consequence of the laws of motion but incorporates in it the *Stosszahlansatz* which, as pointed out earlier, can only be accepted as being true *statistically*, that is, in the sense of an ensemble average. It is precisely the injection of the *Stosszahlansatz* into the argument which imposes an irreversible mode of evolution on the distribution function f.

PROBLEMS FOR CHAPTER 13

13.1. Verify that the Maxwell–Boltzmann distribution function, Equation (6.19a), satisfies the Boltzmann equation identically.

13.2. Show that the Boltzmann collision integral vanishes identically if $f(\mathbf{r}, \mathbf{v}, t)$ has the form of a local Maxwell–Boltzmann distribution function

$$f(\mathbf{r}, \mathbf{v}, t) = n(\mathbf{r}, t)\left(\frac{m\beta(\mathbf{r}, t)}{2\pi}\right)^{3/2} \exp\left\{-\frac{m\beta(\mathbf{r}, t)}{2}[\mathbf{v} - \mathbf{u}(\mathbf{r}, t)]^2\right\}$$

where $n(\mathbf{r}, t)$, $\beta(\mathbf{r}, t)$, and $\mathbf{u}(\mathbf{r}, t)$ may depend on position, $\mathbf{r}$, and time, t.

† G. E. Uhlenbeck and G. W. Ford, *Lectures in Statistical Mechanics*, Am. Math. Society, Philadelphia, Pennsylvania 1963; M. Born, *Natural Philosophy of Cause and Chance*, Oxford Univ. Press (Clarendon), London and New York, 1949. See also P. and T. Ehrenfest, *The Conceptual Foundations of the Statistical Approach in Mechanics* (M. Moravscik, transl.), Cornell Univ. Press, Ithaca, New York, 1959; R. Jancel, *Les Fondements de la Mécanique Statistique Classique et Quantique*, Gauthier-Villars, Paris, 1963. In addition, the preceding references given in this section can also be consulted.

‡ By reversing all velocities in any system it is possible to reach an earlier state from a later state, that is, thermodynamically speaking, a system could reach a nonequilibrium state from an equilibrium state.

13.3. What is meant by the expressions (a) direct collision, (b) restituting collision, (c) inverse collision? Illustrate these collisions with the example of two identical colliding molecules, two spheres, say, as well as with the example of two different molecules, for example, a disc and a triangle colliding in a plane.

13.4. Show that if $f(\mathbf{r}, \mathbf{v}, t)$ satisfies the Boltzmann equation, then

$$\frac{d}{dt}\int d\mathbf{v}\, \psi(\mathbf{v}) f(\mathbf{r},\mathbf{v},t) + \nabla \cdot \int d\mathbf{v}\, \mathbf{v}\, \psi(\mathbf{v}) f(\mathbf{r},\mathbf{v},t)$$
$$= \int d\mathbf{v} \int d\mathbf{v}_1 \int db \int d\phi\, b|\mathbf{v}-\mathbf{v}_1|\, [f'f_1' - ff_1]\cdot\{\psi(\mathbf{v}) + \psi(\mathbf{v}_1) - \psi(\mathbf{v}') - \psi(\mathbf{v}_1')\},$$

where $\psi(\mathbf{v})$ is an *arbitrary* function of $\mathbf{v}$.

13.5. Employing the method of Lagrangian multipliers, prove that the maximum value of the function $\mathscr{H}(t)$, defined by

$$\mathscr{H}(t) = -\mathbf{k} \int_V d\mathbf{r} \int d\mathbf{v}\, f(\mathbf{r}, \mathbf{v}, t) \ln f(\mathbf{r}, \mathbf{v}, t)$$

is attained when f is of the form

$$f = \frac{N}{V}\left(\frac{\beta m}{2\pi}\right)^{3/2} \exp\left(-\frac{m\beta}{2} v^2\right),$$

with $\beta = 3N/2E$. The system contains N particles, has a kinetic energy E and is confined inside a rigid, adiabatic cylinder of volume V.

13.6. Use the divergence theorem of vector calculus to prove that

$$\frac{d\mathscr{H}(t)}{dt} + \int_{\sigma(V)} d\sigma(\mathbf{n}\cdot\mathbf{S}) \geq 0.$$

Here $\mathscr{H}(t)$ is the total H contained in a volume V, $\mathbf{S}$ is the flux in H; $\sigma(V)$ is the surface enclosing volume V, and $\mathbf{n}$ is a unit vector on the surface along the outward normal to the surface.

LIST OF SYMBOLS FOR CHAPTER 13

Latin letters

b	Impact parameter
c_v	Specific heat at constant volume
E	Energy
f	Single-particle distribution function
H	Boltzmann's H-function
$\mathscr{H}$	Total H in a system of volume V
$\mathbf{i}$	Unit vector in x direction
J	Particle flux per unit volume per unit velocity interval
$\mathbf{j}$	Unit vector in y direction

$\mathbf{k}$ Boltzmann's constant
l Mean free path
m Mass of a molecule
N Number of particles
n Number of particles per unit volume
$\mathbf{p}$ Momentum
$\mathbf{r}$ Represents coordinates x, y, z
$\mathbf{S}$ Flux of Boltzmann's H-function
T Temperature
t Time
V Volume
$\mathbf{V}$ Center-of-mass velocity
$\mathbf{v}$ Represents velocity components v_x, v_y, v_z; relative velocity
$\mathbf{u}$ Average molecular velocity

Greek letters
η Viscosity
θ Denotes an angle
λ Thermal conductivity
σ Molecular diameter
$\dot{\sigma}$ Rate of entropy production
ϕ Denotes an angle

Subscript
c Denotes contribution from collisions

CHAPTER 14

FLUCTUATIONS

14.1. Introduction

Except for Chapters 12 and 13, we have, so far, concentrated on equilibrium states and directed most of our energies to the calculation of various thermodynamic properties of macroscopic systems from suitably constructed models. The properties were conceived as the most probable values in an ensemble consisting of a large number of systems, all existing in identical macrostates. The most probable values were computed as averages with respect to a distribution which assured maximum probability. The results were frequently compared with experiments and found to agree with them, the agreement becoming improved with improvements in the details of the models and the accuracy of available measurements. We also ascertained that the microscopic theory agreed with the laws of equilibrium thermodynamics.

It has been stressed many times that a particular equilibrium macrostate in an individual system can be realized by a very large number of microstates, each differing from the most probable microstate by a larger or smaller amount. This dynamic view of the processes which occur within a system even when, macroscopically speaking, it is in equilibrium, means that we must expect that such a system should display *random fluctuations* about its most probable state. In the last chapter of this book, we propose to devote some time to the study of the magnitude and nature of such fluctuations.

Even before we undertake this study, we clearly realize that fluctuations in macroscopic systems must be exceedingly small, for, otherwise, the agreement with experiments would not be as good as it is. In fact, in most cases, lack of agreement—if observed—is attributed either to imperfections in the postulated mechanical models or to inaccuracies in measurements, as already stated. In general, the inaccuracies in the latter are orders of magnitude larger than the fluctuations, and this means that fluctuations can be measured directly only in exceptional cases.

Nevertheless, there exist macroscopic phenomena which are direct results of the existence of fluctuations and which cannot be understood if the latter are ignored. The blue color of the sky, the onset of critical-point opalescence in fluids or the presence of "noise" in electric circuits belong to this group of macroscopically observable effects of fluctuations. We shall see later that the relative magnitude of fluctuations is inversely proportional to the square root of the number of particles per unit volume in the respective system and this leads to the conclusion that fluctuations amplify as the density of particles is reduced. Thus, in very dilute systems fluctuations can be detected directly by suitably designed, sensitive indicators. For example, the very delicate suspension of a precision galvanometer never comes to complete rest and oscillates even when it is in macroscopic equilibrium. Conversely, as the density of particles in a system increases, fluctuations are reduced, and in the limit where the number of particles $N \to \infty$, the ensemble averages of equilibrium thermodynamics turn out to be exact.

A particularly striking and observable effect of fluctuations is afforded by the *Brownian motion* of minute (but still macroscopic) particles suspended in liquids or gases. We shall briefly characterize such motions after having first analyzed the fluctuations in the thermodynamic properties (for example, energy and density) of thermodynamic systems.

14.2. The Probability of a Thermodynamic Fluctuation

The study of fluctuations of the thermodynamic quantities does not require the introduction of any new physical principles beyond those which were formulated in Chapter 5, when the foundations of the statistical theory were laid. At the cost of being repetitive, we shall now briefly summarize those parts of Chapter 5 which are relevant for fluctuations.

We suppose that the macroscopic state of a system is specified by its energy, E, volume, V, number of particles, N, and the set of internal, extensive variables, $z_1, \ldots, z_r$. The variables which we denote by $z_1, \ldots, z_r$, may also be interpreted as the occupation numbers for the various quantum states of the system. For the ensuing argument it suffices to assume that they, together with E, V, and N, completely specify the macroscopic state of the system. For convenience we shall usually think of the quantities $z_1, \ldots, z_r$ as internal, extensive variables, but this need not always be the case.

The probability $\mathscr{P}(E, V, N, z_1, \ldots, z_r)$ of observing the system with specified values of $z_1, \ldots, z_r$, is proportional to the number $\Omega(E, V, N, z_1, \ldots, z_r)$, of microstates associated with this particular macroscopic state. This probability is

$$\mathscr{P}(E, V, N, z_1, \ldots, z_r) = C \exp[S(E, V, N, z_1, \ldots, z_r)/\mathbf{k}], \tag{14.1}$$

where C is a constant and $S(E, V, N, z_1, \ldots, z_r)$ is the entropy associated with the macrostate defined by Equation (5.34a) as

$$S = \mathbf{k} \ln \Omega. \tag{14.2}$$

The most probable values, $z_1^*, \ldots, z_r^*$, of the variables are determined by maximizing $\mathscr{P}(E, V, N, z_1, \ldots, z_r)$ subject to the constraints imposed upon the system. The macrostate with the extensive variables having the values $z_1^*, \ldots, z_r^*$ is identified with the thermodynamic equilibrium state of the system, since it is the one with the largest entropy for fixed values of energy, volume, and number of particles. However, the crucial point in the discussion of fluctuations is the realization that Equations (14.1) and (14.2) provide us with a measure of the probability of the occurrence of macrostates other than the most probable one, that is, states at which the extensive parameters assume values different from those at equilibrium. The fluctuations which we consider here are characteristic properties of systems that contain a finite number of particles. For such systems, the equilibrium state is the most probable state, but there exists a finite probability that other states may also occur. The onset of such different states produces entropy values that differ from that at equilibrium, and a fluctuation results. As the system is allowed to become larger, the probability of the occurrence of a nonequilibrium state decreases, and the relative magnitude of the fluctuations about the equilibrium state decreases accordingly.

By a simple rearrangement of Equation (14.1) we may write the probability $\mathscr{P}$ as

$$\mathscr{P}(E, V, N, z_1, \ldots, z_r) = C' \exp\{[S(E, V, N, z_1, \ldots, z_r) - S(E, V, N, z_1^*, \ldots, z_r^*)]/\mathbf{k}\}, \tag{14.3}$$

where

$$C' = C \exp[S(E, V, N, z_1^*, \ldots, z_r^*)/\mathbf{k}]. \tag{14.3a}$$

Since the probability should be normalized to unity when all possible values of the extensive parameters are summed, it follows that

$$\sum_{z_1, \ldots, z_r} \mathscr{P}(E, V, N, z_1, \ldots, z_r) = 1. \tag{14.4a}$$

If the parameters can vary continuously, the normalization condition becomes

$$\int_{z_1} \cdots \int_{z_r} \mathscr{P}(E, V, N, z_1, \ldots, z_r)\, dz_1 \cdots dz_r. \tag{14.4b}$$

In order to be specific, we consider a perfect gas and identify z_i with the occupation numbers of molecules, n_i, in the single-particle quantum state, i. If, for example, the gas obeys corrected Maxwell–Boltzmann statistics, Ω is given by

$$\Omega(E, V, N, n_1, \ldots) = \frac{1}{\prod_i n_i!} = \exp \frac{S(E, V, N, n_1, \ldots)}{\mathbf{k}}. \tag{14.5a}$$

The most probable state is found by maximizing $S(E, V, N, n_1, \ldots)$ subject to the constraint that the total number of particles is N, or

$$\sum_i n_i = N. \tag{14.5b}$$

It was shown in detail in Section 5.11 that the most probable values for the occupation numbers n_i^* are

$$n_i^* = e^{\alpha} e^{-\beta E_i}. \tag{14.5c}$$

For large values of N, the entropy of this most probable state is

$$S(E, V, N, n_1^*, \ldots) = \mathbf{k} \ln \Omega(E, V, N, n_1^*, \ldots) = -\mathbf{k} \sum_{i=1}^{k} \pi_i^* \ln \pi_i^* \tag{14.5d}$$

where

$$\pi_i^* = \frac{n_i^*}{N}. \tag{14.5e}$$

Thus, it is seen that the quantity

$$\mathscr{P}(E, V, N, n_1, \ldots) = C' \exp\{[S(E, V, N, n_1, n_2, \ldots) - S(E, V, N, n_1^*, n_2^*, \ldots)]/\mathbf{k}\} \tag{14.5f}$$

is the probability of finding the system in a state with any set of occupation numbers $n_1, \ldots$ consistent with the constraint (14.5b).†

14.3. Fluctuations in a Subsystem

Let us now return to the general theory of fluctuations and imagine that the system has been subdivided into two subsystems labeled I and II. The energy, volume, and number of particles of each subsystem are allowed to vary, but

† Further details are contained in Problems 14.1 and 14.2.

the system as a whole is isolated. We stipulate that subsystem I is a very small part of the entire system and that its fluctuations lead to relatively insignificant changes in the equilibrium state of the larger subsystem II. Suppose, now, that the most probable value of the internal, extensive variables of subsystem I are $z_1^*, \ldots, z_r^*$, and that we wish to calculate the probability of finding the subsystem in a state where the variables have the values $z_1^* + z_1'$, $z_2^* + z_2', \ldots$; these now include the energy, volume, and number of particles of subsystem I. To this end, it is necessary to determine the accompanying change in entropy, ΔS_t, where the subscript t refers to the entire system. Since we have divided the system into two subsystems, we may write

$$\Delta S_t = \Delta S_I + \Delta S_{II}. \tag{14.6}$$

Our supposition that the effects of the fluctuations on the equilibrium state of the larger subsystem II are small simplifies the calculation of ΔS_{II}, because the fluctuation produces small changes ΔE_{II}, ΔV_{II}, and ΔN_{II} in the respective quantities. Moreover, since subsystem II is very nearly in equilibrium, we may employ equilibrium relations for it. For this reason we put

$$\Delta S_{II} = \frac{\Delta E_{II} + P\,\Delta V_{II} - \mu\,\Delta N_{II}}{T}. \tag{14.7}$$

However, as the energy, the number of particles and the volume of the entire system remain constant, it follows that

$$\Delta E_I + \Delta E_{II} = 0 \tag{14.8a}$$

$$\Delta N_I + \Delta N_{II} = 0 \tag{14.8b}$$

and

$$\Delta V_I + \Delta V_{II} = 0. \tag{14.8c}$$

Consequently,

$$\Delta S_{II} = -\frac{\Delta E_I + P\,\Delta V_I - \mu\,\Delta N_I}{T} \tag{14.9}$$

and the entropy change produced by the fluctuation is

$$\Delta S_t = -\frac{\Delta E_I - T\,\Delta S_I + P\,\Delta V_I - \mu\,\Delta N_I}{T}. \tag{14.10}$$

The probability of observing a fluctuation of this kind is

$$\mathscr{P} = C \exp\left(-\frac{\Delta E_I - T\,\Delta S_I + P\,\Delta V_I - \mu\,\Delta N_I}{\mathbf{k}T}\right). \tag{14.11}$$

In what follows, we shall omit the subscripts I whenever the context makes it clear that fluctuations about an equilibrium state are studied in a small part of a large system. The quantity $T\,\Delta S_t$, given by Equation (14.10) may be interpreted as the minimum work required reversibly to produce the given change in the thermodynamic quantities in the small subsystem; the remainder of the system is thus regarded as a large reservoir with which the small subsystem is in contact.† It is seen that the probability of observing a fluctuation decreases exponentially with the minimum work, $W_{\min}$, required to produce it. Consequently, we may record that

$$\mathscr{P} = C \exp\left(-\frac{W_{\min}}{\mathbf{k}T}\right). \tag{14.12}$$

If the deviations from equilibrium in the small subsystem are themselves small, we may assume that any three of the four quantities ΔE, ΔS, ΔV, ΔN, are sufficient to determine the fourth, as would be the case in equilibrium. Furthermore, in equilibrium, the temperature and pressure of the small subsystem are equal to those of the large system, that is to T and P, respectively. Since the deviations are small, we may use the expansions

$$\begin{aligned}
\Delta E = {} & \left(\frac{\partial E}{\partial S}\right)^*_{V,N} \Delta S + \left(\frac{\partial E}{\partial V}\right)^*_{S,N} \Delta V + \left(\frac{\partial E}{\partial N}\right)^*_{V,S} \Delta N \\
& + \frac{1}{2}\left[\left(\frac{\partial^2 E}{\partial S^2}\right)^*_{V,N} (\Delta S)^2 + \left(\frac{\partial^2 E}{\partial V^2}\right)^*_{S,N} (\Delta V)^2 + \left(\frac{\partial^2 E}{\partial N^2}\right)^*_{V,S} (\Delta N)^2\right. \\
& + 2\left(\frac{\partial^2 E}{\partial S\,\partial V}\right)^*_N \Delta S\,\Delta V + 2\left(\frac{\partial^2 E}{\partial S\,\partial N}\right)^*_V \Delta S\,\Delta N \\
& \left. + 2\left(\frac{\partial^2 E}{\partial N\,\partial V}\right)^*_S \Delta N\,\Delta V\right] + \cdots,
\end{aligned} \tag{14.13}$$

where the asterisks on the partial derivatives denote their equilibrium values, that is,

$$\left(\frac{\partial E}{\partial S}\right)^*_{V,N} = T \tag{14.14a}$$

$$\left(\frac{\partial E}{\partial V}\right)^*_{S,N} = -P \tag{14.14b}$$

$$\left(\frac{\partial E}{\partial N}\right)^*_{S,V} = \mu. \tag{14.14c}$$

† For a more complete discussion of this thermodynamic quantity, see L. D. Landau and E. M. Lifshitz, *Statistical Physics* (E. Peierls and R. F. Peierls, transl.), Section 20, Addison-Wesley, Reading, Massachusetts, 1958. (A new translation by J. B. Sykes and M. J. Kearsley has become available in 1969 as the second edition.)

Similarly,

$$\Delta S_t = -\frac{1}{2T}\left[\left(\frac{\partial^2 E}{\partial S^2}\right)^*_{V,N}(\Delta S)^2 + \left(\frac{\partial^2 E}{\partial V^2}\right)^*_{S,N}(\Delta V)^2 + \left(\frac{\partial^2 E}{\partial N^2}\right)^*_{V,S}(\Delta N)^2 + 2\left(\frac{\partial^2 E}{\partial S\,\partial N}\right)^*_V \Delta S\,\Delta N + 2\left(\frac{\partial^2 E}{\partial S\,\partial V}\right)^*_N \Delta S\,\Delta V + 2\left(\frac{\partial^2 E}{\partial V\,\partial N}\right)^*_S \Delta V\,\Delta N\right] + \cdots. \tag{14.15}$$

The deviations in entropy, volume, and number of particles produce deviations in the temperature, pressure, and chemical potential; and, again for small deviations, we assume that they may be described in terms of ΔS, ΔV, and ΔN as

$$\Delta T = \Delta\left(\frac{\partial E}{\partial S}\right) = \left(\frac{\partial^2 E}{\partial S^2}\right)^*_{V,N}\Delta S + \left(\frac{\partial^2 E}{\partial S\,\partial V}\right)^*_N \Delta V + \left(\frac{\partial^2 E}{\partial S\,\partial N}\right)^*_V \Delta N + \cdots, \tag{14.16a}$$

$$\Delta P = -\Delta\left(\frac{\partial E}{\partial V}\right) = -\left(\frac{\partial^2 E}{\partial V^2}\right)^*_{S,N}\Delta V - \left(\frac{\partial^2 E}{\partial V\,\partial N}\right)_S \Delta N - \left(\frac{\partial^2 E}{\partial V\,\partial S}\right)^*_N \Delta S - \cdots, \tag{14.16b}$$

and

$$\Delta \mu = \Delta\left(\frac{\partial E}{\partial N}\right) = \left(\frac{\partial^2 E}{\partial N\,\partial S}\right)^*_V \Delta S + \left(\frac{\partial^2 E}{\partial N\,\partial V}\right)^*_S \Delta V + \left(\frac{\partial^2 E}{\partial N^2}\right)^*_{S,V}\Delta N + \cdots. \tag{14.16c}$$

These equations allow us to write ΔS_t as

$$\Delta S_t = -\frac{1}{2T}(\Delta T\,\Delta S - \Delta P\,\Delta V + \Delta N\,\Delta\mu) \tag{14.17}$$

and to express the probability of such a fluctuation about equilibrium as

$$\mathscr{P} = C\exp\left(-\frac{\Delta T\,\Delta S - \Delta P\,\Delta V + \Delta N\,\Delta\mu}{2\mathbf{k}T}\right). \tag{14.18}$$

Using this general result, we can easily compute the effects of fluctuations on the measured values of the thermodynamic functions of a system.

A simple example of the application of Equation (14.18) is the computation of the fluctuations in energy of a system with a fixed volume and fixed number of particles. The energy fluctuations appear in the small subsystem because it is always in contact with the larger subsystem with which it may exchange energy. In this case we set $\Delta V = \Delta N = 0$, and

$$\Delta T = \left(\frac{\partial T}{\partial E}\right)^*_{V,N} \Delta E = \frac{\Delta E}{c_v} \tag{14.19a}$$

$$\Delta S = \left(\frac{\partial S}{\partial E}\right)^*_{V,N} \Delta E = \frac{\Delta E}{T}, \tag{14.19b}$$

because the fluctuations in energy are assumed small. Hence

$$\mathscr{P}(\Delta E) = C \exp\left[-\frac{(\Delta E)^2}{2\mathbf{k}T^2 c_v}\right]. \tag{14.20}$$

The normalization constant, C, can be determined from the condition that

$$\int_{-\infty}^{\infty} \mathscr{P}(\Delta E)\,\mathrm{d}(\Delta E) = 1. \tag{14.21}$$

When we evaluate this integral, we must remember that Equation (14.20) is valid only for small values of ΔE. If, however, we insert it into the normalization condition, Equation (14.21), the resulting error becomes negligible because the largest contribution to the integral is provided by that range of values of ΔE where Equation (14.20) yields an accurate expression for $\mathscr{P}(\Delta E)$. Thus C is determined by†

$$C \int_{-\infty}^{\infty} \exp\left[-\frac{(\Delta E)^2}{2\mathbf{k}T^2 c_v}\right] \mathrm{d}(\Delta E) = C(2\pi\mathbf{k}T^2 c_v)^{1/2},$$

leading to

$$C = (2\pi\mathbf{k}T^2 c_v)^{-1/2}, \tag{14.22}$$

and to

$$\mathscr{P}(\Delta E) = \frac{\exp\{-[(\Delta E)^2/2\mathbf{k}T^2 c_v]\}}{(2\pi\mathbf{k}T^2 c_v)^{1/2}}. \tag{14.23}$$

This expression for $\mathscr{P}(\Delta E)$ is an example of a normal, or Gaussian, distribution function, discussed in detail in Section 4.4. Using Equation (14.23), we can easily compute that

$$\langle(\Delta E)\rangle = \int_{-\infty}^{\infty} (\Delta E)\mathscr{P}(\Delta E)\,\mathrm{d}(\Delta E) = 0 \tag{14.24a}$$

† The definite integral is

$$\int_{-\infty}^{\infty} \exp(-\alpha x^2)\,\mathrm{d}x = (\pi/\alpha)^{1/2}.$$

and

$$\langle(\Delta E)^2\rangle = \int_{-\infty}^{\infty} (\Delta E)^2 \mathscr{P}(\Delta E)\,\mathrm{d}(\Delta E) = \mathbf{k}T^2 c_v. \tag{14.24b}$$

Equation (14.24b) shows that the specific heat of the system may be regarded as a measure of the mean square fluctuation of the energy. The most useful quantity that may be obtained from the expression for $\langle(\Delta E)^2\rangle$ is the mean square of the relative fluctuation in energy which is given by

$$\left\langle\left[\left(\frac{\Delta E}{E}\right)^2\right]_{V,N}\right\rangle = \frac{\mathbf{k}T^2 c_v}{E^2}. \tag{14.25}$$

For large systems both E and c_v are proportional to N and the quantity $\langle[(\Delta E/E)^2]_{V,N}\rangle$ is inversely proportional to N. The root mean square fluctuation defined by

$$\left\langle\left[\left(\frac{\Delta E}{E}\right)^2\right]_{V,N}\right\rangle^{1/2} = \left(\frac{\mathbf{k}T^2 c_v}{E^2}\right)^{1/2} \tag{14.26}$$

becomes proportional to $N^{-1/2}$. For an ideal gas, all of the quantities appearing on the right-hand side of Equation (14.26) are known, and we obtain

$$\left\langle\left[\left(\frac{\Delta E}{E}\right)^2\right]_{V,N}\right\rangle^{1/2} = \left(\frac{2}{3N}\right)^{1/2} \tag{14.27}$$

For example, for $N = 67$, there would be a 10% fluctuation in the energy, but for $N \approx 10^{23}$, the energy fluctuations would be reduced to $\approx 10^{-11}$. In a real gas, the specific heat c_v becomes very large in the vicinity of the critical point. This causes the energy fluctuation to become very large too, regardless of the fact that the number of particles is large. Naturally, Equation (14.26), derived on the assumption that ΔE is small, may still be used since the ratio c_v/E^2 decreases as the number of particles increases.

As the final example of our present theory, we consider the fluctuations in the number of particles in a region of fixed volume, V, and temperature, T. In other words, we propose to analyze the density fluctuations which occur in a region at a fixed temperature. Accordingly, we substitute $\Delta T = \Delta V = 0$ and

$$\Delta\mu = \left(\frac{\partial\mu}{\partial N}\right)^*_{T,V} \Delta N \tag{14.28a}$$

into Equation (14.18) and deduce that

$$\mathscr{P}(\Delta N) = \left(\frac{1}{2\pi\mathbf{k}T}\left(\frac{\partial\mu}{\partial N}\right)^*_{T,V}\right)^{1/2} \exp-\left[\frac{(\Delta N)^2}{2\mathbf{k}T}\left(\frac{\partial\mu}{\partial N}\right)^*_{T,V}\right]. \tag{14.28b}$$

This equation allows us to determine the mean square fluctuation in the number of particles in the region, and we find that

$$\langle(\Delta N)^2\rangle = \mathbf{k}T\left(\frac{\partial N}{\partial \mu}\right)^*_{T,V}. \qquad (14.29)$$

The thermodynamic derivative $(\partial N/\partial\mu)^*_{T,V}$ may be expressed in terms of more familiar functions by the use of Euler's theorem on homogeneous functions. This asserts that if a function $f(x_1, \ldots, x_n)$ is homogeneous of order m, that is, if it satisfies the identity

$$f(\lambda x_1, \lambda x_2, \ldots, \lambda x_n) = \lambda^m f(x_1, \ldots, x_n), \qquad (14.30a)$$

it must also satisfy the identity

$$m f(x_1, \ldots, x_n) = x_1\left(\frac{\partial f}{\partial x_1}\right) + x_2\left(\frac{\partial f}{\partial x_2}\right) + \cdots + x_n\left(\frac{\partial f}{\partial x_n}\right). \qquad (14.30b)$$

The chemical potential μ and the pressure P are intensive thermodynamic variables which means that they are homogeneous of order $m = 0$ in terms of extensive properties, because

$$\mu(T, \lambda V, \lambda N) = \mu(T, V, N) \qquad (14.31a)$$

$$P(T, \lambda V, \lambda N) = P(T, V, N). \qquad (14.31b)$$

It follows that

$$V\left(\frac{\partial \mu}{\partial V}\right)^*_{N,T} + N\left(\frac{\partial \mu}{\partial N}\right)^*_{T,V} = 0 \qquad (14.32a)$$

and

$$V\left(\frac{\partial P}{\partial V}\right)^*_{N,T} + N\left(\frac{\partial P}{\partial N}\right)^*_{T,V} = 0, \qquad (14.32b)$$

or that

$$\left(\frac{\partial \mu}{\partial N}\right)^*_{T,V} = -\frac{V}{N}\left(\frac{\partial \mu}{\partial V}\right)^*_{N,T}. \qquad (14.32c)$$

Referring to the Gibbs equation

$$dF = -S\,dT - P\,dV + \mu\,dN, \qquad (14.33a)$$

and employing Equation (14.32b), we notice that

$$\left(\frac{\partial \mu}{\partial V}\right)^*_{T,N} = -\left(\frac{\partial P}{\partial N}\right)^*_{T,V} = \frac{V}{N}\left(\frac{\partial P}{\partial V}\right)^*_{N,T}. \qquad (14.33b)$$

Thus, finally

$$\left(\frac{\partial \mu}{\partial N}\right)^*_{T,V} = -\frac{V^2}{N^2}\left(\frac{\partial P}{\partial V}\right)^*_{N,T}. \qquad (14.33c)$$

Making use of the thermodynamic relation (14.33c), we can express the fluctuation in the number of particles as

$$\langle(\Delta N)^2\rangle = -\mathbf{k}T\left(\frac{N}{V}\right)^2\left(\frac{\partial V}{\partial P}\right)^*_{T,N} = \frac{N^2\mathbf{k}T}{V}\kappa_T, \tag{14.34}$$

where κ_T is the isothermal compressibility.

In the vicinity of the critical point, the compressibility becomes large and as a result the fluctuation in the number of particles assumes great importance. The effects of these fluctuations are easily observed for they are responsible for the phenomenon of "critical opalescence." As the critical point is approached the gas or liquid becomes very cloudy, or turbid, owing to the large scattering of light by the inhomogeneities in the fluid. Fluctuations in the number of particles in a region of fixed volume produce corresponding fluctuations in the refractive index of the medium. These, in turn, cause incident light to be scattered because the characteristic length of a small inhomogeneous region is of the order of a wavelength, λ, of visible light.

A more spectacular effect of density fluctuations in gases is the blue color of the sky. In the upper atmosphere there exist large fluctuations in the number of particles in a region whose size corresponds to the wavelength of blue light. This explanation of the blue color of the sky was first given by Lord Rayleigh who also proved that under these circumstances the scattering of radiation is proportional to λ^{-4}.† The inverse dependence of the scattering on wavelength may be understood by realizing that the smaller the region, the larger the fluctuations in the number of particles will be .

14.4. Brownian Motion

Minute, but still macroscopic particles, such as specks of dust or colloidal particles suspended in liquids or galvanometric mirrors supported by thin strands and exposed to quiescent air, perform random, oscillatory motions which are accessible to observation through a microscope. Such motions were first observed in 1826 by the botanist Robert Brown (hence the name, *Brownian motion*) who thought that the particles represented the smallest units of living matter. Gradually it became clear that the particles are inanimate and that their motion is due to collisions with the molecules of the surrounding fluid or gas. Thus Brownian motion provides us with a striking, direct, experimental proof of the reality of the fluctuations discussed earlier. It has been aptly said‡ that "Brownian motion is simply kinetic theory

† A complete account of this phenomenon, called Rayleigh scattering, may be found in R. W. Ditchburn, *Light*, p. 472, Blackie, Glasgow and London, 1958.

‡ G. H. Wannier, *Statistical Physics*, p. 476, Wiley, New York, 1966.

becoming visible; it needs only the right circumstances—low mass, weak binding, and small frictional forces—to make its appearance."

Fluctuations in a system arise from the dynamic processes taking place in it and producing local inhomogeneites in the distribution of energy, temperature, and particle density. In the preceding section, we based our discussion on ensemble theory, thus avoiding the complexities of a direct, dynamical description. However, even a simple quantitative study of Brownian motion, first provided by A. Einstein† and later continued by P. Langevin,‡ must be based on a suitably formulated set of equations of motion. In the simplest case, the motion of a Brownian particle is studied on the assumption that it is acted upon by a stochastically varying, time-dependent force, and that its motion is resisted by a Stokesian viscous force. The theory developed by Einstein relates the observable root-mean-square of the displacement of particles to Boltzmann's constant **k**. J. Perrin§ utilized this relation to "measure" **k**. More recent measurements by E. Kappler¶ improved the method to such an extent that it could be used to determine the Avogadro number $\mathbf{N} = \mathbf{R}/\mathbf{k}$ with an uncertainty of 1%. (This is now known from other measurements to 7 ppm.)

Electric ("passive") resistors develop very small internal currents owing to Brownian-type fluctuations in the electron densities. Such currents are the origin of the unavoidable "white noise" in electronic circuits, first analyzed by H. Nyquist.#

PROBLEMS FOR CHAPTER 14

14.1. Compute the mean square fluctuation of the occupation numbers for the quantum states of a gas obeying corrected Maxwell–Boltzmann statistics.

SOLUTION: In order to compute the probability of observing a quantum state i with occupation number $n_i^* + \delta n_i$, where n_i^* is the most probable value, we need to know the entropy associated with the macrostate. The number of microstates, Ω, for a gas obeying corrected Maxwell–Boltzmann statistics is $\Omega = \prod_{i=1}^{k} (n_i!)^{-1}$. The entropy associated with the set of occupation numbers n_i is $S = -\mathbf{k} \sum_{i=1}^{k} (n_i \ln n_i - n_i)$, where Stirling's formula has been used. If we now insert the value $n_i^* + \delta n_i$ for n_i, in this expression, and expand in powers of δn_i, we readily find that

† *Ann. Phys. (Paris)* **17** (1905), 549. See also his *Investigations on the Theory of Brownian Movement*, Dover, New York, 1956.

‡ *Compt. Rend.* 1908, p. 530. See also S. Chandrasekhar, *Rev. Mod. Phys.* **15** (1953), 1, reprinted in *Noise and Stochastic Processes* (N. Way, ed.), Dover, New York, 1954.

§ *Compt. Rend.* **146** (1908), 967.

¶ *Ann. Physik (Paris)* **11** (1931), 233; and *Naturwissenschaften* 649 and 666 (1939).

\# *Phys. Rev.* **32** (1928), 110.

$$S = -\mathbf{k} \sum_{i=1}^{k} (n_i^* \ln n_i^* - n_i^*) - \mathbf{k} \sum_{i=1}^{k} \delta n_i \ln n_i^* - \mathbf{k} \sum (\delta n_i)^2/2n_i^* + \cdots.$$

We may impose the constraints $\sum n_i = N$ and $\sum \varepsilon_i n_i = E$, by adding to S the two quantities: $\mathbf{k}\alpha \sum \delta n_i$ and $-\mathbf{k}\beta \sum \varepsilon_i \, \delta n_i$, each equal to zero, and by setting the coefficients of δn_i equal to zero. Thus S becomes

$$\begin{aligned} S &= -\mathbf{k} \sum_{i=1}^{k} n_i^* \ln n_i^* + \mathbf{k}N - \mathbf{k} \sum \delta n_i(\ln n_i^* - \alpha + \beta\varepsilon_i) - \mathbf{k} \sum (\delta n_i)^2/2n_i^* + \cdots \\ &= -\mathbf{k} \sum n_i^* \ln n_i^* + \mathbf{k}N - \mathbf{k} \sum (\delta n_i)^2/2n_i^* + \cdots. \end{aligned}$$

The probability of observing a macrostate with occupation numbers $n_i^* + \delta n_i$ is given by Equation (14.3a) as

$$\mathscr{P}(E, V, N, n_i^* + \delta n_1, \ldots, n_k^* + \delta n_k) = C \exp\left[- \sum_{i=1}^{k} (\delta n_i)^2/2n_i^*\right]$$

which is valid for small δn_i. The factor C is fixed by normalizing $\mathscr{P}$. Then it may easily be determined that $\langle(\delta n_i)^2\rangle = n_i^*$.

14.2. Compute the mean-square fluctuations of the occupation numbers for the quantum states for a perfect boson and a perfect fermion gas.

14.3. Compute the mean-square fluctuations in temperature and pressure for a system in which both temperature and pressure fluctuations are taking place. Are such fluctuations independent of each other?

SOLUTION: We may apply Equation (14.18) to this problem and write

$$\mathscr{P} = C \exp\left\{\frac{\Delta P\left[\left(\frac{\partial V}{\partial P}\right)_T^* \Delta P + \left(\frac{\partial V}{\partial T}\right)_P^* \Delta T\right] - \Delta T\left[\left(\frac{\partial S}{\partial T}\right)_P^* \Delta T + \left(\frac{\partial S}{\partial P}\right)_T^* \Delta P\right]}{2\mathbf{k}T}\right\}.$$

Using the thermodynamic relations $(\partial S/\partial T)_P^* = c_P/T$ and $(\partial S/\partial P)_T^* = -(\partial V/\partial T)_P^*$, we may write $\mathscr{P}$ as

$$\mathscr{P} = C \exp\left[\frac{(\Delta P)^2}{2\mathbf{k}T}\left(\frac{\partial V}{\partial P}\right)_T^* - \frac{(\Delta T)^2}{2\mathbf{k}T^2} c_p + \frac{2\Delta P \Delta T}{2\mathbf{k}T}\left(\frac{\partial V}{\partial T}\right)_P^*\right].$$

From this, expressions for $\mathscr{P}$, $\langle(\Delta T)^2\rangle$, $\langle(\Delta P)^2\rangle$, and $\langle(\Delta P\, \Delta T)\rangle$ may be computed. If $\langle(\Delta P\, \Delta T)\rangle = 0$, then we say that the temperature and pressure fluctuations are independent.

14.4. A system undergoes simultaneous fluctuations in temperature and volume. Show that the resulting fluctuations in the internal energy satisfy the relation

$$\langle(\Delta E)^2\rangle = \mathbf{k}T[Tc_v - (\partial E/\partial V)_T{}^2(\partial V/\partial P)_T].$$

Show also that $\langle(\Delta V)^2\rangle = -\mathbf{k}T(\partial V/\partial P)_T$ and that

$$\langle \Delta E \Delta V\rangle = \mathbf{k}T\{T(\partial V/\partial T)_P + P(\partial V/\partial P)_T\}.$$

Evaluate all the quantities for an ideal monatomic gas, and provide a physical interpretation of the resulting expressions.

14.5. A system performs simultaneous fluctuations in pressure and entropy. Show that the pressure fluctuations are given by $\langle(\Delta P)^2\rangle = \mathbf{k}T/V\kappa_s$, where $\kappa_s = -(1/V)(\partial V/\partial P)_s$ is the adiabatic compressibility. What are the corresponding entropy fluctuations?

14.6. Show that the energy fluctuations in a canonical ensemble defined by $\langle(\Delta E)^2\rangle = (1/\mathscr{Z}) \sum_i e^{-\beta E_i}(E_i - \langle E\rangle)^2$ agrees with Equation (14.24b). What is the reason for the agreement of these two calculations?

14.7. Evaluate the energy fluctuations at constant volume for a Debye solid. What happens to the amplitude of these fluctuations as absolute zero is approached? Carry out an analogous discussion of the temperature fluctuations at constant volume and examine their behavior as absolute zero is approached. Compare these temperature fluctuations with those of a perfect gas. (*Hint*: Prove that the mean square temperature fluctuations satisfy the general relation $\langle(\Delta T)^2\rangle = \mathbf{k}T^2/c_v$ if the volume is kept constant.)

14.8. Compute the fluctuations in pressure for a perfect gas, by using the expression $P = -\sum_{i=1}^{k} \overline{n_i}(\partial\varepsilon_i/\partial V)$. To what extent is your result independent of statistics?

14.9. Consider a system of N particles which have a Landé factor $g_j = 2$ and are placed in a magnetic field $\mathbf{H}$. Assuming that the particles have angular momentum number $j = \frac{1}{2}$, calculate the mean-square fluctuations in the magnetization $\mathscr{M}$, $\langle(\mathscr{M} - \langle\mathscr{M}\rangle)^2\rangle$.

14.10. Compute the density fluctuation

$$\left[\frac{\langle(n - \langle n\rangle)^2\rangle}{\langle n\rangle^2}\right]^{1/2}$$

for a perfect gas, at $T = 300$ K, and $P = 1$ atm for a volume $V = 1$ cm^3; 1 mm^3; 10^9 Å^3; 10^3 Å^3. Relate the answers to the effects of such fluctuations upon the scattering of light.

14.11. We may use the relation between the probability of a particular fluctuation and the work needed to establish it in the following way. The work necessary to impart a momentum $\mathbf{p}$ to a body is equal to the kinetic energy of the body with momentum $\mathbf{p}$. What then is the probability of observing a system with momentum $\mathbf{p}$, if at equilibrium the system is at rest? What is the mean-square momentum fluctuation for the system?

14.12. Analyze the mean-square angular deviations in a vertical pendulum, which is at rest in equilibrium, and show that $\langle(\Delta\theta)^2\rangle = \mathbf{k}T/mgl$, where g is the acceleration due to gravity. (*Hint*: Use the relation between the probability of a fluctuation and the work needed to establish it.)

LIST OF SYMBOLS FOR CHAPTER 14

Latin letters

C	Normalization constant
c_v	Specific heat at constant volume
E	Energy
F	Helmholtz function
f	Homogeneous function
k	Boltzmann's constant
m	Order of a homogeneous function; mass of a particle
N	Number of particles
N	Avogadro's number
n	Occupation number of a quantum state
P	Pressure
$\mathscr{P}$	Probability
R	Perfect-gas constant
S	Entropy
T	Temperature
V	Volume
W	Work
W_{min}	Minimum work
x	Arbitrary independent variable
z	Internal, extensive variable

Greek letters

α	Constant
β	$(\mathbf{k}T)^{-1}$
κ_T	Isothermal compressibility
λ	Wavelength; constant
μ	Chemical potential
π_i	Fraction of particles in quantum state i
Ω	Number of microstates in a macrostate

Superscript

*	Denotes most probable or equilibrium value

TABLES

TABLE I

Selected Fundamental Physical Constants[a]

Quantity	Symbol	Value
Velocity of light	$\mathbf{c}$	$(2.997\,925\,0 \pm 0.000\,001\,0) \times 10^{8}$ m/sec
Electron charge	$\mathbf{e}$	$(1.602\,191\,7 \pm 0.000\,007\,0) \times 10^{-19}$ C
Planck's constant	$\mathbf{h}$	$(6.626\,196 \pm 0.000\,050) \times 10^{-34}$ J sec
	$\hbar = \mathbf{h}/2\pi$	$(1.054\,591\,9 \pm 0.000\,008\,0) \times 10^{-34}$ J sec
Avogadro's number	$\mathbf{N}$	$(6.022\,169 \pm 0.000\,040) \times 10^{-26}$ 1/kmol
Electron rest mass	$\mathbf{m_e}$	$(9.109\,558 \pm 0.000\,054) \times 10^{-31}$ kg
Bohr magneton	$\mu_B = \mathbf{e}\hbar/2\mathbf{m_e c}$	$(9.274\,096 \pm 0.000\,065) \times 10^{-24}$ J/T[b]
Gas constant	$\mathbf{R}$	$(8.314\,34 \pm 0.000\,35) \times 10^{3}$ J/kmol K
Boltzmann's constant	$\mathbf{k} = \mathbf{R}/\mathbf{N}$	$(1.380\,622 \pm 0.000\,059) \times 10^{-23}$ J/K
Stefan-Boltzmann constant	$\sigma = \dfrac{2\pi^5 \mathbf{k}^4}{15\mathbf{h}^3\mathbf{c}^2}$	$(5.669\,61 \pm 0.000\,96) \times 10^{-8}$ WK^4/m^2
First radiation constant	$\mathbf{c}_1 = 8\pi\mathbf{hc}$	$(4.992\,579 \pm 0.000\,038) \times 10^{-24}$ Jm
Second radiation constant	$\mathbf{c}_2 = \mathbf{hc}/\mathbf{k}$	$(1.438\,833 \pm 0.000\,061) \times 10^{-2}$ mK

[a] Compiled by B. N. Taylor, W. H. Parker, and D. N. Langenberg, *Rev. Mod. Phys.* **41** (1969), 375. Approved by the National Bureau of Standards.

[b] 1 T (tesla) = 10^4 G (gauss) = 1 Wb/m^2.

TABLE II

Characteristic Temperatures for Several Substances

A. Rotational and Vibrational Temperatures for Diatomic Molecules[a]

Substance	Rotational temperature Θ_r (K)	Vibrational temperature Θ_v (K)
Hydrogen, H_2	85.4	6100
Hydrogen deuteride, HD	64.1	5300
Deuterium, D_2	42.7	4300
Hydrogen chloride, HCl	15.2	4140
Hydrogen bromide, HBr	12.1	3700
Hydrogen iodide, HI	9.0	3200
Carbon monoxide, CO	2.77	3070
Nitrogen, N_2	2.86	3340
Nitric oxide, NO	2.42	2690
Oxygen, O_2	2.07	2230

TABLE II (continued)

B. Vibrational Temperatures for Polyatomic Molecules

Substance	Vibrational temperature Θ_v (K)
Carbon dioxide, CO_2	960 (2)
	1900
	3400
Water vapor, H_2O	2294
	5180
	5400
Ammonia, NH_3	1366.9
	2340.9 (2)
	4801.3
	4955.2 (2)
Methane, CH_4	1879.1 (3)
	2207.1 (2)
	4197.0
	4343.7 (3)
Carbon tetrachloride, CCl_4	313.7 (3)
	451.8 (2)
	659.0
	1116.5 (3)
Carbon tetrafluoride, CF_4	625.9 (3)
	909.3 (2)
	1306.4
	1846.0 (3)

[a] From A. H. Wilson, *Thermodynamics and Statistical Mechanics*, Cambridge Univ. Press, London and New York, 1957. The numbers in parentheses give the number of coincident vibrational frequencies.

TABLE III

The Einstein Function and Related Functions[a]

x	$-\mu/\mathbf{R}T$ $= -\ln(1 - e^{-x})$	$h_m/\mathbf{R}T$ $= x/(e^x - 1)$	$s_m/\mathbf{R}T$ $= h_m - \mu/\mathbf{R}T$	$c/\mathbf{R}$ $= (\frac{1}{2}x)^2/[\sinh^2(\frac{1}{2}x)]$
0.01	4.610	0.995	5.605	1.000
0.05	3.021	0.975	3.996	1.000
0.1	2.352	0.951	3.303	0.999
0.2	1.708	0.903	2.611	0.997
0.3	1.350	0.857	2.208	0.993
0.4	1.110	0.813	1.923	0.987
0.5	0.933	0.771	1.704	0.979
0.6	0.796	0.730	1.526	0.971
0.7	0.686	0.691	1.377	0.960
0.8	0.597	0.653	1.249	0.948
0.9	0.522	0.617	1.138	0.935
1.0	0.459	0.582	1.041	0.921
1.2	0.358	0.517	0.876	0.888
1.4	0.283	0.458	0.741	0.852
1.6	0.226	0.405	0.630	0.811
1.8	0.181	0.356	0.537	0.769
2.0	0.145	0.313	0.458	0.724
2.2	0.117	0.274	0.392	0.678
2.4	0.095	0.239	0.335	0.632
2.6	0.077	0.209	0.286	0.586
2.8	0.063	0.181	0.244	0.540
3.0	0.051	0.157	0.208	0.496
3.2	0.042	0.136	0.178	0.454
3.4	0.034	0.117	0.151	0.413
3.6	0.028	0.101	0.129	0.374
3.8	0.023	0.087	0.110	0.338
4.0	0.018	0.075	0.093	0.304
4.5	0.011	0.051	0.062	0.230
5.0	0.007	0.034	0.041	0.171
5.5	0.004	0.023	0.027	0.125
6.0	0.002	0.015	0.017	0.090
6.5	0.002	0.010	0.013	0.064
7.0	0.001	0.006	0.007	0.045

[a] Contributions of a single harmonic oscillator to the several thermodynamic quantities expressed as functions of $x = \mathbf{h}\nu/\mathbf{k}T = \Theta_v/T$.

TABLE IV

The Debye Function

ξ_m	$\mathscr{D}(\xi_m) = \dfrac{c_v}{3\mathbf{R}}$
0.0	1.0000
1.0	0.9517
2.0	0.8254
3.0	0.6628
4.0	0.5031
5.0	0.3686
6.0	0.2656
7.0	0.1909
8.0	0.1382
9.0	0.1015
10.0	0.07582
11.0	0.05773
12.0	0.04478
13.0	0.03535
14.0	0.02835
15.0	0.02307
16.0	0.01902
17.0	0.01586
18.0	0.01336
19.0	0.01136
20.0	0.009741
22.0	0.007318
24.0	0.005637
$\dfrac{c_v}{3\mathbf{R}} = \dfrac{77.92727}{\xi_m{}^3}$	when $\xi_m > 24$

INDEX

The insertion of (prob.) after a page number indicates that the entry is cited in a problem.

D

E

F

T